Sustainable Development Goals Series

The **Sustainable Development Goals Series** is Springer Nature's inaugural cross-imprint book series that addresses and supports the United Nations' seventeen Sustainable Development Goals. The series fosters comprehensive research focused on these global targets and endeavours to address some of society's greatest grand challenges. The SDGs are inherently multidisciplinary, and they bring people working across different fields together and working towards a common goal. In this spirit, the Sustainable Development Goals series is the first at Springer Nature to publish books under both the Springer and Palgrave Macmillan imprints, bringing the strengths of our imprints together.

The Sustainable Development Goals Series is organized into eighteen subseries: one subseries based around each of the seventeen respective Sustainable Development Goals, and an eighteenth subseries, "Connecting the Goals," which serves as a home for volumes addressing multiple goals or studying the SDGs as a whole. Each subseries is guided by an expert Subseries Advisor with years or decades of experience studying and addressing core components of their respective Goal.

The SDG Series has a remit as broad as the SDGs themselves, and contributions are welcome from scientists, academics, policymakers, and researchers working in fields related to any of the seventeen goals. If you are interested in contributing a monograph or curated volume to the series, please contact the Publishers: Zachary Romano [Springer; zachary.romano@springer.com] and Rachael Ballard [Palgrave Macmillan; rachael.ballard@palgrave.com].

More information about this series at http://www.springer.com/series/15486

Vittorio Ingegnoli
Francesco Lombardo
Giuseppe La Torre
Editors

Environmental Alteration Leads to Human Disease

A Planetary Health Approach

Editors
Vittorio Ingegnoli
Environmental Science & Policy
University of Milan
Milano, Milano, Italy

Giuseppe La Torre
Public Health and Infectious Diseases
Sapienza University of Rome
Roma, Roma, Italy

Francesco Lombardo
Department of Experimental Medicine
"Sapienza"
Laboratory of Seminology
Bank of Semen "Loredana Gandini"
University of Roma
Roma, Roma, Italy

The content of this publication has not been approved by the United Nations and does not reflect the views of the United Nations or its officials or Member States.

ISSN 2523-3084   ISSN 2523-3092 (electronic)
Sustainable Development Goals Series
ISBN 978-3-030-83159-2   ISBN 978-3-030-83160-8 (eBook)
https://doi.org/10.1007/978-3-030-83160-8

© Springer Nature Switzerland AG 2022

Color wheel and icons: From https://www.un.org/sustainabledevelopment/
Copyright © 2020 United Nations. Used with the permission of the United Nations.

This work is subject to copyright. All rights are reserved by the Publisher, whether the whole or part of the material is concerned, specifically the rights of translation, reprinting, reuse of illustrations, recitation, broadcasting, reproduction on microfilms or in any other physical way, and transmission or information storage and retrieval, electronic adaptation, computer software, or by similar or dissimilar methodology now known or hereafter developed.
The use of general descriptive names, registered names, trademarks, service marks, etc. in this publication does not imply, even in the absence of a specific statement, that such names are exempt from the relevant protective laws and regulations and therefore free for general use.
The publisher, the authors and the editors are safe to assume that the advice and information in this book are believed to be true and accurate at the date of publication. Neither the publisher nor the authors or the editors give a warranty, expressed or implied, with respect to the material contained herein or for any errors or omissions that may have been made. The publisher remains neutral with regard to jurisdictional claims in published maps and institutional affiliations.

This Springer imprint is published by the registered company Springer Nature Switzerland AG
The registered company address is: Gewerbestrasse 11, 6330 Cham, Switzerland

Foreword

Since the dawn of Medicine, the Hippocratic tradition has emphasized the need for harmony between every individual and the natural environment as the correct philosophy to maintain an optimal state of health.

This approach has been gradually overshadowed over the course of history, focusing on the disease first and then on the single patient. Only in the last decades, a renewed and growing attention has been paid to the environmental issues and their potential impact on human health. To date, environment is estimated to account for almost 20% of all deaths in the WHO European Region. Therefore, the environmental topics and their relationship with health problems have become a priority even on many governments' agendas.

Devastating effects of environmental pollution on global climate change are sadly known, with an increase in Earth temperature and the progressive desertification of many areas. Human beings are responsible for the large-scale fires that are hitting the Amazon rainforest, in particular large livestock and agro-industrial enterprises. Their actions are determining the livability of this place, extremely important and vital for the environmental balance of the whole planet.

In this context, the importance of environmental pollution in large cities is increasing more and more. Today, over 3 billion people live in metropolitan cities and megacities. It is estimated that in 2030 60% of the entire population will live in large urban areas, while in 2050 this number will be 70%. During the last 50 years, this unstoppable trend has been changing the face of our planet. This transformation must be evaluated in all its complexity; not only the socio-demographic aspects, but also the consequences of this migration on individuals and community's health. New habits and new risks changed and are changing the main causes of death in developed countries. In the last century, they faced the great challenges of viral infectious diseases. The important discoveries in the medical field contrasted their spread, leading to an increase in life expectancy and consequently to a rise in the world population, with the following creation of increasingly large and populous urban centers. On the other hand, this transformation has opened up the challenge to chronic noncommunicable diseases, today the main cause of mortality in developed countries. However, the ever-present risk of infectious diseases, which indeed spread much more easily in a globalized world, should never be neglected, as the ongoing COVID-19 pandemic is sadly remembering us.

Today, there is a considerable awareness about environmental pollution and increased risk of cancer. However, little is perceived by the population on the relationship between other diseases, the environment, and its changes. We are observing a steady increase in scientific studies that are investigating the pathophysiological mechanisms of many chronic noncommunicable diseases and the correlation with environmental pollutants although much remains to be clarified. In particular, many studies are focusing on the so-called endocrine disruptors, substances able to impair the normal function of the endocrine system, causing adverse effects on the health of an organism or its offspring. These substances are derived from human activities, mainly from industrial activity, and are ubiquitous in the environment. Consequently, we are continuously exposed to them, even during our fetal life. They are able to bind several hormone receptors, acting as agonists or antagonists. Moreover, they can interfere with hormonal synthesis, secretion, transport, and elimination. Therefore, these compounds can cause severe damage to exposed organisms; their detrimental effects cannot be immediately noticeable because at low dosage they do not cause acute toxic effects. Birth defects and developmental, sexual, and cognitive disorders may be consequence of the exposure. Recently, reproductive disorders, some types of cancer, metabolic diseases such as obesity and diabetes, cardiovascular, skeletal, autoimmune diseases have also been associated with endocrine disruptors. Further studies are surely needed to carry out a complete risk assessment; in particular, low dosage exposure and the cocktail effect (the total effect due to the sum of the individual substances) must be investigated.

In addition, urbanization is also rapidly changing all our lifestyle habits, especially in the latest decades. Indeed, urbanization is producing a harmful "environmental stress": pollutants in air, water, and soil find weakened individuals. The combined action of these factors with other negative outcomes of urbanization, such as traffic and rhythms of life, can generate stress for our communities. As a matter of fact, in last decades stress has become one of the most common disorders in developed countries. Rhythms imposed by urbanization may alter our "biological clock", making subjects more fragile. Recent evidence suggests an association between circadian rhythms and the psychoneuroendocrine system, with a role on the pathogenesis of diseases such as obesity, type 2 diabetes mellitus, tumors, neurodegenerative and immune system disorders. Humans have a complicated organization, which includes social habits, traditions, and culture. Today, we know that healthy eating and adequate sleep are the best tools to ensure a good metabolism. It's not just about what or how much you eat, but also *when*. The stress induced by eating food at the wrong time for our metabolism is enormous. At the wrong moment, the body is unable to correctly metabolize food and therefore stores it as fat. Urbanization has radically changed eating habits. The same applies to sleep, the duration of which has decreased by about 2 h in urban areas over the last 50 years.

Consequently, considering all new emerging data, it is necessary to further investigate the role of the environment in human diseases, with a global approach. Even the current definition of disease seems out of date. It should be remembered that in 1948, in the aftermath of the Second World War, founders of the World Health Organization defined health as a state of complete physical, mental, and social well-being and not merely the absence of disease or infirmity. Today, a more modern and appropriate definition should take into account that our health depends on genetic predisposition for 20% and on environmental factors for 80%. Treatments must necessarily go beyond the single person and the current concept of disease, studying at the same time the surrounding context of life and the community. Therefore, we must move from the concept of a patient-centered medicine to a broader concept of community-centered medicine, up to an even wider approach, **a planetary health approach,** focused on health as a common good of a "planetary" community. The challenge in the coming years is epochal and will necessarily lead to broad reflections in terms of politics and health planning. Personalized medicine certainly remains the future, but we must not neglect the totality of the individuals and the communities. Community-centered medicine means also that doctors and researchers must overcome the current health care system and closely collaborate with other experts, such as those in environmental issues. A biomedical approach is then necessary in order to face the great challenges that phenomena such as globalization, socio-demographic growth, and urbanization will bring in the health care system.

Planetary Health is interdisciplinary, but first of all it must be systemic and it needs a preferential relationship between Ecology and Medicine. This relation is to be upgrading because today both ecology and medicine pursue few systemic characters and few correct interrelations. We need to refer to new principles and methods sustained by the most advanced fields, as Landscape Bionomics and Systemic Medicine.

Thus, we will be able to better discover environmental syndromes and their consequences on human health. Environmental transformations proposed by Planetary Health Alliance, at Harvard University, from biodiversity shifts to climate change do not consider bionomic dysfunctions which can menace human health. On the contrary, finding advanced diagnostic criteria in landscape syndromes can strongly help to find the effects on human well-being. The passage from sick care to health care can't avoid the mentioned upgrading. We must refer to an ecological upgraded discipline, the Landscape Bionomics, emerged following the present shift of scientific paradigm from reductionism to systemic complexity. The first consequence is a new concept of life, not centered on the organism but on the entire "biological spectrum" made of a more defined gradient of living entities, each one strictly interrelated with and interdependent on the inferior and superior ones.

The aim of this book is to analyze the most recent scientific data regarding the complex relationship between environment and human diseases. It is the first book that has this very interesting and extremely difficult goal. I would like to thank all the authors who participated and contributed to its creation, with the hope that it will be the forerunner of a new chapter in the Medical Science.

Andrea Lenzi
President, National Committee for Biosafety
Biotechnology and Life Sciences of the
Presidency of the Council of Ministers (CNBBSV)
President, National Board of Professors in Life Sciences (Intercollegio)
President, Sapienza School of Advanced Studies (SSAS)
Chairman, Department of Experimental
Medicine Chair of Endocrinology
Sapienza University of Rome Policlinico Umberto I
Rome, Italy

Preface

Environmental alteration leads to human health: but the term environment is, today, a misunderstood word, abused, and very often used in a not scientific sense, even in medicine. The concept of environment is mainly intended as hydro-geological and climatic influences, human urbanization and pollution, while it regards firstly the so-called Biological Spectrum (sensu Odum, 1971) that is the hierarchic levels of biological organization: from the ecological communities to the landscape's units and the eco-regions.

Traditional Biology focused on small scales (from biomolecule to the organism) is still mainly reductionist, so marginalizing macro-scales (from community to landscape and biosphere). Medicine seems to be interested in Traditional Biology: the study of pollution and its effects of our health reached today an increasing importance analyzing endocrine disruptors, while very few researches are related with landscape vegetational dysfunctions. Nevertheless, the "rock in the pond" of the systemic turn in scientific paradigms imposes to change our vision: biology does not concern only micro-scales.

We can see that the biological studies on bio-chemical molecules, genetics, viruses, and metabolism led to many successes, but also made insidious errors as, for instance, the statement of DNA as the "central dogma of molecular biology," wrong because the DNA is not a set of formed characters but a set of potentialities. Another tricking error is just the marginalization of macro-scales, which brought to refuse a proper scientific role to the researches in this field.

In Medicine, we can see two reactions: (a) many researchers think even today to the fallacy of ecological aspects in etiology, and (b) some doctors appreciate the problems that come with environmental degradation but, generally, *see them as someone else's problem to solve,* while they focus on *repairing* the damage. Therefore, it is not entirely clear what the medical profession/students are meant to "do" with the ecological problems and how they can use them to help patients.

However, recently, some medical communities recognize that human alteration of Earth's ecological systems threatens humanity's health, as underlined by the recent MACH (Centre for Multidisciplinary Research in Health Science, University of Milan, coordinated by M. Raviglione) and One Health, Global Health, and Planetary Health scientific associations, e.g., PHA (Planetary Health Alliance, Harvard University). All these medical associations are interdisciplinary while, first of all, they must be systemic and pursue a preferential relationship between advanced Ecology and Medicine.

This misunderstanding between Medicine and Ecology is a challenge: we must overcome this impasse! Thus, we cannot discuss the unity of life, but we have to understand better how its scalar interrelations may influence our health. The alterations of life at macro-scales can damage human health, not unlike at small ones. Note that the underestimation of the environment is rooted in Neo-Darwinian's thinking, where the struggle for existence is considered more important that the symbiosis and cooperation. Especially crucial is the *epigenetic* control of gene expression due to DNA methylation. It demonstrates that the phenotype is not directly expressed by the genotype, and part of the genome's methylation pattern can be inherited in the Lamarckian sense.

The dependence of gene expression on the environment is now clear, as confirmed by Psycho-Neuro-Endocrine-Immunology. We move from a mechanistic vision to a complex and systemic one: not only what is written in the *sequence* of the DNA bases matters, but also their modulation due to the information that the *environment* and behavior express. We can see that overcoming the mentioned misunderstanding between Medicine and Ecology needs a theoretical premise on Landscape Bionomics and example of application that correlates bionomics, landscape health, and a disease's incidence. These observations explain the sequence of the first six chapters (and the addition of the 15th one): (1) *The systemic paradigm in Biology* (M. Bertolaso, P. De Felice); (2) *A new Paradigm in Medicine: Psychoneuroendocrineimmunology and Science of Integrated Care (A.G. Bottaccioli, F. Bottaccioli); (3) From general ecology to Bionomics;* (4) *Planetary Health: Human Impacts on Environment* (V. Ingegnoli, E. Giglio); (5) *Landscape Bionomics Dysfunctions and Human Health* (V. Ingegnoli); (6) *Agrofood systems and Human Health* (S. Bocchi, M. Raviglione); and (15) *Some landscape and healthcare considerations comparing European Union and Indian Federation (V. Ingegnoli, E. Giglio).*

Only partially following this premise are Chaps. 7 and 8, anyway, concerning important contents: *Environmental alterations and oncological diseases* (C. La Porta) and *Zootechnical Systems, Ecological Dysfunctions, and Human Health* (L. Bonizzi, F. Campana).

The Planetary Health Approach of this volume is essential to understand the complex relationship between every component of the biology, as well as physics on Earth life.

At the microorganism level, as an example, we underline what is the impact of environmental alterations on human microbiota, the complex and diverse community of bacteria, archaea, fungi, protozoa, and viruses that live on and within human beings. This relationship (Chap. 10: *The impact of environmental alterations on human microbioma and infectious diseases* by D. Barbato, C. Sestili, L. Lia, A. De Paula Baer, and G. La Torre) put in evidence that as human being we are not only what we eat, but also what we breathe and drink in the broadest sense.

Moreover, the relationship between environmental conditions and mental well-being has been acknowledged and has recently garnered additional attention in the face of climate change. In this chapter (Chap. 11: *The relationship between environment and mental health* by R. Cocchiara, A. Mannocci, A. De Paula Baer, and G. La Torre), we try to explain the main potentially associations between the mental illnesses and heavy metals, the climatic factors, and indoor environment. Following this approach, we give an overview on new psychological effect of ecological crises, such as eco-anxiety, ecological grief, and solastalgia.

Finally, the One Health Approach is very consistent with the Planetary Health Approach. In Chap. 12: *Planetary health for clinicians* (G. La Torre, B. Dorelli, and A. De Paula Baer), the One Health Initiative is presented as a worldwide strategy to increase inter-sectorial collaborations in all aspects of health care for humans, animals, and the environment, with the aim to "forge co-equal," all-inclusive collaborations among physicians, veterinarians, nurses, and other scientific-health and environmentally related disciplines.

Further topics with impact on both ecological and biomedical level are discussed in the other chapters. A particular mention is required for the endocrine disruptors (Chap. 13: *Endocrine Disruptors and Human Reproduction* by F. Pallotti, D. Paoli, M. Pelloni, and F. Lombardo), chemicals capable of interfering with human endocrine system at multiple levels. An old story that has become recent news, these chemicals are by-products of human and modern life industry (plastics, flame retardants, cosmetics, pesticides, etc.). Since the last century, these substances have become a relevant public health issue, forcing governments to strictly rule their use. Nonetheless, their widespread use and persistence in the environment have led to measurable consequences in terms of endocrine disease incidence, reproductive health problems, and even cancer.

Endocrine disrupting chemicals also appear to have epigenetic effects capable of being transmitted over the subsequent generations, thus requiring to increase our awareness and vigilance.

Due to extreme interest raised in the scientific community, as one of the most widespread noncommunicable disease, the *Environmental factors in the development of diabetes mellitus* (Chap. 14 by Laura Nigi, Caterina Formichi, and Francesco Dotta) also will be mentioned in this book.

Diabetes mellitus is a global health issue, with a multifactorial pathogenesis. Recently, a great interest has been focused on the environmental factors associated with the onset of the disease, from the best known lifestyle factors, such as diet and physical activity, to environmental pollution and gut microbiome. It is now evident that epigenetic modulation has a pivotal role in determining the phenotype of genetically predisposed individuals in response to environmental factors. Epigenetic changes can occur both in utero and after birth, influencing the onset of chronic diseases, such as diabetes mellitus, in adulthood. Indeed, it has been demonstrated that these new actors in diabetes pathogenesis could represent promising therapeutic targets.

Milano, Italy	Vittorio Ingegnoli
Roma, Italy	Francesco Lombardo
Roma, Italy	Giuseppe La Torre
	31 July, 2021

Contents

1. Understanding Complexity in Life Sciences.................. 1
   Marta Bertolaso

2. A New Paradigm in Medicine: Psychoneuroendocrineimmunology and Science of Integrated Care....................................... 15
   Anna Giulia Bottaccioli and Francesco Bottaccioli

3. From General Ecology to Bionomics........................ 31
   Vittorio Ingegnoli and Elena Giglio

4. Planetary Health: Human Impacts on the Environment 67
   Vittorio Ingegnoli and Elena Giglio

5. Landscape Bionomics Dysfunctions and Human Health 95
   Vittorio Ingegnoli

6. Agrofood System and Human Health 131
   Stefano Bocchi, Simone Villa, Francesca Orlando, Ludovico Grimoldi, and Mario Raviglione

7. Environmental Alterations and Oncological Diseases: The Contribution of Network Medicine 165
   Caterina A.M La Porta

8. Zootechnical Systems, Ecological Dysfunctions and Human Health.. 175
   Luigi Bonizzi, Francesco Campana, and Alessio Soggiu

9. Environmental Pollution and Cardiorespiratory Diseases 195
   Cristina Sestili, Domenico Barbato, Rosario A. Cocchiara, Angela Del Cimmuto, and Giuseppe La Torre

10. The Impact of Environmental Alterations on Human Microbiota and Infectious Diseases........................ 209
    Barbato Domenico, De Paula Baer Alice, Lia Lorenza, Giada La Torre, Rosario A. Cocchiara, Cristina Sestili, Angela Del Cimmuto, and Giuseppe La Torre

11	**The Relationship Between Environment and Mental Health** 229
	Rosario A. Cocchiara, Alice Mannocci, Insa Backhaus, Domitilla Di Thiene, Cristina Sestili, Domenico Barbato, and Giuseppe La Torre
12	**Planetary Health and Healthcare Workers** 241
	Giuseppe La Torre, Barbara Dorelli, Alice De Paula Baer, Domenico Barbato, Lorenza Lia, and Maria De Giusti
13	**Endocrine Disruptors and Human Reproduction** 261
	Francesco Pallotti, Donatella Paoli, and Francesco Lombardo
14	**Environmental Factors in the Development of Diabetes Mellitus**.................................... 275
	Caterina Formichi, Andrea Trimarchi, Carla Maccora, Laura Nigi, and Francesco Dotta
15	**Some Landscape and Healthcare Considerations Comparing European Union and Indian Federation** 319
	Vittorio Ingegnoli and Elena Giglio

Understanding Complexity in Life Sciences

Marta Bertolaso

Abstract

In this introductory chapter I will discuss why in the biological sciences reductionist perspectives often prevent science from making the most of its own empirical evidence and findings. Scientific knowledge in biological sciences, in fact, is mediated by questions that deal with inter-level regulatory processes and emergent complexities that change the causal dynamics relevant to understanding and intervening in living systems' functional states and dynamics. In this contribution I will thus go through some features of systemic thinking and I will discuss some epistemological implications of its use in the life sciences in their wider sense. Overcoming reductionist ontological and epistemological assumptions in scientific practice can open up new possibilities for scientific and technological advancement.

M. Bertolaso (✉)
Research Unit of Philosophy of Science and Human Development, University Campus Bio-Medico of Rome, Rome, Italy
e-mail: m.bertolaso@unicampus.it

Keywords

Complexity · Integration · Mesoscopic way of thinking · Relational epistemology

1.1 Introduction

Philosophers and scientists alike have been discussing for decades what possibilities we have in scientific practice to offer simple explanations of complex phenomena like collective behaviors of animals, cells, and organisms more in general. Such simplifications have often acquired the form of a reductionist way of thinking about biological dynamics and living systems. This reductionist approach is usually articulated at three levels: ontological, for which any biological entity can be eventually represented by its more fundamental elements; epistemological, so that the explanations can be offered at what is considered the most fundamental level of organization of the systems; methodological, so that we can proceed through progressive decomposition of the systems to elaborate a model that can adequately represent the system itself (Brigandt and Love 2017; Bertolaso 2013; O'Malley and Dupré 2009). Despite the expectations of such a reductionist approach, which date back to the earliest complex systems theories, a completely different and systemic trend emerged in the last century for the life sciences. This systemic trend is not merely opposed to reductionism but has been looking for its own epistemological foundations and legitimation in biological sciences, economics, and social sciences (Donati 2009; Aguirre 2013; Bertolaso 2016).

In the same period, empirical evidence challenged both the possibility and the adequacy of a reductionist approach in scientific practice understood as the process by which scientists weighted their evidence and elaborate explanatory models. While, for example and more concretely, some diseases can be clearly traced backward and thus causally explained in terms of a specific molecular function that has been compromised (e.g. diabetes related to genetic mutations), other diseases are still defying such attempts. Clear examples are complex and chronic diseases, such as cancer and Alzheimer's. What prevents cancer, in fact, from being considered a disease reducible to one or more molecular factors and their alteration is the striking heterogeneity of molecular manifestations and characteristics, their heterogeneity over time, and the intrinsic feature of cancer to be able to compromise all levels of the biological organization of a living system. These features clearly prevent scientists from being able to offer simple explanations and a suitable and encompassing definition of cancer. This has consequences at the clinical level in relation to both diagnosis and treatment. Some unanswered questions include: in what sense can biologists and physicians consider a gene, for example, as being sufficiently explanatory of the overall disease? How are ideals of causal specificity simply bad epistemic ideals? Or, how do they hide an ignorance of molecular and genetic features not yet discovered? What does this imply, from a conceptual and explanatory point of view, assuming that the microenvironment plays a role in the failure or reconstruction of proper biological functions? Given these and other related questions, current reflections—both in the scientific and philosophical spheres—are urging a more radical epistemological change in the way of thinking which might be relevant not only for cancer but also for other organizational levels of biological and social entities and organizations (Urbani Ulivi 2011, Green and Batterman 2017, Bizzarri 2020). Forcing science to fit reductionist standards, in fact, seems to prevent science itself from making the most of its own empirical results.

In this chapter I will follow a narrative aimed at familiarizing the reader with some historical reasons and frameworks in both the reductionist and systemic ways of thinking while developing an argument that shows what perspective we could adopt to make sense of complex dynamics and of tensions generated by the previously adopted reductionist approaches in life sciences. In concrete, I will show (Sect. 1.2) how scientific knowledge in biological sciences is mediated by questions that deal with inter-level regulatory processes and emergent complexities that change the causal dynamics that are relevant for understanding and intervening in the functional states and dynamics of living systems. I will then offer (Sect. 1.3) an overview of current systemic views and introduce some main features of a systemic way of thinking, and I will discuss some epistemological implications of its use in the life sciences more in general (Sect. 1.4). Particular attention will be payed to the lack of time scale separation among and within levels of organization in living systems and its main epistemological implication. Overcoming reductionist ontological and epistemological assumptions in scientific practice can open up new possibilities of scientific and technological advancement (Sect. 1.5).

1.2 Where we Are Now: The Systemic Paradigms

Any time we try to understand complexity in biological and life sciences we are faced with a double tension. On the one hand, any concept implies referring to properties of an entity that are the result of correlations among its elements and subsystems at different levels of the entity and also of the interrelations that hold between the entity and its surrounding or (micro)environment. This can be true both in the case that the properties of each factor that plays a role in the emergent complexity remain qualitatively the same (as in the organization of an industry, a group of people, some kind of biological and ecological system, such as a colony of animals) and in the case that properties of the elements are functionally or ontologically (permanently) modified once they are embedded in a specific environment or once their functional integration is compromised. I want to adopt this perspective because I think that a privileged perspective from which we can understand the epistemic and ontological foundation related to our understanding of biological complexities relies very much on our possibility to understand the kind of dependencies we try to modelized and explain. This obviously will determine also the specific approaches we will adopt in trying to deal with and manage the overall emergent dynamics. It will also shed light on other notions and concepts like plasticity and robustness that typically reflect the intrinsic features of a dynamically stabilized functional state that characterizes living systems or entity. In this sense we are thus considering different kinds of complexities widely discussed in literature over the last decades (Weaver 1948; Wimsatt 2007; Mitchell 2009; Bertolaso 2013; Dupré and Nicholson 2018) (Fig. 1.1).

On the other hand, the current coevolution of science and technology is forcing us, from within science itself, to elaborate new multilevel and complex models when we are dealing with complex systems whose adequateness is, nevertheless, often questioned (1) when we are dealing with the organizational and evolving features of living systems, and with their persistence in space and time and when (2) data clearly show that they are not self-evident or neutral in explanatory terms. I am thinking, concretely, of the current difficulties in evaluating social and economic trends, and of the limits that, for example, human genome sequencing has shown in interpreting and predicting cellular and biological behaviors at different levels of biological organization. I have discussed these issues more extensively elsewhere, but what is relevant for us in this chapter is to consider that at the crossroad of these challenges there is a question about what should be considered relevant in explanatory terms when describing and explaining *inter-level regulatory processes*. It is, in fact, clear that a "most fundamental level" in explanatory terms cannot be identified in absolute terms and that an integral and relational epistemology is required to account for biological dynamics and behaviors, requiring new perspectives in order to guarantee a much more fruitful and sustainable development of the digital biotech field as well (Leonelli 2015). A systemic view of nature, based on complexity, seems to have a paradoxical existence in this case because, on the one hand, it seems to merely offer vague metaphors about "holism" and "networks" made all the more abstract by mathematical analysis; on the other hand, emerging technologies tend to become the norm for global and systemic analysis without a real impact on

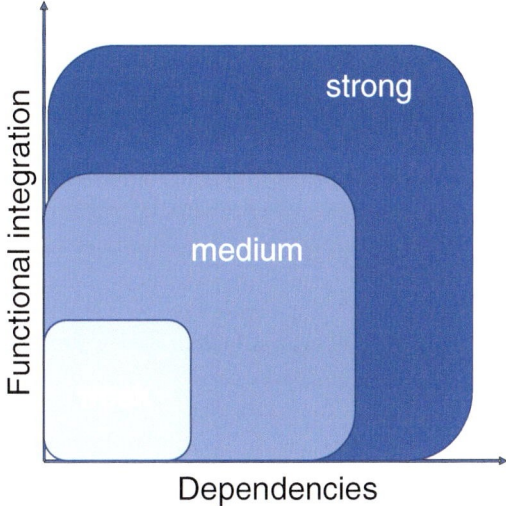

Fig. 1.1 The graphs represent how the progressive integration of functions correspond to different kinds of dependencies. Weak: an example are cells in a petri dish; Medium: an example are the dynamics that determine the normal functioning (topological dependent) of beta-cells; Strong: an example are the dynamics that determine the normal development of the organism and that typically are compromised in cancer, i.e. a genetic drift and genomic instability follow a functional dis-integration over time of the normal activity or differentiation processes of cells at specific levels. Space and time are diversely relevant in these processes as is their combined resultant in terms of microenvironmental features and topological organizations. Similar scales can be easily adopted for social, economic, organizational fields of inquiry

our capabilities of anticipating and managing complexities and therefore of understanding them.

In what follows I will refer to the first tension as "tension about complexity" and to the second as "tension about the explanatory relevance of models." I suggest that in the case of the reductionist expectation, simplicity in science and in life sciences goes with a mereological view of the natural world, and with a reduction *at unum* of the explanatory models, that is, a simplification of the systems in terms of some specific parts or elements, which implies an epistemological reductionism of scientific knowledge usually through a mechanistic understanding of causal dynamics and movements or changes in the living worlds. One example of this is the historical path that the war on cancer has taken. Since U.S. President Richard Nixon declared the "war on cancer" in 1971, various attempts at evaluating the overall progress of cancer research investments have been made. In 2010, cancer was still defined as a key public concern and a tremendous burden while the World Oncology Forum in 2012 asked if we were winning the war on cancer at all. Evidence about the lack of improvement in survival rates 1985–2007 (Pal et al. 2008), the small increases in life expectancy (Langer et al. 2015), and the rare enduring disease-free responses (Hanahan 2014) obliviously posed questions about the strategies we were adopting to understand and manage such complex disease and whether it was adequate to look for the "enemy bases" or molecular entities to be targeted by means of more and more precise drugs (Soto and Sonnenschein 2011) or if "killing undesired cells" is the most promising focus. However, the turn toward a proper understanding of what it means, for example, to think of cancer in processual terms and not merely in terms of selfish genes or crazy cells is taking a huge effort and a long time (Huang 2011). Obviously, it is not surprising that also the question about whether or not high-throughput technology performs better in predicting and managing complex diseases still awaits a satisfactory answer and evidence (Geman and Geman 2016). Are we misguided?

1.3 How we Got Here: Struggling with Organized Complexities

"Something important about complex wholes is lost if they are conceived solely in terms of their least parts of which they are in fact composed" (Grene and Depew 2004, 311). To understand better the meaning and specific contents of these words, let us look at their history and go deeper into the trends and systemic challenges I have highlighted in the previous section to make explicit the main features of a systemic thinking.

As Warren Weaver stated in 1948, it seems clear that "SCIENCE has led to a multitude of results that affect men's lives (…) men's ideas and even their ideals" (p. 536). In particular, decades ago physical sciences had already started developing new conceptual and methodological tools when *problems with simplicity* started: "Living things are more likely to present situations in which a half-dozen, or even several dozen quantities are all varying simultaneously" (ibidem p. 537) resulting in nonlinear and collective dynamics. This challenge became even more evident when new tools like powerful techniques of probability theory and statistical mechanics to deal with many variables were applied to living entities that likewise cannot be "understood as a whole that merely possesses certain orderly and analyzable average properties" (ibidem p. 538). Weaver defined these problems as *problems with disorganized complexity*. Finally, *problems with organized complexity* lead the way toward a more comprehensive understanding of a systemic perspective to account for typical living systems' dynamics. Dealing simultaneously with a "sizable number of factors which are interrelated into an organic whole" asks for a focus on relationships in a "complicated but nevertheless not in helter-skelter fashion" as typically seen in cancer, finance, problems of tactics, and strategies (ibidem p. 541).

Using mantras that have been attributed to Einstein, we are allowed, especially in life sciences, to "make things as simple as possible, but not simpler" and not to stop thinking! Paradoxes, in fact, emerge any time we try to force epistemic reductionisms to account for inter-level regulatory processes, which are, in my view and understanding of living systems, the main feature that asks for a systemic explanation of dynamics and functional states dynamically stabilized in living entities like growth and development (see, e.g. Bertolaso 2016; Bertolaso and Dupré 2018).

Such paradoxes fall mainly into the following categories: (a) Difficulties in accounting for time dependent processes and causalities, that is, for changes that are intrinsically regulated by factors that are established in time later than the elements that constitute them. An example is the influence in regulatory terms of the microenvironment of a tissue that entails non-biological factors like collagen and matrix constituents or the overall production of insulin in beta-cells that is very much related to the topological structure of the islets of Langerhans (Loppini et al. 2014), etc. Examples can be equally derived from ecological and social sciences in relation to the impact of cities and landscapes on their populations, their life styles, or on social behaviors and resilience, etc. (b) Difficulties in understanding the (ir)reversibility of biological processes when they are attributed to intrinsic properties of the elementary constituent of the systems themselves, that is, when the causal explanation relies upon a mechanistic and atomistic (mereological) assumption. Typically, this happens when an attribution of function, which is a typical epistemic operation in life sciences, coincides with an ontological reductionist assumption for which activities belong to fundamental unities or entities as such (in this sense, for example, tumor cells have and retain the properties to proliferate abnormally, to avoid apoptosis, etc. independently of other (micro)environmental factors). This is well represented also by the paradox that emerges in cancer research when accounting for carcinogenesis on the basis of DNA's mutations where quite often the same genes are involved in oncogenetic or tumor reversion processes or even seem to follow more than anticipated neoplastic transformation of the cells. This approach was typical within the so-called central dogma of molecular biology that still sometimes grounds explanations of how biological information flows from genotype to phenotype, that is, the linear causal relationship between a gene and a phenotype through the proteins machine. These explanatory approaches soon showed their pitfalls but a way out from those paradigms was slow to emerge due to their epistemological assumption. As Weinberg (2014), observes: "*Cell* has celebrated the powers of reductionist molecular biology and its major successes for four decades (…) [We] have witnessed wild fluctuations from times where endless inexplicable phenomenology reigned supreme to periods of reductionist triumphalism and, in recent years, to a move back to confronting the endless complexity of this disease."

My thesis is that a way out from these tensions is much easier if we adopt a bionomics perspective. Such bionomics perspective inherits from the traditional systemic views the emphasis of the emergent properties of a system as a result of interactions and interrelations among elements framed in peculiar systemic characters (i.e. measured through specific systemic parameters), more that on elements' activities and properties but also ask for a reflection on time and space-time scale dependencies and integrations which also involve, for example, non-biological factors (e.g. stroma in biology, and geography in ecology, transports and infrastructures in social sciences) like in the case of the tissue constitution and maintenance in an organism or for the establishment or of the sustainability of a wide bio-ecological level at higher scales.[1]

Although paradoxes seem to show limits of scientific and technological approaches and push researchers back in their efforts, we must also

---

[1] I need to disentangle a couple of points on this topic. The first one is that I prefer not to use the concept of ecosystem in this context because it is often ambiguous, not being able to reconcile biotic and functional points of view and dependencies among space-time scales, which characterizes the principle of emerging properties (see also Ingegnoli 2002). Second, in this and other volumes it has been already discussed how *bionomics principles* (Ingegnoli, 2015) underline the difference between what really exists (Life on Earth hierarchically organized in Living Entities) and the different approaches to the study of the environment, transforming the main principles of traditional ecology by being aware that hierarchical levels are types of complex biological systems.

acknowledge that paradox-driven research has always shown itself to be more effective than mere technology driven research in disentangling issues, in showing the compatibility of different models and their possible integration at least at the epistemic level (...). They force us to look at the system as a whole and to focus on the functional capabilities that dynamically emerge when multiple elements interact simultaneously (among other authors we suggest the reading of the work by these authors: Waddington, Needham, von Bertalannfy, Denis Noble, Sui Huang, Gilbert, John Dupré, William Wimsatt, Sandy Mitchell). Moreover they are more capable of making sense of top-down causation and dynamics and of their relationship and possible integration with bottom-up processes, opening up a way from the shortcomings that the two types of paradoxes we have focused on in this section seemed to create.

1.4 The Way out: The Mesoscopic Way of Thinking

As I already discussed elsewhere, paradoxes in explaining biological processes can be better understood by reviewing the principle of *causal identity* in the light of the *mesoscopic principles of causality* and *identity* that work in a synergic way in defining what biological units are relevant from an explanatory point of view and what relational properties should be considered in a first place (cf. Bertolaso 2013, Bertolaso in Greene 2017, Bizzarri et al. 2019). What this basically means is that any time we identify a correlation and make a causal inference we are *understanding by relating*, making a judgment about the dependencies and their regularities and recurrences among entities and facts. When dealing with complex systems and living entities and dynamics, this process takes the form of a *mesoscopic way of thinking* for which we define as explanatorily relevant those levels and interactions that maximize the correlations among dynamics and processes. What actually matter therefore are the relationships among the components of a system and how such interactions reflect or emerge from the inter-level regulatory processes that are typically the object of inquiry in life sciences. This means that the dependencies we described in Sect. 1.2 have ontological foundations in the specific way living entities are constituted and maintained, and epistemic implications in the specific way we have to adopt them in order to understand them and manage with them in terms of prediction and control.

Let us now focus on the first aspect (ontological foundations in the specific way living entities are constituted and maintained) by looking at the kind of dependencies and dynamics that hold the integral functionality of the overall organic system. My discussion here relies on two papers that in my opinion very successfully show the relevance of this approach, beyond what I have already done in a previous volume (Bertolaso 2016) and that are already, in a way, present in the examples I offered above about cancer.

In Gorban et al. (2010), the focus is on the dynamics of correlation and variance in systems under the load of environmental factors, which is described as a universal effect in scenarios of crisis: "in crisis, typically, even before obvious symptoms of crisis appear, correlation increases, and, at the same time, variance (and volatility) increases too. This effect is supported by many experiments and observations of groups of humans, mice, trees, grassy plants, and on financial time series" for which a general approach is developed to explain the effect of individual and collective adaptation. Interestingly, the authors were able to analyze in different areas of practice (from physiology to economics, psychology, and engineering) the behavior of groups of similar systems adapting to similar environments. Groups of humans in hard living conditions, rats under poisoning, enterprises in recession, etc. show similar problems of diagnostics and prediction. Transversally, what the authors discuss is how "the correlations between individual systems are better indicators than the value of attributes. More specifically, in thousands of experiments it was shown that in crisis, typically, even before obvious symptoms of crisis appear, the correlations increase, and, at the same time, the variance (volatility) increases too" (ibidem …). This is represented in Fig. 1.2.

As Fig. 1.2 also shows, "[f]or some systems, it was demonstrated that after the crisis achieves its bottom, it can develop in two directions: recovering (both the correlations and the variance decrease) or fatal catastrophe (the correlations decrease, but the variance continues to increase)." The intriguing problem generated by this evidence is clarified when the authors consider not only the state but also the history of the systems. When ignoring the latter, the only difference between comfort and disadaptation is the value of variance: in the disadaptation state, the variance is larger and the correlations in both cases are low. Qualitatively, the typical behavior of an ensemble of similar systems, which are adapting to the same or a similar environment looks as follows: (a) "in a well-adapted state, the deviations of the systems' state from the average value have relatively low correlations"; (b) "under increasing of the load of environmental factors some of the systems leave the low correlated comfort cloud and form a low-dimensional highly correlated group (an order parameter appears)" (ibidem).

As the authors correctly say, there is "no proof that this is the only scenario of the changes" although the appearance of an order parameter, in their studies, was supported by many experiments as well as by the destroying of the order parameter. However, what most interests us here is that there are similarities among apparently different systems and differences for the same system's evolution depending on its history and under stress or in flourishing conditions that are well represented by the strength and number of the correlations.

Similar evidence was collected by colleagues of mine while looking at the effects of microgravity for cell population phenotypes. Physical factors, which are typically related to environmental conditions do matter, in space and time, in the transition processes of the populations (Po et al. 2019). Moreover, studies carried out on beta-cells equally show the relevance of dependencies among cells that are represented in this case by topological structures and functional correlations. Structure and functions are, as known, strictly related and therefore causally and explanatorily crucial in life sciences. This was also the focus of the project on the morphogenetic models

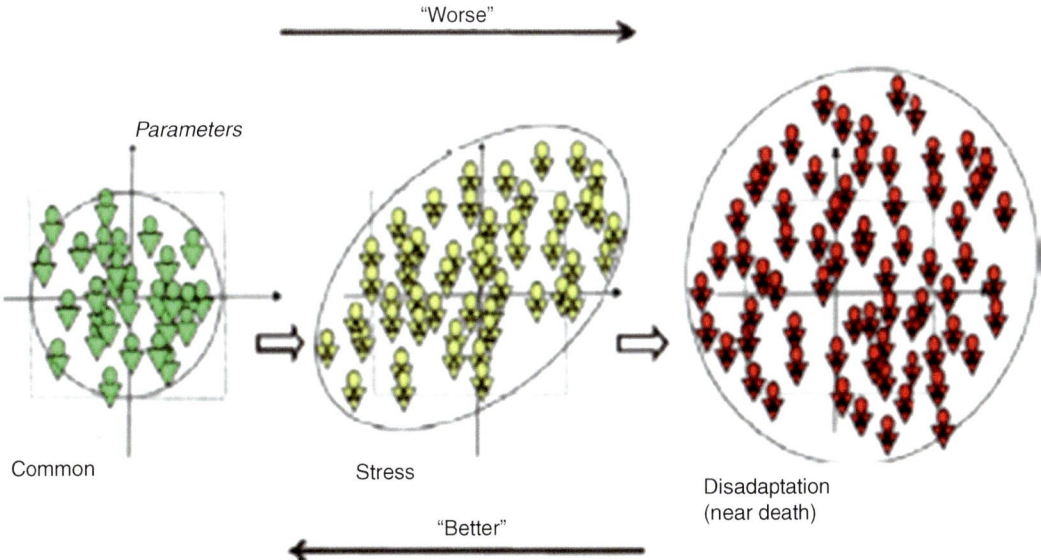

Fig. 1.2 Correlations and variance in crisis. Reproduced from Gorban et al. 2010. Ref. Complete for copyright request: Gorban AN, Smirnova EV, Tyukina T (2010) Correlations, risk and crisis: From Physiology to Finance. Physica A Science Direct

already taken up by the mathematician René Thom (1975) who tried to give mathematical sense to the embryological concept of the morphogenetic field using the theory of topological transformations: "The fundamental problem of biology," he claimed, "is a topological one, for topology is precisely the mathematical discipline dealing with the passage from the global to the local" (1975, p.151). For Thom, biochemical and genetic models failed to address the problem of morphological stability and form.

All these examples imply that what we are eventually pursuing in life sciences is careful *reconstruction of dynamics* and more specifically accounting for the (un)coupling of dynamics in space and time, which typically require an integral systemic approach. This also means that living beings should be understood from the point of view of their spatial-temporal ontology, of their *natural history* usually described in its two most fundamental components, i.e. their (biological) origin and context or environment.[2] Moreover, when we say that part of the ontological and epistemological challenge in life sciences consists precisely of combining the global and the local, the recurrence to a systemic viewpoint—or more widely a relational epistemology—does not merely means taking into consideration interrelations. It asks for a deeper understanding of the *mutual dependencies* between what are considered the local and the global in a given system or model.

Complex systems are nonlinear systems characterized by a multiplicity of interacting components that give rise to higher-order, emergent phenomena. That is, the local behavior of small-scale components gives rise to a global behavior pattern, which in turn affects and restricts the local one. From galaxies to ant colonies—and to human societies—the science of complexity offers a new and computationally powerful lens for understanding the deep structures of nature, human societies, and human minds (cfr. Bertolaso & Sterpetti 2020, Mervich 2020). But there is a specific way in which global patterns affect local behaviors and vice versa depending on the level of integration and dependence through which the specific system and entity is structured.

As Braun and Marom show (2015) in discussing behaviors of microorganisms, gene expression dynamics and neural systems organization, "regardless of the observed level of organization (protein, cell, network, or behavior), above lower boundaries that reflect fairly well understood physical constraints, observed and reported time scales are practically continuous." That is, there is no time scale separation between processes occurring at different levels of organization. As the authors highlight, instead, such separation is possible and even a default assumption in the analysis of physical systems which allows coarse graining "lumping of many microscopic degrees of freedom to a small number of effective system variables. Where such separation does not exist, the path towards complexity is wide open" (ibidem p. 3. See also Bizzarri et al. 2019, Giuliani et al. 2014). As we will see more clearly in the next section, this fact is precisely related to the existence and explanatory relevance for organized complexities of what Weaver (1948) called a *middle region*. "The really important characteristic of the problems of this middle region, which science has as yet little explored or conquered, lies in the fact that these problems, as contrasted with the disorganized situations with which statistics can cope, show the essential feature of *organization*. In fact, one can refer to this group of problems as those of *organized complexity*" (Latin in the original, ibidem p. 4).

1.5 Hidden Simplicity in Biological Complexity

Let us now go back to the second point left opened in Sect. 1.2, that is, the epistemic implication of the dependencies we are focusing on. I said that the convergence between science and technology has been fostering the development of systemic approaches in order to take into

---

[2] This discussion has traditionally also taken the form of the nature-nurture debate that is however beyond the objective of this paper. For more details and reflections on this, see Bertoaso….

account dynamics that often emerge as networks of dynamic relations with elements that acquire a specific explanatory relevance depending on the level of discussion and the scientific question posed. Reaching such an understanding requires a system biology approach "that is defined as the science that deciphers how biological functions arise from interactions between components of living organism. It studies the gap between molecules and life" (Boogerd et al. 2007, p. 6). In Sect. 1.3, however, we have characterized such a gap as an ontological and epistemic feature of living entities that structure themselves through continuous coupling of dynamics that justify also the lack of time scale separability among such processes or dynamics. The global and the local imply each other in non-trivial terms and top-down causality holds as long as the question is precisely related to the dynamic stability that characterizes living entities' behaviors, growth, and development in their most general, but also concrete, sense.

We can thus define a *mesoscopic level* as the scale of network organization at which functionality emerges in responses to higher-level system and environmental constraints (Bertolaso 2017) or the level at which "organizational principles act on the elementary biological units that will become altered, or constrained, by both their mutual interaction and the interaction with the surrounding environment. In this way and in this place is where general organization behavior emerges and where we expect to meet the elusive concept of complexity" (Bizzarri et al. 2011, p. 176, see also Bizzarri et al. 2019). Scientific, methodological, and philosophical questions regarding the adequateness/accountability of the models we develop to account for them rely upon this notion all the time, overcoming tensions generated by reductionist assumptions and making explicit the importance of a relational epistemology in life sciences. Its importance, in fact, lies not only in the identification of a level of the organizations that is successful in descriptive, explanatory, or even predictive terms of a phenomenon—which is already crucial for science and technological applications—but also in helping us understand the relevance of theoretical terms and the abstraction from physical parts and elements of the concepts we usually adopt and apply to different fields of inquiry to describe dynamics and integrated processes. Consider, for example, the already mentioned concept of morphogenetic fields. Such a notion already implies logically and conceptually an explicit reference to inter-level dynamics that are realized in space and time. Similarly, when adopting the language of functional capabilities, instead of mere functions, we are emphasizing the possibility of biological activities to take place not only in time but also over time.

As shown in Fig. 1.3, a mesoscopic way of thinking is able to bridge observations and the identification of the system by making the relationships among the parts at a specific level of a biological organization a proper object of inquiry.

To give a final example of how these epistemic aspects contribute to a scientific question, I will briefly discuss empirical results related to the possibility of predicting Alzheimer's on the metabolic basis of the areas of the brain.

1.5.1 A Case Study: Predicting Time-to-Event Development of Frank Alzheimer

I will follow methodological aspects of the case study as they well illustrate what I have theoretically addressed in the previous section and articulate how a comprehensive approach to predict time-to-event development of frank Alzheimer's can emerge adopting a mesoscopic way of thinking (Pagani et al. 2016; Pagani et al. 2017a, b). Assuming that the question was "how is it possible to find a parameter to predict time to event?" the expected impact would be a better estimation of the time before development of Alzheimer's disease that can allow us to initiate therapeutic interventions (e.g. cognitive therapies) on time, to drastically improve patients' quality of life.

To this end, the scientific problem was clarified in the following way. Brain metabolism (as measured by PET scan) is typically used in the pathology diagnosis of Alzheimer's disease. However, the inherent complexity of the system

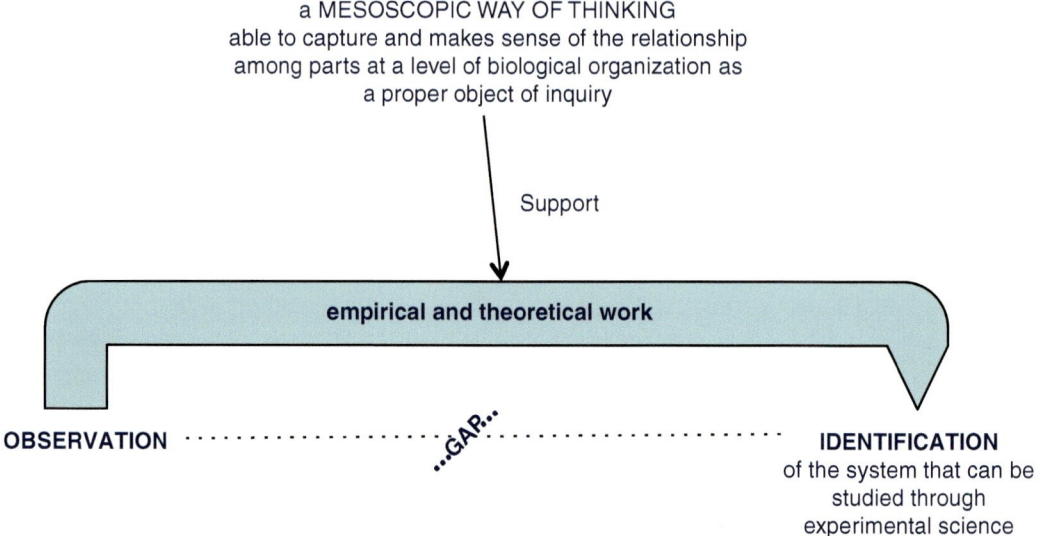

Fig. 1.3 The mesoscopic way of thinking

(brain) and its intermingled network of connections combined with the myriad confounding factors that impinge on the actual PET results make it very difficult to generate an efficient diagnostic criterium from PET data alone.

A reflection followed regarding the adopted strategy for diagnosis that brought it into a new perspective. The intermingled set of connections among brain areas could be switched from a curse to a blessing for diagnostic purposes and treatment: a global statistical measure of the average connectivity of brain metabolism is much more reliable than any single area-based PET Scan. Thus, by means of a classic statistical approach (principal component analysis) the authors measured the relation between the amount of variance explained by PC1 (the first principal component) and the disease severity (time elapsed before development of frank Alzheimer).

What the model offered emerged clearly at that point. The authors of the papers were able to measure the degree of coherence of brain metabolism in different patient groups. By doing so, we discovered that it was very well correlated with the time to development of disease (severity). We found that each brain area had its own specific dynamics in the process of detachment from the main connected component (when each single area stops being coherent with the rest of the brain's metabolism). The finding allowed us to identify the most "promising areas" to develop an easy-to-use model for clinicians to perform diagnosis.

The reliability of the model, therefore, can be traced back to the theoretical assumption that the brain works as a whole as does its metabolism, which is a typical systemic assumption. For this reason, the model proposed works with two measures: the degree of order and the organization of the brain. These measures capture the essential aspect of Alzheimer's disease (from a phenomenological perspective), which is the breaking down of the brain's organization.

As a result of their findings, clinicians diagnosing patients at risk of Alzheimer's have begun working with the information generated by PET scans according to the new model and are making a clear contribution to the knowledge development, that is, that the strength of the mesoscopic methodology lies in this case in changing the correlation structure, thus providing a clearer framework for diagnosis and for brain metabolism research. The authors clarified a very important feature of brain metabolic organization: the decay of a single order parameter governing the entire brain metabolism occurs with a simultaneous increase in local organizations, where sub-clusters of brain areas detach from the "whole" and assume an autonomous local organization.

These findings were made possible by relying on two "complementary" tools: the PCA (principal component analysis) and the ICA (independent component analysis). The first tool captures the "global fields," while the second focuses on "local fields." The loss in the explanatory power of the global order parameter is strictly correlated with the rising explanatory power of local sets of correlations.

Neurologists and Gerontologists (Karolinska University Hospital, Stockholm, Sweden and IRCCS AOU San Martino-IST, Genoa, Italy) who asked for this contribution finally realized that the main mistake was how they framed the problem. What exactly was their (unresolved) problem? This was defined as whether or not it would be possible to predict the individual time-to-outcome (e.g. disease severity) of frank Alzheimer's disease in patients presenting similar levels of impairment at the time of prediction, based on brain PET scan. Their contribution consisted therefore mainly in providing a different interpretation of PET results based on a mesoscopic (systemic) perspective. Clinicians contributed with data and medical expertise, and by selecting patients, based on their identification of the biological role and anatomical disposition of different brain areas, with same degree of cognitive impairment. They followed up on patients and provided treatment based on the mesoscopic model provided. But they also received a ready-to-use method to supplement clinical diagnosis based on cognitive examinations of PET (metabolic) scans that increased the accuracy of their diagnoses. The awareness about the methodological and epistemological potentialities of the mesoscopic approach therefore shows its potentialities in predictive terms as well.[3]

Overcoming reductionist ontological and epistemological assumptions in scientific practice, in fact, opens up new possibilities of scientific and technological advancement.

1.6 Conclusions: A Relational Epistemology

This paper offers an innovative response that seeks to integrate the awareness of the above-mentioned challenges with the opportunities they offer to effectively advance biomedical and engineering sciences, through a relational epistemology to account for complex dynamics and living entities functional stability. Although I think that many of the issues discussed above are very closely related to the difficulty of comprehending a clear-cut notion of life and of living systems, the challenge of doing so opens the possibility of a pluralistic approach to describing and explaining them, that is, a richer understanding of their features and behaviors or dynamics. Thus, we can say that the difficulties associated with appreciating the plurality of models that can be developed to account for different aspects (*explananda*) of a living system and with developing a satisfactory explanatory model of inter-level regulatory dynamics when more or less strong dependencies (Fig. 1.1) are in place follow a wrong understanding of "reductions" and "complexity" within the practice and theories of life sciences.

As we have seen, there is a progressive collective orientation toward new and systemic epistemological assumptions and research practices throughout the scientific community at large. Conversely, as in some examples discussed above, a *multidimensional actionability* emerges and leads interesting future developments in life sciences highlighting, for example, the relevance of the spatio-temporal structures and how *fundamental elements* should be understood differ-

---

[3] Something similar is interestingly discussed in the paper by …. Where showing how the choice of coordinates is strictly related to the problem of invariance: "All indicators of the level of correlations are non-invariant with respect to transformations of coordinates. (….) Dynamics of variance also depends on nonlinear transformations of scales. (…) The observed effect depends on the choice of attributes. Nevertheless, many researchers observed it without a special choice of coordinate system. What does it mean? We can propose a hypothesis: the effect may be so strong that it is almost improbable to select a coordinate system where it vanishes".

ently, and also (explanatory) *relevances*, i.e. what can be considered evidence of some biological phenotype or behavior. Data, in fact, are not self-evident, and little has been invested in research on data evaluation and validation processes as well as on their methodological assumptions. These facts have ethical relevance too if we consider ecological and biomedical scenarios in which right choices shape contemporary and future scenarios as for example in the case of epidemic. Therefore, providing solid data bases for modeling the behavior of living systems can generate an important value.

Acknowledgements I want to thank my co-authors and mentors in many papers I have quoted whose collaboration, discussions, and critical reading of manuscripts have been fundamental in developing the thesis and arguments. In particular I want to thank Alessandro Giuliani, Simonetta Filippi, Sandy Mitchel, and John Dupré.

References

Aguirre MS (2013) An integral approach to an economic perspective: the case of measuring impact. J Markets & Morality 16:1

Bertolaso M (2013) How science works: choosing levels of explanation in biological sciences. Aracne, Roma

Bertolaso M (2016) Philosophy of cancer a dynamic and relational view, vol 18. Springer, Netherlands. ed. no. 1

Bertolaso M (2017) A system approach to cancer. From things to relations. In: Green S (ed) Philosophy of systems biology—perspectives from scientists and philosophers. Springer, Dordrecht. History, Philosophy and Theory of the Life Sciences (HPTL), pp. 37–47

Bertolaso M (2017) "A System Approach to Cancer. From Things to Relations". In Philosophy of Systems Biology, Green S (ed.), 20:37–47. Cham: Springer International Publishing. https://doi.org/10.1007/978-3-319-47000-9_3

Bertolaso M, Dupré J (2018) A processual perspective on cancer. In: Dupré J, Nicholson D (eds) Everything flows: towards a processual philosophy of biology. OUP, pp 321–336

Bertolaso M, Sterpetti F (2020) A critical reflection on automated science—will science remain human? In: Human perspectives in bio-medical sciences and technology. Springer

Bizzarri M (ed) (2020) Approaching complex diseases: network-based pharmacology and systems approach in bio-medicine. vol. 2. human perspectives in health sciences and technology. cham: Springer International Publishing. https://doi.org/10.1007/978-3-030-32857-3

Bizzarri M, Giuliani A, Cucina A, D'Anselmi F, Soto AM & Sonnenschein C (2011) Fractal analysis in a systems biology approach to cancer. Sem Cancer Biol 21, 175–182

Bizzarri M, Giuliani A, Pensotti A, Ratti E, Bertolaso M (2019) Co-emergence and collapse: the mesoscopic approach for conceptualizing and investigating the functional integration of organisms. Front Physiol Front Physiol 10:924

Boogerd FC, Bruggeman FJ, Hofmeyr JHS, Westerhoff HV (eds) (2007) Systems biology. Philosophical foundations. Elsevier, Amsterdam

Braun E, Marom S (2015) Universality, complexity and the praxis of biology: two case studies. Stud Hist Phil Biol Biomed Sci:1–5. *Technion—Israel Institute of Technology, Israel*

Brigandt I and Love A (*First published Tue May 27, 2008; substantive revision Tue Feb 21,* 2017), Reductionism in biology—Stanford Encyclopedia of Philosophy

Donati P (2009) What does subsidiarity mean? The relational perspective. J Markets Morality 12(2):211–243

Dupré J, and Nicholson J (2018) A manifesto for a processual philosophy of biology—chapter of Everything Flows: Towards a Processual Philosophy of Biology

Geman D, Geman S (2016) Opinion: science in the age of selfies. PNAS 113(34):9384–9387

Giuliani A, Filippi S, Bertolaso M (2014) Why network approach can promote a new way of thinking in biology, hosted by Dr(s) Xiaogang Wu, Hans Westerhoff, Pierre De Meyts, Hiroaki Kitano. Front Genet Syst Biol 5(83):1–5

Gorban AN, Smirnova EV, Tyukina T (2010) Correlations, risk and crisis: from physiology to Finance. Physica A Science Direct

Green S, Batterman R (2017) Biology meets physics: reductionism and multi-scale modeling of morphogenesis. Stud Hist Phil Biol Biomed Sci 61:20

Grene M, Depew D (2004) The philosophy of biology: perspectives. An episodic history. Cambridge University Press, Cambridge

Hanahan D (2014) Rethinking the war on cancer. Lancet 383(9916):558–563

Huang S (2011) On the intrinsic inevitability of cancer: from foetal to fatal attraction. Semin Cancer Biol 21:183–199

Ingegnoli V (2002) Landscape Ecology: A widening foundation. Berlin, Heidelberg: Springer Berlin Heidelberg. https://doi.org/10.1007/978-3-662-04691-3

Ingegnoli V (2015) Landscape bionomics biological-integrated landscape ecology. Milano: Springer Milan. https://doi.org/10.1007/978-88-470-5226-0

Langer A, Meleis A, Knaul FM, Atun R, Aran M, Arreola-Ornelas H, Bhutta ZA, Binagwaho A, Bonita R, Caglia JM, Claeson M, Davies J, Donnay FA, Gausman JM, Glickman C, Kearns AD, Kendall T, Lozano R, Seboni N, Sen G, Sindhu S, Temin M, Frenk J (2015) Women and health: the key for sustainable development. Lancet 386(9999):1165–1210

Leonelli S (2015) What counts as scientific data? A relational framework. Philos Sci 82(5):810–821

Loppini A, Capolupo A, Cherubini C, Gizzi A, Bertolaso M, Filippi S, Vitiello G (2014) On the coherent behavior of pancreatic beta cell clusters. Phys Lett A 378:3210–3217

Mervich C (2020) The human infrastructure of artificial intelligence. https://doi.org/10.13140/RG.2.2.20431.10401

Mitchell SD (2009) Unsimple truths: science, complexity and policy. University of Chicago Press, Chicago

O'Malley MA, Dupré J (2009) Varieties of living things: life at the intersection of lineage and metabolism. Philos Theor Biol 1(201306):1–25. ISSN 1949-0739

Pagani M, Giuliani A, Öberg J, Chincarini A, Morbelli S, Brugnolo A, Arnaldi D, Picco A, Bauckneht M, Buschiazzo A, Sambuceti G, Nobili F (2016) Predicting the transition from normal aging to Alzheimer's disease: a statistical mechanistic evaluation of FDG-PET data. NeuroImage 141:282–290

Pagani M, Giuliani A, Öberg J, De Carli F, Morbelli S, Girtler N, Arnaldi D, Accardo J, Bauckneht M, Bongioanni F, Chincarini A, Sambuceti G, Jonsson C, Nobili F (2017a) Progressive disintegration of brain networking from normal aging to Alzheimer's disease. J Nucl Med

Pagani M, Nobili F, Morbelli S, Arnaldi D, Giuliani A, Öberg J, Girtler N, Brugnolo A, Picco A, Bauckneht M, Piva R, Chincarini A, Sambuceti G, Jonsson C, De Carli F (2017b) Early identification of MCI converting to AD: a FDG PET study. Eur J Nucl Med Mol Imaging 44(12):2042–2052

Pal G, Paraz MT, Kellogg DR (2008) Regulation of Mih1/Cdc25 by protein phosphatase 2A and casein kinase 1. J Cell Biol 180(5):931–945

Po A, Giuliani A, Masiello MG, Cucina A, Catizone A, Ricci G, Chiacchiarini M, Tafani M, Ferretti E, Bizzarri M (2019) Phenotypic transitions enacted by simulated microgravity do not alter coherence in gene transcription profile. NPJ Microgravity 5:27

Soto AM, Sonnenschein C (2011) The tissue organization field theory of cancer: a testable replacement for the somatic mutation theory. BioEssays 33(5):332–340

Thom R (1975) Structural stability and morphogenesis: an outline of a general theory of models, Published by W. A. Benjamin Advanced Bk Program

Urbani Ulivi L (ed) (2011) Strutture di mondo. Il pensiero sistemico come specchio di una realtà complessa. Il Mulino, Bologna

Weaver W (1948) Science and complexity Classical Papers—Science and complexity E:CO Vol. 6 No. 3 2004 pp. 65-74

Weinberg RA (2014) Coming full circle—from endless complexity to simplicity and back again. Cell 157(1):267–271

Wimsatt WC (2007) Re-engineering philosophy for limited beings: piecewise approximations to reality, Ch 6. Harvard University Press, Cambridge

A New Paradigm in Medicine: Psychoneuroendocrineimmunology and Science of Integrated Care

Anna Giulia Bottaccioli and Francesco Bottaccioli

Abstract

The industrialisation of medicine has produced the current biomedical paradigm in which health and disease dynamics, as well as individual characteristics would depend on elementary determinants: genes, microorganisms, structural alterations. Health, according to the reductionist paradigm view, would therefore be under the control of science and technique.

However, health and disease depend largely on the organisation of life, grounded by social organisation and individual behaviours. It is particularly evident in pandemic COVID-19 caused by SARS-CoV-2, a novel coronavirus zoonotic origin, that highlighted the serious difficulties and limitations of the pharmacological approach, highlighting the role of the public health service, the prevention and treatment at home and the integrated approach to the patient. The reductionist medical paradigm, pharmaco-centred, is insufficient to effectively combat the pandemic, which is tackled in the territory only with lockdown measures. A new model of care is also needed in the pandemic emergency.

Psychoneuroendocrineimmunology (PNEI) is the discipline that studies bidirectional relationships between psyche and biological systems. Within a single model, PNEI brings together knowledge acquired since the 1930s from endocrinology, immunology, psychology and neuroscience. With PNEI, a model is emerging of research and interpretation of health and disease, which sees the human body as a structured and interconnected unit, where the psychic and biological systems are mutually coordinated.

In an era of great danger, but also of considerable advancement opportunities, the scientific foundations of the new paradigm presented in this chapter are solid: they are the basis upon which it is necessary to construct a new science and a new care.

This chapter is based on Bottaccioli F & Bottaccioli AG (2020) *Psychoneuroendocrineimmunology and Science of Integrated Care. The Manual.* Edra: Milan. Particularly on chap. 23 updated.

A. G. Bottaccioli (✉)
San Raffaele University, Milan, Italy

Italian Society of Psychoneuroendocrineimmunology, Rome, Italy

F. Bottaccioli
Italian Society of Psychoneuroendocrineimmunology, Rome, Italy

L'Aquila University, L'Aquila, Italy

Keywords

PNEI · Science of integrated care
Reductionism · Systemic medicine
COVID-19 pandemic

2.1 Lessons from Pandemic Covid-19

The pandemic coronavirus disease 2019 (COVID-19) was caused by novel coronavirus, named 2019-nCoV, isolated for the first time in human airway epithelial cells by the Chinese Center for Disease Control and Prevention in late December 2019 in familiar cluster of viral pneumonia which was epidemiologically linked to a seafood market in Wuhan, Hubei province (Zhu et al. 2020). The novel coronavirus is the seventh member of the family of coronaviruses that infect humans and it was also named Severe Acute Respiratory Syndrome Coronavirus 2 (SARS-CoV-2) because of the genomic similarity with certain SARS-like coronavirus bat strains. SARS-coronavirus and MERS-coronavirus are two zoonotic coronavirus family members able to infect humans and cause severe and, in some cases, fatal disease, responsible for SARS and MERS outbreaks in Chinese province of Guangdong in 2002–2003 and in 2012 in the Middle East (Zhu et al. 2020).

The COVID-19 pandemic is escalating rapidly and spread in less than 2 months throughout the world. The World Health Organization (WHO) declared the outbreak to be a Public Health Emergency of International Concern on 30 January 2020 and the WHO Regional Director for Europe declared it as a pandemic on 12 March, 2020 (WHO 2020).

Is ongoing as we write an international race for the development of effective antiviral drugs and vaccines. However, within a few months, the SARS-CoV-2 has killed more than 100,000 people worldwide and thrown the world economy into an economic recession that promises to have unprecedented intensity.

The COVID-19 pandemic which, unlike SARS and MERS, was not limited to some eastern areas (China and the Middle East), affecting the heart of the West, has generated astonishment and disbelief among the population. In the imagination of the western citizen, epidemics were a reminder, mostly literary, of the past and, at the same time, a phenomenon of the most miserable areas of the world, which live without hygiene, without food, without drugs. In this case, the epidemic is a dramatic reality in the wealthy, technological West, the cradle of scientific medicine. In addition, the infection usually has a banal onset, with moderate fever, cough, sore throat, asthenia, symptoms that everyone has experienced many times in life without serious consequences. And instead, in the pandemic context, the subject who experiences them can interpret them in a much more threatening way, like the beginning of a chain that, quite quickly, can lead to a condition of serious illness.

For these and other reasons, the new epidemic reality still finds it difficult to be mentally elaborated by the people on whom the restraining measures are incumbent, which, moreover, the governments of the western countries themselves have assumed with much reluctance and wavy attitudes.

Table 2.1 Comparison of old and new paradigm

Reductionist paradigm	PNEI paradigm
• Ignores the complex origin of diseases • Medicalises risk factors • Asymptomatic persons are redefined "ill" to facilitate the treatment with drugs • Failure of prevention • Supremacy of patentable therapies • Unsustainable costs in the face of poor results • Frequent medical errors • Low satisfaction of patients and physicians • Worsening of the disparities of access to care and health	• Knows and cares for the human being in its entirety • Identifies the social and ecological determinants of health • Promotes the knowledge and self-awareness on which to base effective primary prevention policies • Orients molecular research within the systemic vision of the human being • Promotes effective integrated care and with a low impact on the human network and on the terrestrial systems • Promotes social solidarity and real equality of access to care and health • Promotes empathy between patients and health professionals

The concepts expressed in the left column are taken from: Fani Marvasti F, Stafford RS (2012) From sick care to health care—Reengineering prevention into the U.S. system, *New Engl J Med* 367(10): 889–891

The first investigations on the mental states during the lockdown among populations revealed a strong incidence of distress, anxiety, depression, sleep disturbances (Pfefferbaum and North 2020).

Similar phenomena are recorded among health professionals, especially the front-line doctors and nurses, who registered numerous deaths due to the infection contracted, in hospital and at home, in treating COVID-19 patients (Lai et al. 2020).

In the absence of effective drugs, the medical system can only offer emergency respiratory assistance, which is certainly a life-saving activity, but which is insufficient to effectively combat the pandemic, which is effectively tackled in the territory only with lockdown measures.

The majority of individuals who came into contact with SARS-CoV-2 exhibit few symptoms or often no symptoms. In a proportion of those infected, we now do not know exactly in what proportion, the infection can evolve into interstitial pneumonia which can give rise to acute respiratory distress syndrome (ARDS) and systemic inflammation, with possible fatal outcome. If we want to effectively tackle the pandemic, it is in the sea of the infected that we need to intervene with an integrated treatment program.

Clinically, in a first phase, fever, cough, dyspnoea, immune alterations occur, such as a high neutrophil: lymphocytes ratio, an overproduction of IL-1β, IL-6 and, however, a low production of IFN-γ (Qin et al. 2020). If the disease evolves into a more severe form, other signs of systemic inflammation, particularly of the vessels, occur alongside these alterations: a condition produced by the so-called cytokine storm, that is, by the significant concentration of inflammatory cytokines released by immune cells and also from other damaged cells. The virus, in patients who develop more severe forms of the disease, evades the immune response that could block it and which is based on the cytotoxic Th1 and T lymphocytes. In place of this antiviral immune circuit, a hyperactivity of neutrophils intervenes. Neutrophilia and lymphopenia appear to be a constant trait of COVID-19. Neutrophil activity, if not accompanied by the action of B lymphocytes and cytotoxic T and T helper lymphocytes, not only does not eliminate the infection but can also be at the origin of hyper-inflammation, which characterises the transition to ARDS, with overproduction of inflammatory cytokines and extracellular traps of neutrophilic derivation (Neutrophil Extracellular Traps NET)(Van Avondt and Hartl 2018).

2.2 Support the Resilience of the Population

The great majority of the population has endogenous resources to counteract the infection which can be silent or with few symptoms. The psychoneuroendocrine-immunological (PNEI) approach, which studies the two-way relationships between the psychic dimension and biological systems in the environmental and social context (Bottaccioli and Bottaccioli 2020; Ader 2007), provides an adequate model for identifying risk factors and resistance to infection. It also illuminates the way to understand the effects of the infection on the overall health of the affected person, including his mental state (Bottaccioli et al. 2019).

As we described above, the balanced immune response to SARS-CoV-2 is central. A number of factors can regulate or unbalance the antiviral response, which are individual and collective. Individual factors related to nutrition, physical activity, stress and mental states. Collective factors related to state of environment and social conditions.

2.2.1 Environment

Recent research by the European Environment Agency has estimated that PM2.5 pollution in 2016 was responsible for approximately 412,000 premature deaths in Europe, caused by heart attacks, strokes and lung diseases (EEA 2019).

We have known for a long time that chronic exposure to air polluted by fine particulates, PM10, PM2.5 and ultrathin PM <0.1, coming from industrial exhausts, domestic heating and vehicular traffic, can cause damage to the respiratory system. In particular, the PM2.5 particulate penetrates into the bronchial and pulmonary levels, while the ultrathin one can pass directly into the blood and spread to the organs.

Polluted air can contribute to the development of serious lung diseases, but also cause low-grade inflammation that favours the progression of the viral infection. Research in progress at Harvard University, Department of Biostatistics, School of Public Health, has recorded, in the USA, a direct relationship between exposure to air pollution from PM2.5 particulates and mortality from COVID-19. Harvard epidemiologists have found that the increase of only 1 μg/m^3 in PM2.5 is associated with a 15% increase in COVID-19 mortality rate (Wan et al. 2020).

In addition to pollution, the immune system is also affected by other factors, including nutrition, physical activity and psychological status.

2.2.2 Nutrition

A low protein diet is one of the main causes of immunodeficiency in the elderly population (Salazar et al. 2017) and the lack of an adequate amino acid pool has been associated with poor production of immunoglobulins, thymic atrophy, reduced proliferation of naïve and poor lymphocytes maturation of cells with lytic activity (Natural Killer, Lymphocytes with cytotoxic activity).

As is known, the food style selects and deeply shapes the microbiota, a complex set of resident microbial populations (bacteria, viruses, fungi) that form colonies in contact with the body's mucous membranes and therefore also in the respiratory mucosa. A state of dysbiosis, which can arise as a result of various conditions including the use of drugs (i.e. antibiotics, antacids), an inflammatory diet, surgery and hospitalisation, can be associated with various infectious and life-threatening clinical pictures. The clinical features of COVID-19 critically ill patients show widespread malnutrition.

The timely start of nutritional therapy is therefore vital, in particular in patients with organ failure and septic state, but it could significantly change the disease process even in non-critical patients, hospitalised in ordinary wards or treated at home (Liang 2020).

2.2.3 Physical Activity

Reduced mobility is one of the main effects of forced pandemic quarantine. Although all sections of the population may be adversely affected by a prolonged period of almost total physical inactivity, the elderly population can once again pay the highest price. In fact, the reduced mobility in the elderly (Jiménez-Pavón et al. 2020) raises dangerously the fragility index upwards, rapidly depletes the muscle reserve and accelerates the bone turnover, promoting sarcopenia, osteo-articular degeneration, falls and osteoporotic fractures, worsens respiratory function by increasing the risk of acute seasonal diseases of the airways and exacerbations of chronic bronchopathies, alters the metabolism and the regulation of blood pressure, increasing the use of specific drugs and therefore health expenditure. Regular physical activity is also a trophic stimulus for the brain and immune system. Regular physical activity also modulates immune function, making the response against viruses and cancer cells (Th1 circuit) more efficient by counteracting immunosenescence (Abd El-Kader and Al-Shreef 2018) and age-related low-grade inflammation (inflammaging).

2.2.4 Stress

A condition of prolonged stress, such as during the COVID-19 pandemic, brings profound adaptive changes to the psycho-neuro-endocrine-immune network (Bottaccioli and Bottaccioli 2020): the psychological state is predominated by anxiety, depression, alteration of the sleep-wake rhythm and anhedonia; the biological side is characterised by impaired activation of the hypothalamic-pituitary-adrenal (HPA) axis and the release of circadian and stress-induced cortisol, imbalance of the autonomic nervous system and adrenergic hypertonicity, pathological changes in metabolic and cardiovascular functions, dysregulation immune, systemic and central inflammatory state.

Therefore, the link between psychological condition, immune dysregulation and inflammatory state, which in turn worsens the psychological picture in a deleterious pathological loop, is now sufficiently clear from a biological point of view (Bottaccioli et al. 2019) and there it allows us to understand how fragile this balance is if there are sudden and destructive environmental factors, such as the pandemic and the consequent state of isolation from quarantine, and how much the elderly population is at risk in this historical era.

2.2.5 Conclusions on COVID-19 Pandemic

The COVID-19 pandemic, due to its exceptional level of involvement of the populations of the richest and most technologically advanced nations—who experience unprecedented experiences of widespread mortality, fear and social isolation—due to the considerable difficulties encountered by health services to cope with, due to the emergency and uncertainty about the evolution of the pandemic and its predictably heavy economic repercussions on a global scale, requires a change in the approach to the prevention and treatment of infection, based on the integration of biomedical and psychological sciences and professions. It requires a change of the reductionist biomedical paradigm, whose effectiveness in prevention and treatment is below the needs and possibilities provided by current scientific knowledge.

2.3 Effectiveness of Standard Care

We often read and hear that the life expectancy increase in rich countries is the result of the medical achievements while neglecting the crucial role played by the great structural changes in social life that occurred in the course of the twentieth century: the removal of sewage from the houses and the urban environment, the reduction of the time and physical intensity of work, the protection of children from early fatigue, the climatic conditioning of homes, the elimination of hunger and undernourishment. In this general structural change, which has only taken place for a part of humanity, the one that has achieved a higher average life expectancy, the improvement of the prevention and treatment of certain infectious diseases (but not for emergent and unknown infective disease outbreaks), the improvement of emergency treatments in the cardio-cerebrovascular field, the drastic reduction of infant mortality and childbirth mortality has certainly brought their important contribution to the increase of average longevity, even though it is difficult to quantify. However, it is rare to encounter, in the contemporary age, a definitive treatment, i.e. that heals ill people: when therapies work, which is not always the case, they mostly have short-term or chronicising effects on the pathological state. This conclusion, which rests upon the analysis of efficacy trials, is also supported by independent investigations on the efficacy of the drugs that are most currently used. As shown in a meta-analysis, carried out by a group of researchers of the Technische Universität of Munich that are particularly skilled in evidence-based medicine, few drugs have a high effect size. Among these, only one (proton pump inhibitors) has an effectiveness related to the care of a pathology (reflux oesophagitis); the others (painkillers, antiparkinson drugs, antidiabetic agents) are effective in reducing the symptoms, but very poor at healing, as it is clear from the case of metformin, definitely useful in lowering the blood glucose concentration, but with a minimal effect on the reduction of diabetes mortality. Similarly, statins reduce by 30% the blood cholesterol levels, but despite the propaganda of the producers and researchers affiliated to them, the difference with the placebo, with respect to the reduction of cardiovascular events, is 4%, which becomes a miserable 1.2% for mortality at 5 years from the event (Leucht et al. 2015). But what is the reason for this poor efficacy of pharmacological therapies? We might think of insufficient effort in biomedical research. But the data on the financial resources used for research say the opposite.

2.4 A Sea of Money for Biomedical Research

In the year 2012 (last standardised data available), the (public and private) global investment in research and development in the biomedical sector at world level was approximately 268 billion dollars. By calculating the 2007–2012 period, approximately 1500 billion dollars were invested; of these, a little more than half (793 billion) was invested in the USA. In recent years there has been a growth of investments in Asia, both because the Western industries are relocating their investment in India, South Korea and Singapore due to the more favourable conditions of exploitation of the work and for the state facilitations, and because of the strong growth in China and Japan (Chakma et al. 2014). The funds to the main public biomedical research institutions in the USA (and worldwide) are impressive. The National Institutes of Health (NIH), despite some contractions in the years of economic crisis, have had a budget of over 30 billion dollars per year since 2014. Even if it is chilling to read that the USA spend twice that amount (60 billion per year) for defence, with an increase in 2016 decided by the Peace Nobel Prize winner Obama of 8.4% on the 2015 Budget, health research however remains the second financing for the expenditure entity that is approved every year by the United States Congress. Calculating only from 2000 to 2016, the NIH received from the American Congress over 500 billion dollars of financing (roughly the entire annual budget of the Italian State), of which 83% went to finance basic and applied research. The budget for cancer research, which is the most substantial and has been practically stable for several years, is around 5 billion per year. To get an idea of the order of funding priorities, we need to consider that the National Institute in charge of environment received little over 600 million dollars for 2016 and the newly born National Center for Complementary and Integrative Health, which deals with integrated care, had to settle for a little more than 100 million dollars (National Health Institute 2016). The problem does not therefore lie in the money allocated to research to improve health which, as we have already noted, represents imposing amounts, although always too little compared to that spent on instruments, people and the means to kill each other. The cause must be identified in the paradigm that directs research. But how can we correctly define the reductionist biomedical paradigm?

2.5 The Paradigm: A Vision Based on Concepts, Practises, Interests and Apparatuses

In 1962 the American philosopher with a training in physics Thomas Kuhn (1923–1996), proposed for the first time in the book The Structure of Scientific Revolutions the concept of paradigm. Yet Kuhn's most famous creature, the "paradigm", at least at a linguistic level, is alive and well and the quotation from The Structure of Scientific Revolutions belongs to the group of classic quotations, whenever reference is made to a significant change in any field, not only scientific.

2.5.1 A Paradigm to Learn and Know

Without entering into the details of the American philosopher's deep and articulate reflection, it is enough for the purposes of our discourse to remember a double distinction made by Kuhn. The science philosopher distinguishes between a paradigm as an "exemplary case" and a paradigm as a "disciplinary matrix". In the first sense, the paradigm is like a painting, a single work of art that serves as a model, which the student in the painter's workshop uses to learn to represent reality and that the student at school uses to solve in a similar way the problems that arise in the individual disciplines. In this sense, the manuals in which we study physics, biology, medicine and other scientific disciplines are a collection of models, a series of "solutions to concrete problems that the profession has come to accept as paradigms". These texts "then ask the student to solve for himself problems that are closely modelled in method and substance upon those through which the text has led him". From here the role of science education "as a relatively dogmatic initiation into a pre-established problem-solving tradition that the student is neither invited nor equipped to evaluate"(Kuhn 2000). The paradigm in the limited sense, as an exemplary case, therefore forms part of a wider and more structured cognitive system, which Kuhn calls "disciplinary matrix", which is not called into question and that contemplates symbolic generalisations, metaphors and similarities, values. The first are the laws, we can say the universal assertions of the discipline. The second constitute the "shared dogmas". Finally, the values relate to common rules, to research goals in a certain field. The sociopsychological context in which researchers are formed and work is therefore fundamental in orienting research that, for Kuhn, is not the work of isolated individuals, however brilliant, but of groups of professionals who have received the same education and that follow the same research models. All these cognitive apparatuses that have their own peculiar specialised language, practices and interests constitute "normal science", in which problems not solved by the paradigm are classified as "exceptions", which time, i.e. scientific progress in the vision of "normal science", will bring back to the rules. Cognitive apparatuses are based on educational institutions, on world scientific networks and on research programs of which industry has been the protagonist over the past two centuries.

2.5.2 The Paradigmatic Role of Industry

Historically, the industry of manufacture of drugs and medical devices has played a very pervasive and incisive role with doctors, researchers and policymakers. It did so without hesitation, as shown by the numerous examples of corruption in public administration, scientific institutions and individual operators, which have involved all the main rich countries whose details the reader will find in two books cited in the Recommended reading, by two important authors: Jerome Kassirer and Marcia Angell, both former directors of the New England Journal of Medicine. This is probably also the reason why the pharmaceutical branch of the industrial sector has crossed the great economic crisis unharmed and has increased its turnover and monopolistic concentration: in 2015 the 15 leading companies that constitute the so-called Big Pharma, recorded a turnover of over 500 billion dollars, but generally all pharmaceutical industries in the world cash in more than 1000 billion dollars each year. However, in our opinion, the most important and dangerous aspect of the industrialisation of medicine does not lie so much in the outpouring of money, which still has the unedifying aim of creating a favourable context for the sale of drugs, as in the effects of a serious distortion of research and of the vision of health and disease. The industrialisation of medicine has produced a standard "medical mind", a paradigm with which to watch health and disease that has uniformised researchers, health workers and citizens. At the base of the paradigm there is the following syllogism: health, disease and individual characteristics depend on elementary factors which are genes, microorganisms, structural alterations. Medicine can locate and solve the problems that they create. Health is in the hands of science (pharmacology and surgery) and of the physician. The effects on research are highly distorting, because most of the answers to the scientists' questions originate from the questions proposed by the reductionist paradigm path oriented by industry. The questions are of this type: which is the molecular switch that I can switch on or off with a molecule synthesised ad hoc, to influence the activity of the cell and therefore the genesis of a disease? Asking oneself exclusively these questions cuts away research in one go, because it is not reducible to a molecular switch and therefore deletes the investigation on the complex effects of substances (plants, food, natural products), of complex methods (diets, acupuncture, meditation, psychotherapy and other methods) and of behaviour (physical activity) and, a fortiori, excludes research on integrated care programmes. Or, even better, these domains of scientific research are relegated to the margins in terms of funding, publication in high impact magazines, dissemination of scientific ideal (the "shared dogmas" mentioned by Kuhn) within the research community.

2.6 Integrated Care

The clinical damage caused by the reductionist paradigm in terms of prevention and care is huge, because, as we have seen in this book, the standard "medical and psychological mind" deprives the therapist of the possibility of realising a quality leap in care centred on the integration and involvement of the patient in the first person. At the same time, the therapist is deprived of the possibility to establish themself as an autonomous subject that benefits of an evolved, non-dogmatic scientific culture, who knows the effectiveness, the limits and risks of each of their therapeutic proposals and therefore knows that not only drugs, but also social support, psychotherapy, meditation, power, physical exercise, plants, needles, body manipulations can affect care and that their combination, adapted to the individual patient seen in their entirety, may have surprising synergistic effects. Integration, for us, means first of all seeing the person who asks for help in their entirety. Therefore, every patient, whether they meet a doctor, a surgeon, a psychologist or another therapist, should receive an integrated diagnosis, i.e. that combines examinations and biological and psychological evaluations, in the context of an examination of their biography, that covers the main events of life and not simply those concerning health, as is currently done in the traditional medical history.

2.7 The Promotion of Health in the Neoliberal Era

As Hippocrates recalled 2500 years ago, if you want to try to cure, it is necessary to study the person in their environmental, physical and social context: physis kaí nómoi, the nature and the rules, the customs. In the contemporary age, the philosopher Gregory Bateson (1904–1980) has described the human being as "imprisoned in a frame of epistemological and ontological premises"(Bateson 1976). Our epistemology (conscious and unconscious beliefs on the world) affects our behaviour and also our biological being, ontology, which in turn affects our epistemology. Reductionist science finds it very hard to insert the social and cultural data in the study of the human being and it is laughable as well as infuriating to read in important neuroscience magazines that, for example, the resounding failure of the pharmacological cure for addictions perhaps derives from the fact that "we have neglected the context in which drug addiction develops" and that therefore it is necessary to change the route, by "adding the social factors to research on addictions" (Heilig et al. 2016). With this example we are faced with the admission of the failure of the pluridecennial research paradigm centred on the brain dopamine of the little mouse in a cage. This has built prestigious university careers which have split the hair on dopamine receptors and their genetic polymorphism: hard science, founded on the genetic binome and animal research, that was tossed as a weapon against those who saw drugs as a prominent social phenomenon, to be fought with studies and the contrast of the social and economic reasons that feed it. This is also why animal research needs to be rethought. For ethical reasons but also for strictly scientific reasons: a human being is not a 70 kilo rat, our psyche-brain system and the human social context are not comparable to those of other animals. This does not mean, from our point of view, the sudden abolition of animal experimentation, because its necessary overcoming will be a long process, an integral part of the emerging of the new scientific paradigm. There is also a need to pursue with determination and with the necessary safety measures the idea that scientific research for human health should be done on and with the human being in its entirety. The more we advance along this road, the more we will spare animal suffering and gain scientific knowledge.

2.7.1 Physis and nómoi Today

Contemporary human beings are living a historical era of major upheavals, linked to increasingly threatening environmental changes and unsettling social changes. It is the end of classic modernity, which had its "glory" in the early decades after the second world war. An era in which the nineteenth-century ideal of progress seemed to come true. Of course, at the price of hard social and cultural battles that strove upon the social structure, but which at the same time gave a unified image of it, inside which the worker and the manager, the student and the professor, the shopkeeper and the bureaucrat, the intellectual and the politician, the woman and the priest, played clearly defined roles in their respective organisations. A stability of antagonism that directed state policies of welfare, of extension of civil rights and which constituted the basis of identity and of individual lives. From the point of view of public health, the decade 1975–1985, which culminated with the so-called Ottawa Charter on health promotion (World Health Organization 1986), was the attempt to spread health policies to the whole population based mainly on the change of social determinants. This is the era in which we tried to build networks of "healthy cities", of "health-promoting schools" and also the dissemination of structures such as family counsellors and tools for early diagnosis, such as the Pap test. Starting from the end of the last century we definitely passed from that phase to a condition dominated on one side by corporate health economism (hospitals and territorial health services have become companies lead by managers whose goal is the control of the budget), and on the other, by the closing of the individual on himself. Marketing, privatisation, individualisation of health are the words that summarise some underlying characteristics of the current condition. The instruments of the previous step, as we recalled, were of an institutional type: the individual dimension of health came second, was so to say overdetermined by the collective, social and political dimension. In that model, the individual was recognised freedom of treatment, but the health game was played outside the subject, in the so-called social determinants, which were the subject of public policies. With the current phase, the focus of the health promotion action has fallen on the individual, whose capacity to take care of himself has been strengthened.

2.7.2 Neoliberal Self-Care as a Deception and as Bad Conscience of the State

It is quite clear that the emphasis on the individual as health promoter serves to cover the guilty obliteration of the institutions, if, as it has been established by a survey (Censis-RBM Assicurazione Salute 2016), 11 million Italian citizens in 2016 were not able to receive adequate care for economic reasons, because now, in contempt to the public health service, you must reach into your purse if you want to receive health care. Among the excluded citizens, 2.4 million are elderly (with starvation pensions) and 2.2 million are young people under 30 years (without work). This deception opens the way to private health care and to the growing wellness industry while obscuring the dramatic and increasing inequality of access to support and health renewal sources. In addition, by separating the public conditions on which individual health is built from the social and economic context, we prevent health care sciences from understanding the peculiarities of the troubles and the risks of the current era. The second half of the last century, with respect to the first, has seen a fundamental change in the health epidemiology of the population, which has passed from a prevalence of infectious diseases to a prevalence of acute cardiovascular diseases. But in the phase that has opened now, the main pathologies will be cancer and dementia for the older groups of the population, autoimmune pathologies for the middle-aged, in particular women, disorders of neurodevelopment and psychiatric disorders for children, adolescents and young adults. The forecast results from the analysis of current trends, but above all is based on what is happening and will happen in human society.

2.7.3 Automation, Unemployment, Isolation, Virtual World

Since the 1980s of the last century, the so-called third industrial revolution founded on microelectronics combined with information technology has begun in rich countries. "It has allowed to automate to a very high extent both the physical processes typical of industrial production and the collection, compilation and distribution of large volumes of information: the so-called big data" (Gallino 2015). The consequences have been and will be imposing: there will be a massive reduction of the workforce required for production processes in the broad sense. Not only workers but also the traditional professional figures, from the architect to the lawyer and in a near future the doctor and the psychologist, are and will be increasingly redundant, and replaced by artificial intelligence programmes. The capitalist dream of reducing employees to a minimum, up to the delirium of imagining doing without the labour force, that has the disadvantage of being composed of human beings with needs, ideas, claims, diseases, is happening before our eyes. Google has about 60,000 employees and has a 15 billion dollar declared annual profit. Apple has 115,000 employees, has a 700 billion capital and records an annual revenue of 40 billion, but also the young company Facebook, founded in 2004, with 12,000 employees realises an annual revenue close to 4 billion dollars. A traditional company, for example, the automotive giant General Motors, has almost twice as many employees as Apple, but its profits are 6–7 times lower, not to mention Fiat Chrysler (FCA), which has three times more employees than Apple with ridiculous profits compared to the Cupertino company (Wikipedia 2016). The way is pointed by Apple, Google, Facebook, Amazon, which are able to combine the use of the most advanced technology and a reduced component of human workforce with an image of captivating modernity, youth, innovation and freedom.

In fact, under their "liberal" patina, these companies pursue, with a determination that is not inferior to that of the old "steam masters", the objective of paying the lowest taxes possible and of getting the most out of their workers, as demonstrated by the actions of the European Union toward Apple for the recovery of several billion euro circumvented taxes and as documented by accurate journalistic surveys on the working conditions at Apple and Amazon (Kantor and Streitfeld 2016). Proto-industrial working conditions in terms of time (60–80 h a week on average) and gruesome in human terms, as the company stimulates ruthless competition without excluding low blows between their employees. It is expected that shortly, in 2020, half of the U.S. workers will be freelance, i.e. will not have a stable work relationship and will therefore continually need to seek for and negotiate their own employment with different employers, in systematic competition with their own kind. This condition, which is typical of all "developed" countries, places the human being in a continuous state of alert, like a hounded animal that tries to interpret the signals of a hostile environment, according to the effective definition of the psychiatrist Miguel Benasayag (Benasayag and Del Rey 2016).

In particular young people, the so-called millennials (born between the Eighties and the year 2000), according to the sociologist Zygmunt Bauman, are not supported by the stable and authoritative values of the past. They are therefore in the condition of "continually redefining their identity to be put on the market to find its value"(Bauman 2015) and then to verify if it has a market, if it receives the necessary "likes" or if it is instead ignored or even pilloried by the media. Isolated, alert, hunted down, insecure, but incapable of rebelling, indeed even of imagining to possibly contest the national and world order that continues to accumulate wealth, prosperity and security in a minimum proportion of the population. Here lies the main novelty of the current condition compared to the dramatic economic and social crisis of the past: collective inertia. What is it due to? To the dismantling of the traditional collective action organisations (political parties and trade unions), but also and above all for young and middle-aged people, to epistemological frameworks, described by Bateson, that imprison considerable masses of people in a virtual reality that, in spite of the real conditions of life, makes you feel free, self-determined, "entrepreneur of yourself". Under this profile, the ideology and the new apparatuses of neoliberalism have hit the mark: Facebook (that in July 2016 reached 1 billion and 650 million people, with approximately 1 billion people who use it daily) makes you believe that you have extensive social relations; it gives you the illusion that you exist as an autonomous subject equipped with a free and conditioning power when you click "like" or when you "post" a disagreeable or slanderous comment; it also enables you to participate, while staying sheltered, in what is called by a refined British neologism shitstorms (literally, storms of shit), i.e. bloodthirsty pillory rites and mediatic killing of a designated victim, which then, as it sometimes happens, really kill themselves. In addition, the ideology of self-determination, summed up in the formula "everyone is an entrepreneur of himself", becomes the means of structuring a self-exploitation without any rights, with a person perpetually in business, day and night, to keep up with the cage of commitments that it has built for itself.

2.7.4 Artificial Intelligence at the Service of Psychopolitics

The development of artificial intelligence in sophisticated form of machine learning (operating systems, algorithms, able to learn alone on the basis of the data available) will increase to the maximum the control over individuals, since the systems already know their tastes, preferences, habits, weaknesses, thus opening the way to an unprecedented control and possible manipulation of the human psyche (Han 2014). At the same time, artificial intelligence will manage human loneliness with virtual reality and robotic tools that will address the poverty of relationships and social support by making people live in an unreal bubble, feeding and gigantifying the distortions of the human psyche. This is the case with the new generation of robotic dolls with human features equipped with language, which we imagine will be increasingly more evolved to meet the perverse male dream of an obedient, loving, helpful wife, who stays in her place and is always sexually available.

2.8 A Phase of Extreme Danger and Opportunity

All those involved in the third industrial revolution, which is also called the "second machine age"(Brynjolfsson and McAfee 2015), agree in indicating that with the arrival of machine learning (Domingos 2016) we are close to an epochal change that will leave its mark in the social organisation and therefore also in the psychobiological structure of human beings. We foresee: a strong increase in unemployment also of the intellectual professions; an unprecedented acceleration of the trend that is already well underway of social inequality increase; a more rigid social control; a modification, which we can only guess today, of the psychic features and the brain of human beings. It is indeed not science fiction to imagine that through the processes described, it would be possible to produce a really epochal evolutionary pressure on the human species. When 8–10,000 years ago the human populations started to become sedentary and to till the soil, our species underwent some peculiar changes: our ancestors became shorter than their hunter-gatherer ancestors, but proliferated at a speed that had never been seen before and that progressively allowed the diffusion of the species over the whole planet.

When technological jumps are pervasive and disruptive, they change the conditions of life reproduction forever. It is not possible to go back, as it happened for industrialisation, motorisation, the computerisation of life over the last 150 years. We can only think, soon enough, of opening a phase of collective awareness on where we are going, in order to place robust corrections on the processes underway, with the aim of using them to improve the quality of life of the great majority of people who would otherwise not only miss the exceptional opportunities in terms of cultural growth, effective freedom of health and wellness, offered by the second machine revolution, but would instead be literally crushed by it. Just think what increase of current wars and terrorist campaigns the full maturity of the machine learning technology could lead to: the human species could be literally devastated by devices that would only obey their own algorithm, which would be self-sustaining starting from the simple objective of destroying the enemy.

2.9 Care Sciences in the Face of the Epochal Changes under Way

Medicine and care sciences have always been very closely linked to industrial development and even military technology. The first pharmaceutical industries, at the end of the nineteenth century, were chemical industries that produced dyes; the first anticancer drugs were the result of research to produce chemical weapons during the First World War 1914–1918; the first mass experiment of penicillin was made in the trenches of the Second World War (1939–1945). Atomic physics that led, with the contribution of Enrico Fermi and the best world physicists, to the construction of the first weapon of mass destruction, dropped on Japan more than 70 years ago, with hundreds of thousands of deaths and with the unspeakable suffering of successive generations, on orders from the Democratic president Harry Truman who had just succeeded to the legendary Democrat Franklin D. Roosevelt, changed the medical diagnosis and therapy of certain cancers.

Over the centuries, biomedical sciences have also been supplied with technological metaphors. From Descartes until the mid-eighteenth century, the metaphor most frequently used by doctors and physiologists to represent the functioning of the human organism was the machine: the clock, the mechanical fountain, the industrial machine. Devices with a wonderful network of springs and levers which allowed the activity of the instrument without human intervention, except in their programming, control and maintenance. A famous doctor and philosopher, Julien Offroy de La Mettrie (1709–1751), wrote, in 1747, a beautiful treaty entitled "L'Homme Machine", which described human physiology as a fruit of the activation of mechanical devices. One hundred years later, in the steam machine era, German physiologists repeatedly resorted to the concept of energy produced by fuel to describe the human metabolism. Almost three centuries after La Mettrie, the risk is to pass from the metaphor to the reality of the man–machine. In the sense of the replacement of man by machine, which mimics him by eliciting our peculiarity, the ability to learn and then evolve under the cognitive profile and that of performance, but also in the sense of the absolute domination of a mechanistic vision of the human being and of the universe.

The reductionist paradigm in biomedicine, which is going through a serious crisis at the beginning of this century, could find in the "second machine age" a new lifeblood, which would be detrimental to the care sciences and professions. This would worsen all its negative traits, already identified and summarised in a comment of New England of Medicine and presented in the left column of Table 2.1 (see below). In particular, the foreseeable dissemination and adoption by state and scientific authorities of diagnostic and therapeutic algorithms, built on the reductionist paradigm, will produce such constraints on the therapeutic activity that the current one dominated by "guidelines" will appear as an age of freedom. It is expected that in this scenario, the therapist, independently from his or her specialisation, will be subservient to the protocols built by machine learning. This will lead to a worsening of the limited satisfaction for care which already now unites many patients and a growing number of therapists. There will be an attempt to mask this dissatisfaction by increasing what another comment of the New England Journal of Medicine defines the "therapeutic illusion" that, on the one hand, works as a self-justification for the use of useless drugs and medical devices and, on the other hand, as a protection against the pressures and threats of legal action on the part of the patient's family members (Casarett 2016).

Health and care capacity in the so-called third millennium (the thirteenth millennium, having the agricultural revolution as reference) may instead take a spectacular leap forward by combining the systemic paradigm of psychoneuroendocrinoimmunology with a technology that is not enslaved to potentates, but to the development of humanity. A systemic vision of scientific imprint can indeed only rely on an advanced technology, which allows the development of molecular research and which greatly improves the knowledge of human complexity and abandons reductionism which does not mean doing without molecular research, but doing without the idea that the complexity of biological phenomena is reducible to simple determinants. An extremely difficult challenge that however also represents the hope of building a restraining and opposing force to the thrust toward catastrophic results that, in the world that we have described, is active and powerful. We can imagine the objection: but this is a cultural and political dimension that reaches beyond science and care professions!

In fact, this dimension has always existed in the sciences that deal with the human being. It used to exist in philosophy until recent times, until the collapse of the "Great Stories" was replaced by "unstable forms of rationality" and by the "weak thought" of postmodern philosophy (Lyotard 1984; Vattimo and Rovatti 1983), whose cultural climate has objectively favoured the imposition of the neoliberal hegemony. It used to exist in medicine, which joined in large waves, from the nineteenth century up to the seventies of the twentieth century, social progress and the conquest of civil rights for all with the growth of scientific knowledge and care capacity, and then had to endure a long period of cultural restoration. We need to reconnect to those experiences of partnership between science and philosophy and care professions and structure it upon new bases. The individual dimension of health, of self-care, needs to be subtracted to the rampant neoliberal ideology.

We think that health is an attribute of the subject in a socially and politically determined context. "Health is a common good because it is also a good of the individual" (Ingrosso 2013). Individuals can therefore intervene actively in the determination and management-renewal of their health by using social and environmental conditions which favour their efforts. This implies a structured plan for health practises, whose interdependent areas are self-care and the care of others, practices of health promotion on a mass scale and health intervention based on integrated care. By care of others we mean not only the exercise of the medical profession which, to be exercised at its best, needs an operator who is able to take care of himself and therefore can recognise, in the other whom they are helping, themself and their life and health management problems (empathy). We also intend care practices of social relations and of environmental ecosystems, whose influences on individual health we have widely documented.

For this reason, in order to proceed in the renewal of care sciences and professions, it seems essential to build a critical thinking trend in science, culture and politics, encouraging desires and collective projects to bring changes at all levels of the human health structure.

References

Abd El-Kader SM, Al-Shreef FM (2018) Inflammatory cytokines and immune system modulation by aerobic versus resisted exercise training for elderly. Afr Health Sci 18(1):120–131. https://doi.org/10.4314/ahs.v18i1.16

Ader R (2007) Psychoneuroimmunology, 4th edn. Academic Press, San Diego

Bateson G (1976) Verso un'ecologia della mente. Adelphi, Milano

Bauman Z (2015) "I giovani 'liquidi'. Una, nessuna, centomila identità", la Repubblica 21 maggio, p. 39

Benasayag M, Del Rey A (2016) Oltre le passioni tristi. Dalla solitudine contemporanea alla creazione condivisa. Feltrinelli, Milano

Bottaccioli AG, Bottaccioli F, Minelli A (2019) Stress and the psyche-brain-immune network in psychiatric diseases based on psychoneuroendocrineimmunology: a concise review. Ann N Y Acad Sci 1437(1):31–42. https://doi.org/10.1111/nyas.13728

Bottaccioli F, Bottaccioli AG (2020) Psychoneuroendocrineimmunology and science of integrate care. The manual. Edra, Milano

Brynjolfsson E, McAfee A (2015) La nuova rivoluzione delle macchine. Lavoro e prosperità nell'era della tecnologia trionfante. Feltrinelli, Milano

Casarett D (2016) The science of choosing wisely—overcoming the therapeutic illusion. N Engl J Med 374:1203–1205

Censis-RBM Assicurazione Salute (2016) Dalla fotografia dell'evoluzione della sanità italiana alle soluzioni in campo, www.censis.it, accessed 25.09.2016

Chakma J, Sun GH, Steinberg JD et al (2014) Asia's ascent—global trends in biomedical R&D expenditures. N Engl J Med 370(1):3–6. https://doi.org/10.1056/NEJMp1311068

Domingos P (2016) L'algoritmo definitivo. La macchina che impara da sola e il futuro del nostro mondo. Bollati Boringhieri, Torino

(EEA. European Environment Agency) 2019. Air quality in Europe—2019 Report. ISSN 1977–8449, Luxembourg: Publications Office of the European Union)

Gallino L (2015) Il denaro, il debito e la doppia crisi. Einaudi, Torino

Han B-C (2014) Psychopolitik. Neoliberalismus und die neuen Machttechniken. Fischer Verlag, Frankfurt am Main

Heilig M, Epstein DH, Nader MA, Shaham Y (2016) Time to connect: bringing social context into addiction neuroscience. Nat Rev Neurosci 17:592–599

Ingrosso M (2013) Attualità e riorientamento della promozione della salute nello scenario sociale contemporaneo. Sistema Salute 57:249–264

Jiménez-Pavón D, Carbonell-Baeza A, Lavie CJ (2020) Physical exercise as therapy to fight against the mental and physical consequences of COVID-19 quarantine: special focus in older people. Prog Cardiovasc Dis. https://doi.org/10.1016/j.pcad.2020.03.009. [published online ahead of print, 2020 Mar 24]

Kantor J, Streitfeld D (2016) Inside Amazon: wrestling big ideas in a bruising workplace. The New York Times. 15 August

Kuhn T (2000) La funzione del dogma nella ricerca scientifica in Kuhn T. Dogma contro critica, Cortina, Milano, pp 3–32

Lai J, Ma S, Wang Y et al (2020) Factors associated with mental health outcomes among health care workers exposed to coronavirus disease 2019. JAMA Netw Open 3(3):e203976. Published 2020 Mar 2. https://doi.org/10.1001/jamanetworkopen.2020.3976

Leucht S, Helfer B, Gartlehner G, Davis JM (2015) How effective are common medications: a perspective based on meta-analyses of major drugs. BMC Med 13:253. https://doi.org/10.1186/s12916-015-0494-1

Liang T (ed) 2020. Handbook of COVID-19 Prevention and Treatment Zhejiang University School of Medicine. http://www.zju.edu.cn/english/2020/0323/c19573a1987520/page.htm

Lyotard J-F (1984) The postmodern condition: a report on knowledge. University Of Minnesota Press, Minneapolis, Minnesota

National Health Institute (2016) The data are taken from the website of the National Health institutes and specifically from the "Office of Budget" page, officeofbudget.od.nih.gov/spending_hist.html accessed 18.09.2016

Pfefferbaum B, North SC (2020) Mental health and the covid-19 pandemic. N Engl J Med 15:2020. https://doi.org/10.1056/NEJMp2008017

Qin C, Zhou L, Hu Z et al (2020) Dysregulation of immune response in patients with COVID-19 in Wuhan, China. Clin Infect Dis 2020:ciaa248. https://doi.org/10.1093/cid/ciaa248. [published online ahead of print, 2020 Mar 12]

Salazar N, Valdés-Varela L, González S, Gueimonde M, de Los Reyes-Gavilán CG (2017) Nutrition and the gut microbiome in the elderly. Gut Microbes 8(2):82–97. https://doi.org/10.1080/19490976.2016.1256525

Van Avondt K, Hartl D (2018) Mechanisms and disease relevance of neutrophil extracellular trap formation. Eur J Clin Investig 48(Suppl 2):e12919. https://doi.org/10.1111/eci.12919

Vattimo G, Rovatti PA (1983) Il pensiero debole. Feltrinelli, Milano

Wan Y, Shang J, Graham R, Baric RS, Li F (2020) Receptor recognition by the novel coronavirus from Wuhan: an analysis based on decade-long structural studies of SARS coronavirus. J Virol 94(7):e00127-20. Published 2020 Mar 17. https://doi.org/10.1128/JVI.00127-20

Wikipedia (2016), it.wikipedia.org, accessed 24.09.2016

World Health Organization (1986) WHO Ottawa Center for health promotion: an international conference on health promotion. The move towards a new public health. WHO, Geneva

WHO access to: http://www.euro.who.int/en/health-topics/health-emergencies/coronavirus-covid-19/news/news/2020/3/who-announces-covid-19-outbreak-a-pandemic, 07 april, 2020

Zhu N, Zhang D, Wang W et al (2020) A novel coronavirus from patients with pneumonia in China, 2019. N Engl J Med 382(8):727–733. https://doi.org/10.1056/NEJMoa2001017

Suggested Reading

Bottaccioli F, Bottaccioli AG (2020) Psychoneuroendocrineimmunology and science of integrated care. The Manual. Edra, Milan chap. 23

http://www.euro.who.int/en/health-topics/health-emergencies/coronavirus-covid-19/news/news/2020/3/who-announces-covid-19-outbreak-a-pandemic, 07 april, 2020.

From General Ecology to Bionomics

Vittorio Ingegnoli and Elena Giglio

Abstract

Background: The necessity to follow a new scientific paradigm (shifting from reductionism to systemic complexity) leads to the emergence of a new concept of life, not centered on the organism, but on the entire "biological spectrum." Limits to traditional ecology have emerged yet, and the aim of this chapter is mainly to present an upgrading ecological discipline.

Theory and Method: Bio-hierarchical systems interact and govern Culture, but men do not respect nature. This violence against nature damages human health, so biologists must upgrade traditional ecology to rehabilitate our environment following the System Theory. Thus, Ingegnoli proposed the Bionomics discipline and the Landscape Bionomics or biological-integrated Landscape Ecology. A synthesis of the main concepts is presented: the complete Biological Spectrum, Structures, and State Functions, Diagnostic Evaluation, Vegetation Science and Agroecology, Territorial Governance, and Planning and role of Urban and suburban Parks.

Findings: Observe that epigenetics confirms bionomics principles, allowing a crucial linkage between genomic scale and environment, enhancing the interrelations among space-time-information scales and the possibility to deepen the relationships between environmental health and human health. The etiology due to environmental stress assumes vast importance.

Discussion and Conclusion: Landscape syndromes damage human health independently from pollution, and this changes the relations between Bionomics/Ecology and Medicine. Even medicine doctors should have a formation on landscape bionomics, at least who is interested in Public Health and Planetary Health.

Keywords

System complexity · Bionomics · Landscape syndrome · Diagnostic evaluation · Health

---

V. Ingegnoli (✉)
Department of Environmental Sciences and Policy, University of Milan, Milan, Italy

PHA (Planetary Health Alliance, member of), Harvard University, Cambridge, MA, USA

SIPNEI (It. Soc. of Psycho-Neuro-Endocrine-Immunology), Rome, Italy
e-mail: vittorio.ingegnoli@guest.unimi.it

E. Giglio
PHA (Planetary Health Alliance, member of), Harvard University, Cambridge, MA, USA

Environmental Science: man and the environment in the Appennine, PhD, Teacher of Geography at High School, Milan, Italy

3.1 Paradigm Shift and a New Concept of Life

3.1.1 Premise: Effects of Life Alterations on Human Health from Small to Broad Scale

The passage from reductionistic biology to a systemic one needs the revision of the concept of life. Biology is primarily concerned with the concepts of cell and the organism. In the last century, biologists investigated its components arriving at bio-molecular scales while leaving the biological levels above the organism to generic and often confusing descriptions.

Billions of years ago, when opportune environments emerged, complex sets of macromolecules were led to maintain themselves separately from the mother-environment, self-manufacturing in cells supported by a *metabolic* process. Note that the cell separation does not signify to cut out the environment, but, on the contrary, to develop more linkages enhancing the creative potentialities of this hyper-complex system. Thus, we cannot discuss the unity of Life. The alterations of Life at these broader scales can damage human health, not unlike from small ones. As underlined by Bottaccioli, in the previous Chap. 2, health and disease depend primarily on Life's organization, and we must affirm "on the *full organization*," from small to broad scales. The study of biology must be upgraded even to the broader scales, trying to understand their anatomical components (structure), physiological processes and functions, transformation processes, clinical-diagnostic evaluation, pathology, and therapy.

Dotta et al. (Chap. 14) noted that genetic variability alone could not explain the profoundly different risk of developing chronic diseases. Genome-Wide Association Studies (GWAS) revealed a limited causal effect (estimated less than 20%) of genetic susceptibility on phenotypic variance. Consequently, environmental exposure plays a crucial role in disease development, especially in non-communicable diseases (NCDs), such as cancer, asthma, allergy, cardiovascular and endocrine diseases.

In reality, we have to underline that environmental exposure and exchanges, and their interaction with the body's biological systems and apparatuses, play an essential role in disease development. Note that the concept of exposure [e.g., the exposome, sensu Wild (Wild 2012)] may be necessary but not sufficient because of the complex structures and interrelations of life. We can distinguish at least four categories of environmental alterations capable of influencing human health through exposure to them:

(a) internal processes (e.g., metabolism, hormonal balance, gut microbiota, aging, etc.),
(b) specific external factors (e.g., infections, pollutants, smoking, drugs, etc.),
(c) general external factors (e.g., socioeconomic status, technological behaviors, climate change, etc.), and
(d) landscape structural and functional alterations (e.g., concerning hierarchical relations, the biological territorial capacity of vegetation, vital space per capita, ratio human/natural habitats, etc.).

Medical communities are recognizing that human alteration of Earth's ecological systems threatens the health of humanity. This fact has given rise to the field of Planetary Health (Almada et al. 2017), which is interdisciplinary, but, first of all, it must be systemic: it needs a special relationship between Ecology and Medicine. This relation is to be upgrading because, today, both ecology and medicine pursue few systemic characters and few correct interrelations. It is the aim of this book: this upgrading is essential when we have to study the four categories here mentioned, particularly the fourth (d). It requires real systemic medicine and, still more, a new branch of systemic ecology. It needs to follow the new scientific paradigm shift and Bionomics foundation, as we will see in this chapter, theory, and applications, prolonging in Chap. 4 and presenting preliminary research on environment and health in Chap. 5.

3.1.2 Paradigm Shift from Reductionism to Systemic Complexity

Many epistemological studies (Agazzi 2014; Bottaccioli 2014; Ingegnoli 2002; Urbani-Ulivi 2019) indicate that the modern scientific method, deductive and experimental, although not yet been passed, shows severe limitations. Therefore, the limitations of Galilean methods, the basis of modern science today, are no more suitable? Let us deepen.

First of all, Galilei removed a cognitive openness from modern natural science: the concept of teleonomy, that is, the sense within nature. As well expressed by Agazzi (Agazzi 2014), the elimination of the *final causes* or *fundamental goals* of natural entities from natural sciences appears as an arbitrary constraint, especially in the study of actions and living organisms too: the radical empiricism is linked with determinism, which gives no importance to time and puts in question the human freedom and the knowledge of the real world and bring to reductionism.

The second and third limitations of the Galilean method are the necessity of quantification by mathematics and the repeatability of experiments, following which people affirm that General Ecology is not a Science: even if math is undoubtedly the language of Nature and Science, math and physics are not able to describe all the aspects of Nature. Note that the second principle of thermodynamics is not able to describe the characters of living entities. Furthermore, some aspects of Nature cannot be submitted to repeatable experiments (especially at higher Spatio-temporal scales). Conversely, complex behaviors of Nature may be model through opportune complex equations, following a systemic, integrated approach.

While Physics has been eliminating the cage of linear causality even before Einstein's times (1905) (Einstein 1944) and has been beginning to act by the systemic approach, Biology (all branches) has been remaining reductionist for decades. In Biology, the dominion of the dogmatic Neo-Darwinian theory of evolution and the so-called Central Dogma of Molecular Biology (Crick 1970) reduced the capacity to follow a systemic approach. The insistence to consider the DNA able to respond to every question on life is well commented by the Nobel David Baltimore: "It is clear that we do not gain our undoubted complexity over worms and plants by using more genes" (Baltimore et al. 2001).

Here is a classical question, related to man/environment health alteration, which can help us focus on the problem: people living in an urban neighborhood with 20% of green areas present more precarious health than people living with 40% green areas? The answer can be complicated and even wrong if ecology cannot make a valid diagnosis of green areas and the regional complex system related to the landscape unit. The reductionist analysis used by general ecology (e.g., photosynthetic area, species number, wood biomass, distance from housing, etc.) is insufficient. The problem needs a systemic clinical-dysfunctional analysis, to investigate landscape pathology, and an exact etiology linking the state of the environment to human health. Remember that it is possible to register a high ecological efficiency on the 20% green area and a lower value on the 40%; the mean efficiency value of the two examined green areas may be near the same: e.g., $6.70 \times 0.20 = 1.34$ vs. $4.0 \times 0.4 = 1.60$. If the regional environment presents an efficiency green value, respectively, of 2.00 and 2.70, we will have an ecological dysfunction in the second case (40% green) wider than in the first (20% green): $1.34/2.0 = 0.67 > 1.6/2.7 = 0.59$. It is impossible to find the hypothetical function of the ecological efficiency of green areas presented here within General Ecology: we will see forward that, that is only thanks to the new scientific paradigms that we found a function like this, named BTC. In addition, Bionomics enlighten that what common language named "Urban Green" correspond in reality to at least the following ten different types:

1. Urban forest (Allochthonous plants: 0–4%; Tree cover: >60%; BTC[1]: > 6.0 Mcal/m^2/year).
2. Urban parks with natural core green (Allochthonous plants: 5–6%; Tree cover: 50–70%; BTC: 3.5–4.5 Mcal/m^2/year).
3. Urban parks (Allochthonous plants: <20%; Tree cover: 30–40%; BTC: 2–3 Mcal/m^2/year).
4. Public garden (Allochthonous plants: <30%; Tree cover: 25–35%; BTC: 1.3–2 Mcal/m^2/year).
5. Private garden (Allochthonous plants: 30–40%; Tree cover: 20–45%; BTC: 1.1–3 Mcal/m^2/year).
6. Remnant fields (Allochthonous plants: 15%; Tree cover: 0–10%; BTC: 0.7–1 Mcal/m^2/year).
7. Vegetable garden (Allochthonous plants: <20%; Tree cover: 0–10%; BTC: 0.9–1.3 Mcal/m^2/year).
8. Square garden (Allochthonous plants: <40%; Tree cover: 10–20%; BTC: 0.4–2 Mcal/m^2/year).
9. Road green (Allochthonous plants: <30%; Tree cover: 5–20%; BTC: 0.4–1.5 Mcal/m^2/year).
10. Sport green (Allochthonous plants: 35%; Tree cover: 10–20%; BTC: 0.8–1.5 Mcal/m^2/year).

As previously highlighted, the necessity of a scientific *paradigm shift*, the so-called Systemic Turn (Urbani-Ulivi 2019), is thus impellent, even in biological fields and emerges from many reasons: the old Scientific Paradigm is mainly reductionist, anchored to the concept of process reversibility, to the Darwinian struggle for existence, to Newtonian physics, to the division among knowledge, to analytical quantifications, etc.

The new Paradigm is nearly the opposite: it is mainly holistic, better systemic, able to admit process irreversibility and give more importance to symbiosis and cooperation, to the Einsteinian physics of relativity, to information theory, Trans-disciplinarity, to consider systemic quantifications, etc. Trans-disciplinarity is an activity which may exceed the boundaries (often artificial) separating various disciplines, due to the need to overcome inter-disciplinarity, in the sense of M. Vitruvius Pollio *"Encyclios enim disciplina uti corpus unum ex his membris est composita"* (Vitruvius 1931), giving them a different importance in relation to the whole contest of the examined problem/situation. Therefore, the new scientific Paradigm Shift is hardly followed by Biology!

General ecology is the first to suffer from the reductionist limits, derived from this vision. Thus, it can be fallacious when it does not recognize the entire systemic organization of Life on Earth (i.e., the complete Biological Spectrum) and cannot frame the concept of environment and its complex implication. Today, physicians consider a few aspects of the environment, e.g., pollution, nutrition, climate condition, physical, and sport behavior, and often separately. On the other side, Medicine prefers to attend *sick care* than *health care* and rarely adhere to a right systemic view (Bottaccioli 2014), because it follows a particularistic and symptomatologic approach, not recognizing a wider one based on complex living systems. These views are not acceptable: an honest comparison with reality brought the most advanced medical communities to recognize that human alteration of Earth's Ecological Systems threatens humanity's health.

Nevertheless, both Ecology and Medicine reject such systemic concepts considering them as imprecise, belonging to the superficial level of common-sense language, that should be banned from the rigorous (reductionistic) discourse of science. As again underlined by Agazzi (Agazzi 2004, 2014), this attitude was in keeping with the *positive,* inspired scientific culture still predominant in the first half of the twentieth century but, up to today, too frequently followed.

Instead, studying complex systems, the Principle of Emergent Properties assumes prominent importance (Fig. 3.1). This principle affirms that system's behavior depends not only on the characters of its components and their reciprocal position but on the unpredictable properties

---

[1] BTC = Biological Territorial Capacity of vegetation. See Sect. 3.3.1

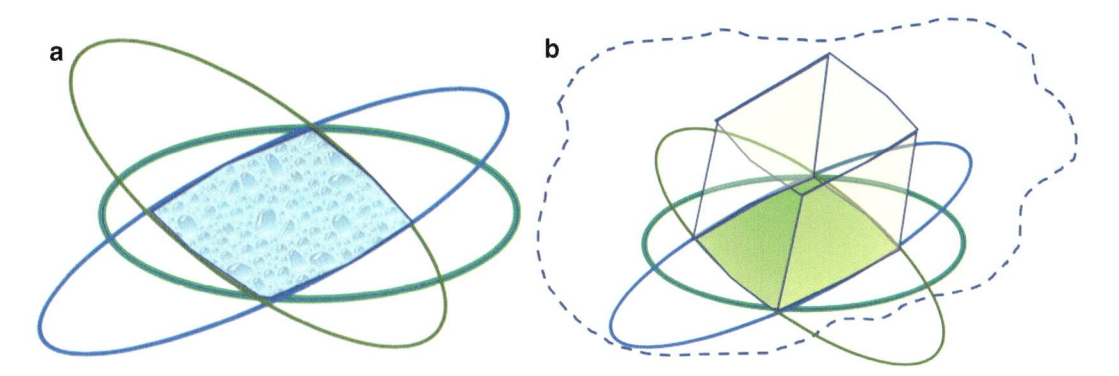

Fig. 3.1 Differences between an "ecological space" in conventional ecology (**a**) and in upgraded ecology (Bionomics), following the new scientific paradigm (**b**) (Ingegnoli, 2002)

acquired in its structuring as a system, properties that its parts do not have on their own. So, even if the "constitutive elements" of a system are disposed to enter complex compositions, the program (i.e., the design) is not evident in their structure (Lehman-Dronke 2007). Konrad Lorenz (Lorenz 1978) used to affirm that a system is always more than the sum of its parts.

Figure 3.1 shows the sharp difference between general ecology and a new one able to follow the new scientific paradigms. An "ecological space," i.e., the set of characters of a species community living in a local habitat, is represented by conventional ecology in Fig. 3.1a as an overlapping of the niche of each species. This overlapping was used by phytosociology (Pignatti 1976). However, it occurs also in general environmental studies, when the quality level of a place, or the zonation within an urbanistic plan, is defined through the overlapping of maps concerning: level of pollution, hydrogeologic risks, distance from disturbances, adequate climate conditions, etc. On the contrary, following the systemic point of view, the "ecological space" of the studied community derives by the emergent properties of this system, here expressed by the green volume (Fig. 3.1b).

3.1.3 The Emergence of a New Concept of Life

We have underlined that the passage from reductionistic biology to a systemic one needs the revision of the concept of life. We stressed that general ecology is mainly concerned with the concept of the organism. In the last century, biologists investigated its components arriving at biomolecular scales while leaving the biological levels above organism to generic and often confusing descriptions. So, the study of biology must be upgraded to wider scales (Fig. 3.2), trying to understand how they perform fundamental living processes and functions. We can define as their anatomical components (structure), how their transformation and evolution processes act, and how it is possible to gain a clinical-diagnostic evaluation of them, identify their pathology, and the possible criteria of therapy.

Only following these paths, the discipline of Biology may proceed more organically, consenting to its main fields to be more efficient and operative. Indeed, Life is a complex self-organizing process, operating with a continuous exchange of matter and energy with its environment: the system can perceive, process, and transfer information, can follow rules of correspondence among independent worlds (coding), can reach a target, can reproduce itself, can have a history and participate in the process of evolution (Ingegnoli 2001; Ingegnoli 2002; Ingegnoli 2015; Naveh and Lieberman 1984). Moreover, observe that, in an evolutionary view, structure and function become complementary aspects of the same evolving whole. Life is an open system. Consequently, life cannot exist without its environment: both are the necessary components of

Fig. 3.2 Hierarchical levels of life organization: examples at small space-time scale (**a**) from cell to ribosomes and at broader scales (**b**) from an Alpine landscape to a forested ecocoenotope (see Table 3.1) (Ingegnoli 2015)

the system, because life depends on the exchange of matter, energy, and information between a concrete entity, like an organism or a community, and its environment.

However, the world around life (an organism, a community) concerns other life systems; so, the integration reaches new levels again. That is why the concept of life is not limited to a single organism or a group of species. Life organization consists of hierarchic levels [i.e., the "biological spectrum" sensu E.P. Odum (Odum 1971, 1983)]. Moreover, biological levels cannot be limited to cell, organism, population, communities, and life support systems, as asserted in general Ecology. Life also includes ecological systems such as ecocoenotopes, landscapes, ecoregions, and the entire ecosphere (ecobiogeonoosphere) (Table 3.1). These hierarchies are not new to human culture, rather a confirmation, from the scientific community, of what our ancestors already sensed through the myths and religions, e.g., the Greek-Roman and Celtic sacred woods. As all remember, the Gaia Theory (Lovelock and Margulis 1974) has asserted that the Earth can be considered a living entity. Deepening, this new principle stated by Bionomics (i.e., the Theory of Life Organization on Earth), exposed in Table 3.1, underlines the difference between what exists (Life on Earth organized in Living Entities) and the different approaches to the study of the environment (viewpoints). They completely transform the main principles of traditional ecology

Table 3.1 Hierarchic levels of Biological Organization on the Earth

Scale	Viewpoints				
	SPACE[1] CONFIGURATION	BIOTIC[2]	FUNCTIONAL[3]	CULTURAL-ECONOMIC[4]	REAL SYSTEMS[5]
Global	Geosphere	Biosphere	Ecosphere	Noosphere	Geo-eco-bio-noosphere
Regional	Macro-chore	Biome	Biogeographic system	Regional Human systems	Ecoregion
Territorial	Chore	Set of communities	Set of Ecosystems	District Human systems	Landscape
Local	Micro-chore	Community	Ecosystems	Local Human systems	Ecocoenotope
Stationary	Habitat	Population	Population niche	Cultural/Economic	Meta-population
Singular	Living space	Organism	Organism niche	Cultural agent	Meta-organism

1= not only a topographic criterion, but also a systemic one;   2= Biological and general-ecological criterion;
3= Traditional ecological criterion;   4= Cultural intended as a synthesis of anthropic signs and elements;
5= Types of living entities really existing on the Earth as spatio-temporal-information proper levels

by being aware that hierarchical levels are types of living complex systems.

Focusing on the Earth, for millions of years, diverse geophysical environments had been forming, and about 3.5 billion years ago, the first cells appeared. Then, after the emergence of O2, the eukaryotic cells, multicellular systems, and, finally, the organisms occurred. When opportune environments emerged, complex sets of macro-molecules had been led to maintain themselves separately from the mother-environment self-manufacturing in cells through a *metabolic* system. The metabolism is a vast and complex web, depending on *enzymes* acting on nutrient inputs derived from the environment, which also allows a protective habitat: cell separation does not signify to cut out the environment. *Enzymes* are polypeptides composed of ordered amino-acid sequences synthesized by ribosomes. They acquire functionality when suitably folded by the internal conditions of the protoplasm. This intracellular milieu is led to the opportune condition by electrolytes, which allow other proteins to maintain a difference of homeostatic balance between inside-outside the cell.

This behavior activates inter-scalar complex relationships capable of *changing* the *environment* and *evolving it together with the cells*, which meanwhile evolve into organisms that continue to change the environment in a mutual exchange of information, matter, and energy. *The landscape is the information system essential for co-evolution and group selection* because the genetic characterization is linked to three scale levels: cell, population, *and landscape*. The extension of Life to large scales, as the landscape ones, permits a first understanding of the linkages among hierarchic levels of biological organization, because constraints are imposed from above (Hierarchical systems theory).

Deepening, the laws that underlie the behavior of territory as *landscape—the peculiar place for the evolution of man in nature*—are mostly the same ones that govern every other living entity, although if declined suitably. So, to speak of metabolism and order maintenance even at *a larger space-time scale*, e.g., in a forest ecotope or a landscape unit, is possible. In cases like these, there are two (or more) separate and distinguishable functioning levels, which depend strictly on each other for proper operation. Thus, the actions directed to the change of landscapes lead to the emergence of rules of correspondence between the complex structure of landscapes and new structures. During these transformations, the high complexity of the landscape expresses relationships allowing the maintenance or the increase of the right level of metastability through cybernetic processes. Similarly, chains of interacting organisms and communities behave as information networks in an ecotissue (which represents the structure of a landscape, see Fig. 3.4 further on), so that it is possible to maintain a certain level of metastability.

3.1.4 The Control of Bio-Hierarchical Systems on Men and Culture

Even human landscapes present a modality of transformation led by ecological laws, which may cause a change in the culture and the ethology of man. Many phenomena of landscape reorganization, both in the design of the territory and in population movement, seemingly decided by man, are controlled at the highest hierarchical level, and man remains a mere executor of an unconscious life necessity [e.g., land abandonment, natural-shaped gardens, natural remnants among the fields, interface plots, etc. (Ingegnoli 2015)].

For instance, let us examine the change in garden/park planning criteria during the industrial revolution. We may observe that, when rural landscapes were structured as "gardens" with remnant forest patches and a web of hedgerows in a very heterogeneous field matrix, the theory of garden design was strictly "formal," that is geometric, terraced and with leisure buildings (French or Italian gardening). However, when the presence of industries and agriculture changed, rural landscapes began to increase monoculture, a new theory—that of English gardening—became dominant, following natural landscape criteria. Figure 3.3 shows the sharp difference between the vast formal garden of the Villa Gallarati Scotti at Oreno (Vimercate, Milan) in the eighteenth century (above) and the transformation of that garden in an English Park, since the nineteenth century (below). The study has been elaborated following landscape bionomics principles, as we will see in the next paragraphs.

The biblical texts are to be read in the claim following which "our Mother Earth can sustain and *govern* us" has been expressed by St. Francis of Assisi (di Assisi 1224), in the well-known Canticle of the Creatures (1226). The reductionist objectivism, still followed by the most conservative scientists, suggests that all the ancient interpretations of nature were wrong, but that is not true. Rather, the current concept of nature is limited: for the ancient Greeks, the *Physis* was not a world of objects but a world of vital processes, as pointed out by Israel (Israel 2005). There is no doubt that man's position in nature involves problematic aspects, as he has to submit to the laws of nature that govern complex biological systems and can manage many aspects of their components, including humans. Similarly, the man should preside over, and control, the organization of the natural environment that allows him to live and develop: this position implies a problematic role, because it concerns a domain, even if in compliance. So, the role of man in nature involves a creative way, but also a clear responsibility. We must recognize that today, the duty of accountability is not heard, often even forgotten.

Furthermore, traditional western ethical perspectives are anthropocentric, so that they assign intrinsic value only to human beings. Aristotle (Politics, Vol.1, Chap. 8) maintains that "nature has made all things specifically for man." In a note essay on the historical roots of the environmental crisis, historian Lynn White (White 1967) argued that the main features of Judeo-Christian thinking had encouraged the overexploitation of nature by maintaining the superiority of humans over all other forms of life on Earth and by depicting all of nature as created for the use of humans. Nevertheless, influent traditions within Christianity (e.g., St. Francis) might provide an antidote to the "arrogance" of a mainstream tradition steeped in anthropocentrism. The Encyclical Laudato-Sì affirms (Francesco 2015):

> The biblical texts are to be read in their context, with an appropriate hermeneutic, recognizing that they tell us to "till and keep" the garden of the world (cf. Gen 2:15). *"Tilling"* refers to cultivating, plowing or working, while *"keeping"* means

Fig. 3.3 Transformation of the landscape unit dominated by Villa Gallarati Scotti, Oreno. Above, the formal garden in the eighteenth century; below, a photo of the English park in XIX cent. Note the island configuration of the park in an agricultural landscape without any more hedgerows and woods. The demonstration of the buffer effect due to the Oreno park (right) is based on the g-LM (general landscape metastability). The blue segment is the real transformation (1805–2005), the red one the potential development without the English park

caring, protecting, overseeing and preserving. This implies a relationship of mutual responsibility between human beings and nature. Each community can take from the bounty of the earth whatever it needs for subsistence, but it also has the *duty* to protect the earth and to ensure its fruitfulness for coming generations. *"The earth is the Lord's"* (Ps 24:1); to him belongs *"the earth with all that is within it"* (Dt 10:14).

From the preceding, in practice, the greatest threat to nature, conservation biology, landscape planning, one health, and sustainable development or even to the medical clinic, is represented by the betrayal of the role of man, expressed as *violence against life* in all its aspects in the form of pseudo-scientific arrogance. A betrayal claims a principle of justice: it speaks of environmental ethics as a necessary reference in the relationship between man and nature. As Gandhi asserted, "Justice is the truth in our acts"; therefore, harmony with nature is a question of justice. This fact has been confirmed ever since the roots of our western civilization, in the first book of "De Legibus," by M.T. Cicero (Cicero 1973). Thus, deaf at the miracle of life, man uses violence to life at all the scales (see forward Fig. 3.8).

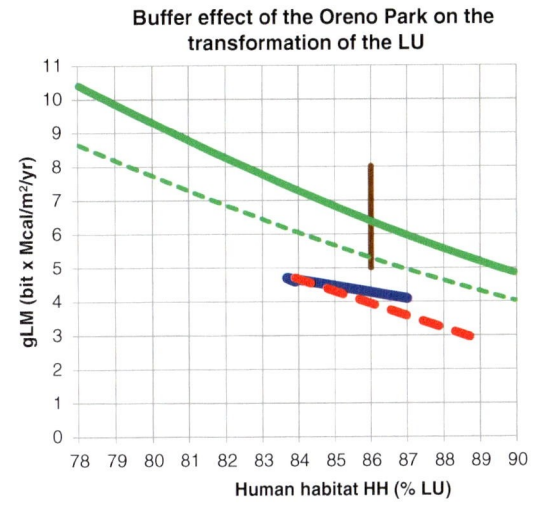

3.2 Environmental Alteration Needs a New Ecology

3.2.1 Environmental Alterations and the Damage of Human Health: The PHA

Man's violence against nature cannot remain without consequences, properly because of this ontological interrelation. Remember that land-use changes on the Emerged Lands in the last 125 years (1890–2015) were produced by forest cuts (diminishing global forest cover from 36.7 to 26.8%), and agrarian areas increase (from 14.1% to 27.7%). From the beginning of the third millennium, the strong population growth (from 3.3 to 7.0 billion in the period 1970–2012) and the Climate Change (with the mean T increase from +0.01 to +0.60 °C in the same period) enhanced the environmental alterations and the ecological alarm. WHO Environmental Burden of Disease (Prüss-Üstün et al. 2003; Prüss-Üstün et al. 2016) estimated 17.06% deaths in 2002, but those values became 21.80% in 2012 (+127.8%)! Attribution of the burden of disease to environmental risks highlights the importance of environmental protection for people's health and the study of "Global Health."

More specifically, it could be interesting to estimate the fraction of the total global burden of disease in DALY (Disability-Adjusted Life Year). Remember the WHO definition: one DALY can be thought of as "one lost year of healthy life." The sum of these DALYs across the population, or the burden of disease, can be thought of as a measurement of the gap between current health status and an ideal health situation where the entire population lives to an advanced age, free of disease, and disability. DALYs for a disease or health condition measures the sum of (a) the Years of Life Lost (YLL) due to premature mortality in the population and (b) the Years Lost due to Disability (YLD) for people living with the health condition or its consequences. The environmental fraction of the burden of selected diseases (percentages relate to the environmental share of the respective disease) reported, for example, by Prüss-Ustün et al. (Odum 1983) show that:

- Ischemic heart disease is estimated to have 35% of environmental causes and 6.5% of DALY;
- Cancers have at least 20% of environmental causes and 8.5% of DALY;
- Malaria has 42% of environmental causes and 2% of DALY.

Therefore, in 2014 the Rockefeller Foundation-Lancet Commission on Planetary Health, composed of experts in medicine and ecology, first met in Bellagio (Lake Como, Italy), reaching a dominant consensus around key messages and the urgency of planetary health as "an idea in jeopardy." Launched with the Rockefeller Foundation's support in 2016 (Rockfeller Foundation and Lancet Commission 2015), PHA has been integral to the rapid growth of planetary health through its engagement in advancing research, education, and policy. Now a highly effective coalition of 180+ member institutions from over 36 countries, the PHA is supported by a Secretariat based at Harvard University and a Steering Committee of international experts.

Planetary Health is interdisciplinary, better transdisciplinary, but first of all, it must be systemic. Its relationship between Ecology and Medicine has to be upgraded. Refer to new principles and methods sustained by the most advanced fields, as Bionomics and Systemic Medicine is necessary. Thus, a better understanding of environmental syndromes and their consequences on human health is possible. Environmental transformations proposed by PHA (from biodiversity shifts to climate change) do not consider bionomic dysfunctions, which can menace human health. On the contrary, finding advanced diagnostic criteria in landscape syndromes can actively help to find the effects on human well-being. The passage from sick care to health care cannot avoid the mentioned upgrading.

3.2.2 Limits of Traditional Ecology and the Birth of Bionomics

People affirming that Ecology is not a Science is increasing, due to an over-exposition of not well-defined ecologist figures on the media. However, it is nevertheless true that the limits of Traditional/General Ecology are impressive (Figs. 3.4 and 3.5).

Therefore, the birth of a new discipline, able to upgrade the conventional principles and methods that lead to the distortions set out below, is undelayable. Landscape Ecology was the first to begin an upgrading of Ecology, turning to a systemic view (Lovelock 2006), but it was mainly limited to the discovery of a spatial context (Forman and Godron 1986; Forman 1995): a critical approach, but the study of a complex system needs more.

A new course began in 1999, based on first statements of the entire biological spectrum and of its ontological properties and functioning (Ingegnoli 2001, 2002): thanks to suggestions and discussions with Zev Naveh, Richard Forman, and Sandro Pignatti, it brought to a re-founding of General Ecology with Bionomics and a widening foundation of the scientific studies on man-landscape relationships with Landscape Bionomics. A comparison between Traditional/General Ecology and Landscape Bionomics is shown in Table 3.2, in which the most important concepts are synthetically exposed: the differences are generally severe.

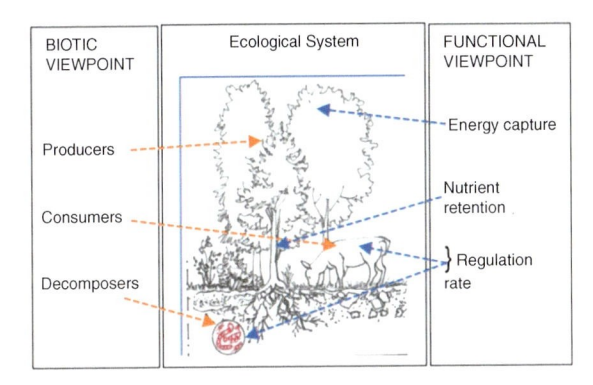

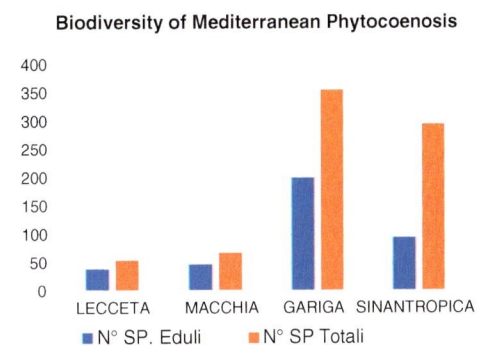

Fig. 3.4 (Left) The concept of an ecosystem is intended to be independent of space-time scales and incapable of integrating the biotic and the functional views. It is an ambiguous concept (O'Neill et al. 1986; Ingegnoli 2002, 2015), conflicting with reality, the new scientific paradigm, and the Emerging Property Principle. (Right) The value of biodiversity does not correspond to the value of the biological organization. Note that from Lecceta (Sclerophyll Oak forest) to Gariga (Chaparral scrubland), we pass from 40–50 to 250–350 species; but the top level of organization is the Lecceta

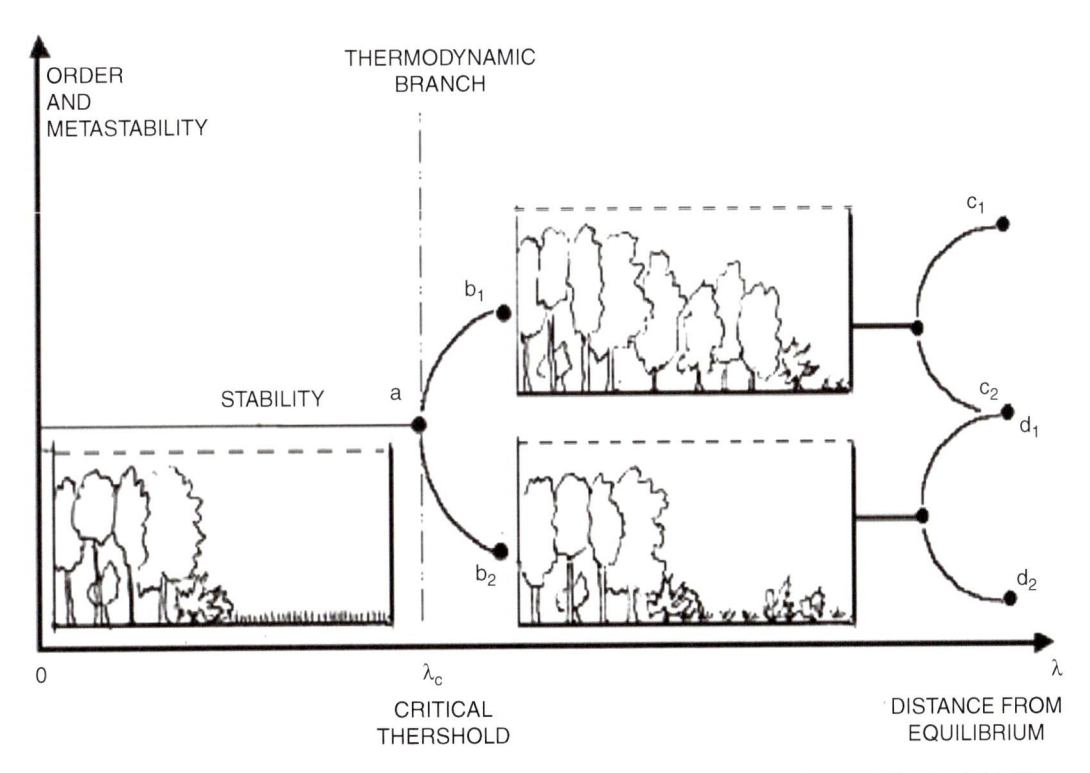

Fig. 3.5 A macro-fluctuation (e.g., land abandonment) leads to a bifurcation point with two possible ecological transformations: b1 and b2, which have different metastabilities. Note that after another macro-fluctuation, the successive transformation might lead to d1/c2, due to an increasing organization and metastability level (b2-d1) or to a degradation process (b1-c2). Examples like these show the need to revise the concept of succession, leading to significative changes in vegetation science

Table 3.2 Comparison between some fundamental biological concepts following General Ecology vs. adhering to Landscape Bionomics

GENERAL ECOLOGY	LANDSCAPE BIONOMICS
LIFE, Tab. 3.1	
Life is a characteristic, a condition that distinguish.... This concept is limited to an incomplete Biological Spectrum and investigated only through the biotic or the functional viewpoints, impossible to be overlapped! The Emerging Properties Principles is enunciated but not applied, so reductionism remains the rule.	Life is a complex self-organizing open process, operating with a continuous exchange of matter, energy, and information with its environment. At present, Life on Earth is organized in a hierarchy of Living Entities, that are six types of complex systems existing, linked together and interacting. Each specific biological level emerges, expressing a process personally (as they are scale-dependent), although processes allowing the definition of life are ontological.
ADAPTATION	
It is the capacity of an organism to become more suited to an environment only through the process of natural selection.	It is the capacity of every biological system to modify itself and to specify the admissibility of the environmental constraints, so as to define its proper domain of environmentally pertinent perturbations.
EVOLUTION	
Based on the process of natural selection, the so called "struggle for existence" is due to the environmental selection of casual changes in the hereditary characters (caused by 'mistakes' within the process of molecular copying) acting on organisms.	Based on the Principle of Emergent Properties, it implies at least natural selection (linked to the process of copying, which is related to biological information) and natural convention processes (due to the process of coding, related to biological meaning, acting both at molecular scale and on collective set of objects); it is linked to both energetic order and information order and involves all the biological spectrum proceeding through the force of mutual cooperation and symbiosis.
ENVIRONMENT	
Today, the definition of the environment is "the set of physical, chemical and biotic factors *that act upon an organism* or an ecological community and ultimately determine its form and survival." It is generally intended to be scale independent and studied for separate sets of elements (water, air, soil, species, pollution). So, only climate, geological components, pollution, or limited biological indicators are taken into account.	In reality, the components cited in traditional ecology pertain to complex and scale-dependent systems. The environment is characterized *by a reciprocity of relationships*, which must be intended as systemic integration of biotic and abiotic components. We have to intend these interrelations as comprise within the 'biological spectrum', except for *out-of-scale* calamitous disturbances produced by geophysical forces and/or devastating anthropic actions.
ECOSYSTEM, Fig. 3.4-a	
Even leaving aside the abuse of the term committed every day by journalists, politicians, etc. or the misuse within the universal language as the contraction of "ecological system," the concept of "ecosystem" (Tansley, 1935) is scale-independent, so presenting the same structure and functions "from a temporary pond to an entire alpine valley."	It is an ambiguous concept (O'Neil et al. 1986), due to the conflict between the biotic and the functional view. Moreover, this term implies a definition based on structure and functions that cannot remain the same in a small ecotope and a vast landscape (system of ecosystems), due to the Emergent Property Principle. At the local scale, the ecosystem has to be changed with the concept of ecocoenotope.
LANDSCAPE	
The landscape means an area, as perceived by people, whose character is the result of the action and interaction of natural and/or human factors (European Landscape Convention, 2004)	The landscape means a complex system, existing as a biological entity, whose character and behavior is the result of the action and interaction of natural and human components.
BIODIVERSITY, Fig. 3.4-b	
Biodiversity is intended near exclusively as specific, and it has to be always high in numbers. Usually, it's strictly related to the concept of resilience.	Biodiversity concerns all the levels of the Biological Spectrum (except the highest), but its role and importance is inversely proportional to them: every biological system admits a proper biodiversity, so both a decrease or an increase of biodiversity are destabilizing. Indeed, the most evolved natural systems follow the resistance stability, not the resilience one.
TRANSFORMATION (Fig. 3.5) & ENVIRONMENTAL BALANCE	
Transformation acts following the concept of ecological succession, which consists of a linear series of steps. The concept of environmental balance follows classic thermodynamics and reversible processes, e.g., degradation and recovery or "universal" energy flux models.	The transformation of a complex system takes place according to a process of macro-fluctuation, which leads to bifurcation points, presenting two possible paths with different levels of metastability. Not out-of-scale disturbs can play a positive role. Extensive mutual relations occur among the components: so the process is neither linear nor deterministic. The new concept of environmental balance depends on non-equilibrium thermodynamics, irreversibility, and the concept of metastable equilibrium.

(continued)

Table 3.2 (continued)

QUALITY DEGREE	
It is referred to the concept of 'distance from the climax', as a mature, relatively stable community within an area, representing the top of ecological succession, considered as linear and based on the frequency/presence/absence of key species. In add, the system behavior is found merely through the characters of its components.	The concept of Quality degree is based on a condition of normality for a set of integrated parameters, the range of which depends on the spatio-temporal-information scale involved.
HEALTH STATE	
The idea is limited to that of plants, animals, and human beings. Considerations mainly concern Climate Changes and Pollutions.	The idea concerns each living system on the Earth pertaining to every level of the Biological Spectrum, in a proper way and in relation with the other levels. The pathology of the environmental systems of the upper scales is crucial both for landscape rehabilitation and for the study of human health damages.
ROLE OF URBAN AND PERI-URBAN 'GREEN'	
Green is intended as place to rebalance environmental stress and to 'reset' our level of fitness; place of positive influence on urban meso-climate; area of refuge for flora and fauna. Formalism and originality, walkability and easy management, presence of exotic colored species are the dominating criteria.	Green can be divided in typologies, each one with proper characters and functions within its specific contest. It acts as sub-regulatory system of the metastability and/or with a protective function (within the PRT Landscape Apparatus); its ecotopes may contribute to the formation of ecological networks. Rehabilitating role within the LU and rebalancing of Human Health are to be improved and need completely different criteria.

3.2.3 Upgrading Landscape Ecology: Landscape Bionomics

Within Bionomics principles, the real living system involved in the right relationship between man and nature is the Landscape.

The discipline of Biological-Integrated Landscape Ecology (Ingegnoli 2010), re-named Landscape Bionomics (Ingegnoli 2011a, 2015), has been proposed just to deepen the investigations at the light of these new scientific paradigms (Ingegnoli 2001, 2002; Ingegnoli e Pignatti 2007; Ingegnoli et al. 2017). Its bases are:

1. The enlargement of the definition of life, recognizing the landscape as a peculiar biological level, performing its birth, growth, development, evolution, and death in an own way;
2. The study of the Landscape units (LU) as real systems integrating the different environmental viewpoints today considered separately: spatial configurations, biotic characters, functional characters, cultural/economic aspects;
3. Consequently, the changing of the structural model based on a mosaic of ecosystems, with the concept of ecotissue;
4. The diagnosis of the systemic health through opportune state functions concerning the complex system forming a landscape unit, through a proper quality-quantitative clinical-diagnostic methodology;
5. The suggestion of therapeutic criteria and methods of its strategic rehabilitation.

If we are going back to Table 3.1, it is necessary to consider two hierarchic levels in the middle "biological spectrum": (1) the ecobiota, composed of the community, the ecosystem, and the microcore [i.e., the spatial contiguity characters, sensu Zonneveld (Zonneveld 1995)], which we will name ecocoenotope and (2) the landscape, formed by a system of interacting ecocoenotopes (the "green row" in Table 3.1). Therefore, the biological organization of a territory as a "complex living system formed by natural and human components" must be named "landscape."[2]

No doubt that some community and ecosystem characters are also available at the landscape level, and even the inverse is true: only reductionism pretends to separate all the charac-

---
[2]Going on within the chapter, the old reductionist term "territory" will be substituted by the upgraded concept of "landscape," to be intended as the real living complex system at the Spatio-temporal scale of the territory.

ters related to each level. The theoretical corpus of this new discipline is quite complex, so here only an extreme synthesis of the essential principles and functions, needed to understand the applications of the present chapter, is reported suggesting to refer to the most recent published book "Landscape Bionomics, Biological-Integrated Landscape Ecology" by Ingegnoli, Springer publ. (Ingegnoli 2015) for a deeper understanding.

3.3 Landscape Bionomics: Some Key Concepts

3.3.1 Main Structures and State Functions in Landscape Bionomics

As one has to relate to real systems, within the biological level of Landscape, we identify 16 mayor types of it (Table 3.3), based on the presence/absence of specific functions, each one performed by a proper Landscape Apparatus. A Landscape Apparatus (L-Ap) is constituted by a proper functioning system of ecocoenotopes[3] (even not connected), forming a specific configuration within the ecotissue. It is possible to distinguish many types of landscape apparatuses, the most important being:

1. HGL = Hydrogeologic (emerging geotopes or elements dominated by geomorphic processes).
2. RNT = Resistant (elements with high metastability, e.g., forests).
3. RSL = Resilient (elements with high recover Capacity, e.g., prairies or shrublands).
4. PRT = Protective (elements which protect other elements or parts of the mosaic).
5. PRD = Productive (elements with high production of biomass).
6. SBS = Subsidiary (systems of human energetic and work resources).
7. RSD = Residential (systems of human residence and dependent functions).
8. CON = Connective (elements with essential connective functions in the mosaic).
9. EXR = Excretory (the fluvial web as landscape catabolite processing).
10. DIS = Disturbance (elements with a range of non-incorporating disturbances).

Rarely the previous functions have been linked to some basic concept of landscape ecology, such as to its structure and dynamic!

Each landscape type and sub-type is concretely constituted by a different number of simple or complex **Landscape Units (LU)**. A Landscape Unit (LU) is characterized by peculiar structural (i.e., the recurrence of a specific configuration) or functional aspects, which identifies it as regard to the entire landscape. This organization is not always immediately recognizable, needing proper studies: indeed, while general ecology approaches a landscape through the concept of eco-mosaic, the fundamental structure of a landscape is systemic; so, we have to pass to the concept of "ecological tissue," similar to the weaving or the histological tissue. The **Ecotissue** (Ingegnoli 2001) is a multidimensional conceptual structure representing the hierarchical intertwining, in the past, present, and future, of the ecological upper and lower[4] biological levels and of their relationships in the landscape: it is constituted by an underlying mosaic (usually the vegetation one for its importance, due to photosynthetic capacity) and a hierarchic succession of correlated structural and functional patchworks and attributes (Fig. 3.6). As a consequence, in the operative chart of integration, it is possible to subdivide a LU in *ecotopes*, each one being the smallest *multidimensional* element of the LU that owns all the structural, functional, and informative characters of the concerned landscape: concretely, it is the minimum system of interdependent ecocoenotopes (not less than two) as determined by, i.e., the topographical recurrence or the geomorphological origin or the role within the LU.

---

[3] Remember that the **ecocoenotope** is a *multifunctional* entity in a definite geographic locality: it is the "tessera" of the underlying mosaic of an ecotissue.

[4] Ecoregions and ecocoenotopes, definitions in (Giglio 2011, 2015)

Table 3.3 Main composition of landscape apparatuses forming landscape types

Geo	Exr	Rnt	Rsl	Con	Prt	Prd	Rsd	Sus	Landscape types
++	-	-	+-	-	-	-	-	-	Desert
++	-	-	++	-	+-	-	-	-	Semi-desert
-	+-	-	++	++	+-	+-	-	-	Prairie
+-	+-	+-	++	+-	++	+-	-	-	Shrub-prairie
++	+-	+-	+-	+-	+-	+-	-	-	Shrubby
+-	++	++	++	+-	++	++	-	-	Open forested
-	+-	++	+-	++	++	+-	-	-	Closed forested
+-	+-	++	++	+-	+-	++	-	+-	Semi-natural. > bmass
+-	+-	+-	+-	+-	+-	++	-	+-	Semi-natural. < bmass
+-	++	+-	++	++	++	++	+-	-	Cultivated, protective
+-	+-	-	++	-	+-	++	+-	-	Cultivated, productive
+-	+-	-	+-	-	+-	++	+-	+-	Rural
+-	+-	-	+-	-	+-	+-	+-	+-	Suburban, rural
+-	+-	-	-	-	+-	-	+-	++	Suburban, industrial
+-	++	-	-	+-	++	-	++	+-	Urban, open
-	+-	-	-	-	+-	-	++	++	Urban, closed

Geo geologic. *Exr* excretory. *Rut* resistant. *RSL* resilient. *CON* connective. *PRT* protective. *PRO* productive. *RDS* residential. *SUS* subsidiary. ++ full presence. + – partial presence. – – absence, *bmass* biomass

To describe the landscape as a complex system, we need some state functions. The essential state functions are the Standard Habitat per capita, the Carrying Capacity, the Human and Natural Habitat, and the Biological Territorial Capacity.

The **Standard Habitat (SH)** is the state function intended as *the set of portions of landscape apparatuses* (within the examined LU) *indispensable for an organism to survive*, also known as the **vital space per capita** [m²/ab]. It is available for an organism (man or animal), divisible in all its components, biological, and relational. In the case of human populations (idem), we will have SH_{HH}, that is, an SH referred to the human habitat (HH):

$$SH_{HH} = (HGL + PRD + RES + SBS + PRT)\,areas\,/\,N° \text{ of people } \left[m^2/\text{inhabitant}\right]$$

The connected Minimum Theoretical Standard Habitat per capita (SH*) is the state function estimated as dependent on

(a) the minimum edible Kcal/day per capita [1/2 male + female diet];
(b) the productive capacity (PRD) of the minimum field available to satisfy this energy for 1 year, taking into account the production of major crops;
(c) an appropriate safety factor for current disturbances;
(d) the need of natural or semi-natural protective vegetation for the cultivated patches

(Table 3.4). It is estimable for each type of animal population too.

The ratio SH/SH*, named **Carrying Capacity** (σ) of a LU, is the state function able to *evaluate the self-sufficiency of the human habitat (HH)*, a basic question for sustainability and ecological territorial planning.

The **Human Habitat (HH)** is the surface evaluation (% of LU) of *the human ability to affect and limit the self-regulation capability of natural systems*. Ecologically speaking, the HH cannot be the entire territorial (geographical) surface (Fig. 3.7): it is limited to the human ecotopes

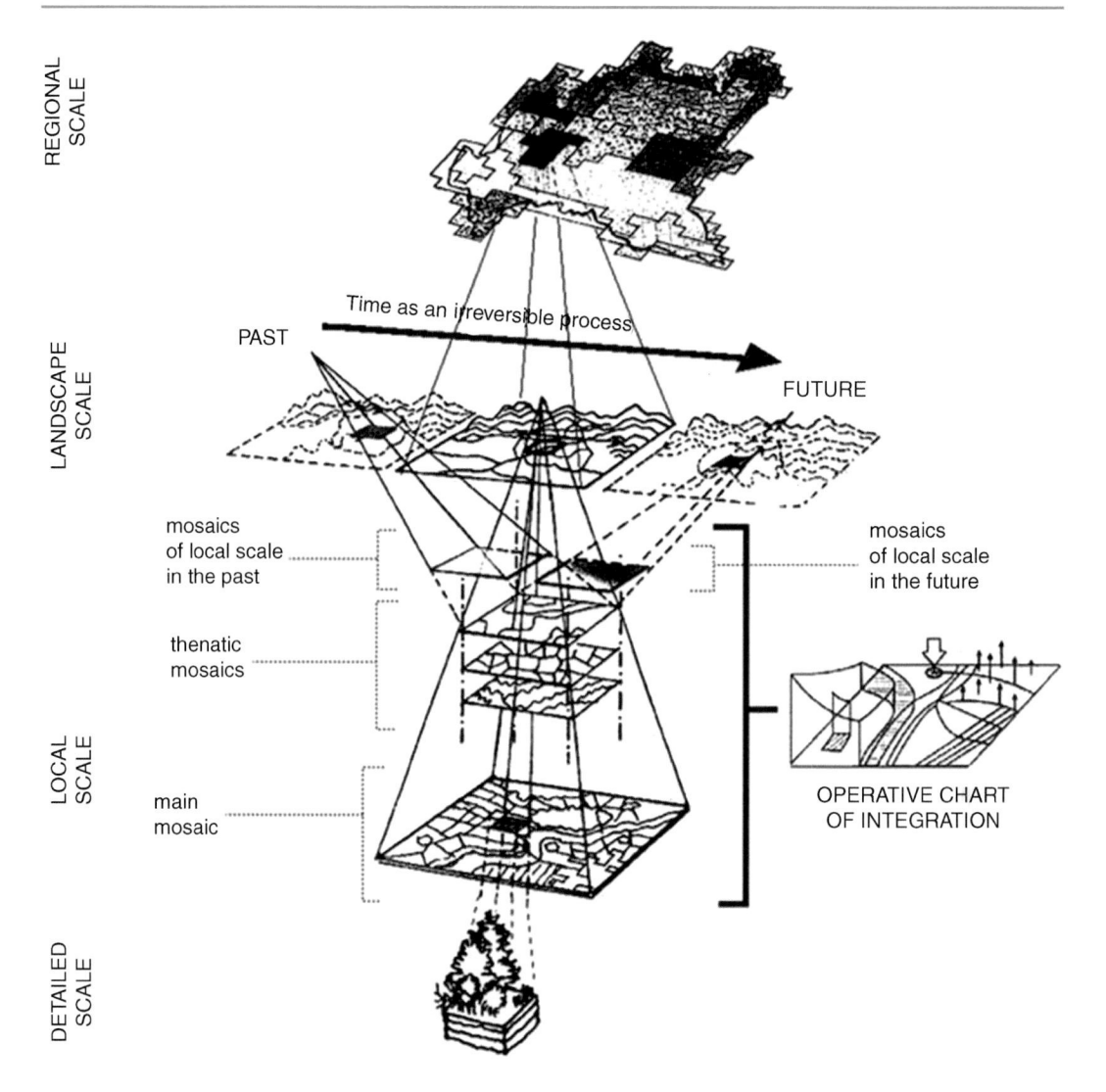

Fig. 3.6 The ecotissue model gives the right importance to the landscape and integrates the fundamental dimensions of the landscape: (1) a range of spatial scales, from the regional to the local configuration of elements, (2) a set of thematic mosaics on species (biomass) and resources (energy) components, (3) a range of temporal scales on developing processes, which permits the evolutionary dynamic of the landscape to be forecasted and reconstructed, (4) a set of information contents, which permits to evaluate the level of the systemic organization (Ingegnoli 2015)

and landscape units (e.g., urban, industrial, and rural areas) and the semi-human ones (e.g., semi-agricultural, plantations, ponds, managed woods). The NH concerns the natural ecotopes and landscape units, with the dominance of natural components and biological processes, capable of healthy self-regulation. In Fig. 3.7, it is exposed to the drastic difference between the reductionist concept of natural and human components vs. the systemic one, in which the landscape apparatus and the human (HH) and natural (NH) habitats are integrated: two non-linear state functions express them. The absence of HH or NH can be possible only as an exception, e.g., in a wide industrial-urbanized area.

The **Biological Territorial Capacity of vegetation (BTC)** is related to the physiology of vegetation, which leads to the concept of the *latent*

3 From General Ecology to Bionomics

Table 3.4 Theoretical minimum standard habitat/capite in Landscape Units with presence of human populations

Climatic Belts	Kcal/inhab[a]	SH* m²	Agricultural surface/capita
Arctic	3,500	2,500	1670
Boreal	3,100	1,850	1250
Cold-Temperate	2,850	1,480	1050
Warm-Temperate	2,750	1,360	980
Sub-Tropical	2,550	1,250	870
Tropical	2,350	1,020	730

[a]Minimum edible Kcal/day per capita

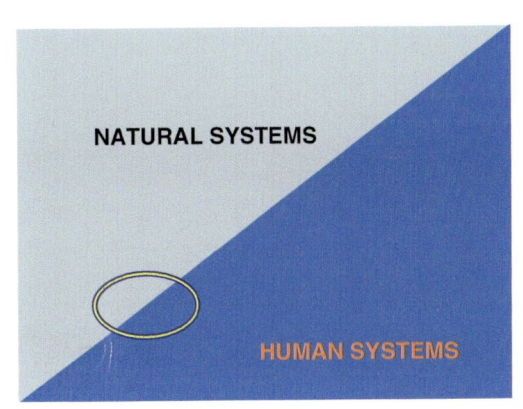

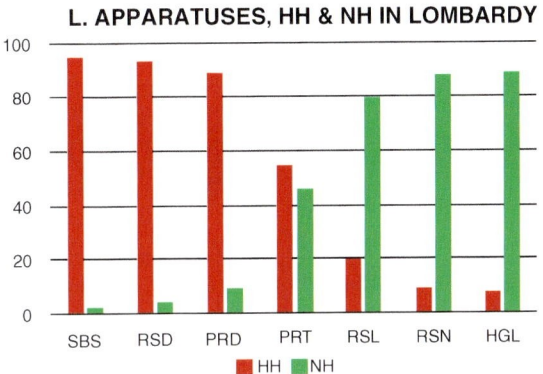

Fig. 3.7 Comparison between the reductionist concept of Natural vs. Human components in standard Ecology (left) and the systemic concept (right). The difference is very sharp

capacity of homeostasis of a phytocoenosis (Ingegnoli 1991, 1999, 2011b). It can be studied on the basis of: (a) the concept of resistance stability; (b) the type of vegetation community; (c) its metabolic data (biomass, net or gross primary production, respiration, B, NP, GP R); (d) their metabolic relations R/GP (respiration/gross production); and (e) their order relations R/B (respiration/biomass) = dS/S (anti-thermal maintenance). We have to elaborate on two coefficients:

$$a_i = (R/GP)_i / (R/GP)_{max}$$
$$b_i = (dS/S)_{min} / (dS/S)_i$$

a_i measures the degree of the relative metabolic capacity of principal vegetation communities;

b_i measures the degree of the relative anti-thermal (i.e., order) maintenance of the same main vegetation communities.

The degree of the homeostatic capacity of a *phytocoenosis* (vegetation community) is proportional to its respiration. It is expressed as the flux of energy that the phytocoenosis must dissipate to maintain its condition of order and metastability [Mcal/m²/year].

$$BTC_i = (a_i + b_i) R_i \; w \; \left(Mcal/m^2/year\right)$$

Therefore, *the BTC function is essential* because it is systemic and *can evaluate the flux of energy available to maintain the order reached by a complex system*.

Note that a high level of energy flux expresses the complexity of a highly ordered system: e.g., a Tropical rain forest (high resistance) may arrive up to BTC = 17–18 Mcal/m²/year, a Tropical seasonal forest to BTC = 12.1, a *Querco-Carpinetum* to BTC = 8.50–9.50, while a *cultivated meadow*

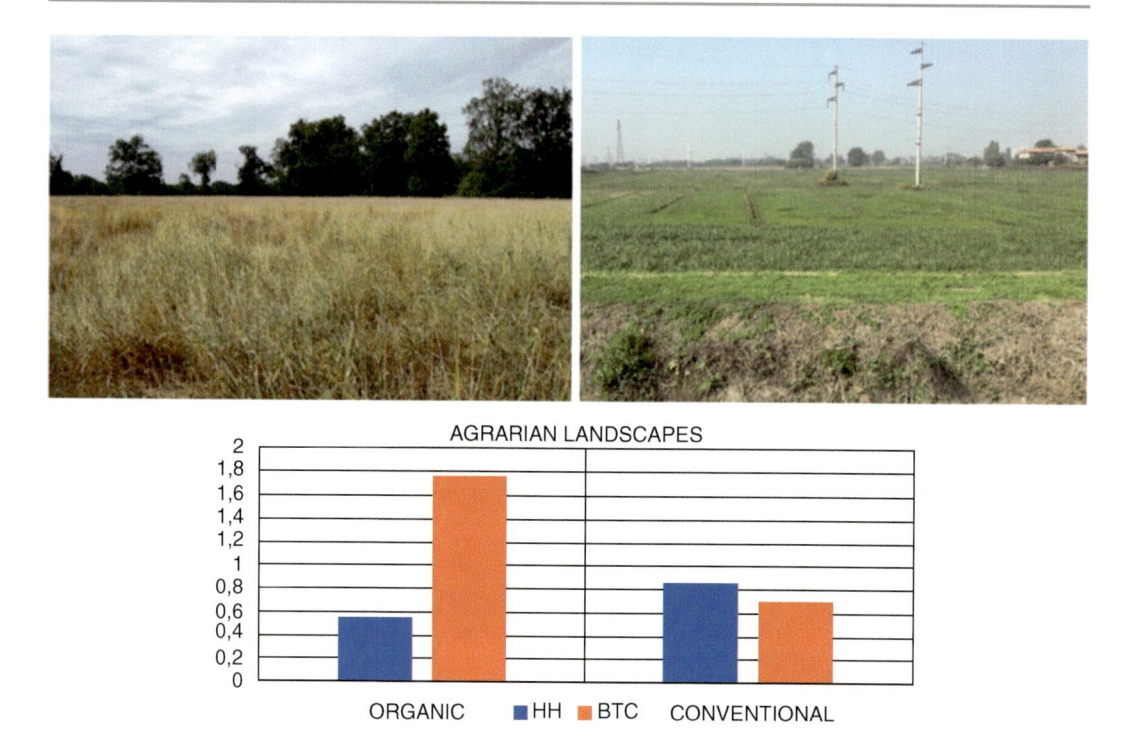

Fig. 3.8 Comparison between two very different agrarian landscapes near Milan. The difference in BTC level is very sharp: Albairate-organic farming presents BTC = 1.75 Mcal/m^2/year vs. Chiaravalle-conventional farming, with BTC = 0.73 Mcal/m^2/year. In the case of Chiaravalle, the violence of man against nature should be evident

(low resistance) to BTC = 0.45–0.55 Mcal/m^2/year.

In Fig. 3.8, we can show the comparison between two agrarian landscapes units (LU) near Milan: note that Albairate-organic farming presents a BTC level near three times the BTC of Chiaravalle-conventional one, even if the human habitat value is not so different. This example may prove the possibility to arrive at a systemic diagnostic valuation of a complex system like a LU.

3.3.2 Main Syndromes Concerning Landscapes

As the landscape is a specific level of life organization on Earth, therefore it is a *living entity* and may present many *syndromes*, as synthesized in Table 3.5. As one can see, it is a simple framework of categories, but it is evident that the *majority* of landscape syndromes are *not* due to pollution.

3.3.3 Diagnostic Evaluation in Landscape Bionomics

The difficulty to understand the concept of *pathology* lies in its inextricable link with *physiology*, within which, to understand a function, we need its alteration and vice versa, with continuous feedback. Taking into account that the environmental formations, from the ecocoenotope to the entire landscape, are living systems, and, as all the living systems, they can fall ill, we need to know the whole system behavior, so its main state functions. The analysis of the components is necessary but not sufficient, because they are not essential, not able to synthesize the main characters of the complex system, not capable of controlling the movement of the system and its limits. Moreover, as in medicine, environmental evaluation needs comparisons with *standard* patterns of behavior of the investigated system. The main problem becomes how to know the normal state of a landscape unit and the levels of

Table 3.5 Landscape health [from Ingegnoli (2002)]. Main landscape syndromes categories and sub-categories

Main landscape syndrome categories and sub-categories	
Main landscape syndrome categories	**Sub-categories of syndromes**
A- Structural alterations	A1- Landscape element anomalies A2- Spatial configuration problems A3- Functional configuration problems A4- Multiple structural degradation
B- Functional alterations	B1- Geobiological alterations B2- Structurally dependent dysfunctions B3- Delimitation problems B4- Movement and flux dysfunctions B5- Information anomalies B6- Reproduction problems B7- Multiple dysfunctions
C- Transformation syndromes	C1- Stability problems C2- Changing process dysfunctions C3- Anomalies in transformation modalities C4- Complex transformation syndrome
D- Catastrophic perturbations	D1- Natural disasters D2- Human-made destruction
E- Pollution degradations	E1- Direct pollution E2- Indirect pollution
F- Complex multiple syndromes	F1- Acute F2- Chronic

alteration of that system. Nevertheless, with the BTC function, we can provide it.

The System Theory affirms that the scale capable of maximizing the importance and the quantity of relations among the components of a system is the scale that consents to discriminate the different forms, especially the relational ones. The territorial scale is the best one capable of maximizing the importance of the relations among the elements, both natural and human (Ingegnoli 2011a; Ingegnoli 2015)—the principal methodologies capable of expressing a correct diagnosis of an entire landscape are:

1. The congruence with the normal function correlating the human habitat (HH) with their levels of BTC, resulting in the *Bionomics Functionality ratio (BF)*, at Local and Regional scale.
2. The evaluation of the distances from the normal range of *N* parameters, resulting in a Probe Spectrum, quantifiable with a *Diagnostic Index (DI)*.
3. The control of transformations obtained evaluating the ratio of the BTC of forest and agrarian components with the regional BTC level referred to a favorable period, resulting in the *Transformation Deficit (TD)*.

In this work, we have to limit the diagnostic methodologies to the first point (1).

3.3.3.1 The Bionomic Functionality (BF)

After the study of 45 landscape units, it was found that their Biological Territorial Capacity of Vegetation (BTC) and their Human Habitat (HH) present an excellent correlation between *the* BTC and HH. The correlation consisted in an $R^2 = 0.95$ and a Pearson's correlation coefficient of 0.910.[5]

Thus, as we can see in Fig. 3.9 (right), it was possible to build the simplest mathematical model of bionomic normality, available for *a first, but fundamental, framing of the dysfunctions* of landscape units. Below normal values of *bionomic functionality (BF = 1.15–0.85)*, with a tolerance interval (0.10–0.15 from the curve of normality), we can register three levels of altered *BF*: altered

---
[5] Excellent: about three times the minimum value of significance for a sample of 45 elements.

Fig. 3.9 (Left) The HH/BTC model, able to measure the bionomics state of a LU. Dotted lines express the BF level, which is the bionomics functionality of the surveyed LU. (Right) The correlation HH/BTC at Regional-National scale (blue line), with an example of land transformation between India (red) and Europe

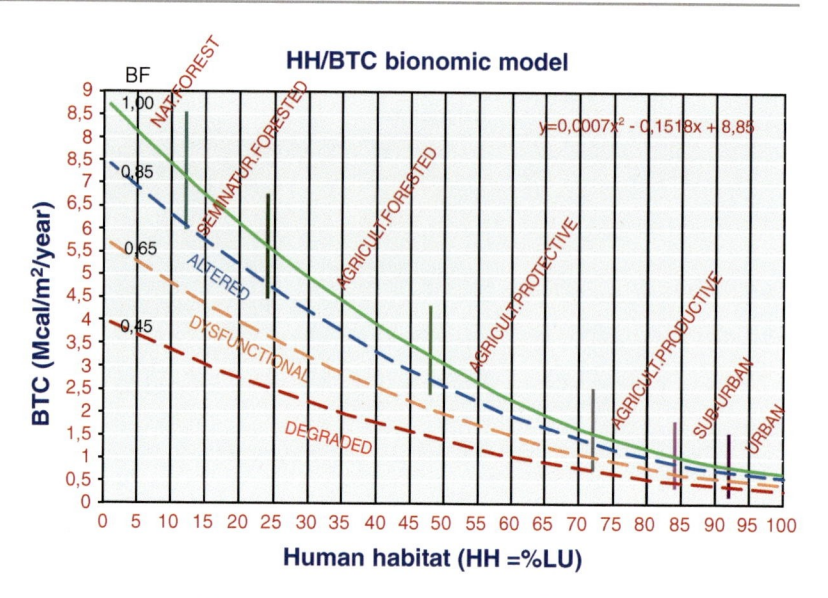

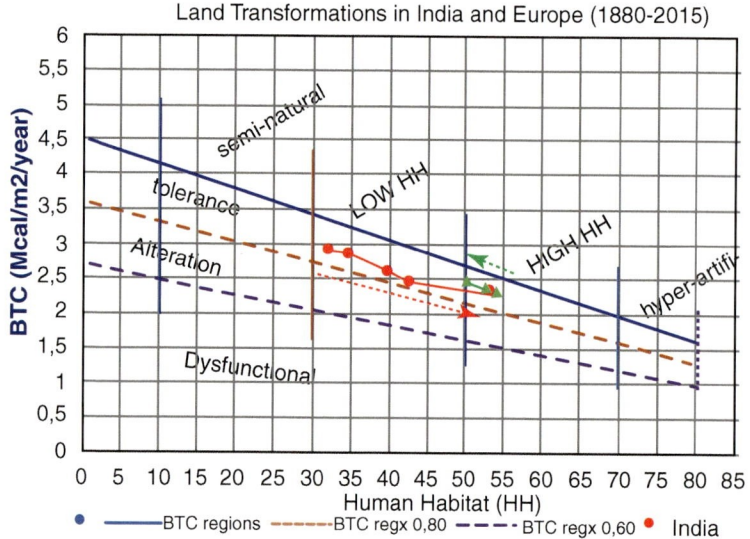

($BF = 0.85$–0.65), dysfunctional ($BF = 0.65$–0.45), and highly degraded ($BF < 0.45$). The vertical bars divide the main types of landscapes, from Natural-Forest (high BTC natural) to Dense-Urban: each of them may present a syndrome.[6]

As already underlined, *complex systems are scale-dependent*. Changing scale from local to regional-national one, the diagnostic model HH/BTC changes, because we pass from 10^1–10^4 km² to 10^4 a 10^6 km², so the curve of normality is different (Fig. 3.9 left).

Note that at a regional scale, HH has severe limits, because over HH = 70% the HH values become typical of local scale. So, *Hyper-anthropization is an environmental syndrome* because the level of HH is decidedly out-of-scale. Consequently, on a regional-national scale, the most functional-systemic parameters are not BF but BTC + BF.

3.3.4 Role of Vegetation Science and Agroecology

Landscape Bionomics and Vegetation Science. Vegetation is the key-system at the landscape bio-

---

[6] For the articulation of landscape pathologies, see (Ingegnoli 2015), Sec. 4.5, pp. 100–110.

logical level, playing at least the triple role of "managing energy" (so relating to metastability and the efficiency concepts), organizing the landscape and structuring the chronotopes (three spatial dimensions + the temporal one) (Ingegnoli 2015). Thus, the analysis of vegetation usually taught in Italy, but also still in France, Spain, and Germany, mainly based on Phytosociology, cannot be useful.

Studies of Phytosociology, sensu Braun-Blanquet (Braun-Blanquet 1926), have apparent advantages regarding the description of vegetation, and its safe limitations. Let us take a quick overview of them (Naveh and Lieberman 1984; Pignatti et al. 2002; Ingegnoli 2002, 2015; Ingegnoli and Giglio 2005): (a) reference to a concept of naturalness that excludes humans in any event; (b) dynamics based on the concept of ecological succession mainly understood as linear and deterministic; (c) reference to an "ecological space" that does not consider the principle of emergent properties (remember Fig. 3.1); (d) use of the concept of "potential vegetation" that does not consider the role of disturbances in ecological systems; (e) ignorance of the complex system and scale-dependent functions, claiming to be able to study the landscape with the deterministic approach of the concept of "sygmetum" and "geo-sygmetum" by Tüxen (Tuxen 1956) and Rivas-Martinez (Rivas-Martinez 1987). Despite synthetic, these premises enhanced the urgent need to develop new concepts and a method to study the vegetation following the landscape bionomics principles. Here we will only focus on two key points:

- The first one had been the proposal of the new concept of *"the fittest vegetation for..."* (Ingegnoli 2002, 2015; Ingegnoli and Pignatti 2007), to overcome the one of *potential vegetation*, which indicates "the vegetation most fitting in climatic and geomorphic conditions, in a limited period, in a certain defined place, in the function of the history of the same place and with a certain set of incorporable disorders (including those humans) in natural and not natural conditions." Hence, also anthropic vegetation is involved;
- The second one consisted in the proposal of a new methodology, called LaBiSV (Landscape Bionomics Survey of Vegetation) (Ingegnoli 2002, 2011a, 2015; Ingegnoli and Giglio 2005; Ingegnoli and Pignatti 2007), whose theories are summarized as follows: (a) reference to the concepts of ecocoenotope and ecotissue as structural entities of the landscape; (b) use of bionomics territorial capacity of vegetation (BTC) as the main integrative function; (c) drawing up of development models "time-BTC" for different types of vegetation based on a logarithmic and exponential functions[7]; (d) possibility of comparison between the ecological status of natural and human-made vegetated tesserae; (e) ability to determine the state of normality of the ecological parameters of different types of vegetation; (f) ability to measure the efficiency[8] of vegetation (CBSt) in providing ecological services; (g) ability to measure the concept of biodiversity at the landscape level (diversity of biological organization of the context). Since their first tests, this concepts and methods have been confirming their validity.

Bionomics and Vegetation at Earth scale.
The mentioned "Gaia Theory" (Lovelock 2006; Lovelock and Margulis 1974) asserts that living organisms and their inorganic surroundings have evolved together as a single living system that greatly affects the physics, chemistry, and conditions of Earth's surface. In the past 30 years, many of the mechanisms

---

[7] Remembering the well-known relationships among gross productivity, net productivity, and respiration in vegetation systems, the development of a vegetation community may be synthesized in (a) the growing phases from young-adult to maturity, expressed by an exponential process; (b) the growing phase from maturity toward old age, expressed by a logarithmic process (Ingegnoli 2002, 2011a, 2015).

[8] Efficiency relates the maturity level (MtL) of a vegetation coenosis and its bionomic quality (bQ). It is fundamental for the comparison among different types of vegetation.

by which Earth self-regulates have been identified. The photosynthesizing plants release water vapor into the atmosphere; the air rises and eventually condenses into clouds. For instance, it has been shown that cloud formation over the ocean is almost entirely a function of the metabolism of *oceanic algae* that emit a large sulfur molecule that becomes the condensation nuclei for raindrops. Previously, it was thought that cloud formation over the ocean was a purely chemical/physical phenomenon.

In the Emerged Lands, forest systems act in a similar way but with *more potent* effects. According to recent research, vegetation may contribute as much as 90 percent of the moisture in the atmosphere derived from land surfaces, far more than earlier estimates. Trees produce water vapor flows that are typically more than ten times greater than that from herbaceous vegetation per unit of land area, surpassing those produced by wet ground or open water. Transpiration "is an active biological process" that is not fully reflected in climate models (Sheil and Murdiyarso 2009). The theory of "Biotic Pump" (Makarieva and Gorshkov 2012) confirms this view, specifying that forest landscapes create and control even the oceanic winds, enriched with wet air, condense it in the cloud, and then in the rain. Moreover, trees influence cloud formation by emitting carbon-based chemicals called volatile organic compounds (VOCs) into the atmosphere. Some of those compounds are deposited on tiny airborne particles such as dust, bacteria, pollen, and fungal spores. As the particles grow with the deposition of VOCs, they promote condensation and gather the resulting moisture, hastening cloud formation.

The Importance of Agroecology. The problem is that the regulation capacity of Gaia, first of all, depending on forest systems, is today strongly decreased, because of the forest destruction in the last century. There is no doubt that the present climate change is due not only to the greenhouse gases but also to forest destruction! A recent research of Ingegnoli evaluated the Planetary Health, through the measure of the transformation deficit (TD) related to the Emerged Lands. Remember that TD is the loss of energy use to maintain the level of order reached by an ecological system at its best recent period: in our study, the end of the nineteenth century (1880–1885), when World Population was only 1.5 billion. In Fig. 3.10, forest, and agrarian data from FAO (FAO 2010), we can note:

1. The decrease of forest cover from 36.6% in 1882 to 26.8% in 2015 (−26.9%), but (worst) the loose of bionomics efficiency (−33.0%)!
2. The increase of agrarian areas (green and crop) from 14.09% to 27.65% (+192.2%), but (worst) followed by a poor increase of bionomics efficiency (+175.7%), due to the degradation of agricultural landscapes.
3. TD, with a 4% tolerance, passed from 88% to 65%: a difference of −26.14%, which brings to a total TD waste of 7.974×10^{18} kcal.

From this research, the drastic resizing of the "principle of Borlaug" (Borlaug 1974) is evident, linking agrarian industrialization to forest preservation (i.e., crop increase limits the need for deforestation). In reality, forests have decelerated their destruction, but also their bionomics efficiency, while agricultural landscapes BTC flux remained too low. Thus, TD grew impressively.

The forest destruction may be reduced. In some continent, forests are increasing (e.g., Europe), but their negative global trend would continue (e.g., Brazil) and, in any case, the times for their regrowth is too slow. To reforest desert and urban landscapes is not enough: also, it needs financial efforts and long times. In any case, this action would not be sufficient, being their land cover only 14.56% and 0.62%. The most available remedial action should be the *rehabilitation of Agrarian Landscapes*, 27.68% of land cover,

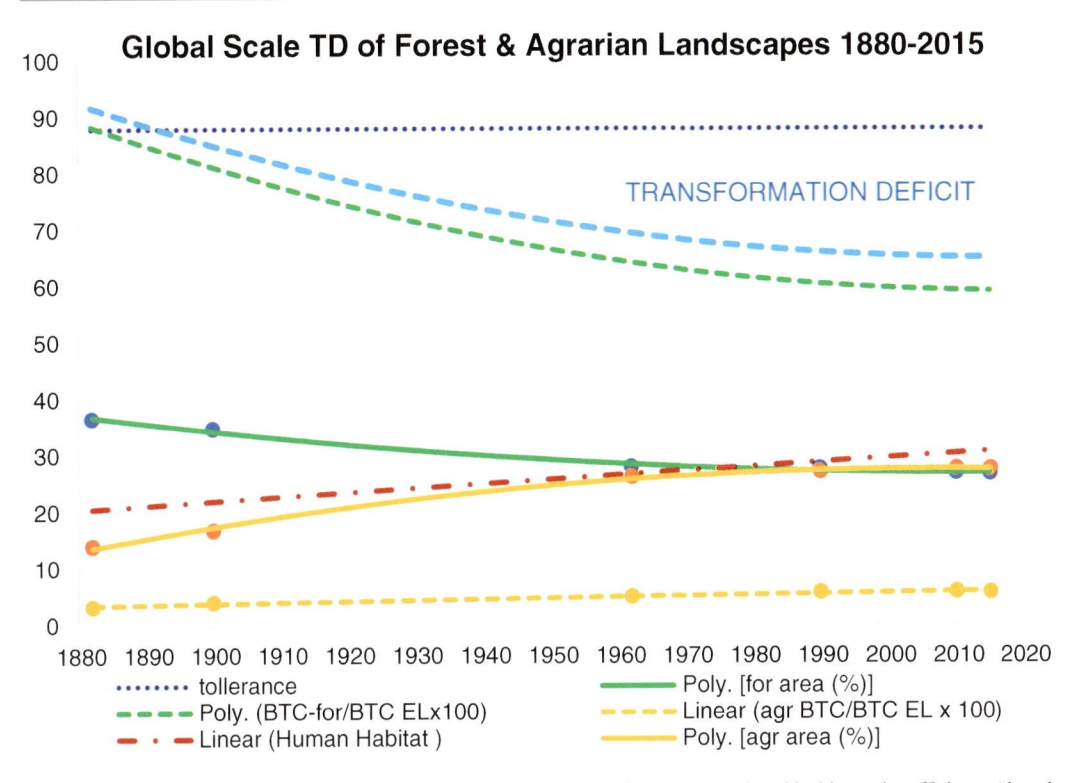

Global Scale TD of Forest & Agrarian Landscapes 1880-2015

Fig. 3.10 Evaluation of TD (blue) of the Emerged Lands. Forest cover decrease (green) and its bionomics efficiency (dotted green); Agrarian increase (yellow) and its bionomics efficiency (dotted yellow). BTC EL = BTC of Emerged Lands

more natural to transform, and useful for full agroecology goals.

In summary, we have to underline that a few decades ago in human history, almost all agriculture was traditional and agrarian systems were interspersed across the more significant natural landscapes. Managed habitats maintained the integrity of ecological systems while diversifying the landscape. This condition is in open contrast with today's situation, due to industrialized agriculture (Bocchi 2015; Gliessman 1984). The failure of the "green revolution," not able to save the natural resources because of an excess of environmental simplification and landscape destruction, brings an enlarging agreement, towards the concept of agroecology.

Note that agrarian scientists recently proposed the new discipline of Agroecology, but it is still limited to update farming systems *without consider* the *entire agricultural landscape* or only mentioning the landscape in the generic way.

However, the necessity to rehabilitate the agricultural landscape is truly impellent, as shown in Table 3.6. In the plain landscape we can have near 80% of industrial farming, vs. less than 20% of organic, while their differences in BTC and CBSt are severe: the biological territorial capacity of vegetation of industrial agricultural $BTC_{IND} = 0.73$ Mcal/m^2/year results only 64% of $BTC_{ORG} = 1.14$, where their difference in agrarian land use is only 5%. Moreover, the bionomic efficiency of vegetation of industrial agricultural $CBSt_{IND}$ is 9.03 in comparison with $CBSt_{ORG} = 13.55$: no comment! To deepen these arguments, see the publications of Ingegnoli and Ingegnoli & Bocchi (Ingegnoli 2011a, 2015, 2018; Ingegnoli and Bocchi 2017). See also Chaps. 4 and 5.

Table 3.6 Comparison among different agrarian managements. Average values in North Italy

Landscape Element	INDUSTRIAL FARMING					ORGANIC FARMING					ALPINE FARMING				
	%	HH'	HH	BTC'	BTC	%	HH'	HH	BTC'	BTC	%	HH'	HH	BTC'	BTC
Grass and Pasture	10	60	6	0,65	0,065	14	60	8,4	0,65	0,091	59	60	35,4	0,7	0,413
Meadow	15	80	12	0,45	0,0675	8	80	6,4	0,5	0,04	6	80	4,8	0,5	0,03
Crop Field	60	90	54	0,74	0,444	58	88	51,04	0,78	0,4524	14	88	12,32	0,78	0,1092
Tree hedgerow	1	50	0,5	2,85	0,0285	3	50	1,5	3	0,09	1	50	0,5	3	0,03
Forest Patch	2	12	0,24	5,5	0,11	7,5	10	0,75	6	0,45	17	10	1,7	7	1,19
Urbanized	8	98	7,84	0,15	0,012	7	98	6,86	0,15	0,0105	2	98	1,96	0,15	0,003
Roads and subsidiary	4	100	4	0,05	0,002	2,5	100	2,5	0,05	0,00125	1	100	1	0,05	0,0005
Total LU	100		84,58		0,73	100		77,45		1,14	100		57,68		1,78
Mean CBSt			9,03					13,55					22,17		

3.3.5 Territorial Governance and Planning

Territorial Governance. This term recently introduced is still an object of debate. Substantially, it is the process of organization and coordination of actors to develop territorial capital at a local-regional scale through the sustainable exploitation of it, in a non-destructive way to improve territorial cohesion at different levels. The territory is intended as "territorial capital," constituted of "factors," which could also be grouped, though to some extent overlapping, as natural features, material and immaterial heritage, and fixed assets; as infrastructure, facilities and relational goods; as cognitive, social, cultural, and institutional capital (Davoudi et al. 2008). For its vital role in decisions and actions to be taken concerning Human Health, it should have needed a deepen discussion, but let us focus at least on some key points.

First of all, it is necessary to accept and adopt that the "territory" is, in reality, a part—or a group—of Landscape Units, to be intended as the real living complex systems at the Space-time-information scale of the territory, as already underlined. So, it is not a warehouse!

Second, a territory/landscape is constituted by components (not factors) and processes that interact following the EPP, by the principles of trans-disciplinarity, hierarchy, and scale dependence: so it is neither sufficient nor always correct to simply grouping or overlapping them.

Third, the aims of the management of the territory—according to Bionomics (Ingegnoli et al. 2017) and Planetary Health necessities, taking into account population growth, too—must be to improve "environmental quality," reaching new metastable equilibria, to let the territory able to improve human health (as we will see forward ahead) and support a sustainable development: note that, to gather them, it could be necessary to go against people consensus, as a physician must do. Remember that man must act in compliance with the laws of nature that govern all biological systems and their component (including man himself): he must also preside over and safeguard the organization of life that allows him to live correctly. Territorial planning can be a first field of challenge.

Territorial Planning.[9] The first modern methods of eco-design, or "Design with Nature," begun in 1968–1969, primarily due to the studies of Mc Harg (Mac Harg 1969), also cited by E. P. Odum in his famous treatise "Fundamentals of Ecology" (Odum 1971). These methodologies were divided into four main phases, as follows:

(a) Ecological analysis of the site (site analysis) with the use of "overlapping maps."
(b) Professional qualification for the *objectives of intervention* (*capability*).
(c) Responsiveness better (*suitability*).
(d) Choices feasibility (*feasibility*).

Could this method be available even today? The answer is "Yes," but only following what previously presented, in accordance to the prin-

---
[9] A more in-depth discussion in (Ingegnoli 2015, Chap. 10)

ciples and methods of Landscape Bionomics (Ingegnoli 2015; Ingegnoli and Giglio, 2005; Ingegnoli et al. 2017). However, please pay attention: the use of bionomics methodologies to substitute the overlapping maps is not sufficient, we have to note that the objectives of intervention (capability) are dual: (i) ecological-bionomics recovery, (ii) socioeconomics use.

Note that the "recovery objectives" represent the primary necessity to rehabilitate the examined landscape units, through opportune therapeutic criteria: since a landscape is a living system and it can be altered or dysfunctional, actions must be able to reactivate dynamics of ecological balance (in a metastable sense), bringing it able to evolve within the present given conditions, but also to adapt if they should change. In other words, the term "recovery" does not mean returning to conditions similar to that of a century ago, but to be able to gain a higher level of metastability. Only after the decisions on recovery, we may verify the fitness to the territory's capability, the objectives regarding health, recreation, tourism, economics, etc. These actions follow the necessity to choose the best correspondence (suitability) to the socio-economic use of the territory, elaborating the preliminary Systemic Plan and its verification (e.g., bionomic SEA, Strategic Environmental Assessment). After this stage, we can pass to the executive planning (feasibility).

3.3.6 Role of Urban and Suburban Parks

Urban and Suburban Parks. Since the beginning of 1900, due to the gradual destruction of the environment in big cities and their surroundings, the urban and suburban park (being a portion of a living entity, usually an ecotope, formed by at least two different ecocoenotopes) have been assuming, consciously or not, prominence in an ecological function, first as: place to rebalance environmental stress and to "reset" our level of fitness; a place of positive influence on urban meso-climate; a sub-regulatory system of the metastability; ecotope contributing to the formation of ecological networks; ecotope with a protective function (within the PRT Landscape Apparatus); area of refuge for flora and fauna. Thus, to design an urban or a suburban park, it is essential to understand the bionomics laws and follow a clinical-diagnostic method to act on it (Ingegnoli e Giglio, 2005; Gliessman 1984; Ingegnoli 1998, 2002, 2015). The design criteria of a park must necessarily follow such a theorem. If a park design remains anchored to the principle of pure *formalism* and research of the maximum *originality*, it forgets the crucial importance of the quality of human life itself and the necessity of fitting the environment.

On the contrary, an urban park design has to follow the suggestions of an *ecoiatra*, dealing with the pathology of the landscape units in which it is inserted, at least at three Spatio-temporal scale (punctual, local and of the contest; two temporal threshold in the past plus present time plus two scenarios in the future) and due to the relationships with human health, well beyond the problems of pollution.

Some of the most frequent aims underlined by conventional and bionomics views present astonishing differences and show the importance of the new aspects added by bionomics to park design. For instance, in conventional design, maximum walkability is very important, while in bionomic design, it is necessary to avoid this pervasive walkability. Similarly, easy management everywhere is to avoid in favor of self-maintenance.

A fundamental role remains in the *design of the form*. However, the formal verification criteria have to change: they are no longer reducible only to the visual perception and to the caprices of the creative but must be somewhat related to bionomic and ecological conditions. It is necessary to underline that today the whims of a designer can reach hallucinating formalisms: man creates for himself, in total freedom (understood as the absence of constraints) the criteria for his life, and its landscape, aided by technology! For example vertical walls, "green" lawns of pure geometric shapes (cones, prisms, etc.), trees from opposites habitat held together by technical devices, iron trees, benches of branches of boxwood (but plastic!), colored lights, abstract art and squares concrete with water jets instead of tree-lined spots (but the surface always counted as park!)! We have to underline this because to fully understand an innovative criterion, it is not

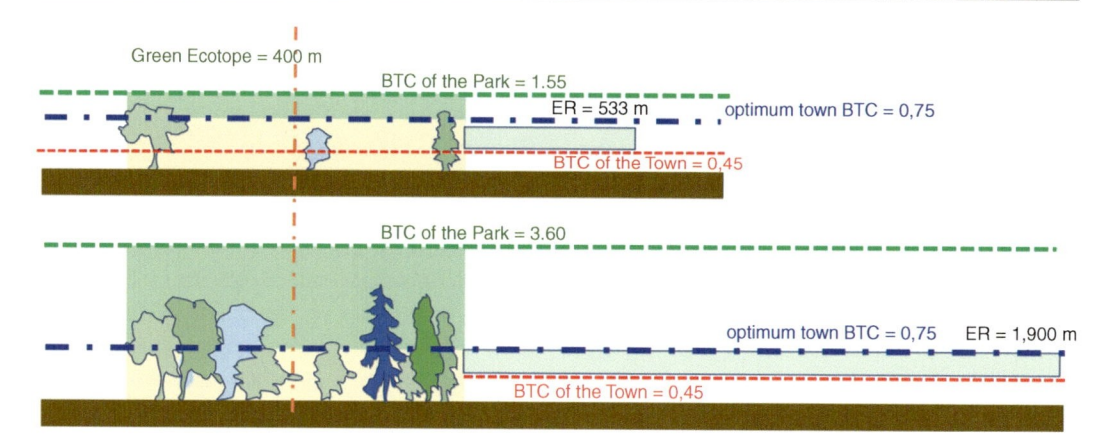

Fig. 3.11 Example of the ecological range of influence (ER) to evaluate two park designs to build in an urban landscape. Note that the mean $BTC_B 3.60/BTC_A 1.55$ Mcal/m²/year = 2.32, while $ER_B / ER_A = 3.58$. The process is not simply proportional

enough to know what to do: we must also know what not to do.

Again, bionomics characters assume crucial importance also in improving human health in a heavily urbanized town. Landscape Bionomics principles and methods can evaluate the difference among various park designs, controlling even the surroundings with the **ER** = *ecological range of influence*, available to balance the lack of BTC of the landscape unit around the park (Fig. 3.11).

3.4 Bionomics and Scales, from Genome to Region

3.4.1 Epigenetics Confirmation of Bionomics Principles

In the first paragraph, we underlined that the old Scientific Paradigm is mainly reductionist, anchored to the concept of process reversibility, to the Darwinian struggle for existence, etc. and that the new Paradigm is nearly the opposite: holistic, systemic, able to admit process irreversibility and to give more importance to symbiosis and cooperation, etc. However, Biology has been a reductionist for decades, e.g., because of the dominion of the dogmatic Neo-Darwinian theory of evolution and the "Central Dogma of Molecular Biology" (Crick 1970), which pretends a direct flux of information DNA-RNA-Protein, so the univocal and direct expression genotype-phenotype. Today, Bionomics principles, following the new paradigm, are marginalized because of both the Neo-Darwinist hypothesis of evolution and the Central Dogma, which count many supporters again!

Therefore, we must underline, with Agazzi (Agazzi 2004, 2014) that the basilar factor of Darwinist evolution is the natural *selection*: after very tiny variations (random but transmissible), some individuals survive while other disappear. Pay attention: (a) *no adaptation* is required, but only the external action of the environment as simple life condition, (b) *no design* is admitted in evolution. Moreover, many Neo-Darwinism supporters pretend that their theory of evolution was unique, and this, is another severe logical mistake!

To get over the Central Dogma it is indispensable to see that the information written on the chromosome basis is not sufficient to sustain and evolve the complex entity of life, as demonstrated by Baltimore (and others) (Baltimore et al. 2001) studying the human genome: it is even necessary the *modulation* of *information* due to the relations with the environment and the behavior. So, the interaction *gene-environment* is crucial!

We have underlined that the new concept of life, at the base of bionomics principles, is not lim-

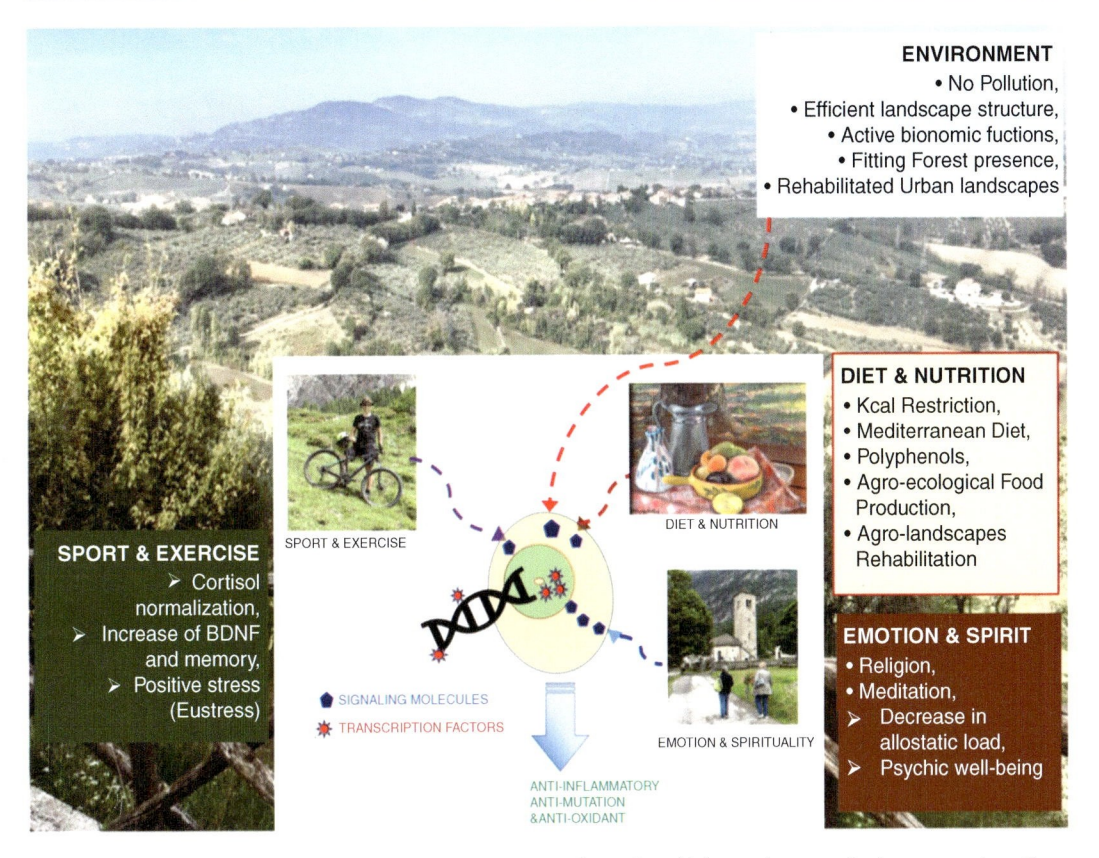

Fig. 3.12 It is not sufficient the information written on the chromosome bases; we also need the modulation of information given by the environment and our behavior. Note that nutritional factors are linked with the population of cereals, which may improve the immune system. Even active bionomics functions are linked with environmental signals and transcription factors. Note the importance of the environment, emerging in all the activities

ited to a single organism or a group of species and, therefore, life organization can be described in hierarchic levels: this implies the necessity to interrelate all the levels of biological organization, thus confirming the bionomic principles. However, only after the acquisition of the importance of epigenetics, these interrelations were possible. Differences between the old view (DNA, RNA, Protein) and the modulation of information linked to the environment (Signaling Molecules, Transcription Factors, DNA, RNA, Protein) are crucial. The new concept of life, at the base of bionomics principles, is not limited to a single organism or a group of species, and therefore the linkage due to epigenetic processes is very significant.

Moreover, epigenetics allowed a braiding of interrelations with other new disciplines, particularly in environmental fields (Fig. 3.12). Food, health, and environment are strictly linked with gene expression and the bionomic state of landscape units, especially where man lives.

We have to add another note to this paragraph: widening the scales to the entire biological spectrum, we must enhance the recognition of the processes of *symbiosis and cooperation* as more critical, even for evolution, of the Darwinian struggle for survival. The dominance of symbiosis does not remain at small scales, e.g., cell endosymbiosis, but it exists even at large ones, as exposed in Fig. 3.13: a village and its fields and forests in a specific landscape unit, e.g., in a Pre-Alpine valley, as the example of Pannone (Trento). Remember that, in Latin, the term *civitas* refers to *urbs* (city) plus *territorium* (its related surroundings).

Fig. 3.13 Agriculture as mutualistic symbiosis near the Ticino river (left) and landscape unit symbiosis (right): a village and its fields and forests (Pannone, Trient)

3.4.2 Bionomics and Scale Interferences

The central concept of the Theory of Hierarchical System is that the organization of a system results from differences in process rates, changing with the scale. Levels in the hierarchy are isolated from each other because they operate at distinctly different rates. Boundaries, which are not only the physical ones, separate the set of processes from components in the rest of the system. So, the inferior level components *explain the origin* of the level of interest (i.e., allow an inner description). In contrast, a superior level system *explains the significance* of the level of interest (i.e., allowing the understanding of the characters derived from transferred conditions).

One of the most important consequences of the hierarchical structure of a system is the concept of *constraint*. It is correct than the non-systemic concept of "limiting factor," and it shows the behavior of an ecological system as limited: (a) by the behaviors of its components and (b) by environmental bonds imposed by superior levels of the organization. Remember: there is a linkage between constraint and information.

This premise is necessary because only the study of the relations among different scales can eliminate the dominance of reductionism: for instance, Agroecology only updates farming systems, *without considering* the *entire agricultural landscape*. In fact, from the interactions among vegetation, fauna, and bacteria at different space-temporal scales, it is possible to avoid a broad input of fertilizers and pesticides. This situation is dangerous facing the prospect of a revival of the Green Revolution today, e.g., with the manipulation of the transcription factor OsGRF4 to improve the efficiency with which some high yielding cereal crops use nitrogen (Li et al. 2018). Therefore, we must understand the focal reasons by which agroecology must rehabilitate the *entire agricultural landscape*! These reasons are mainly concerned with two disciplines: Epigenetics and Bionomics and their relationships with the field-crop scale, i.e., ecocoenotope, ecotope (Fig. 3.14).

Genome and Cell Scales: **Epigenetics** and the environment. The definition of Epigenetics is the study of changes in gene function that are mitotically and meiotically heritable, and that do not entail a change in the DNA sequence. The epigenetic modifications described in current literature generally (Maroon 2009) comprise histone variants, post-translational modifications of amino acids on the amino-terminal tail of histones, and covalent modifications of DNA bases. The information written on the chromosome bases is insufficient; we also need the modulation of information given by the *environment* and men's behavior. So, the interaction gene-environment is crucial. This fact implies a direct linkage between cell scale and environmental scale. That is why Gene Activation improves health and longevity through many directions, from gymnast exercises to emotional perceptions. Signaling molecules and transcription factors modulate genetic information producing anti-inflammatory cytokines, anti-oxidant, anti-mutation, a decrease

3 From General Ecology to Bionomics

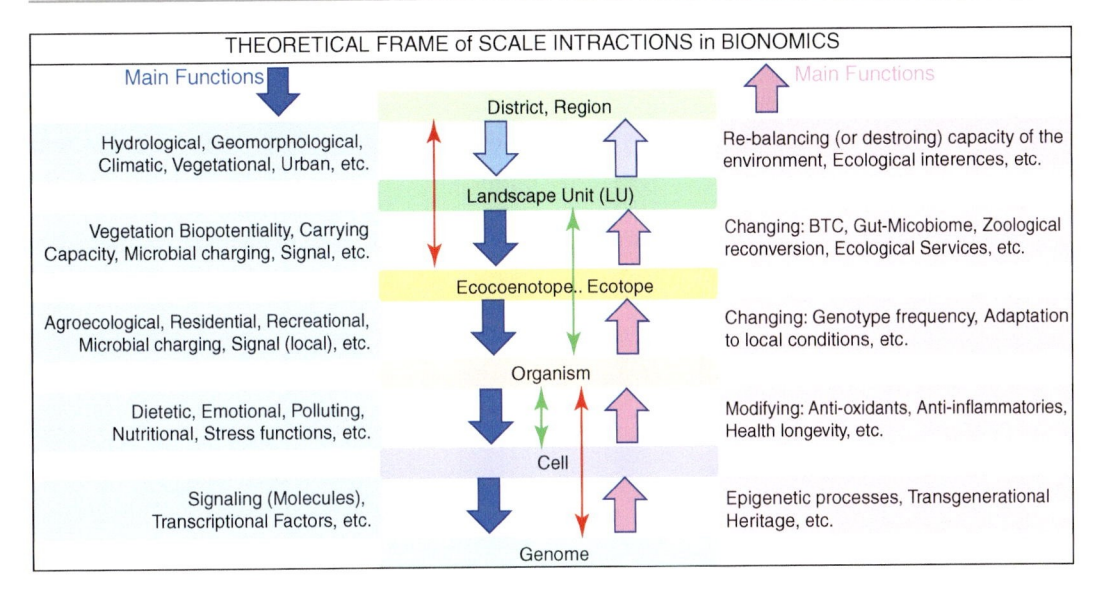

Fig. 3.14 An example of scale interactions, following bionomics principles. Note that (a) some interactions result even with a gradient jump, (b) the directions are both top-down and vice versa

of cortisol, etc. Critical are nutritional factors and environmental parameters, both deeply interesting bionomics and agroecology.

Organism and Ecocoenotope Scales: **Population** and the environment. The complete homogeneity of monocultural industrialized crops, due to cultivars with the same genetic characters, is dangerous for our immune system and mental health, disrupting gut microbiota. Therefore, agroecology suggests to plant populations of cereals, e.g., mixed varieties of *Triticum durum* (Ceccarelli 2013). These plant populations evolve under different types of agronomic management and in the face to specific combinations of *landscape unit characters* (LU) as microclimate, insects, weeds, trees and shrubs, BTC, HS/HS*, TD, DI, BF, etc. In this way, the frequency of genotypes that have adapted to local conditions gradually increases. Epigenetic processes lead to the mentioned evolution of the population "Evolutionary Plant Breeding" (EPB): local environmental characters guide DNA expressions and make them transmitted. Note that these populations do not need to be treated with pesticides and herbicides.

Landscape Unit and Region Scales: **Bionomics** and the environment. The extension of Life to large scales, as the landscape ones, permits a first understanding of the linkages among hierarchic levels of biological organization because, in hierarchic systems, constraints are imposed *from above*. Let us list the main processes demonstrating the influence of broader scales on detailed ones: (a) LU characters are indispensable to the evolution of populations; (b) in the MR/BF (mortality rate/bionomic functionality) correlation, the health damages depend on living complex system (i.e., LU) dysfunctions, independently from pollution (Ingegnoli 2015; Ingegnoli and Giglio 2017); (c) the so-called pruning effect, which improves the environmental fitting in the development of brain, extends the influence of large scale on organisms, especially on man; (d) the *protective* effects of landscape structures, as tree corridors networks and forest patches with high BTC, have to influence the detailed scales; (e) the zoological reconversions, due to LU structure changes, lead to an *exacerbation of virus* and parasites influencing man and animals; (f) the interactions "landscape apparatuses/vital space per capita" express other meaningful linkages among scales, permitting the assessment of Carrying Capacity and the measure of heterotrophy of a LU.

Note that all these inter-scale influences may have many consequences on human Health, becoming very important in Planetary Health studies. Comparative Ethology and Psycho-Neuro-Endocrine-Immunology (Bottaccioli and Bottaccioli 2017) may help better understand the influences of the landscape scale on human (and animal) health.

3.5 Environmental Alteration and Human Health

3.5.1 Landscape Pathology Damages Human Health

The relations human health-environment are usually connected to pollution: globally, 1 in 8 deaths depended on pollution in 2012 (WHO research found—Geneve, 03/24/2014). It is the most significant environmental health problem. However, we also observe a steady increase in landscape alterations (Table 3.5): the essential landscape syndromes derive from structural and functional disorders. Therefore, it is necessary to check the eventuality that even these non-toxicological landscape pathologies should be dangerous for human health. It is a fundamental question, because any scientific demonstration of these threatening linkage may change, genuinely, our responsibility and our actions to protect our health and the importance of landscape bio-structural rehabilitation would become imperative.

Moreover, the recreational factors of the environment changed in the last decades, being today generally far from residential areas and expensive to be reached. Many environmental components are today altered at a wide-scale (e.g., an entire landscape unit, LU), and a dangerous stress condition is more diffuse, often in an unconscious way. Thus, the spontaneous rebalance of stress has become more complicated, and many illnesses are growing.

Oncological epidemiology indicates that in metropolitan suburban LU, the illness frequency is higher than in country LU. Suburbia generally presents many landscape pathologies, but it is difficult to distinguish between the prominent role of pollution and environmental stress due to landscape structural dysfunctions. Insufficient epidemiologic research emerges in this direction. So, psychic disorders (Kessler 2005) may be more indicative: for instance, major depression disorder (MDD) in metropolitan LU is known to be higher than in country LU (Ingegnoli 2015), and its frequency is near twice the one in suburban LU.

The etiology of coronary syndrome demonstrated that, when the sympathetic response exceeds the limits of control, it may generate a sequence of dangerous episodes (Baroldi 2003). A sympathetic alteration can lead to medial neuritis, post-atherosclerotic spasms, regional asynergy, and increase of ventricular pressure, extra-vas compression, hematic block, and necrosis infraction, heart–brain–endocardial reflex interaction, catecholamine myonecrosis of the infraction/sudden death, arrhythmogenic hypersensitivity, ventricular fibrillation/cardiac arrest. This sequence is significant because it demonstrates another way to cardiac infraction besides the classic thrombogenesis. The catecholamine myotoxicity has been found at the histological level with high frequency (78%).

We have to underline that the above-exposed etiology results in a sympathetic response exceeding the capacity of control in dependence on environmental stress. Even if it should be impossible to quantify how much of this stress should be due to landscape pathologies, it is also impossible to exclude this component. No doubt that the main origin of the stress is due to the working environment, but the role of the natural environment remains crucial both in rebalancing or in the aggravation. Consequently, if the adrenergic stress may be partially caused by a landscape structural and bionomic dysfunction, our habitat's bionomic rehabilitation reaches a new level of primary importance.

Our organism is related to the environment even through a system of stress alarm described in Physiology, after pioneer Selye studies (Selye 1956), in 1992 by Berne and Levy (Berne and Levy 2005). As shown in Fig. 3.15, the human body presents very efficient ways to stress

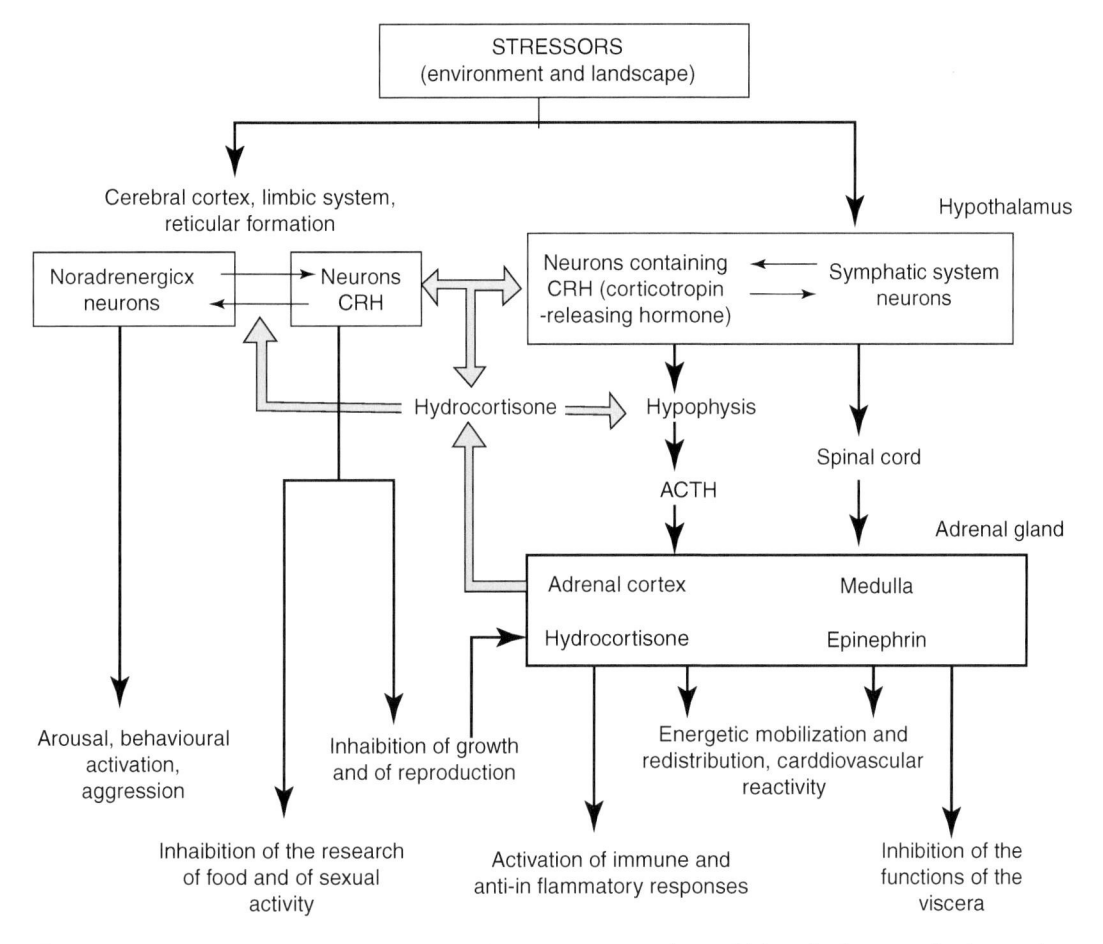

Fig. 3.15 The sympathetic nervous system and the hypothalamus–pituitary–adrenal axis mediate the integrate answers of the organism to stress. The negative feedback acted by cortisol (hydrocortisone) functions to limit too strong reactions, which can be dangerous for the organism (from Berne and Levy (Baroldi 2003) modified). Many of these stressors are due to landscape structural dysfunctions, even in absence of pollution

alarm, both via the sympathetic nervous system and the hypothalamus–pituitary–adrenal axis. The negative feedback acted by cortisol works to limit too strong reactions, which can be very dangerous for the organism. Many of these stressors are inevitably due to landscape structural dysfunctions, even in the absence of pollution. Remember that cortisol follows a circadian rhythm.

The increase of illness depending on landscape pathologies is theoretically available to reach levels of danger, mainly because the capacity of natural rebalancing of natural ecotopes is strongly reduced.

3.5.2 New Relationships between Bionomics, Ecology, and Medicine

Ingegnoli gave a preliminary demonstration that the increase of the mortality rate correlates to the landscape alterations, as shown in Fig. 3.16. The demonstration concerns research on the Monza-Brianza province, N-E Milan hinterland. These 55 municipalities were divided into nine groups, from agricultural to dense-urban *landscape types* and plotted in the HH/BTC model, in which the BF curves measure from BF = 0.50 to BF = 1.10. Four to five of these landscape types (suburban-

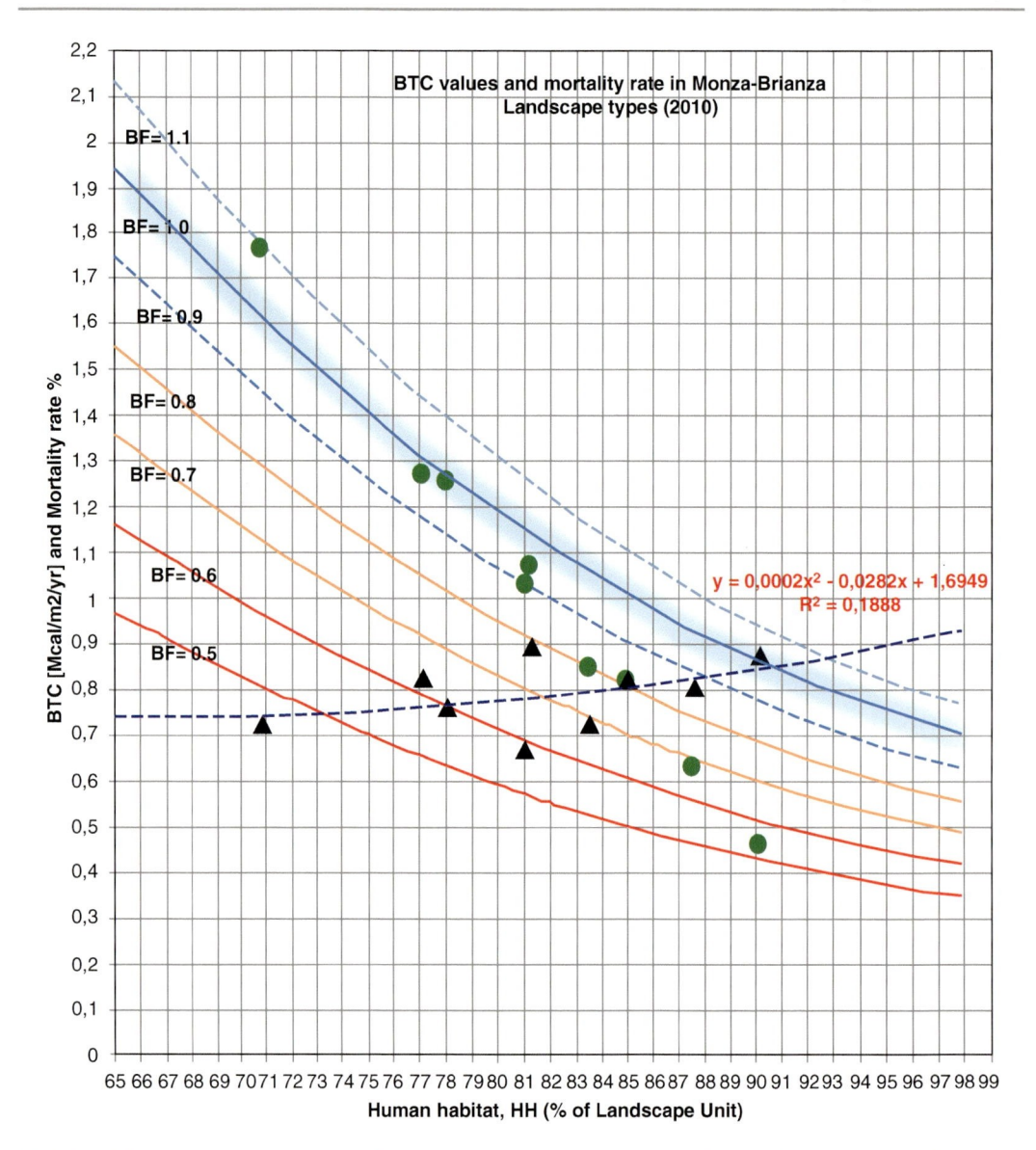

Fig. 3.16 The 55 municipalities of the Monza province were divided into nine landscape types (from agricultural to dense-urban) and plotted (green dots) in the HH/BTC model: only four-five of them can be considered in a normal Bionomic Functionality state (blue zone, i.e., 0.9 < BF >1.1), the others resulting altered (orange and red curves, 0.8 < BF >0.5). Their mortality rate MR (black triangles) increases from 0.74 to 0.78 when the relative LU lies within the tolerance BF band (blue zone), but from 0.78 to 0.91 out of this BF band. Correlation between a BF value and the relative MR value for a LU can be found following the HH values. The violet curve represents the trend of MR.

rural, suburban-technologic, urban-suburban, urban, and dense-urban) do not follow the curve of normality presenting heavy structural dysfunctions (Fig. 3.16).

The tolerance of normality is considered here in the band BF = 0.90–1.10. Relating the mortality rate MR with the Human Habitat HH, we can note that the increase within the tolerance zone (HH from 65% to 81%) results in +5.4%, while the MR increase in the second interval (HH = 81–97%) is +16.7%. This exceeding increase in MR is due to bionomic alterations, an evident correlation between human health and landscape structural degradation. A more in-depth investigation is following, as we will see in Chap. 5.

The strategic environmental assessment and territorial governance must be obliged to enclose health evaluation, even in the absence of pollution, in the direction of bionomic rehabilitation. The figure of the "*ecoiatra*," the doctor of ecological systems and their syndromes relating to human health, must be seriously considered.

Note that without the principles and method of bionomics, this study on MR/BF correlation is merely impossible. Therefore, in medical formation, and especially in Public Health faculties, we strongly suggest adding a course on Landscape Bionomics. This course may also contribute to open up Medicine to the Systemic paradigm. Remember that if both ecology and medicine do not consider life as a unitary biological spectrum, they are *challenging to engage* with the field of planetary health.

Physicians appreciate the problems that come with environmental degradation but, *generally, see them as someone else's (i.e., environmental policymakers') problem to solve*, while they focus on repairing the damage (Fig. 3.17, left).

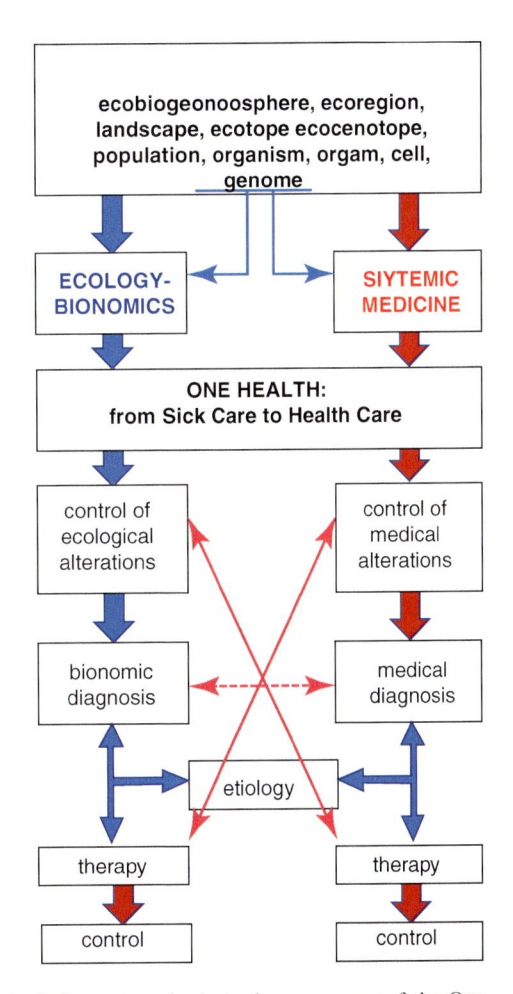

Fig. 3.17 Comparison between the relationships Ecology-Medicine before the Systemic Paradigm (left) and after (right). Note that to consider Life as a unitary biological spectrum leads to the emergence of the One Health concept, so it forges closer ties between the two disciplines

Similarly, better environmental protection for the benefit of human health is recognized as a benefit, but again, *medics do not "do" preventative health in this way*—public health policymakers do. With this in mind, it is not entirely clear what the medical profession and/or medical students are meant to "do" with this Planetary Health knowledge and how they can use it to help patients.

In Fig. 3.17, we underline the differences between the mentioned position of doctors related to ecology

(a) when the systemic paradigm was not recognized, so the reductionism maintains Ecology and Medicine strictly separated, and the new one,
(b) when they admit the new definition of life as a unitary biological spectrum.

Note that the first viewpoint did not consider the concept of "One Health," but only a "Pressure-Response assessment" of ecology that had nothing to do with "Sick care" of medicine. New linkages between the two disciplines emerge following the new systemic paradigm, both in diagnostic and therapeutic fields and especially in etiology.

3.6 Conclusion

We have affirmed that the landscape syndromes damage human health independently from pollution, and this changes the relations between bionomics/ecology and medicine. Consequently, medicine doctors must have a formation on landscape bionomics, at least who is interested in Public Health and Planetary/Global Health. This aim needs more than a single chapter, even mainly dedicated to bionomics, because we need a synthesis related to theory and its inferences on medicine, for instance, to demonstrate the cited affirmation.

So, it is necessary, at least, a sequence of three chapters related in prevalence to (a) bionomics theory, (b) environmental alterations, (c) etiological paths on human diseases: respectively, Chap. 3, Chap. 4, Chap. 5.

Thus, the next chapter will concern environmental alterations, which have to be upgraded following bionomics' principles. A stimulant Concept Note has been distributed among PHA members at the PHA Meeting of Edinburg (2018): It represents the network of drivers relating to environmental alterations to dangerous health effects. Upgrading traditional ecology brings to upgrade these Man–Environment drivers, becoming the main aim of next Chap. 4.

References

Agazzi E (2004) Right, wrong and science. The ethical dimension of the techno-scientific enterprise. Amsterdam-New York, Rodopi

Agazzi E (2014) Science, metaphysics, religion. F. Angeli, Milano

Almada AA, Golden CD, Osofski SA, Myers SS (2017) A case for planetary health/geo health. Geohealth 1:75–78. https://doi.org/10.1002/2017GH000084

Baltimore D et al (2001) Frontiers of life. Academic Press, Boston

Baroldi G (2003) Patologia cardiovascolare: ruolo nell'iter diagnostico-terapeutico. Edizioni Primula, Pisa

Berne RM, Levy MN (2005) Principles of physiology, 4th edn. Mosby

Bocchi S (2015) Zolle, storie di tuberi, graminacee e terre coltivate. R. Cortina Ed, Milano

Borlaug NE (1974) In defence of DDT and other pesticides. Pesticides 8(5):14–19. Bombay, India

Bottaccioli F (2014) Epigenetica e Psiconeuroendocrinoimmunologia, le due facce della rivoluzione in corso nelle scienze della vita. Edra spa, Milano

Bottaccioli F, Bottaccioli AG (2017) Psiconeuroendocrinoimmunologia e scienza della cura integrata. Il manuale. Edra spa, Milano

Braun-Blanquet (1926) Pflanzensoziologie: Grundzüge der Vegetationskunde, Berlin

Ceccarelli S (2013) Produrre i propri semi. Manuale per accrescere la biodiversità e l'autonomia nella coltivazione delle piante alimentari. Libreria Editrice Fiorentina, Firenze

Cicero MT (1973) (52 a.C). *De Legibus libri tres*. Reprint by Mondadori

Crick F (1970) Central dogma of molecular biology. Nature 227:561–563

Davoudi S, Evans N, Governa F, Santangelo M (2008) Territorial governance in the making. Approaches, methodologies, practices. Boletín de la AGE 46:33–52

Francesco di Assisi (1224) Laudes Creaturarum. In: Codice 338, f.f. 33r - 34r, sec. XIII - Biblioteca del Sacro Convento di San Francesco, Assisi

Einstein A (1944). *Einstein archives* (pp. 61–574)

FAO (2010) Global forest resource assessment: Main report. FAO

Forman RTT (1995) Land mosaics: the ecology of landscapes and regions. Cambridge University Press, Cambridge, UK

Forman RTT, Godron M (1986) Landscape ecology. John Wiley & Sons, New York, p XIX+619

Francesco I (2015) Lettera Enciclica Laudato Sì del santo Padre Francesco sulla Cura della Casa Comune. Vaticano, Roma

Giglio E (2011) Glossario. In: Ingegnoli V (ed) Bionomia del paesaggio. L'ecologia del paesaggio biologico-integrata per la formazione di un "medico" dei sistemi ecologici. Springer-Verlag, Milano, pp 315–326

Giglio E (2015) Synthetic glossary. In: Ingegnoli V (ed) Landscape bionomics. Biological-integrated landscape ecology. Springer, Heidelberg, Milan, New York, pp 425–431

Gliessman SR (1984) An Agroecological approach to sustainable agriculture. In: Jackson W, Berry W, Coleman B (eds) Meeting the expectation of the land: essays in sustainable agriculture and Sturadship. North Point Press, S. Francisco, CA, pp 160–171

Ingegnoli V (1991) Human influences in landscape change: thresholds of metastability. In: Ravera O (ed) Terrestrial and aquatic ecosystems: perturbation and recovery. Ellis Horwood Ltd. Chichester, England, pp. 303–309

Ingegnoli V (1998) Landscape Ecological criteria as a basis for the planning of a suburban park in Milan. in: Breuste J, Feldmann H, Uhlmann O (eds). Urban Ecology. Springer-Verlag, Berlin Heidelberg. pp. 657–662

Ingegnoli V (1999) Definition and Evaluation of the BTC (Biological Territorial Capacity) as an indicator for landscape ecological studies on Vegetation. In: Sustainable Land Use Management: the challenge of ecosystem protection. EcoSys: Beitrage zur Oekosystemforschung, Suppl Bd 28: 109–118

Ingegnoli V (2001) Landscape ecology. In: Baltimore D, Dulbecco R, Jacob F, Levi-Montalcini R (eds) Frontiers of life, vol IV. Academic Press, New York, pp 489–508

Ingegnoli V (2002) Landscape ecology: a widening foundation. Springer, Berlin, New York, p XXIII+357

Ingegnoli V (2010) Ecologia del paesaggio: L'ecologia del paesaggio biologico-integrata. In: Gregory T.(ed.) XXI secolo, vol. IV. Istituto della Enciclopedia Italiana Treccani, Roma. pp. 23–33

Ingegnoli V (2011a) Bionomia del paesaggio. L'ecologia del paesaggio biologico-integrata per la formazione di un "medico" dei sistemi ecologici. Springer-Verlag, Milano, p XX+340

Ingegnoli V (2011b) Non-Equilibrium Thermodynamics, Landscape Ecology and Vegetation Science. In: Moreno-Pirajan JC (ed.) Thermodynamics: systems in equilibrium and non-equilibrium. IntechWeb.org, Rjeka, Croatia, pp.139–172

Ingegnoli V (2015) Landscape bionomics. Biological-integrated landscape ecology. Springer, Heidelberg, Milan, New York, p XXIV + 431

Ingegnoli V (2018) L'agricoltura industrializzata è oggi una minaccia per la salute e per il pianeta. PNEI News n°6, pp: 18–22

Ingegnoli V, Bocchi S (2017) The Crucial Contribute of Bionomics to Agroecology from Farming-Systems to Agricultural Landscapes. In: ResearchGate, Preprints

Ingegnoli V, Bocchi S, Giglio E (2017) Landscape bionomics: a systemic approach to understand and govern territorial development. WSEAS Trans Environ Dev 13:189–195

Ingegnoli V, Giglio E (2005) Ecologia del Paesaggio: manuale per conservare, gestire e pianificare l'ambiente. Sistemi editoriali SE, Napoli, p 685+XVI

Ingegnoli V, Giglio E (2017) Complex environmental alterations damages human body defence system: a new bio-systemic way of investigation. WSEAS Transactions on Environment and Development, Pp. 170–180

Ingegnoli V & Pignatti S (2007) The impact of the widened Landscape Ecology on Vegetation Science: towards the new paradigm. Springer, Rendiconti *Lincei Scienze Fisiche e Naturali*, s.IX, vol.XVIII, pp. 89–122

Israel G (2005) The science of complexity: epistemological problems and perspectives, vol. 18. Science in Context

Kessler RC, Chiu WT, Walters EE et al (2005) Prevalence, severity, and comorbidity of twelve-month DSM-IV disorders in the national comorbidity survey replication (NCS-R). Arch Gen Psychiatry 62(6):617–627

Lehman-Dronke J (2007) Risposta alla relazione del card. C. Schonborn. In: Creazione ed Evoluzione; Convegno di Papa Benedetto XVI a Castelgandolfo, pp. 158–163

Li S, Tian Y, Wu K et al (2018) Modulating plant growth-metabolism coordination for sustainable agriculture. Nature: 595–600. https://doi.org/10.1038/s41586-018-0415-5

Lorenz K (1978) Vergleichende Verhaltensforschung: Grundlagen der Ethologie. Wien, Springer-Verlag, Berlin

Lovelock J (2006) The revenge of Gaia: why the earth is fighting back and how we can still save humanity. Penguin Books

Lovelock J, Margulis L (1974) Atmospheric homeostatsis by and for the biosphere: the Gaia hypothesis. Tellus XXVI

Mac Harg I (1969) Design with nature. Natural History Press, New York

Makarieva AM, Gorshkov VG (2012) Revisiting forest impact on atmospheric water vapor transport and precipitation. Theor Appl Climatol. https://doi.org/10.1007/s00704-012-0643-9

Maroon JC (2009) Epigenetics living smarter. Tri-State Neurosurgical Associates-UPMC, Pittsburg

Naveh Z, Lieberman A (1984, 1994) Landscape ecology: theory and application. Springer-Verlag, New York, Inc. pp. XXVII+360

O'Neill RV, De Angelis DL, Waide JB, Allen TFH (1986) A hierarchical concept of ecosystems. Princeton Univ. press, Princeton, NY

Odum EP (1971) Fundamentals of ecology. Saunders, Philadelphia, PA

Odum EP (1983) Principles of ecology. Saunders, Philadelphia, PA

Pignatti S (1976) Geobotanica. In: Cappelletti, Botanica, vol. ii, UTET Torino

Pignatti S, Box EO, Fujiwara K (2002) A new paradigm for the XXIth century. Annali di Botanica II:31–58

Prüss-Üstün A, Mathers C, Corvalán C, Woodward A (2003) Introduction and methods. Assessing the Environmental Burden of Disease

Prüss-Üstün A et al. (2016) *Preventing disease through healthy environments: a global assessment of the burden of disease from environmental risks.* WHO, iris

Rivas-Martinez S (1987) Nociones sobre Fitosociolog´ıa, biogeograf´ıa y bio-climatolog´ıa. In: La vegetacion de Espanā. Universidad de Alcala´ de Henares, Madrid, pp 19–45

Rockfeller Foundation & Lancet Commission (2015) Report on Planetary Health. The Lancet

Selye H (1956) The stress of life. Mc Graw- Hill, New York

Sheil D, Murdiyarso D (2009) How forest attract rain: an examination of a new hypothesis. BioScience 59(4):341–347. https://doi.org/10.1525/bio.2009.59.4.12

Tansley AG (1935) The use and abuse of vegetational concepts and terms. Ecol. 16: 284–307

Tuxen R (1956) Die heutige potentielle naturliche Roma Vegetation als Gegenstand der Vegetationkartierung. Angew. Pflanzensoziologie Stolzenau/Weser 13:5–42

Urbani-Ulivi L (ed) (2019) The systemic turn in human and natural sciences. A rock in the pond. Springer-Nature, Switzerland

Vitruvius MP (1931) De Architectura/On Architecture, Books I-X. William Heinemann Ltd. London

White L (1967) The historical roots of our ecological crisis. Science 155:1203–1207. reprinted in Schmidtz and Willott 2002

Wild CP (2012) The exposome: from concept to utility. Int J Epidemiol 41:24–32. https://doi.org/10.1093/ije/dyr236

Zonneveld IS (1995) Land ecology. Academic Publishing, SPB Amsterdam

Planetary Health: Human Impacts on the Environment

Vittorio Ingegnoli and Elena Giglio

Abstract

Background: A stimulant Concept Note has been distributed among PHA members at the PHA Meeting of Edinburg (2018): It represents the network of drivers relating environmental alterations to dangerous health effects. Upgrading traditional ecology brings to upgrade these Man–Environment drivers too, becoming the main aim of this chapter.

Theory and Method: After a brief synthesis of the main principles of Landscape Bionomics (LB), we add to the drivers the arguments derived from LB: ecological economy, living systems dysfunctions, LB degradations, meditation, diseases independently from pollution. Moreover, all principal aspects of environmental alterations are concisely considered.

Findings: The result of this work led to (a) upgrade the "ecosystem services," (b) integrate pollutants with biologic landscape parameters, (c) change the concept of biodiversity, (d) consider the climate change less linked to CO_2 and more to forest destructions, (e) express functions of landscape unit (LU) change, (f) recover a LU changing configuration, not land use, (g) underline the therapeutic help of meditation.

Discussion and Conclusion: The upgrading of Man/Environment drivers brings to widening the etiological paths. Pollution, infective agents, and agrofood dysfunctions remain the most known, while environmental stress, lack of defense aspects, and lack of hierarchical relations need more attention and more studies.

Keywords

Human Impact · Landscape Bionomics Man/Environment drivers · Environmental alteration · Etiological paths

---

V. Ingegnoli (✉)
Department of Environmental Sciences and Policy, University of Milan, Milan, Italy

PHA (Planetary Health Alliance, member of), Harvard University, Cambridge, MA, USA

SIPNEI (It. Soc. of Psycho-Neuro-Endocrine-Immunology), Rome, Italy
e-mail: vittorio.ingegnoli@guest.unimi.it

E. Giglio
PHA (Planetary Health Alliance, member of), Harvard University, Cambridge, MA, USA

Environmental Science: man and the environment in the Apenninens, PhD, Teacher of Geography at High School, Milan, Italy

4.1 Human Impact on the Environment: Main Drivers

4.1.1 The Planetary Health Alliance Concept Note

Since the famous book of Rachael Carson (Carson 1962) "Silent Spring," ecological and medical communities have been recognizing that human alteration on the environment threatens human health, but their reactions have been remaining quite limited. From the beginning of the third millennium the strong population growth, from 3.3 to 7.0 billion in the period 1970–2012, and the Climate Change from +0.01 to +0.60 °C in the same period, greatly enhanced the environmental alterations and the ecological and medical alarm, as mentioned in the previous Chap. 3. WHO Environmental Burden of Disease (Prüss-Üstün et al. 2003; Pruss-Ustun et al. 2016) estimated 17.06% deaths in 2002, but those values became 21.80% in 2012!

Everything is connected—what we do to the world comes back to affect us and not always in ways that we would expect. Understanding and acting upon these challenges call for massive collaboration across disciplinary and national boundaries to safeguard our health. Therefore, in 2014 the *Rockefeller Foundation-Lancet Commission on Planetary Health* was formed to explore the scientific basis for creating this new transdisciplinary field at the intersection of accelerating global environmental change and human health.

As published by Almada et al. (Almada et al. 2017), they released its "Safeguarding Human Health in the Anthropocene Epoch" white paper in mid-2015 (The Lancet Commission 2015). With unplanned but powerful *convergence*, Pope Francis produced his "*Laudato Sì*" encyclical on environment and health around the same time (Pope Francis 2015), and Ingegnoli published "Landscape Bionomics" (Ingegnoli 2015), the test book presenting the new discipline of Bionomics able to begin an upgrading of ecology for the challenges of the third Millennium, following a holistic approach. In the meantime, the Welcome Trust has developed a significant research funding initiative called "Our Planet, Our Health: Responding to a Changing World" to fund pilot projects investigating the connections between environmental change and health; the University of Sydney has recently named its first Professor of Planetary Health, and other universities around the world are rapidly developing curriculum and training opportunities in Planetary Health; the Rockefeller Foundation launched a "Planetary Health Alliance" to build a community of practice in planetary health education, research, and policy. The United Nations Framework Convention on Climate Change has recently announced a Planetary Health track, and the United Nations Environment Program and the World Health Organization have been collaborating on using a Planetary Health lens to address the Sustainable Development Goals.

In 2019 the University of Milan and the Centre for Multidisciplinary Research in Health Science (MACH) had decided to propose the Global Health Master. The Master Course in Global Health (MGH), coordinated by Prof. Gori and Prof. Raviglione and organized at MACH with the support of Intesa San Paolo, is a professional, specialty master of 1-year duration and is offered by the University of Milan in close collaboration with experts from major institutions worldwide.

A very stimulant Concept Note (PHA 2018) has been distributed among PHA members at the end of the PHA Meeting at the University of Edinburg (2018), the main scheme of which is reported in Fig. 4.1. It represents the network of driver interrelations that emerged from the integration of underlying drivers and ecological processes, proximate causes, and mediating factors, leading to dangerous health effects.

New scientific paradigms are systemic; so, we need to upgrade ecology and medicine, turning to systemic direction. Planetary Health is interdisciplinary but, first of all, it must be systemic, and it needs a special relationship between Ecology and Medicine. This relation is to be upgrading because, today, both ecology and medicine pursue few systemic characters and few correct interrelations. Therefore, we need to refer to *new* principles and methods sustained by the most advanced fields, as

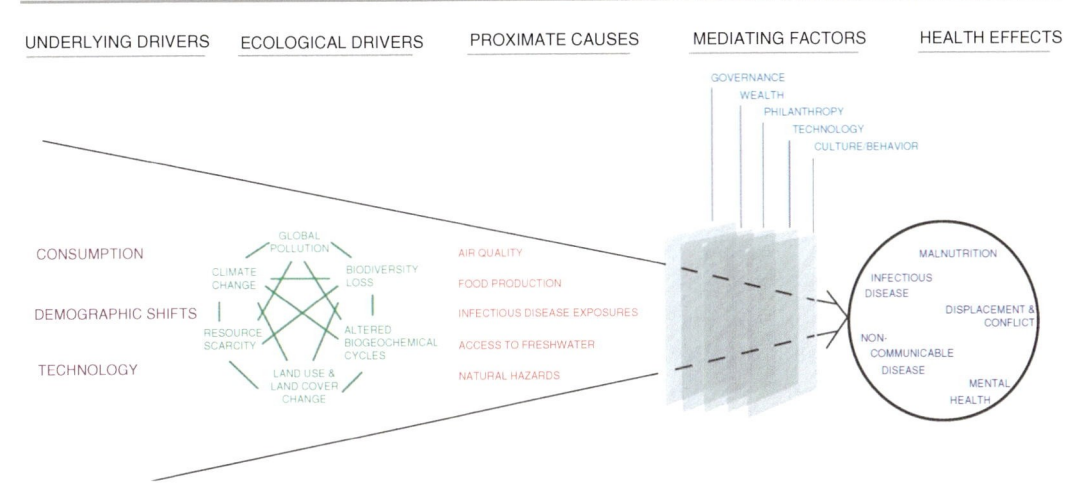

Fig. 4.1 Schematic illustrating impacts of anthropogenic change on human health proposed by PHA

Landscape Bionomics and Systemic Medicine. Following these new principles, we will see that a robust upgrade is needed in two directions: (a) traditional ecology (Chap. 3) and (b) the human impact on the environment (Chap. 4).

4.1.2 Upgrading Traditional Ecology

In this chapter the presentation of Landscape Bionomics, showing the sharp differences with traditional ecology will be limited to some crucial arguments: (a) life definition, (b) environment, (c) ecosystem, (d) land structure, (e) land transformation and evolution, (f) biodiversity, and (g) health state. See Chap. 3 to deepening and the book of Ingegnoli (Ingegnoli 2015) published by Springer.

4.1.2.1 Life

Traditional Ecology. This concept is limited to an incomplete Biological Spectrum and investigated only through the biotic or the functional viewpoints, impossible to be overlapped! No linkages among hierarchic levels of biological organization are taken into consideration, the Emerging Properties Principles is enunciated but not applied, so reductionism is the rule.

Bionomics. Life on Earth is organized in a hierarchy of Living Entities that are six types of complex systems concretely existing, the complete Biological Spectrum (see Table 3.1): it is a complex, self-organizing open system, operating with a continuous exchange of matter, energy, and information with its environment. Processes allowing the definition of life are ontological, but each specific biological level emerges, expressing a process in its own way, by the Principle of Emerging Properties.

4.1.2.2 Environment

Traditional Ecology. Today, it is defined as "the set of physical, chemical and biotic factors that act upon an organism or an ecological community and ultimately determine its form and survival." It is generally intended to be scale-independent and studied for separate sets of elements (water, air, soil, species, pollution). So, it considers only climate, geological components, pollution, or limited indicators.

Landscape Bionomics. The components cited in traditional ecology need to be integrated into complex and scale-dependent systems. The interrelation among the biological levels of life and their environment can be generally intended as comprised within the "biological spectrum" except for out of scale disturbances produced by geophysical forces (earthquake, eruption, climatic catastrophe, etc.) or anthropic ones (soil cementing of large areas, agricultural industrialization, deforestations, dams and large artificial lakes constructions, etc.): also, it is not the environment that "ultimately determine form and survival of an organism": the process implies a continuous interchange between the organism (deepen, between each biological level) and its environment, both of one needing to admit the reciprocal constraints.

4.1.2.3 Ecosystem

Traditional Ecology. Even leaving aside the abuse of the term committed every day by journalists, politicians, etc. or the misuse within the universal language as the contraction of "ecological system," the concept of "ecosystem" (Tansley 1935) is scale-independent, so presenting the same structure and functions, both dealing with "a temporary pond or an entire alpine valley."

Landscape Bionomics. The concept of ecosystem is ambiguous (O'Neill et al. 1986), the biotic and the functional view being unconceivable. Moreover, the real living systems need a conceptual definition based on structure and functions that cannot remain the same in a small ecotope and a vast landscape (system of ecosystems), due to the Emergent Property Principle. At the local scale, the ecosystem must be changed with the concept of ecocoenotope.[1]

4.1.2.4 Land Structure

Traditional Ecology. In a spatio-temporal scale, the basilar structure of the territory is the concept of *eco-mosaic*, composed of different ecosystems covering a geographic territory.

Landscape Bionomics. The common term "territory" corresponds to the underlying mosaic of a multidimensional conceptual structure representing the hierarchical intertwining, in the past+ at present+ in the future, of the ecological upper and lower biological levels and their relationships in the landscape: the *ecotissue* (see Fig. 3.6), referring to the cells within a histologic tissue (Ingegnoli 1993; Ingegnoli 2002; Ingegnoli 2011). Note that the tesserae of the essential mosaic are not ecosystems, but ecocoenotopes[1]. The same multidimensional concept of ecotissue can also be applied at the ecoregion level.

4.1.2.5 Land Transformation

Traditional Ecology. Ecological succession is stated occurring through the serial steps, while the transformations of complex systems are not linear. Even the concept of environmental balance follows classic thermodynamics and reversible processes, e.g., degradation and recovery or "universal" energy flux models, not available in System Theory.

Landscape Bionomics. A good model for a complex system transformation should be based on non-equilibrium thermodynamics (see Fig. 3.5), irreversibility, and metastability concept. Extensive mutual relations occur among the components. Superior biological levels of the organization impose environmental constraints and/or bonds to the lower ones.

4.1.2.6 Biodiversity

Traditional Ecology. Biodiversity is intended near exclusively as specific, and it has to be always high in numbers. Usually, there is a dominance of the concept of resilience.

Landscape Bionomics. Biodiversity of ecocoenotopes, landscapes, and ecoregion are also essential. The most evolved natural systems follow the *resistance* stability, not the resilience. The increase of biodiversity stabilizes an ecological system if it presents too low diversity, but (Pennekamp et al. 2018) destabilizes it if it has already reached its proper diversity.

4.1.2.7 Health State

Traditional Ecology. No idea of Health State outside that of plant, animal, and human beings is conceived. Considerations mainly concern Climate Changes and Pollutions.

Landscape Bionomics. The pathology of upper-scale environmental systems is crucial for landscape rehabilitation and the study of human health damages.

---

[1] Remember that the **ecocoenotope** is a *multifunctional* entity in a definite geographic locality, an ecobiota, composed of the community, the ecosystem, and the microcore (i.e., the spatial contiguity characters, see Chap. 3, Fig. 3.2 and Sect. 3.2.2).

4.1.3 Trying to Upgrade the Man/Environment Drivers

At the light of the bionomic principles, we will find how to complete human and ecological drivers' system, leading to threatening our environment, hence our health. The passage from sick care to health care cannot avoid the mentioned upgrading. Let us refer again to Fig. 4.1, the PHA network diagram, adding to each driver some notes derived from bionomics principles and systemic medicine. We will see in Fig. 4.2 the synthesis of all these adjuncts. Here are some concise notes: the following paragraphs will join some suggestions:

1. UNDERLYING DRIVERS: Addition of "*Neoclassical Economy,*" because it is primordially indebted to Nature: the exchange values derived from labor cannot be the economic measure of everything and ecology cannot be dependent on economy, allowing the degradation of natural systems and their components! Even the so-called Ecosystem Services have to be upgraded.

2. ECOLOGICAL DRIVERS: Addition of "*Living Complex Systems Dysfunctions,*" not present within the six ecological functions, even if these conditions are more than land cover change, implying the dynamic of a hierarchical organization of living entities, acting as complex systems, well farther populations and communities. Even each standard ecological driver has also to be upgraded to the principles and methods of Bionomics.

3. PROXIMATE CAUSES: Addition of "*Landscape Bionomics Degradation and Artificialization*"; it implies the Clinical Diagnosis of the state of the Landscape Units (LU). Note that the level of Bionomics Functionality (BF) is correlated with the Mortality Rate (MR), so with the increase of morbidity. Extremely dangerous is also the degradation of agrarian landscapes.

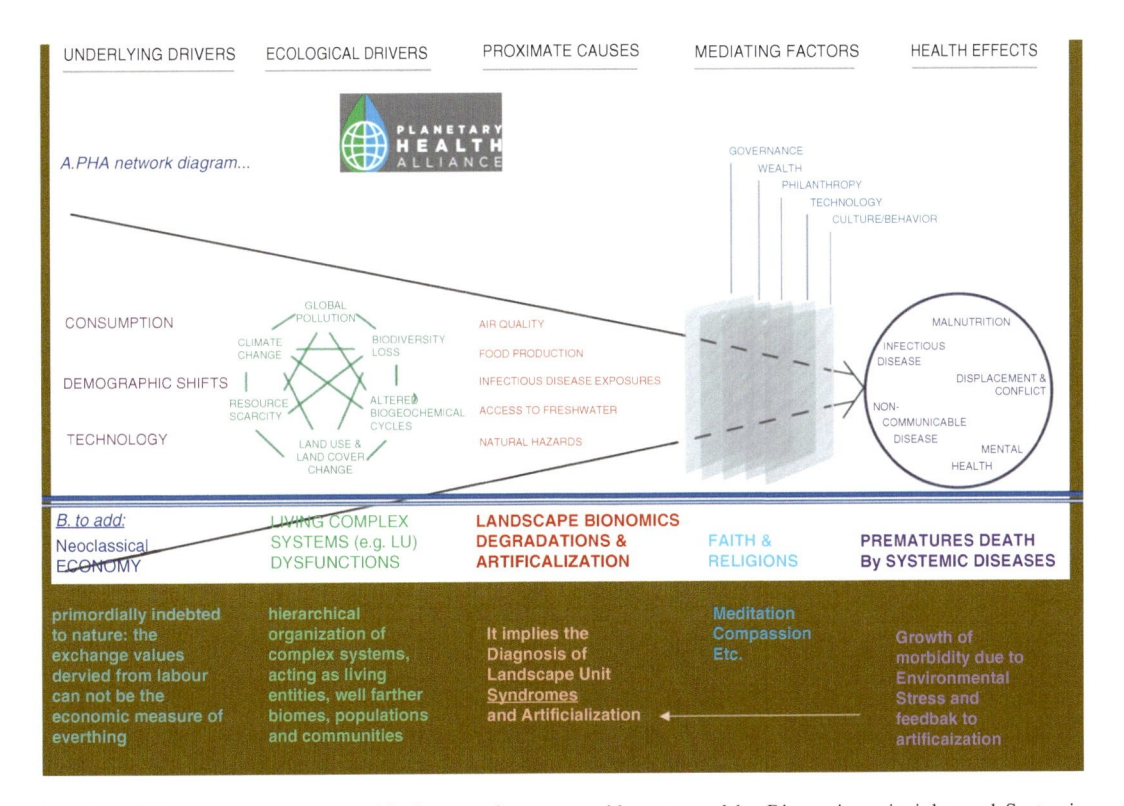

Fig. 4.2 PHA Network Diagram, with the most important adds suggested by Bionomics principles and Systemic Medicine. This upgrading of PHA scheme (Fig. 4.1) opens new theoretical and practical views

4. MEDIATING FACTORS: Having forgotten *Epistemology* and *Faith/Religions*, mediating factors eliminate (a) a correct interpretation of facts, (b) the aid of meditation (also considered in Medicine) (Bottaccioli and Bottaccioli 2017), and (c) the consciousness related to the most significant ethical threat to nature: the betrayal of the role of man expressed as violence against life (Chap. 5, Sect. 5.6).
5. HEALTH EFFECTS: Addition of "*Premature death by Systemic Diseases,*" the main consequence of previous lacks and "*mortality due to excessive drug treatment*" due to the refusal of systemic medicine.

4.2 Underlying Drivers: The Distortions due to Economy

4.2.1 Neoclassical Economy Is Indebted with Nature

In the neoclassical Economy (Morgan 2016), today still dominating our societies, the interaction between nature and the economy is not regulated by the law of *equal exchange*. Economy is founded upon the assumption of *impossible* exchange: so, it is *primordially* indebted to nature (Bjerg 2016). In this vision, the measure of the value of the soil and the land is the labor exerted on them.

The neoclassical economy is in open contrast with the reality because the exchange values derived from labor cannot be the economic measure of everything, the value of each component being linked to its *role* in *complex natural systems* (Ingegnoli 1993). Our *survival* depends on this kind of linkage; therefore, we need to exceed this contrast. However, today a convergence between the natural role and the concept of labor is possible to be considered, completing the concept of Ecological Services.

The real problem is that all the economists sensitive to ecological principles have not a sufficient ecological background to propose alternative actions of a broader cultural perspective. The relationship between ecology and economy is complicated because both are generally misunderstood and separated. Summarizing, in economy:

1. Exchange-value is thought to be the measure of richness and to be independent of nature. *By contrast*, we note that the value of something depends on its role in natural systems.
2. Our culture rejects any ethical evaluation of means and ends. So, the end justifies the means. *By contrast,* ends and means are strictly linked in nature, like a seed to its plant. Therefore, they must not be separated, and we have to choose among different ends.
3. Today absolute scarcity is not seen as a limitation because the low entropy is abundant for our necessities. *However,* these necessities must contain the proper functionality of all the ecological systems.

The relations between ecology and economy show another significant paradigm: the ecological-economics one. It is easier to explain their vast differences by referring to Fig. 4.3. In the first part (a) the scheme represents the ecological-economics. The economy is inserted in the complex system of ecological systems, and re-cycles and ecological services are considered. In contrast, the second part (b) represents the neoclassical economy. The ecological systems, considered as subsystems of the paradigm of economy, postulate that ecology represents merely the extractive and waste disposal sector of the economy. Even if these services become scarce, growth may continue without limits, since technology allows us to "grow around" the ecological sector by the substitution of natural capital with human-made capital. It results in the dictates of market prices, if and when prices of natural capital rise. Thus, nature is nothing but a supplier of the economy. The only limit to growth is technology, which is supposed to have no limits.

The new concept of "Ecosystem Services" may be used to turn the economy to approach the real world. The neoclassical "Labor Theory" can finally recognize that even living entities (e.g., an ecotope or a Landscape Unit) do a labor in the

Fig. 4.3 (**a, b**) The scheme of the two main paradigms on economy-ecology relationships. (From Daly (1999): redrawn). (**a**) Ecological-Economics, (**b**) Neoclassical Economy

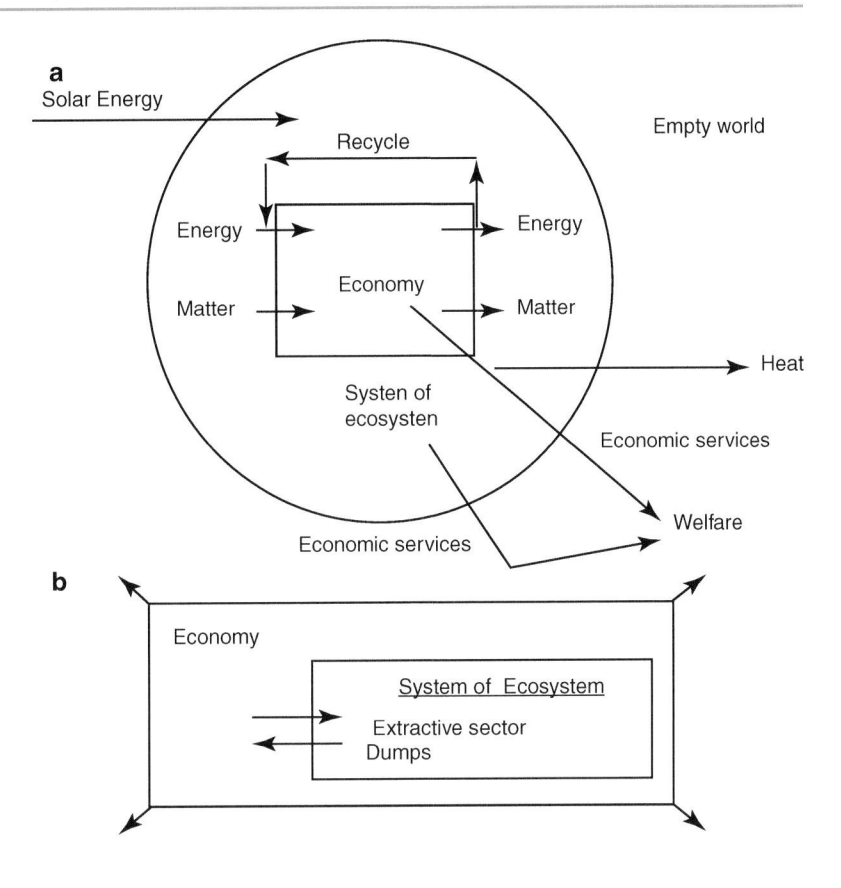

form of the eco-bionomics' services in favor of its components, first of all, our society. For instance, the "Protective Land" compensating the "Transformation Deficit" (TD) develops necessary labors, like:

1. maintaining a proper level of BTC,
2. protecting human health,
3. purifying fine dusts,
4. controlling temperature,
5. regulating rainfall, and,
6. producing food and wood, etc.

These natural labors signify that the land's value consists of two components: the labor of a farmer plus the labor of the ecotope. Remember that the labor of farmer is widening: (a) traditional food production and (b) ecological protection. The problem is to find the correct methodology available to express this labor. That was the aim of a study done by Ingegnoli, Bocchi, and Giglio (Ingegnoli et al. 2018). To reach this goal, we need (a) to follow the principles and methods of Landscape Bionomics and (b) to upgrade the concept of Ecosystem Services.

4.2.2 Upgrading "Ecosystem Services"

Therefore, it is appropriate to promptly identify some crucial changes, both of the terms and the concepts, in order to reduce the risks of insufficient consistency in relation to the systemic approach and possible misunderstandings generated by the too general use of the terms. For these reasons:

1. As underlined (Sect. 4.1.2.3), the concept of ecosystem is at least ambiguous; so, "Ecosystem Services" should be more correctly called "Ecological Services"; this incorrectness would determine a limited ability to recognize all the same services, due to

the lack of integration already mentioned, that is, it would not be easy to recognize the different jobs that natural systems perform.

2. The "Ecological Services" should not refer only to the human population, but to all the components of a Landscape Unit (LU), as Landscape Bionomics suggests on the basis of the study of the components of the landscape as living entities: the human population is one of the components, but not the only one. The landscape performs services necessary for all its components.
3. The ecological functions of Health and Protection should appear explicitly among the Services' guiding concepts (Table 4.1). In the Millennium Ecosystem Assessment (Millennium Ecosystem Assessment 2005) health is mentioned, but subordinate to the four canonical categories (supporting, regulating, provisioning, cultural) and only for the human population, but the concept of protection is not directly expressed.

On the contrary, we have to note:

- health preserving must be placed as a further primary category, as it must be understood as twofold, concerning pathologies of the landscape and human pathologies, obviously with multiple reciprocal interactions;
- bionomic-protective functions (Ingegnoli 2011, 2015) are of primary importance too, because they are indispensable for the measuring of the transformation deficit (TD) concerning changes in the system over time; so, we arrive at the recognition of the equivalence of ecological and economic costs.

Moreover, we have to note that if the labor made by ecotopes or by landscape units is not considered within the land evaluation, the market prices remain low, but the difference with their eco-service labor value is shifted to the local/district society, in a way not dissimilar to what happens with a new urban settlement. Planners know that most of the Nations had to develop a tax on "infrastructure costs" inherent to new urbanizations.

In the case of ecotope services, we must note that their eco-bionomics monetary value is 2–4 times higher than the mentioned costs of urbanization, and, in some aspects, our society is not able to make all these kinds of labors. Thus, such eco-bionomic services are needful but irreplaceable. Therefore, we suggest dividing the costs of this ecological labor by raising the price of land (cropland and forest lots) in the market exchanges and by adding a tax on "eco-bionomics costs" (the economic value of which can be pondered following bionomic criteria) (Ingegnoli et al. 2018) to be paid when new urbanizations regard cropland and forest lots, even in addition to the tax on "infrastructure costs."

We insist! It should be impossible to avoid the continuous destruction of our environment because of the vast abusive differences among built lots vs. agrarian and forest ones. The sprawl of urbanization on the agricultural landscape

Table 4.1 Eco–bionomic services following bionomics principles

Eco–Bionomics SERVICES	CONTROL ACTIONS	BIONOMICS References
1. Supporting	Primary production, soil formation, etc.	Bionomics *payment of ecologic services (PES)* method
2. Regulating	Water and air purification, etc.	Water and air monitoring/economy
3. Provisioning	Food and fibers production, etc.	Bionomics *PES* method
4. Cultural	Architectural heritages and recreations, etc.	Bionomics/culture relations
5. Health preserving	*Geo-health/human-health; risk factor control*	*LU diagnostic evaluation, MR/BF ratio*
6. Protecting	*Transformation deficit; equivalence: Bionomics and economics costs*	*SH*, TD, and bionomics methods (BTC flux, etc.)*

around a city is like cancer on healthy tissue as underlined by the Nobel Konrad Lorenz near 50 years ago (Lorenz 1978).

4.3 Ecological Drivers: An Upgrading

Chapter 3 shows the crucial importance of the paradigm shift, which brought to the upgrading of ecology with the development of the new discipline of bionomics. This shift from ecology to bionomics changed the concept of life, which must be intended as a complete "biological spectrum" from micro to macro scales.

If we concentrate our biological research on the classical range of biochemistry-cell-organism, leaving the medium-macro scales only to obsolete studies, the ecological drivers are the six reported in Fig. 4.1. However, if we follow bionomics and landscape bionomics, putting in evidence the anatomy (structure) and physiology (function) of ecocoenotopes, ecotopes, landscape units (LU), landscapes, and regional bio-systems, we must add "*Living Complex Systems Dysfunctions*" (Fig. 4.2).

Note that the standard ecological drivers may influence the health state of the environment too, but what we purpose to add permits to arrive to *intrinsic syndromes* of the biological levels of life organization. We have also to upgrade each standard ecological driver to the principles and methods of bionomics and landscape bionomics. Consequently, the *ecological* drivers are:

1. Global pollution,
2. Biodiversity loss,
3. Altered biogeochemical cycles,
4. Land use and land cover changes,
5. Resource scarcity,
6. Climate change,
7. Living complex systems dysfunctions.

In the following sections, a short synthesis of their contents and some notes on the systemic/bionomic interpretation are stated.

4.3.1 Global Pollution

The relations between landscape pollution and human health are well known, even if underestimated. Landscape pollution is known to produce toxins, chemical substances that cause any of a vast number of adverse effects in living organisms. The effects may be acute and chronic, including changes in living tissues, physiology, mutations, and consequently, cancer (Kampa and Castanas 2008). Figure 4.4 shows the distribution of air pollutants in Europe, from a study of ESA. Ruhr, Benelux, North Italy (Po valley), and England present the most polluted landscapes. The impact of air pollution was estimated in 22 months, detracted from local life expectancy and € 31.5 billion of sanitary and social costs!

The possibility to reduce many air pollutants is concrete (Gimeno et al. 2001), and we reached a success, e.g., for calcium and sulfate ion emissions in both EU and the USA, where in 10 years the decrease reached about 2/3 of the previous level.

The alterations of the territorial-bionomic system are not usually limited to pollution: structural and functional disorders of the landscape unit (LU) in question are often present (see Table 3.5). These disorders usually lead to systemic pathologies in the human body, that is, to alterations in the control systems starting from the nervous system (central and peripheral), the immune system, and the hormonal system.

The International Association for Impact Assessment (IAIA) correctly indicates the "*cumulative impacts*" of the essential health/environment studies. The effects of landscape pathologies on human health bring *not only* to the integration among different types of pollutants and geophysics parameters *but even* among the biologic-structural parameters of the landscape, as we see in Fig. 4.5, comparing the two flow charts. Note, also, the difference between pollution-direct organic alterations-health damages (partial linear linkages) and the network linkages due to structural and functional disorders, which add new health damages.

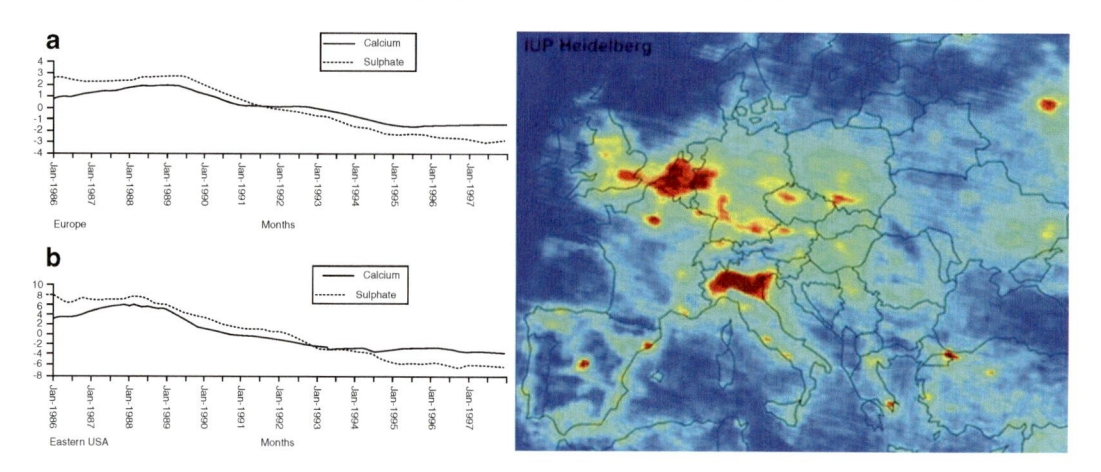

Fig. 4.4 The reduction of calcium and sulfate emissions in the EU (**a**) and USA (**b**) (left, from Gimeno et al. 2001) and the distribution of air pollutants in Europe [from ESA (2012)]

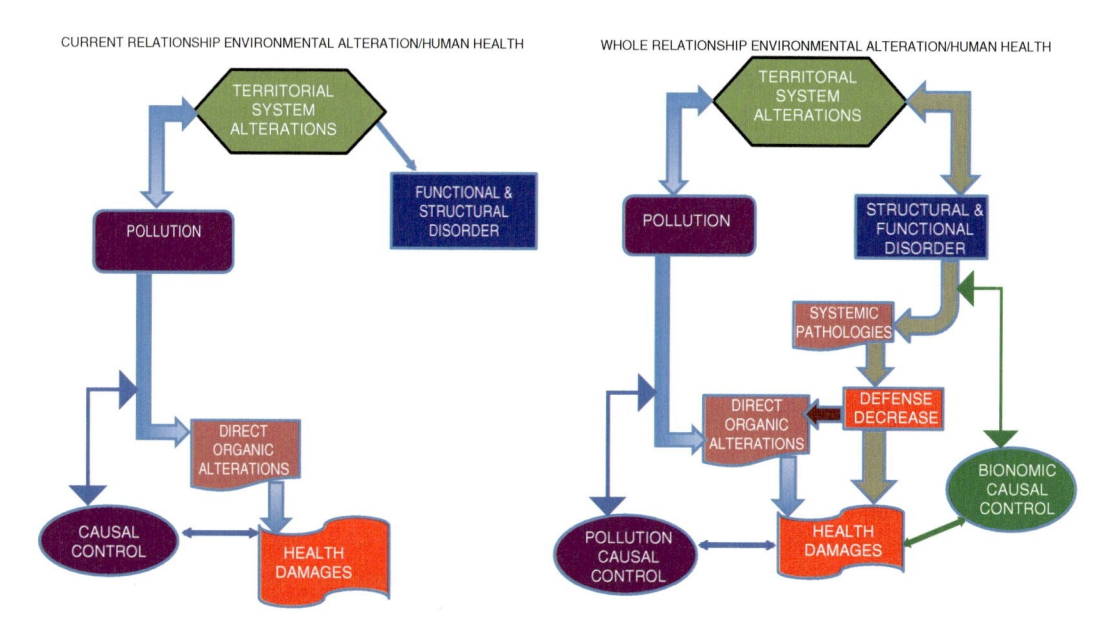

Fig. 4.5 The effects of landscape pathologies on human health bring not only to integration among different types of pollutants and geophysics parameters but even among the biologic-structural parameters of the landscape. Study related to the carbon air pollution at Vado Ligure (Ingegnoli). Comparison between current and bionomic ecology

Therefore, the checks must be carried out not only at the Pollution-Health level but also at the Environmental Disorder-Human Health level, for example, with bionomic studies on the state of the territorial unit in question and in particular on the vegetative subsystems, compared with similar systems but little or not at all attacked by pollution (Chap. 3, Sect. 3.5).

Furthermore, in situations of known full-blown and long-term air pollution, a further investigation of lichens becomes necessary, even if the confirmation of a lichen desert or the toxicological sensitivity to heavy metals would not be sufficient to affirm the level of risk, because within the center of large metropolitan areas one is sure to find similar situations. As an eco-specialist, it is suggested to combine the bionomic monitoring of the territory with the chemical and lichen monitoring surveys.

4.3.2 Biodiversity Loss

The System Theory and the Principle of Emerging Property suggest revising the concept of Biodiversity because many ecologists tend to overestimate its importance. As already pointed out with the example of Chap. 3, Fig. 3.4, we noted that the increase in Biodiversity stabilizes an ecological system if this system has too low a diversity, but destabilizes it when the system had already achieved a right diversity!

However, it is above all the land use for agricultural reasons that causes the decline of Biodiversity on a global level, generating the degradation or loss of habitats. Many estimations indicate that 80% of all species of birds and land mammals are threatened by agricultural practices, which reduce the surface available for animals and fragment populations in small isolated areas threatening them with pesticides. "The history of industrial agriculture is a story of outsourced costs and exploitation of nature," said Raj Patel (University of Texas) (Patel 2006) in an interview with The Independent. Furthermore, again: "Extinction is the elimination of diversity, and that is what happens in Brazil and other areas of the world: green deserts are obtained—soybean or corn monocultures and nothing else."

Although there are over 7000 types of plants cultivated for food, in the surveys carried out on 2014 (Khoshbakht and Hammer 2008), the species that make a significant contribution to production globally would be less than 200, i.e., about 3.0%. Just nine species (0.15%) represent more than 66% of all agricultural production (sugar cane, corn, rice, wheat, potatoes, soybeans, palm oil, sugar beets, and cassava). Besides, food production is responsible for one-third of all greenhouse gas emissions and it is the main defendant for the loss of biodiversity and depletion of water resources, impacting heavily on our planet. At the same time, bad eating habits cause higher health risks for tobacco and unprotected sex and alcohol, while malnutrition still affects many populations.

Deforestation, urbanization, mining activities, and the construction of roads and other transport corridors that interrupt the continuity of habitats also cause the loss of biodiversity: many naturalists note that agriculture and cutting of logs in the forest are the most important reasons, due to consequent habitat loss, even if the human habitat (HH) is not more then 30–32% at the global scale (Emerged Lands).

We have to underline that biodiversity cannot be considered only at the scale of species. It is imperative even at the scale of the landscape. When we affirm that the agricultural landscapes are going to be destroyed, we find principally the destruction of the diversity of their ecocoenotopes and landscape elements, which cannot be only cropped but must be mixed with natural infrastructures.

The main types of agrarian landscapes in Europe, from the least to the most ecologically efficient, are:

1. industrial agriculture, with BTC = 0.60–0.80 Mcal/m^2/year,
2. suburban-rural, with BTC = 0,80–1,00 Mcal/m^2/year,
3. agricultural-productive, with BTC = 0.90–1.20 Mcal/m^2/year,
4. agricultural-protective, with BTC = 1.30–1.80 Mcal/m^2/year,
5. agricultural-forestry, with BTC of around 1.90–2.40 Mcal/m^2/year.

To have a comparison measure, consider that for a couple of centuries the Lombard agricultural landscapes had been remaining constant with a BTC equal to 1.70–1.80 Mcal/m^2/year, while in the last 50 years the average BTC of these landscapes had fallen sharply below the sustainability limit (BTC = 1.20) recording values of 0.80–0.90 Mcal/m^2/year. In Fig. 4.6 we show the vast differences among landscape biodiversity, measuring their landscape-ecological components (average) in comparison with the differences in BTC levels (Mcal/m^2/year).

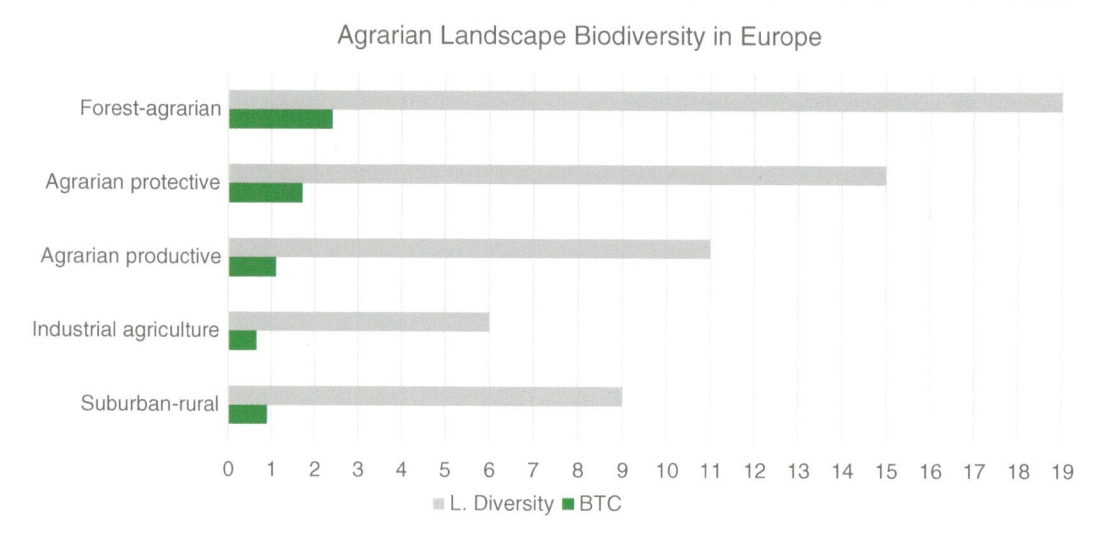

Fig. 4.6 Landscape biodiversity in the main types of Agrarian landscapes in Europe: L. Diversity = n° of landscape-ecological components; BTC = Mcal/m^2/year × 10

4.3.3 Altered Biogeochemical Cycles

In this sector, we will have to limit ourselves to a few quick hints, because this paper is to be understood as indicative of problems of alteration due to man.

We know that the water cycle is altered due to forest destruction (see also Sect. 3.3.4 and Sect. 4.3.5), to the destruction of the river margins, to the increase in water collection from the subsoil, to the exasperated waterproofing of the soils, etc. with repercussions on both climate change and the scarcity of resources.

Let us try to bring a much less known example: the influence of man on the nitrogen cycle.

We read in the publications of ecological interest (Balzani and Venturi 2014) that the nitrogen cycle has been profoundly disturbed by the man in the last 100–110 years due to the development (a) of industrial processes (Fig. 4.7) that use molecular nitrogen, N_2, to obtain ammonia and subsequently fertilizers and explosives, (b) the intensive production of cereals, and (c) the extensive use of fossil fuels. The consequences are different:

1. increase in the amount of nitrogen oxide (N_2O) produced by agricultural fertilization, biomass combustion, and industrial activities, with serious adverse effects because N_2O contributes to destroy ozone and increase the greenhouse effect (it is a greenhouse gas three hundred times more potent than CO_2),
2. a substantial increase in nitrogen oxides (Nox) in the atmosphere, generated by high-temperature combustion processes, responsible for photochemical smog, acid rain, and the onset of severe diseases in humans,
3. increase in nitrogen compounds that end up in rivers, lakes, and seas, generating the phenomenon of *eutrophication*, damage to ecological systems and reduction of biodiversity,
4. contamination of groundwater by the nitrate ion, a dangerous pollutant for drinking water.

Note the importance of considering the landscape structure and components in the nitrogen cycle, not limited to separate processes, from industrial to natural (Ingegnoli 1993).

4.3.4 Land Use and Land Cover Changes

Human actions toward the environment give rise to one of the most significant environmental degradations in this sector, for causes that can always

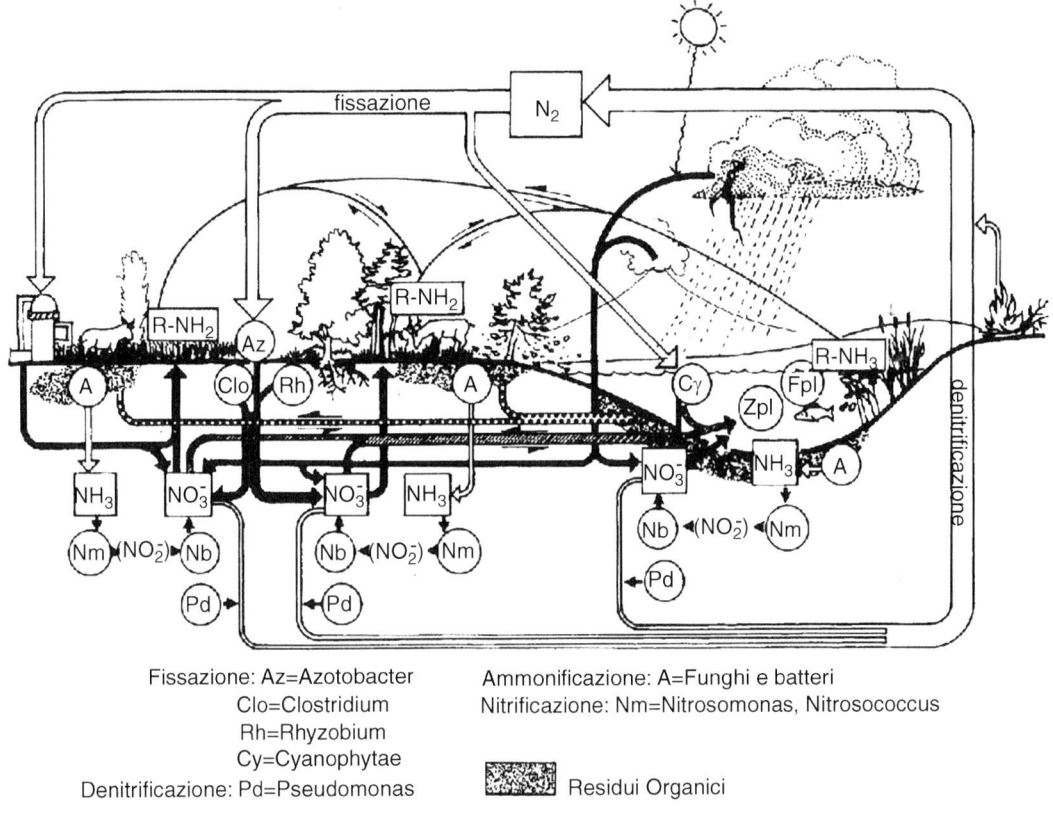

Fissazione: Az=Azotobacter
Clo=Clostridium
Rh=Rhyzobium
Cy=Cyanophytae
Denitrificazione: Pd=Pseudomonas

Ammonificazione: A=Funghi e batteri
Nitrificazione: Nm=Nitrosomonas, Nitrosococcus

Residui Organici

Fig. 4.7 Schematic representation of the nitrogen cycle in a landscape unit. [from Ingegnoli (1993)]. It should be noted that this cycle also added the sources of industrial and volcanic origin, sources that can destabilize the natural balance

be traced back to senseless exploitation of resources (see the equivalence "territory-territorial capital" Sect. 3.3.5). However, another aspect, never considered is that *use and cover changes* within a territory give rise to profound alterations in Landscape Apparatuses[2] structure and functioning within the relative landscape units and, thus, in the metastability of the all ecotissue, due to the complex interlacements and redundancies among all the components, as Bionomics underlines. Moreover, this is true also for the landscape units' respect to the ecoregion scale. So, land use and cover changes must be pondered and also evaluated in relation to their proper context through a landscape bionomic approach before their realization and cannot be projected through a mechanistic approach.

Concerning this situation of alteration of the environment, we can recall that forests are fundamental for the maintenance of health, as will be seen in Chap. 5, but they must be at their maximum ecological-bionomic efficiency. As for agricultural landscapes, we will note that the need to drastically decrease products for plant protection and ensure a proper nutraceutical function would oblige to keep the crops' natural infrastructures integrated. As expressed by FAO (FAO 2015), the emerging lands present both net gains and net loss: e.g., the USA and India registered a net gain

---

[2] A Landscape Apparatus (L-Ap) is constituted by proper functional systems of ecocoenotopes (even not connected), forming a specific configuration within the ecotissue, performing functions among the protective, productive, subsidiary, residential, connective, excretory, resilience, resistance, ... ones (see Sect. 3.3.1 and Ingegnoli 2002; Ingegnoli 2011; Ingegnoli 2015).

Fig. 4.8 Change of land cover in the metropolitan area of Milan 1969–2001. Note that these data can be impressive and give the idea of the strong increase in urbanization: but they cannot evaluate the possible ecological dysfunctions. *URB* urban, *SURB* suburban, *AGR* agrarian, *S-NAT* semi-natural

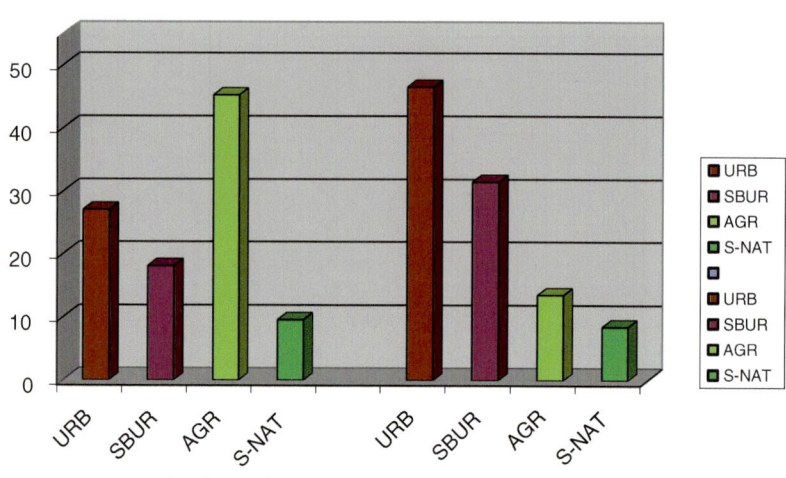

of about 250–500,000 ha, while Brazil and Indonesia have a net loss of about 500–750,000 ha. Note that the grown back of forests is generally linked with a less ordered phytocoenosis, so the ecological efficiency decreases.

Another example at the regional scale is shown in Fig. 4.8, exhibiting (Ingegnoli 2002) the change of land cover in the metropolitan area of Milan from 1969 to 2001. Note that these data are undoubtedly impressive and measure the substantial increase in urbanization: a great transformation! However, remember that they cannot evaluate the possible ecological dysfunctions. That is why we introduced the new section of "Living Complex Systems Dysfunctions."

In Chap. 3 (Sect. 3.3.4), we underlined that our Earth's regulation capacity first depends on forest systems, which is today intensely decreased because of the forest destruction in the last 135 years. Recent research of Ingegnoli was made to evaluate the Planetary Health, through the measure of the transformation deficit (TD) related to the Emerged Lands. Remember that TD is the loss of energy necessary to maintain the level of order reached by an ecological system at its best recent period: in our study, the end of the nineteenth century (1880–1885), when World Population was only 1.5 billion. In Fig. 3.10, forest and agrarian data from FAO (FAO 2015), we can note:

(a) We can note the decrease of forest cover from 36.6% in 1882 to 26.8% in 2015 (−26.9%), but (worst) the loss of bionomics efficiency (−33.0%)!
(b) The increase of agrarian areas (green and crop) from 14.09% to 27.65% (+192.2%), but (worst) followed by a poor increase of bionomics efficiency (+175.7%), due to the degradation of agricultural landscapes. Therefore,
(c) The decrease of TD, passed from 88% to 65% even with a 4% of tolerance: a difference of −26.14%, which brings a total TD waste of 7.974×10^{18} kcal.

Note that today is not sufficient to contain this process: we must also restore the agricultural landscapes, increasingly degraded.

4.3.5 Resource Scarcity

In ecology, the problem of scarce resources concerns mainly natural resources, due to an excessive level of exploitation, with a low and decreasing availability/consumption ratio. In this short essay, we will limit ourselves to a mention of the most crucial problem: water scarcity.

As FAO argues, widespread degradation and growing scarcity of land and water resources are putting a large number of critical food production systems around the world at risk, posing a severe threat to the possibility of being able to feed a predicted world population, reaching 9 billion people by 2050 says a new FAO report released today. The World State of Water and Land Resources for Food and Agriculture (SOLAW) points out that although there has been a significant increase in world production in the past 50 years, on too many occasions, these improvements have been accompanied by management practices of resources that have degraded the terrestrial and water ecological systems on which food production itself depends.

The availability of water resources, particularly for civil (drinking water and sanitation) and agricultural uses, constitutes a significant crisis factor on a global scale, which is continuously increasing. The Treccani Encyclopedia (natural resources) believes that water removals may exceed 5000 km^3 per year in 2025: according to this calculation, and due to the natural uneven geographical location of resources, some forecasts for 2025 indicate that 1.8 billion people will have to face a condition of absolute water scarcity. This alarming prospect has led the international community to set up a permanent forum (World Water Forum), which promotes actions for the sustainable exploitation of water resources while respecting the integrity of ecological systems.

The degrees of water stress in the world are measured as the ratio between the quantity of water extracted each year and the total of long-term renewable freshwater resources. As reported by FAO Aquastat (FAO 2012), also in high-income countries (e.g., West USA and South Spain and Italy), high water stress (40–80%) may be estimated. Italy, Spain, and Germany have medium stress indeed, on the order of magnitude of 35%. However, areas with high water stress (>70%) are populated by over half a billion people, so the problem is not to be overlooked.

4.3.6 Climate Change

Two alternative theories, in the last 35–40 years, try to explain the rise in global temperature (Mariani 2017): (a) the Solar Theory (TS) which attributes this growth to the increased solar activity; (b) the Anthropogenic Global Warming (AGW) theory which attributes the increase to the increment in CO_2 induced by human activities.

Both these theories have some strengths but different elements of weakness; the weakness concerns the increase in ground energy induced by both increased CO_2 forcing and higher solar activity. It is intrinsically too contained to justify such significant global warming: + 1.3 °C in the last 150 years. So, the followers of the two theories have started hunting for amplifiers (positive feedback). Expressly, it can be noted that:

1. followers of the solar theory have identified stratospheric interactions with solar UV as the most accredited amplifiers, the effect of which would propagate in the troposphere (top-down theory) or the effect of galactic cosmic rays (GCR) on the clouds;
2. followers of the AGW theory have identified water vapor and cloud cover as the most trusted amplifiers.

However, both these theories show other weaknesses. The solar theory states that, since the end of the small ice age (PEG, sixteenth to nineteenth century), from 1850 to today, solar activity has been dramatically increasing. However, this increase seems to stop in 1980 because, from this date, there has been a constant decrease in solar radiation (Data: SILSO, Royal Observatory of Belgium) (SILSO 2019).

Since the highest increase in temperature occurred precisely in the last 35–40 years, solar theory alone does not seem to be acceptable. Conversely, the AGW produced an unbelievable climate model, because (i) the IPCC (Intergovernmental Panel on Climate Change) (IPCC 2001) had not considered neither the Medieval climate optimum nor the small ice age in the northern hemisphere, notwithstanding their demonstrations (Fig. 4.9), and (ii) it predicted a 1980–2020 temperature rise of approximately 1.1 °C, while the increase was 0.6–0.7 °C. Furthermore, the growth of CO_2 is constant, but the growth of temperature is not (Fig. 4.10). Therefore, the two phenomena are

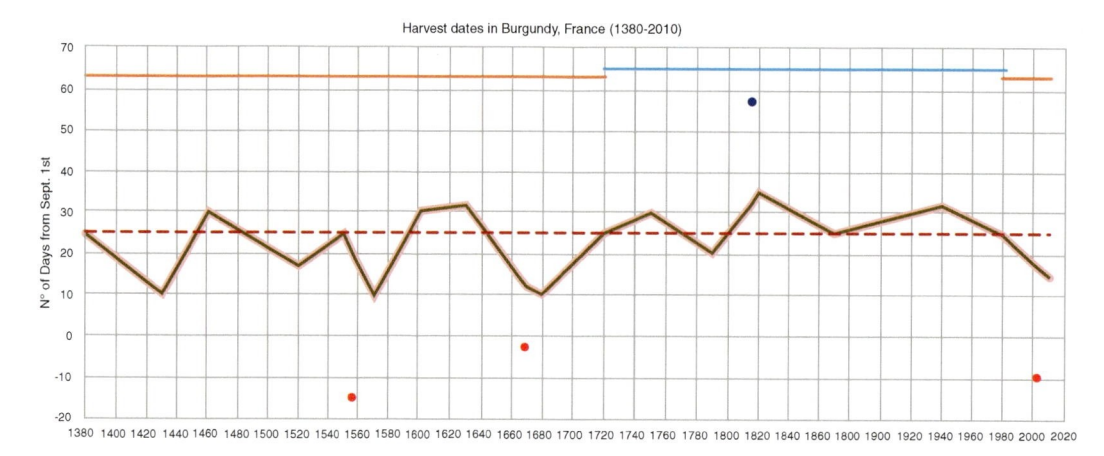

Fig. 4.9 Diagram of the harvest dates in Beaune (Burgundy) 1380–2010. The ordinate indicates the number of days from September 1st. The earliness of the harvest depends above all on the average temperatures of the months of April–May and June. Note the substantial variability and the presence of multidecadal cycles related to the sea temperature and the intensity of the current western degrees. The dotted line indicates the average harvest date, which is September 23rd. The isolated dots (red and blue) represent some exceptional cases [source Labbé and Gaveau 2013]. The blue line shows the coldest period

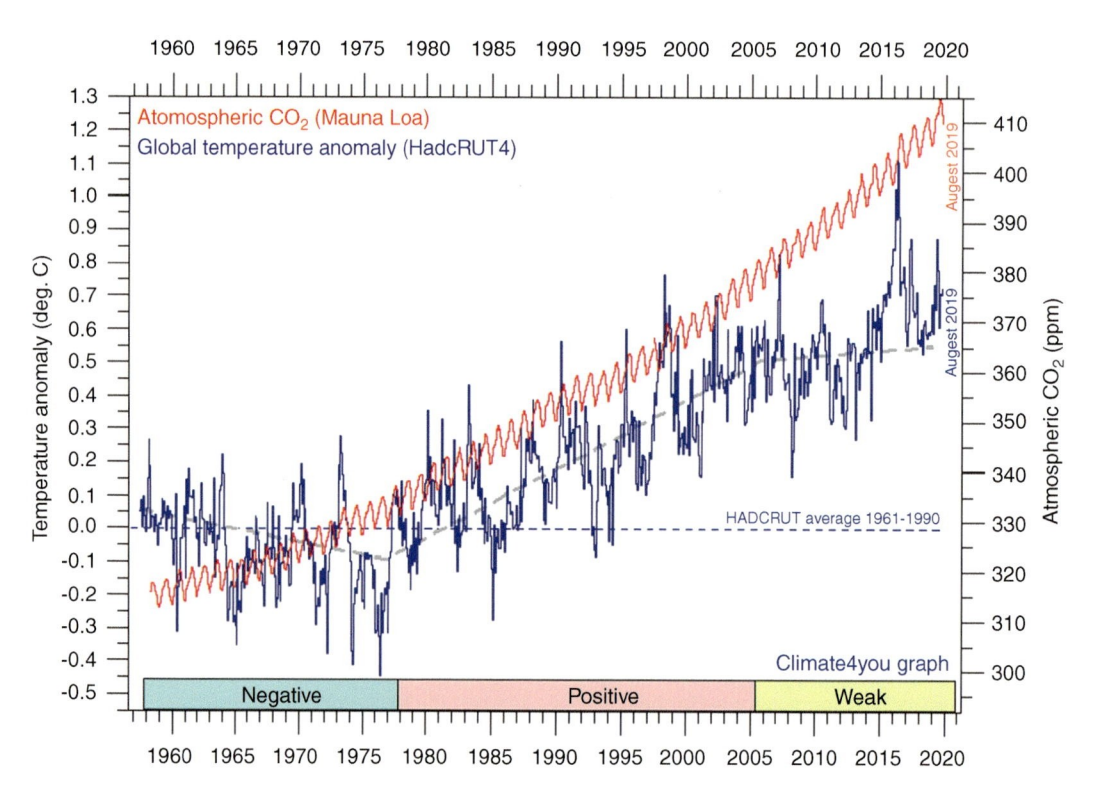

Fig. 4.10 See that CO_2 (Carbon Dioxide) has a much more limited absorption than water vapor. On the right, the growth in CO_2 increases steadily from 320 to 395 ppm, while the temperature has a cooling phase (1957–1977), an intense heating phase (1978–2003), and a more contained heating phase (from 2004 to date) [from Humlum et al. (2012)]

not as closely linked as one would have us believed.

Now consider the recent bio-geophysical theory of the "Biotic Pump" (Makarieva and Gorshkov 2012). This theory is based on the observation that forest landscapes create and control oceanic winds which, passing over them, are further enriched by humidity, condensed here in clouds, then in the rain and subsequently distributed.

When the steam condenses, the number of gas molecules dramatically decreases, and the pressure drops, forming a relatively persistent depression. The water vapor in the atmosphere is unstable on condensation and, when a sufficient quantity is reached in the lower atmosphere, condensation begins. This phenomenon is much faster and more consistent in the forest than on the sea because evapotranspiration occurs on a three-dimensional scale. Forest landscapes, therefore, become areas of depression that recall ocean winds and form clouds. The hydrological cycle is completed with rainfall on land areas and then returned to the sea utilizing the river network.

The temperature gradient (due to the shade and the height of the trees) and the distribution of evapotranspiration by layers, often with the addition of volatile chemical components (VOC, capable of providing condensation nuclei), trigger the formation of clouds and subsequent precipitation. We know that the greenhouse effect is mainly determined by water vapor and cloud cover: the absorption of CO_2 molecules covers less than 25% of the spectrum of thermal radiation, while atmospheric humidity uniformly absorbs 95% of the spectrum.

To better explain the process of the biotic pump, Fig. 4.11 shows (a) the Amazonia-Congo transept and (below) (b) the Norway-East Asia transept, (c) the short transept over Australia. Average annual rainfall (red) is highlighted. Note the difference between the rain forests (a), the temperate ones (b) with negative winters (black hatching), and the effect of the Australian desert (c).

From a study on Borneo (Hance 2019), *deforested* areas have an average temperature higher than 1.7 °C compared to forests and *agricultural* areas even a higher temperature, equal to 2.8–6.5 °C, even when it comes to oil palm plantations, that is arboriculture, because the local landscapes do not comply with bionomic requirements.

Note that in the light of the Biotic Pump theory, the climate change process should also be reviewed. In fact, the increase in CO_2 on the greenhouse effect should have been compensated by a minor change in the hydrological cycle, i.e., significantly reducing the destruction of forests (and changing their management) and rehabilitating agricultural landscapes. On the contrary, we have seen in Chap. 3 (Fig. 3.10) that the transformations of forest and agricultural landscapes in the last 135 years have influenced the hydrological cycle because of the "transformation deficit" (TD), which reduced the possibility to compensate the CO_2 increase. The TD we mentioned, evaluated as TD = 7.974×10^{15} Mcal, signifies the non-assimilation of an incredible quantity of CO_2 equal to 6140 billion tons, estimating Kg_{CO2}/Mcal = 0.76–0.78 at Emerged Lands scale.

The consequences of Climate Change should also include storms (cyclones, hurricanes, etc.) made more frequent by human responsibility for Climate Change (Giorgi and Mearns 2003). However, this component is partial to be scaled down, as the increase due to the higher temperature is verifiable, but less long than expected, about 7–10 storms/year and 10–14 hurricanes/year in the period 1850–2015 in the Atlantic Area.

As regards the Extra-Tropical Cyclones (ETC) in Europe there have been 12 in the last 66 years (Fig. 4.12), with a slight increase in wind speed: however the damages have been very remarkable, as in the case of the ETC Vaia (October 2018), which destroyed nearly 42,000 ha of forests in Trentino, Veneto, Lombardy, and South Tyrol in just a few hours.

It should be noted that the destruction of forests also depends on their management. If man favors spruce trees instead of larches (more resistant and deciduous) and if it facilitates formations of the same ages and without undergrowth (mono-stratification), it is evident that a lower

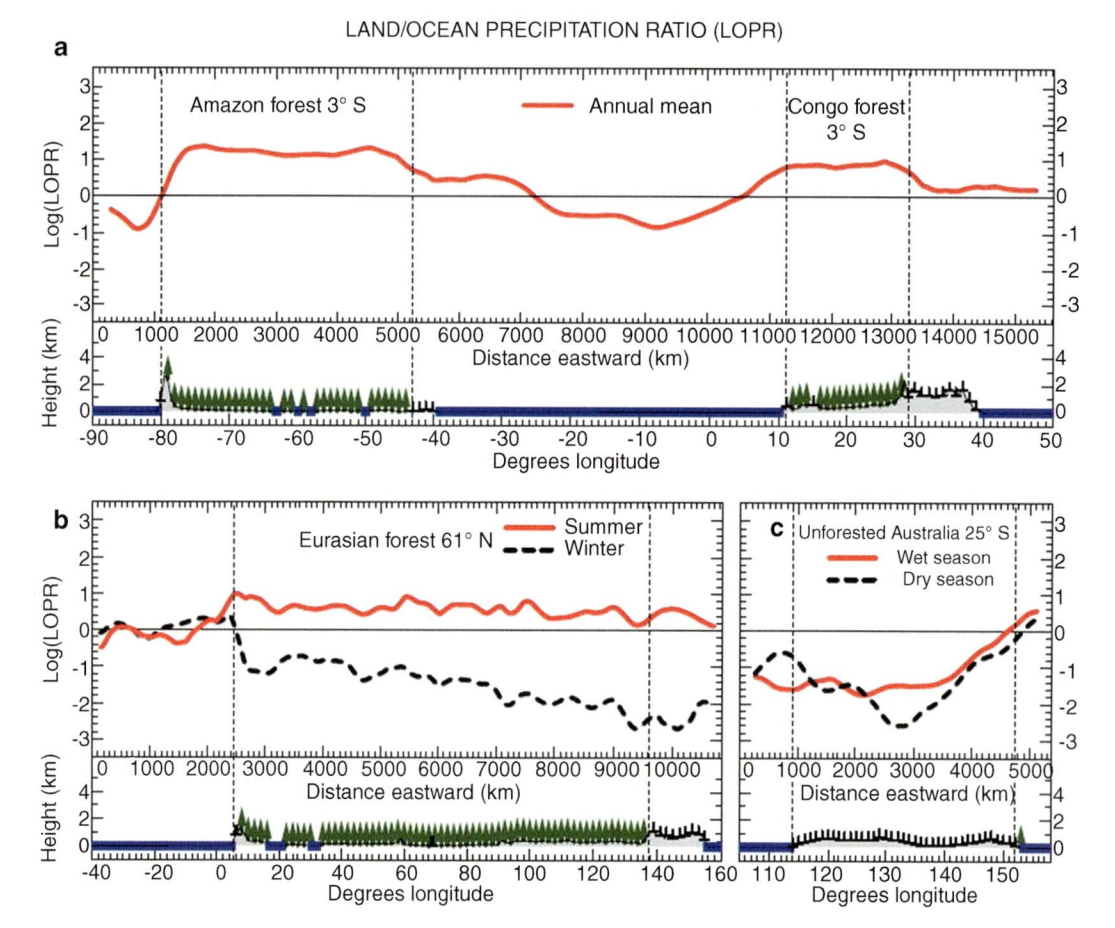

Fig. 4.11 (**a**) Amazonia-Congo transept, (**b**) Norway-East Asia transept, (**c**) Australia. Average annual rainfall (red) is highlighted. The vertical scales indicate the land/ocean precipitation ratio, measured in Log (LOPR). Note the difference between rainforests (**a**), temperate ones (**b**) with negative winters, and the effect of the Australian desert (**c**); from Makarieva (Makarieva and Gorshkov 2012)

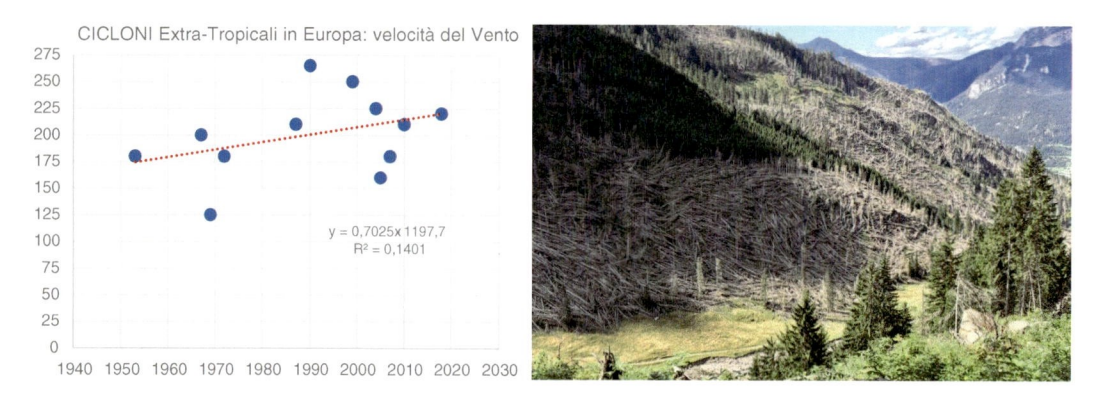

Fig. 4.12 Extra-Tropical Cyclones (ECT) in Europe and wind speed (Km/h) from 1953 to 2018. On the right, destruction of forests of Picea abies in Val di Fiemme caused in a few hours by the ECT of October 2018

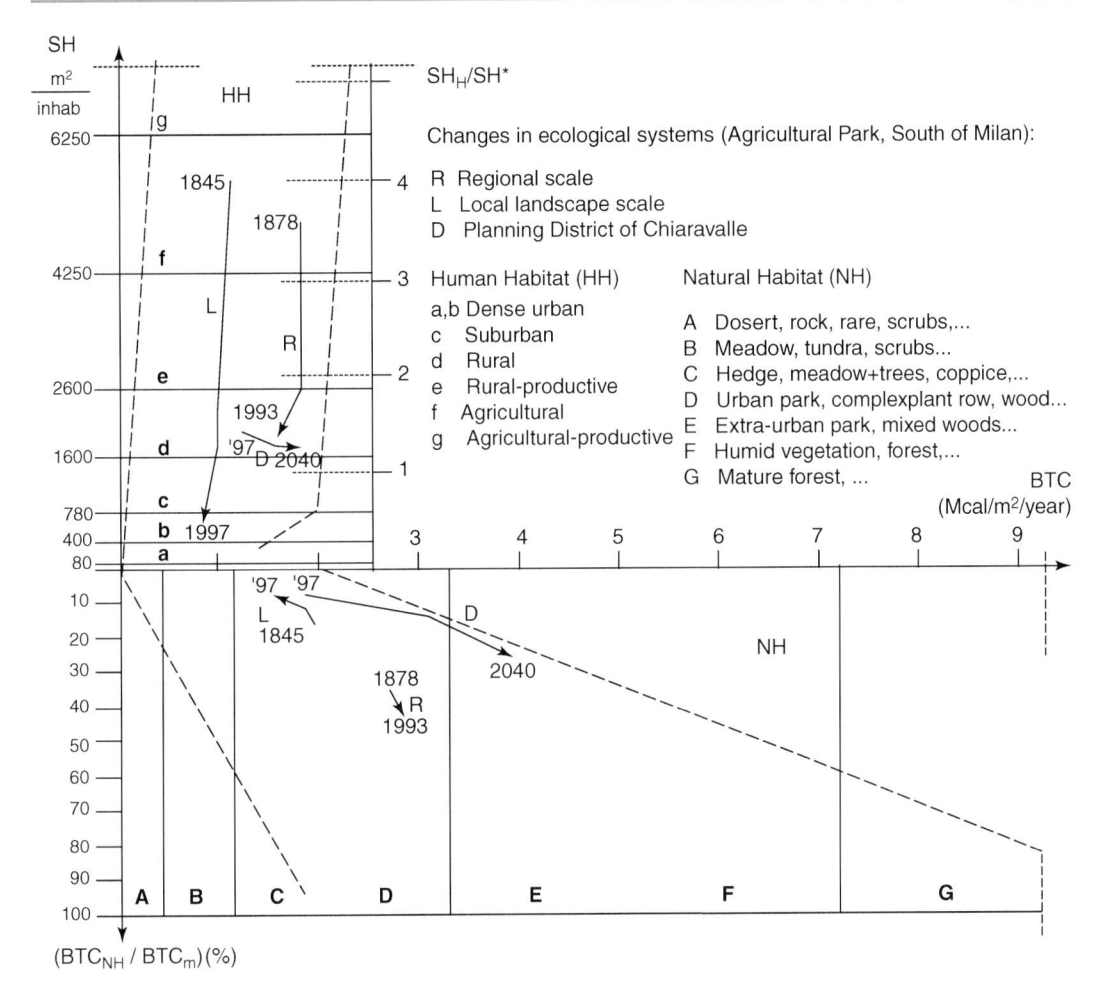

4 Planetary Health: Human Impacts on the Environment

Fig. 4.13 Transformation modalities of a landscape: a model of control based on the BTC function. We can see three spatial levels of change: R regional scale (Lombardy), L local landscape unit scale (suburban landscape of South Milan), D district of Chiaravalle Abbey; dotted lines define the more frequent movement field of the system. Note the different directions of District arrows vs. L and R within the HH field and all the three arrows within NH (from Ingegnoli, 2000: redrawn)

ecological efficiency adds a lower ability to withstand severe weather.

4.3.7 Living Complex Systems Dysfunctions

This section is different from the simple land cover change, implying a hierarchical organization of complex systems, acting as living entities, well farther populations, and communities (see Bionomics principles, Chap. 3). It is a problem of complex system behavior.

It is possible to visualize the regional thresholds of landscape replacement in two coordinated fields, monitoring changes in anthropic and natural sets of landscape types, measured on the same scale of BTC (Fig. 4.13). The possibility of analyzing broad-scale changes allows us to study the past of a landscape, control the present state, and guide future management. Note that estimation procedures can also measure the BTC. As we may observe in Fig. 4.13:

- In the HH, the transformation modalities generally follow a cluster of parabolic functions crossing a series of thresholds, from the semi-natural protective type of agricultural landscape to the most urbanized one. The mosaic sequence remains that expressed by Richard Forman (Forman 1995). Even if the opposite is theoretically possible, in human landscapes, these transformations tend to be unidirectional.

- In Natural Habitat, the transformation modalities are more complex. An ecological succession from a near-desert type of landscape to a high BTC mature forest type is theoretically possible, but yes, it is not the primary modality and does not follow a straight line. The role and the range of disturbances generally lead the way to change.

The following Fig. 4.14 (Ingegnoli 2011) presents land-use transformations from 1878 to 2015 in Lombardy. There are phenomena related to each other such as, for example, (a) forests regrowth that was in detriment of meadows-pastures, (b) urbanization that increases enormously mainly at the expense of crops (arable and permanent).

The human habitat and the BTC have little changed, except between the two wars (1915–1950). So, one would think that the regrowth of the forests (from 20 to about 25%) has improved the regional state of the environment. However, this is not the case. The population in this period has tripled (from 3.5 to 10 million), while the level of the urbanized landscape grew by ten times (from 1.5% to 15%). This fact shows the aggressiveness and waste of soil of urban sprawl. The Carrying Capacity, following bionomy, indicates that today the region has entered a phase of heterotrophy, having diminished from SH/SH* = 2.65 in 1878 to 0.99 in 2015. These changes mean that, even if considering a tolerance of 5%, already in 2020–25 SH/SH* would reach a value of 0.93, and despite the high agricultural productivity of the region, this vast territory is no longer able to guarantee the food needs for its population.

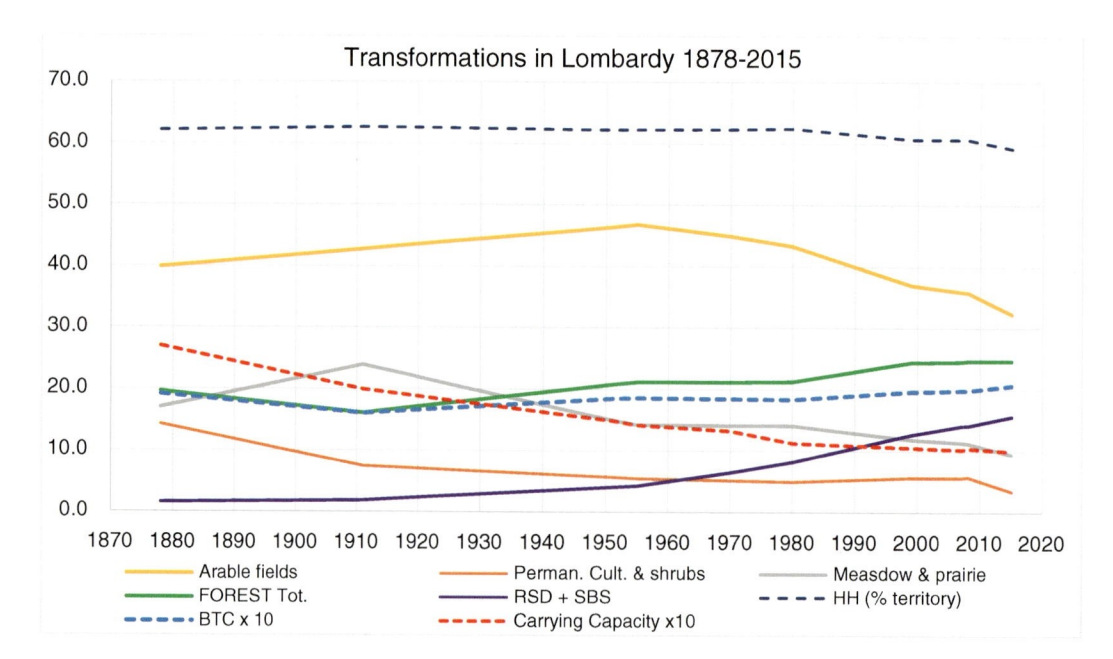

Fig. 4.14 Transformations in Lombardy (1878–2015). The bionomic state functions of the human habitat HH (black), the Biological territorial Capacity BTC (blue), and the Carrying Capacity (red) are shown in dashed lines. Note that the forests are re-growing in detriment of the meadows-pastures, the urbanized area has increased enormously mainly at the expense of the crops; from Ingegnoli (Ingegnoli 2011). Note: RSD = residential apparatus; SBS = subsidiary apparatus; Carrying Capacity = the state function able to evaluate the self-sufficiency of the human habitat (see Sect. 3.3.1)

Furthermore, we know that the regrowth of forests has not avoided a transformation deficit (TD), because both forest and agricultural landscapes have lowered their ecological efficiency in these 137 years, respectively, from BTC = 6.8–7.0 Mcal/m^2/year to 5.7–5.8 (−17%) for forests, which are managed by increasing coppices, and from BTC = 1.25–1.30 Mcal/m^2/year to 0.8–0.9 (−33%) for agricultural landscapes, which are increasingly industrialized.

4.4 Proximate Causes and Landscape Unit Syndromes

4.4.1 Landscape Unit Syndromes

We suggested to add to this driver "Landscape Bionomics Degradation and Artificialization"; it implies the clinical Diagnosis of the health state of the Landscape Units, concerning both their Syndromes and the excess in surface cover by anthropic elements (e.g., infrastructures). Note that the level of BF (Bionomics Functionality, see Sect. 3.3.3) is correlated with the mortality rate (MR), so with the increase of morbidity. Extremely dangerous is also the degradation of agricultural landscapes.

Some considerations on landscape pathology. The definition of landscape as a specific level of life organization becomes a challenge for environmental evaluation, first of all because man has to pass from a discipline related to technology, economy, sociology, urban design, visual perception, and ecology, to another related to biology, natural sciences, medicine, and traditional disciplines. Consequently, as we underlined in Chap. 3, analysis, evaluation (and intervention or planning) of the territory require changed methodologies: from engineering, economical, and aesthetical rearrangement to biological diagnosis and therapy.

Let us remember that the study of the pathology of any biological system, independently from the levels of scale and of the organization, needs a basic diagnostic methodology, which cannot be avoided. This valid is also for landscape dysfunction, and it consists of six phases:

1. Survey of the symptoms.
2. Identification of the principal causes.
3. Analysis of the reactions to pathogen stimuli.
4. Risks of ulterior worsening.
5. Choice of therapeutic directions.
6. Control of the interventions.

Like in medicine, environmental evaluation needs comparisons with "normal" patterns of behavior of a system of ecocoenotopes. The main problem becomes how to know the normal state of an ecological system and the levels of alteration of that system. In medicine, it is the physiology/pathology ratio that permits a clinical diagnosis of an individual. Here it is the same ratio which permits a clinical diagnosis of an ecocoenotope or a landscape.

For example, dysfunctional landscapes will have less patchiness than normal ones: any remaining patches will have lower concentrations of soil nutrients, lower water infiltration rates, lower levels of biological activity, and lower production cycles, as sustained by Tongway and Ludwig (Ludwig et al. 1997). We may add (Ingegnoli 1993, 2002, 2011): higher transformation deficit, decreasing BTC, natural habitat loss, the incorrect ratio between HH and NH, the decreasing correlation between heterogeneity and information, higher fragmentation, loss of connectivity, incongruent landscape apparatuses, higher landscape resistance, etc. Landscape bionomics principles and methods have a wide capability to characterize landscape syndromes, as shown in Fig. 4.15: a pathogenetic scheme of the agrarian industrialization syndrome in temperate regions.

Nevertheless, it is sometimes challenging to reach a correct diagnosis, because some pathologies have "low" symptoms or not directly alarming ones. For instance, a forested landscape may present a high biomass volume and quite high BTC, but it may have a too low reproductive rate, too few patches far from the dominant state, and a too homogeneous structure of the landscape central mosaic thus presenting hazardous senescence. Conversely, we have just seen the considerable capacity of disturbance incorporation on a regional scale, which seems reassuring.

Pathogenic scheme of the agrarian industrialization syndrome of temperate agricultural landscape of the plains

Original State Permanence : 3 to 18 centuries	**Traditional agricultural landscape "a Bocage"** (BTC = 1.2-2.1 Mcal/m^2/yr; HS/HS* = 2.5-6.0; HH = 50-75%) heterogeneity, coneectivity and circuitry = good	
Main human cause of alteration	Socio-economic pressure due to growing agrarian production	Increase of help for agrarian technologies
	Specialization of cultivations	Canalization of small rivers
	Increase of arable land and of chemical fertilizers	Cutting of tree lines and hedgerows, mechanical irrigation
Positive feedback processes	Destruction of forest patches, Increase of crop pests, Increase of chemical pesticides, Decrease of fauna, Soil depletion	
	Enlarging field area, increase road network	Rupture of geomorphologic constraints,
Altered State Permanence : 20 to 80 yr	**Open mono-cultural landscape** (BTC = 0.9-1.3 Mcal/m^2/yr; HS/HS* = 4-9; HH = 70-85%) heterogeneity, connectivity and circuitry = weak, partial	
	Structural weakness, Increase system fragility, Attraction for highways, Attraction for industrial areas, Increase of fragmentation	
Disordered State Permanence ?	**Suburban-rual landscape** (BTC = 0.7-1.1 Mcal/m^2/yr; HS/HS* = 0.8-2.7; HH = 80-90%) heterogeneity = increasing; connectivity and circuitry = disrupted	
	Loose of functionality Decrease of agrarian production	

Fig. 4.15 The agrarian industrialization syndrome (Ingegnoli and Giglio, 2005; Ingegnoli 2015). Note that landscape bionomics functions can evaluate each level of transformation. HS, HS* = Standard Habitat and Minimum Theoretical Standard Habitat per capita (see Sect. 3.3.3)

The identification of the leading causes producing landscape syndromes needs a good knowledge of the anatomy and the physiology of the complex system of ecocoenotopes, of their pathologic disturbances, and a good anamnesis. Semeiotic must be added, in which even perceptive studies may help.

Summarizing, the landscape is a specific level of life organization on Earth; therefore, it is a *living entity* and may present many *syndromes*, as synthesized in Chap. 3, Table 3.4. Most landscape syndromes are *not* due to pollution or land cover change, so we must investigate if *these* non-toxicological landscape pathologies can also *influence* human health. If this hypothesis were proven, the importance of landscape bio-structural rehabilitation would become imperative.

4.4.2 Recover with no Land Cover Change

An example of the application of "Landscape Bionomics Degradation and Artificialization" concerns the transformation of an ecotope brought to an altered condition of landscape health. This fact is the case study of the

Environmental Assessment for the Intermodal Station HUPAC, in Gallarate (Ticino Park) for Termi spa, Chiasso (Switzerland). First, it is necessary to define the field of normality of this specific type of Landscape unit, concerning at least the more involved dimensions (here the BTC and the Fractal Dimension D) applying Landscape Bionomics criteria and methods. Then, it is necessary to evaluate the magnitude of the alteration of the ecological state: here, the shift from $BTC_{1856} = 3.65$ to $BTC_{2000} = 1.85$ and from Fractal dimension of the forest cover $D = 1.92$ to $D = 1.3$.

Figure 4.16 shows the comparison between the two possible criteria of recovery of the ecotope (about 250 ha): A = bionomic rehabilitation (green arrow) and B = traditional mitigation and compensation (red arrow). Note that B should be improved in BTC (from 1.85 to 2.20), but its *direction* is misleading on the field of normality of the Landscape Unit (LU); that is, it is not directed toward the *field of normality* for an ecotope of agricultural-forested type. This example is significant because the two proposed ecotope recovery criteria differ in their configuration and the vegetational components. However, their land cover/land use elements remained the *same*: no changes in the percentage of cropland, grassland, urbanized infrastructure, forest patches. However, we know that, following the Emergent Property Principle, the different dispositions on the territory of the same elements (number and surface) can affect the system's different behaviors. Observe that Following Landscape Bionomics methods, it is possible to calculate the reachable value of the involved parameters and monitor them along time, as in regular rehabilitation therapy.

What we exposed for remediation is available in the evaluation of the landscape diagnosis: even if we have no change in land cover/land use elements, we can find different environmental health conditions because these conditions mainly depend on *bionomics* functions!

Fig. 4.16 Comparison between two criteria of recovery for an agrarian-forested ecotope, near Gallarate, North to Milan. Note the different change directions

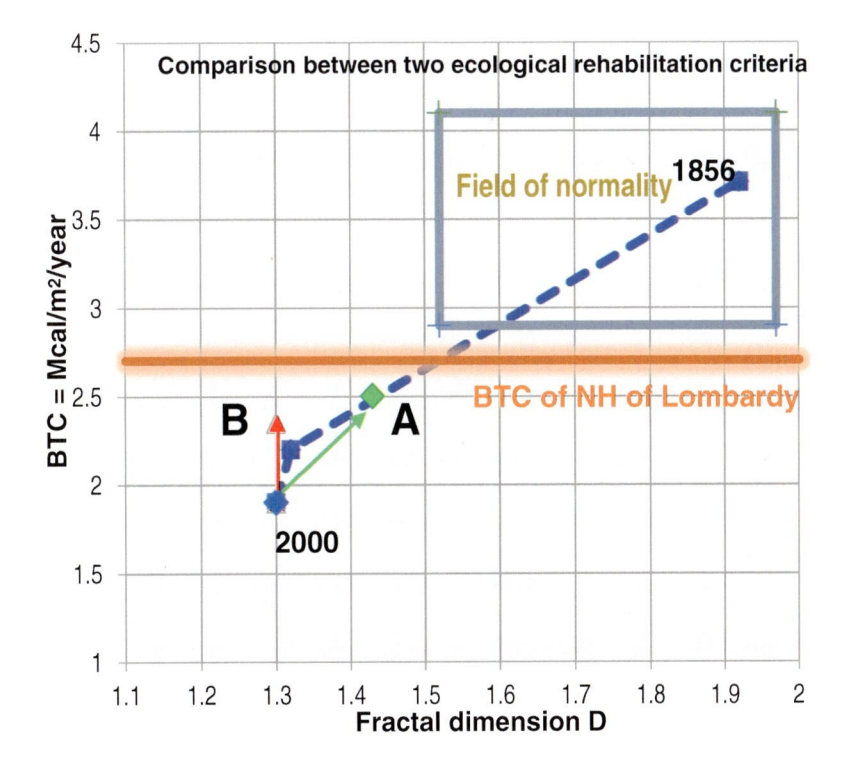

4.5 Mediating Factors, Epistemology, and Meditation

Having forgotten Epistemology and Faith/Religions, mediating factors eliminate (a) a correct *interpretation* of facts, (b) the contribution of *meditation* (also considered in Medicine), and (c) the *consciousness* related to the most significant *ethical* threat to nature: the betrayal of the role of man expressed as violence against life (for this third point see Sect. 3.5).

4.5.1 Epistemology and Bioethics

Being the landscape a living entity, represented by a hyper-complex system, it is difficult to study. Therefore, we need to follow the best epistemological criteria to mediate factors from environmental health to human health.

This epistemological approach, however, in the world of biology is highly controversial and limited because it follows a great relief, as well explained by Lorenz (Lorenz 1978 and Israel 2010): the destruction of any notion of subjectivity, design, and finality, the key to biological thought.

Einstein wrote in his "Scientific Autobiography" that even scholars of relief (such as Mach) could be hampered by prejudice in the interpretation of the facts: "Prejudice, which still is not gone, is the belief that the facts can and should result in scientific knowledge by itself, without free conceptual construction." The same great scientist (Einstein 1944) wrote that independence from the prejudices is determined by philosophical analysis, and it is "the mark of distinction between a mere artisan or specialist and a real seeker after truth." Again, Einstein used to underline that he learned more from Dostoevsky than from his theoretical physics colleagues.

In the third chapter we underlined that the study of biology must be upgraded even to the broader scales, trying to understand their anatomical components (structure), physiological processes and functions, and transformation processes, clinical-diagnostic evaluation, pathology, and therapy. Life's alterations at these broader scales can damage human health, not dissimilarly to small ones. As underlined by Bottaccioli, health and disease depend primarily on the organization of life, and we must affirm on the *full organization*, from small to broad scales. However, today the discipline of bioethics follows the limits of biology: rarely it is widened to the broad life scales. The linkage environmental alteration/health damages must change this limitation of bioethics.

Remember: disciplines like Bionomics enlighten that the Technological World leads to a syndrome reducing Gaia self-regulation capacity. Having to do with a living entity, to remember that man is a part of a more complex living entity produces a completely different control on man's activities on a geophysical territory! But, the alternative to the Technological World, proposed by greens and traditional ecologists, seems to be the return to the state of nature. This proposal is impossible, but not because the true environment of man is the Technological one. The reason is the withdrawn of traditional ecology, still anchored to reversible processes (degradation and recovery), while the System Theory and so bionomics enlighten that, due to Time arrow, processes are irreversible and the environment does not stop modifying: thus, we need environmental pathology and a systemic vision of the problem. So? How to indicate the main orienting principles capable to readdress man-nature relations?

Environmental Bioethics needs to be based at least on six assertions:

- Life cannot exist without its environment.
- Life on Earth is organized in hierarchic levels, each one being a complex system.
- Each level is unique because of its emerging properties.
- Each level has proper characters but, through opportune constraints, upper-scale processes explain the significance of it, while the inferior scale processes explain its origin.
- Man is a part of this biological hierarchy.
- Man and the Environment co-evolve.

Remember: Man must take care of Rhythm and Logic of Gaia and of systemic approach. This fact leads us to be governed by Nature co-evolving along the time arrow and assuming the responsibility of rehabilitating our Mother Earth role and functionality for the future generations. Environmental structural and functional alterations lead to human health alteration, we cannot forget this fact! The recent birth of Planetary Health studies, correlating environmental syndromes with human health pathologies confirms it. Environmental Ethics is strictly linked with Bioethics and a renewed ecology needs a Hippocratic Oath of the Ecoiatra, as shown within Giglio and Ingegnoli (Ingegnoli and Giglio, 2005).

4.5.2 Faith and Meditation

In the face of the materialist theories, scientism, and neoclassical economy, which influence the society on a global scale, we would need a revolution. Without a revolution seems to be impossible to eliminate the present ecological crisis. What kind of revolution? We have two possibilities: (a) violence, e.g., French or Russian revolution, or (b) non-violence, as demonstrated by the Mahatma Gandhi (Gandhi 1927):

"Revolution is the return to the beginning and the Lord. The ones they stick to the forms of the past and the memory of the dead, and live like the dead, the others fly themselves into new crazy until they fall on deaf ears. But I'll go ahead and I don't lose my way, because I return to the most ancient traditions through the complete revolution which is the total overthrow, but natural, and willed by God, and which happens at the right time."

This direction, underlined by Gandhi, brings to enhance meditation. It is at the basis of human spirit, so it comes before the same religion.

Meditation is a way of connecting with a natural state of mind that is spacious and clear. It is not eliminating thoughts but noticing when our mind is busy or racing. Meditation can help to connect with the breath and bring calmness to mind. There are many different types of meditation. Most involve being still and quiet. Some involve movements, such as tai chi, chi gong, or walking meditation (Cancer Research UK).

In a hypothesis-generating study, a prospective study measured and compared 12-hour urinary 6-sulfatoxy-melatonin between women who regularly meditated versus those who did not. Women within the meditation group had increased levels of physiological melatonin compared to the non-meditation group. This fact is significant since melatonin has been shown in multiple studies to have anti-cancer properties as well as other biologic functions important in maintaining health and preventing disease such as immunomodulation and hematopoiesis (Srinivasan et al. 2019).

We have to underline how meditation changes the activity of the immune system. Bottaccioli and Bottaccioli (Bottaccioli and Bottaccioli 2017; Slavich et al. 2010) indicate that the activity of the Nk (natural killer cells) increases in the group of meditating patients, so to become similar to people without cancer: an increase evaluated +50%.

In Fig. 4.17 it is reported the IFN-γ (the primary signal of the immune circuit Th1), which increases in 1 month from 7 to 15 ng/mL in the group of meditating patients, while it decreases in non-meditating ones. Andersen et al. (Andersen et al. 2008) underline the consequences: more than 80% of the patients of the meditating groups did not die of cancer vs. 60% of the other: but, even if also meditation promoted by Psychical therapies and Oriental philosophies is effective, nothing more than deep religious practice (like spiritual prayers, retreats and fasting) can reach the maximum of efficaciousness in recovery.

Fasting is particularly essential, and Medicine has studied how to develop fast-mimicking diets: they have shown that intermittent fasting reduces markers of inflammation, such as C-reactive protein (CRP), tumor necrosis factor-alpha (TNF-α), interferon-gamma (IFN-γ), leptin, interleukin one beta (IL-1β), and interleukin 6 (IL-6) (Brandhorst and Longo 2019). In a study, people practicing alternate-day fasting for religious holi-

Fig. 4.17 Meditation (MBSR) and mental well-being increase the interferon-gamma, produced by the B and T lymphocytes, which at the time T4 (1 month) returns to being almost normal and also manages to counteract any oncological diseases. [Redesigned, from Bottaccioli and Bottaccioli (2017)]

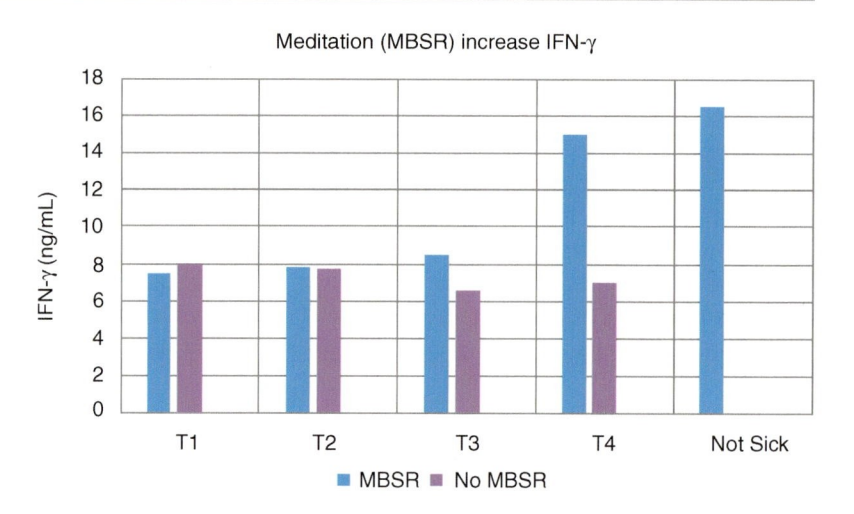

days, proinflammatory cytokines were significantly lower during the alternate-day fasting period than the weeks before or after.

4.6 The Effects of Anthropogenic Alterations on Human Health

4.6.1 Main Etiological Processes

The environmental leading disruptive causes (proposed in the *ecological drivers* by Planetary Health Alliance at Edinburgh University 2018 (PHA 2018) Annual Meeting) were six (remember Fig. 4.1), plus the new one that has been suggested in this study (Fig. 4.2): these are no doubt the main sections. However, their interactions and their subsections may produce many more subsections. Some examples: Climatic Alterations, Infectious and Parasitic Diseases, Direct and Cumulative Pollutions, Resource Scarcity, Wide Land Use Changes, Biochemical Cycles Alterations, Large Biodiversity Loss, Degradation of Protective Areas, Rural Landscape Destruction, Urban Landscape Alteration, etc.

These drivers are usually acting in cumulative convergences, often challenging to be analyzed. Moreover, their influences may change on the space-time scales, so at different levels of life organization. Bionomics principles and methods can help to study the syndromes given by the mentioned ecological drivers to the environment, as we may understand, following the previous Chap. 3. In that chapter we underlined the availability of bionomics to make the systemic diagnosis at the scale of interest. The diagnosis can be positive or negative: if negative, it is possible to arrive to a correct environmental syndrome and consequently to propose a therapy.

We know, it is the aim of this book, that the presence of a syndrome in our environment, e.g., a landscape unit (LU), leads to alter human health. These damages to human health are ranked by PHA as follows: (a) Non-Communicable Diseases, (b) Infectious Disease, (c) Mental Health, (d) Malnutrition, (e) Civil Strife and Displacements.

That is good, undoubtedly, but we think it is necessary to add the principal paths regarding the main processes of etiopathogenesis environment/health, as exposed in Fig. 4.18.

Obviously, in Fig. 4.18, the damage to health due to natural catastrophes is not considered. Note that in this figure are exposed six processes, each with two subdivisions, to simplify the figure: each section may have a deeper etiological subdivision and may present a cumulative impact; a complex "combined and cumulative" impact (and interference) is possible. This situation indicates that in many cases, it becomes challenging to arrive at an understandable etiology. However, efforts should be made to clarify these processes as far as possible if the damage to human health is addressed and contained.

ENVIRONMENTAL MAIN DISRUPTIVE CAUSES	BIONOMIC PRINCIPLES AND METHODS
CLIMATIC ALTERATIONS INFECTIOUS & PARASITIC DISEASES DIRECT & CUMULATIVE POLLUTION RESOURCE SCARCITY WIDE LAND USE CHANGES BIO-CHREMICAL CYCLES ALTERATIONS LARGE BIODIVERSITY LOSS DEGRADATION OF PROTECTIVE AREAS RURAL LANDSCAPE DESTRUCTION	BIONOMICS DEFINITIONS THE SISTEMIC PARADIGM BIONOMIC FUNCTIONS LANDSCAPE ANATHOMY LANDSCAPE PHISIOLOGY VEGETATION & FAUNA ANALYSIS HUMAN POPULATION ANALYSIS LANDSCAPE PATHOLOGY DISTURBANCES ANALYSIS DIAGNOSTIC EVALUATION OF LAND

UNFAIR ECOLOGICAL & BIONOMIC DIAGNOSIS

Tab. 1. HEALTH/ENVIRONMENT main etiopathogenesis paths

POLLUTION		INFECTIVE AGENTS		AGROFOOD DYSFUNCTIONS		ENVIRONMENTAL STRESS		LACK OF DEFENCE CONTRIBUTIONS		LACK OF HIERARCHICAL RELATIONS	
direct toxicity	endocrine disruptor	viral & bacterial	fungal & protozoa	OGM cultivars	Hyper-homogeneous Crops	neural path	hormone path	gut microbiome	phytoncydes	lack of disturbances incorporation	hierarchical disruptions
cumulative impact		cumulative impact		cumulative impact		cumulative impact		cumulative impact		cumulative impact	
complex combined and cumulative impacts and interferences											

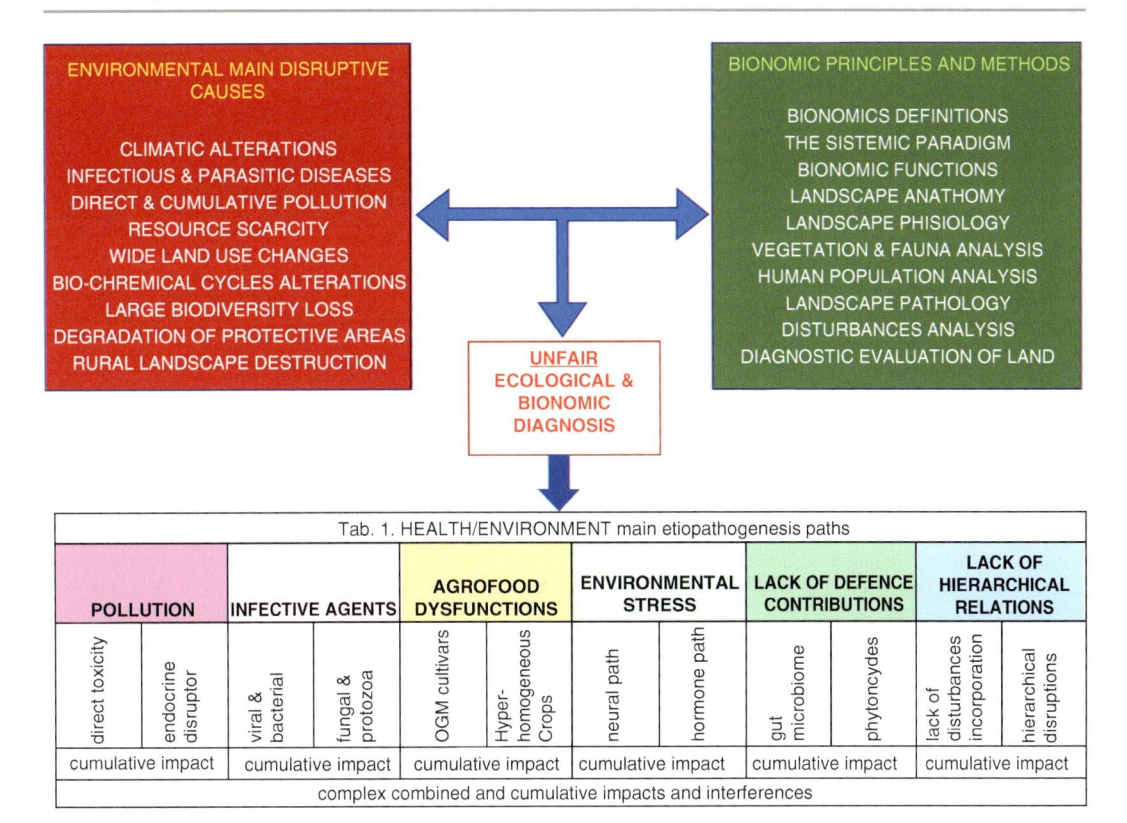

Fig. 4.18 The most critical etiological processes are linking environmental syndromes and human health. These processes have to be checked when the bionomics diagnosis is definite (unfair)

It is interesting to observe that: (a) every etiological path section may lead to a wide range of morbidity, implying near all the disease categories mentioned before and (b) not all the six pathological sections are commonly known. Pollution, infective agents, and agrofood dysfunctions remain the most known, while environmental stress, lack of defense contribution, and lack of hierarchical relations need more attention and new studies.

References

Almada AA, Golden CD, Osofski SA, Myers SS (2017) A case for planetary health/geo health. Geohealth 1:75–78. https://doi.org/10.1002/2017GH000084

Andersen BL, Yang HC, Farrar WB et al (2008) Psychological interventions improve survival for breast cancer patients. Cancer 113:3450–3458

Balzani V, Venturi M (2014) Limiti e confini planetari, Il ciclo di azoto e fosforo. Ecoscienza n°6, Bologna

Bjerg O (2016) Parallax of growth, the philosophy of ecology and economy. Polity Press, MA, USA

Bottaccioli F, Bottaccioli AG (2017) Psiconeuroendocrinoimmunologia e scienza della cura integrata. Il manuale. Edra spa, Milano

Brandhorst S, Longo VD (2019) Protein quantity and source, fasting-mimicking diets, and longevity. Adv Nutr 10(Suppl_4):S340–S350. https://doi.org/10.1093/advances/nmz079

Carson R (1962) Silent spring. Penguin books

Daly HE (1999) Uneconomic growth and the built environment. In: Theory and in fact. In reshaping the built environment. Island Press, California, pp 73–86, a c. di C.J. Kibert

Einstein A (1944) Einstein archives (pp. 61–574)

ESA (2012) Reorganisation of the SCIAMACHY Level 2 data set at D-PAC, Earth Online

FAO (2012) Water Report 37 Aquastat 12, Rome

FAO (2015) Global forest resource assessment: main report. FAO, Rome

Forman RTT (1995) Land mosaics: the ecology of landscapes and regions. Cambridge University Press, Cambridge

Francis I (2015) Laudato Sì. Encyclical, Vatican, Rome

Gandhi MK (1927) My experiments with truth. Yerwada Central Jail, Pune, Maharashtra

Gimeno L, Marin E, del Teso T, Bourhim S (2001) How effective has been the reduction of SO2 emissions on the effect of acid rain on ecosystems? Sci Total Environ 275(1–3):63–70

Giorgi F, Mearns LO (2003) Probability of regional climate change based on the Reliability Ensemble Averaging (REA) Method. Geophysical research letters.—Wiley Online Librar

Hance J (2019) Less rainforest, less rain: a cautionary tale from Borneo. Mongabay Series

Humlum O, Stordahl K, Solheim J (2012) The phase relation between atmospheric carbon dioxide and global temperature. Glob Planet Chang 100(2013):51–69

Ingegnoli B, Giglio (2018) Agricultural landscapes rehabilitation suggests 'ecosystem services' updating. WSEAS Trans Environ Dev 14:233–242

Ingegnoli V (1993) Fondamenti di ecologia del paesaggio: Studio dei sistemi di ecosistemi. CittàStudi (poi UTET-Cittàstudi) Milano

Ingegnoli V (2000) Ecologia e pianificazione: metodi e criteri di pianificazione ecologica del Parco Sud di Milano, settore Chiaravalle. In: Biondi E, Colantonio R (eds) La pianificazione del paesaggio tra ri-naturazione e iper-antropizzazione. Atti Con. Naz. Ancona, Accademia Marchigiana di Scienze, Lettere, Arti. pp. 209–240

Ingegnoli V (2002) Landscape ecology: a widening foundation. Springer, Berlin, New York, p XXIII+357

Ingegnoli V (2011) Bionomia del paesaggio. L'ecologia del paesaggio biologico-integrata per la formazione di un "medico" dei sistemi ecologici. Springer-Verlag, Milano, pp. XX+340. Foreword by Giacomo Elias

Ingegnoli V (2015) Landscape bionomics. Biological-integrated landscape ecology. Springer, Heidelberg, Milan, New York, p XXIV + 431

Ingegnoli V, Giglio E (2005) Ecologia del Paesaggio: manuale per conservare, gestire e pianificare l'ambiente. Sistemi editoriali SE, Napoli, pp. 685+XVI

IPCC (2001) TAR climate change 2001: the scientific basis. REPORT

Israel G (2010) Per una medicina umanistica. Apologia di una medicina che curi i malati come persone. Lindau, Torino

Kampa M, Castanas E (2008) Human health effects of air pollution. Environ Pollut 151(2):362–367

Khoshbakht K, Hammer K (2008) How many plant species are cultivated? Genet Resour Crop Evol 55:125–928

Labbé T, Gaveau F (2013) Les dates de vendange à Beaune (1371–2010). Analyse et données d'une nouvelle série vendémi- ologique. Rev Historique 666:333–367

Lorenz K (1978) Vergleichende Verhaltensforschung: Grundlagen der Ethologie. Springer-Verlag, Berlin, Wien

Ludwig J, Tongway D et al (1997) Landscape ecology, function and management. CSIRO, Australia

Makarieva AM, Gorshkov VG (2012) Revisiting forest impact on atmospheric water vapor transport and precipitation. Theor Appl Climatol. https://doi.org/10.1007/s00704-012-0643-9

Mariani L (2017) Cambiamenti climatici e viticultura. Museo Lombardo di Storia dell'Agricoltura- DISAA, Università di Milano

Millennium Ecosystem Assessment (2005) *"Ecosystems and human well-being: the assessment series"* (4 vol + summary). Island Press, Washington DC

Morgan J (2016) What is neoclassical economy? Routledge

O'Neill RV, De Angelis DL, Waide JB, Allen TFH (1986) A hierarchical concept of ecosystems. Princeton Univ. press, Princeton, NY

Patel R (2006) International agrarian restructuring and the practical ethics of peasant movements solidarity. J Asian Afr Stud 41:71–93

Pennekamp F et al (2018) Biodiversity increases and decreases ecosystem stability. Nature 563. https://doi.org/10.1038/s41586-018-0627-8

PHA (2018) Meeting at the University of Edinburgh. Concept Note

Prüss-Üstün A, Mathers C, Corvalán C, Woodward A. (2003) Introduction and methods: assessing the environmental burden of disease at national and local levels. Geneva, World Health Organization. (WHO Environmental Burden of Disease Series, N° 1)

Pruss-Ustun et al (2016) Preventing disease through healthy environments: a global assessment of the burden of disease from environmental risks. WHO, iris

SILSO (2019) Solar Cycle 25. World data Center Royal Observatory of Belgium

Slavich GM, O'Donovan A, Epel ES, Kemeny ME (2010) Black sheep get the blues: a psychobiological model of social rejection and depression. Neurosci Biobehav Rev 35(1):39

Srinivasan V, Spence DW, Pandi-Perumal SR, Trakht I, Cardinali D (2019) Therapeutic actions of melatonin in cancer: possible mechanisms.P. Integr Cancer Ther 7:189203. [PubMed]. 2008

Tansley AG (1935) The use and abuse of vegetational concepts and terms. Ecology 16:284–307

The Lancet Commission (2015) Safeguarding human health in the Anthropocene epoch: report of The Rockefeller Foundation-Lancet Commission on planetary health. The Lancet

5. Landscape Bionomics Dysfunctions and Human Health

Vittorio Ingegnoli

Abstract

Background: Landscape bionomics (LB) affirms that geobiological systems are living entities, not sets of separate issues (water, air, soil, species, and pollution), in which to find some interrelations. LB leads to significant changes in the understanding of reciprocal influences between environment and health.

Theory and Method: Today, the relations environment/health are mainly concerned with pollution, infective agents, partially with agrofood dysfunctions and environmental stress, but rarely with lack of defense contributions or lack of hierarchical interrelations. It could be necessary to remember that the processes leading to diseases can be due to excesses and deficiencies, underlining their peculiar characters.

Findings: Following LB, it is possible to show clear correlations between: (a) landscape structural alterations and health damages; (b) bionomic functionality and COVID-19 insurgence; (c) Cancer incidences in Europe and correlation with carcinogen parameters; (d) bionomic functionality (BF) and mortality rate (MR). The reasons for the importance of bionomic parameters in cancer incidence and the complex relationships brain-environment lead to developing etiological webs and flowcharts on the relations environment/health in cancer onset.

Discussion and Conclusion: Human activities drive fundamental biophysical change at the steepest rates in the history of our species leading to dangerous health effects. It must be clear that the Technological World cannot be considered as the real environment of man.

Keywords

Landscape Bionomics · Environment/health
Bionomic functionality · Cancer incidence
Deep vein thrombosis

5.1 Landscape Alterations and Human Health

5.1.1 Bionomics: the Need of a New Ecological Discipline

As underlined in Chap. 3, bionomics (Ingegnoli 2002; Ingegnoli 2015) profoundly upgrades the main principles of traditional ecology. In reality, *Life on Earth is organized in a hierarchy of*

V. Ingegnoli (✉)
Department of Environmental Sciences and Policy, University of Milan, Milan, Italy

PHA (Planetary Health Alliance), Harvard University, Cambridge, MA, USA

SIPNEI (It. Soc. of Psycho-Neuro-Endocrine-Immunology), Rome, Italy
e-mail: vittorio.ingegnoli@guest.unimi.it

hyper-complex systems (often indicates as *levels*), *each one being a Living Entity which cannot exist without its proper environment.*

Life, in all its form, and the related environment are the necessary components of each system, because life depends on exchange of matter, energy, and information between a concrete entity, like an organism or a community, and its environment. That is the reason why the concept of life is not limited to a single organism or to a group of species and, consequently, life organization can be described in hierarchic levels [i.e. the so-called biological spectrum sensu Odum (Odum 1971, 1983)]. The world around life is made also by life itself; so, the integration reaches again new levels.

Thus, bionomics (Ingegnoli 2001, 2002, 2015) underlines the difference between the various approaches to the study of the environment (viewpoints) and *what really exists*, the six scale of Living Entities (Table 5.1), each one definable through *ontological properties* common to all the levels of the biological spectrum, even if each specific biological level expresses a process in a *personal way*.

The System Theory affirms that the scale capable to maximize the importance and the quantity of relations among the components of a system is the scale which consents to discriminate the different forms, especially the relational ones: so, bionomics underlines that the territorial scale is the best one capable of maximizing the importance of the relations among the elements, both natural and human. Consequently, a theoretical corpus has been developed to study the real systems (Ingegnoli 2002, 2015; Ingegnoli and Giglio 2005), particularly concerning the central levels of Table 5.1. Here is a brief synthesis of the main principles proposed by landscape bionomics (see, also, Sect. 3.3):

1. It stated that Life on Earth is subjected to "time arrow," thus no return to the prior state (restoration) is possible: intervention must be intended in the sense of structural and/or functional rehabilitation.

2. Each Living Entity, from the local to the upper scales, manages *a flux of energy to reach and maintain a proper level of organization and structure, through its vegetation communities,* their metabolic data and order functions (biomass, gross primary production, respiration, B, R/GP, R/B); a landscape systemic function, named **Bionomics Territorial Capacity of Vegetation (BTC)** (**Ingegnoli** 1991, 1999, 2002, 2011b, 2015) linked to metastability (based on the concept of resistance stability) gives us *a quantitative evaluation of this flux of energy.*

3. Humans affect and limit the self-regulation capability of natural systems. An evaluation of this *aptitude* brings to the concept of **Human Habitat (HH)**.

4. The health state of a territory/landscape/region can be investigated through a proper quali-quantitative clinical-diagnostic methodology, being Living Entities; therapeutic criteria and methods of its strategic rehabilitation can be suggested and monitored.

5. Like in medicine, environmental evaluation needs **comparisons with "normal" patterns** of behavior of a system of ecocoenotopes. So, the main problem becomes how to know the normal state of an ecological system, and/or, at the same time, how to know the levels of alteration of that system.

6. The fundamental structure of a landscape is clearly systemic, an *"ecological tissue"* as the weft and the warp in weaving or the cells in a histologic tissue: so the ***Ecotissue*** concerns a multidimensional conceptual structure representing the hierarchical intertwining, in past, present, and future, of the ecological upper and lower biological levels and of their relationships in the landscape: it is constituted by a basic mosaic (the territory with the vegetation cover enhanced) and a hierarchic succession of correlated structural and functional patchworks and attributes.

7. The ***Landscape Unit (LU)***, intended as a sub-landscape, is a part of a landscape, the peculiar structural or functional aspects of which characterize it as regards to the entire landscape.

Table 5.1 Hierarchical levels of life organization on the earth

Scale	Biotic Viewpoint[a]	Functional Viewpoint[b]	Spatial configuration Viewpoint[c]	Human cultural Viewpoint[d]	Living entities[e]
Singular	ORGANISM	Organism niche	Living space	Cultural agent	Meta-organism
Stationary	Population	Population niche	Habitat	Cultural site	Meta-population
Local	Community	Ecosystem	Micro-chore	Historic-cultural district	Ecocoenotope
Territorial	Set of communities	Set of ecosystems	Chore	Historic-cultural landscape	LANDSCAPE
Regional	Biome	Biogeographic system	Macro-chore	Historic-cultural region	Ecoregion
Global	Biosphere	Ecosphere	Geosphere	Noosphere	Ecobiogeosphere[f]

[a]Biological and general-ecological criterium
[b]Traditional ecological criterium
[c]Not only a topographic criterium, but also a systemic one (Crf. Emergent Property Principle)
[d]Cultural, intended as a synthesis of anthropic signs and elements
[e]Types of living entities really existing on the Earth as spatio-temporal- information proper levels
[f]remember the "Gaia Hypothesis"

The meaning of this hyper-complex system is to be a *living entity*, not an inconsistent set of separate issues and themes (water, air, soil, species, and pollution) in which some interrelations can be found! This leads to very significant changes in the understanding of reciprocal influences among man (both as individual and/or community) and his environment and suggests a human health risk factor due to bionomic alterations (even without pollution). Landscape bionomics put in evidence new landscape structure and functions, landscape behavior, transformation and evolution laws, landscape pathologies. Its disciplinary methods are clinic-diagnostic and concerns landscape health state analysis, evaluations and diagnosis, rehabilitation strategies, etc. The creation of a new figure of man-environmental physician, the Ecoiatra, is hoped.

To strengthen the principles and methods of landscape bionomics you may refer to the book of Landscape Bionomics (Ingegnoli 2015).

5.1.2 Environmental Alteration and Human Health

In the Premise of Chap. 3 we have underlined that health and disease depend primarily on the organization of life, and we must affirm: on the *full organization*, from small to broad scales. Consequently, we distinguished four categories of environmental alterations capable of influencing human health:

(a) internal processes (e.g., metabolism, hormonal balance, gut microbiota, aging, etc.),
(b) specific external factors (e.g., infections, pollutants, smoking, drugs, etc.),
(c) general external factors (e.g., socioeconomic status, technological behaviors, climate change, etc.), and,
(d) landscape structural and functional alterations (e.g., hierarchical relations, the biological territorial capacity of vegetation, vital space per capita, ratio human/natural habitats, etc.)

Note that these categories can receive a new contributions and new linkages, following landscape bionomics principles.

5.1.2.1 Green Disposition in Wide Cities and its Effects

Let us take as an example the criticisms of the Milan green plan (Fig. 5.1), which propose again the structural scheme of the Green Belt, a scheme like that adopted by London in the immediate post-war period.

However, Sukopp (University of Berlin) (Sukopp and Hejny 1990) criticized this scheme because it worsens the fallout of fine dust in the central town, as seen in the figure. If the temperature difference (T °C) between the center and the periphery increases, the ordinarily centripetal airflow and, therefore, the concentration of dust also increases. The combined effects of the heat island (UHI) from urbanization and peripheral cooling from peri-urban reforestation can exacerbate the environment in the center. This model (Green Belt) is not good as it increases air pollution. Similar problems arise from landscape bionomics because this model does not permit to have a right ecological range (ER, see Chap. 3, Sect. 3.3.6) diffusion in the urbanized ecotopes.

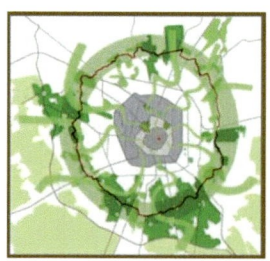

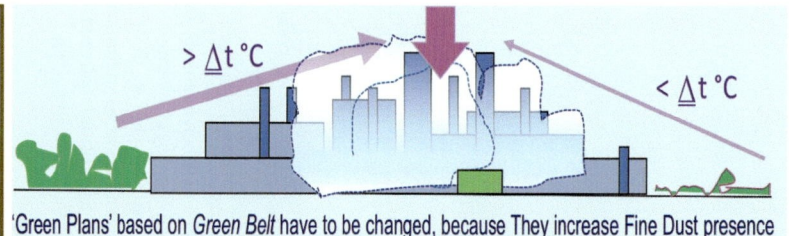

Fig. 5.1 Note that the increase of green parks around the city increases $\Delta T°C$ and fine dust transport and deposition in the central town. The UHI effect enlarges this process, which becomes very dangerous to human health

At this point, we must observe that the increase in air pollution from PM_{10} and $PM_{2.5}$ is particularly harmful to health, as it does not just produce damage to the lungs, but can cause several other damages, including serious ones. An example linked to the above is still from Milan.

A team of medical doctors published a study in the USA entitled "Exposure to Particulate Air Pollution and Risk of Deep Vein Thrombosis," which is based precisely on the hinterland of Milan (Baccarelli et al. 2008). This study highlights a clear relationship between the risk of deep vein thrombosis (DVT) and the level of atmospheric particulate matter <10 μm (PM_{10}) in the years preceding the diagnosis.

This correlation begins at the level of $PM_{10} = 12$ μm/m^3, while the legal limit for United Europe is $PM_{10} = 40\text{–}45$ μm/m^3 per year.

The motivation of the DVT formation process due to PM_{10} pollution is exciting because it assumes that fine particles interfere with the immune system, where monocytes or macrophages release different pro-inflammatory cytokines. Endothelial and blood cells can be induced to express "Tissue Factor" (TF, coagulation initiator) in the presence of cytokines such as IL-1beta, TNF (tumor necrosis factor), C-reactive protein, i.e. under inflammatory conditions.

Gilmour et al. (Gilmour et al. 2005) have shown that TF expression increases in macrophages exposed to PM_{10}. Even if the pro-inflammatory effects produced by PM_{10} fine dust (often in the presence of CO and NO2) have not yet been sufficiently investigated, we observed that they are capable of promoting the onset of various pathologies if they persist over time. The ascertainment that chronic inflammation leads to an increase in morbidity is hardly surprising.

But, among the above listed category, the fourth (d) is the less studied, so it is necessary to synthesize the main processes concerning environmental alteration and human health eminently focalized on this category.

5.1.2.2 Ethological Information and Environmental Stress

Stress needs a closer examination, because it can be direct and indirect. *Direct* stress may be due to climatic disturbances (heat, cold, rain, wind, etc.), to high noises (traffic, airport, train, etc.), to industrial food production (too high transformation and homogeneity), to high difficulties and contrasts on work, etc. *Indirect* stress may be due to behavioral and psychic inputs, consequently more abstract and difficult to study, even if after the research of Selye (Selye and Timiras 1949; Selye 1956, 1979) it has been demonstrated that the body's reaction to stress is the same both in direct and indirect cases.

The *behavioral* inputs are studied in comparative *Ethology* (Eibl-Eibesfeldt 2007; Lorenz 1973). All the organisms are capable to follow the so-called concept of *"value judgment"*: the correspondence between the perception of reality and the passage between what is more probable and disordered and what is more ordered and unlikely. It is an innate process, at the base of survival and largely *unconscious*. This ethological process can have:

- *Recreational meaning*, if the environment transmits information of good health; this activates prevention and well-being (psychic and physiological).
- A*larm meaning*, if the environment transmits information of altered environment; this activates the stress responses. Chronic stress leads to getting sick more easily.

The *psychic* inputs are recently considered in a more systemic vision, passing from agent of diseases due to unconscious conflicts expression (sensu Freud) (Freud 1923) to psychosomatic medicine, mostly depending on physiopathology. An emotional activation indicated by an increase in signal in dACC, *Anterior Cingulate Cortex*, is correlated with an increase in the plasma levels of the soluble TNF-alpha type II receptor, a basic inflammatory cytokine, as shown in the graph (Fig. 5.2). Therefore, a direct relationship between psychic stress and inflammation is enlighten (Slavich et al. 2009).

Regarding this argument, we have shown that: (a) landscape pathologies are not limited to pollution (Table 5.2), (b) the most critical landscape syndromes derive from structural and func-

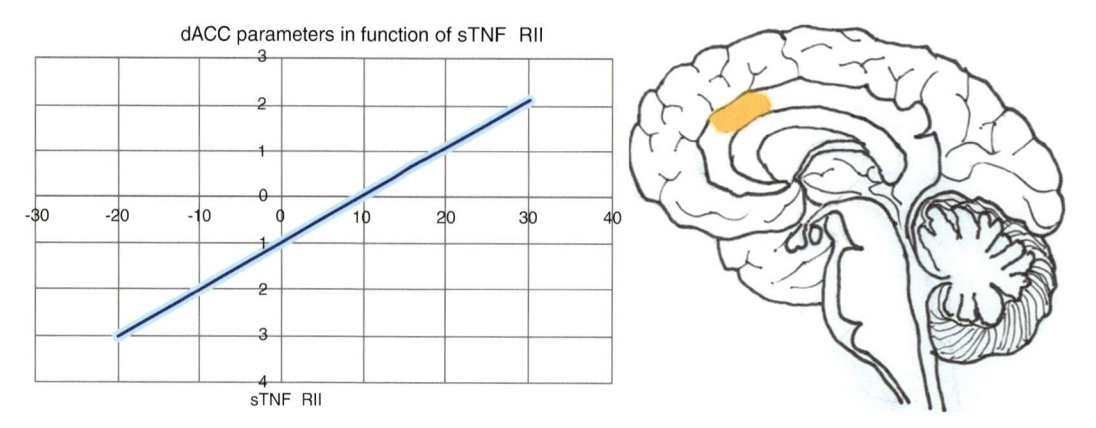

Fig. 5.2 Neural activity in the dorsal Anterior Cingulate Cortex (dACC), the yellow patch to the right, during social exclusion that correlated positively with inflammatory responses to the Trier Social Stress Test. Values derived from changes in levels of soluble receptor for tumor necrosis factor (sTNFαRII). Note that greater activity in the dACC was significantly associated with greater sTNFαRII responses mentioned Test. (data from Slavich et al. 2009)

Main landscape syndrome categories	Sub-categories of syndromes
A – Structural alterations	A1 – Landscape element anomalies
	A2 – Spatial configuration problems
	A3 – Functional configuration problems
	A4 – Multiple structural degradation
B – Functional alterations	B1 – Geobiological alterations
	B2 – Structurally dependent dysfunctions
	B3 – Delimitation problems
	B4 – Movement and flux dysfunctions
	B5 – Information anomalies
	B6 – Reproduction problems
	B7 – Multiple dysfunctions
C – Transformation syndromes	C1 – Stability problems
	C2 – Changing process dysfunctions
	C3 – Anomalies m transformation modalities
	C4 – Complex transformation syndrome
D – Catastrophic perturbations	D1 – Natural disasters
	D2 – Human-made destruction
E – Pollution degradations	E1 – Direct pollution
	E2 – Indirect pollution
F – Complex multiple syndromes	F1 – Acute
	F2 – Chronic

Table 5.2 Landscape health (from Ingegnoli, 2002). Main landscape syndromes categories and sub-categories

tional alterations, and (c) it is necessary to check if these landscape pathologies should be dangerous for human health (Chap. 3, Sect. 3.5), (d) stress agents can be registered by the ethological concept of *"value judgment."*

We underlined that this last judgment is a natural process at the base of survival, and it is mostly *unconscious*. This ethological process can have *recreational* meaning or meaning of *alarm*. It could be added that the reality of the subconscious is an emotionless database whose function is linked to the decoding of environmental signals and the activation of behaviors. The power of image processing of the conscious and subconscious mind is very different: the subconscious mind processes up to 2000 environmental stimuli/sec (both external and internal to the body) while the conscious mind only 40–50 (about 2.25%).

Therefore, a corollary has to be observed from these considerations: an alarm signal registered in *conscious* mind processes generally derives from an environmental alteration patently evident, impossible to be overlooked or not well understood (Fig. 5.3), while the most of environmental degradations seems to be not relevant; but here emerges the importance of subconscious processes.

Note that the subconscious mind frequently proceeds from comparison with normal or optimal conditions of a similar environment, a process followed even by the conscious mind, but after critical analysis, generally induced by opportune disciplines, e.g., ecology and bionom-

Fig. 5.3 (left) Example of patently evident environmental alterations, immediately traduced as alarm signal even in a conscious mind, also from people not specialized in ecology. (right) Common aspects of our environment, seen in a suburban agrarian landscape: in a case like these, the environmental alterations are rarely noted by the conscious mind, while the subconscious can receive massive stress alarm

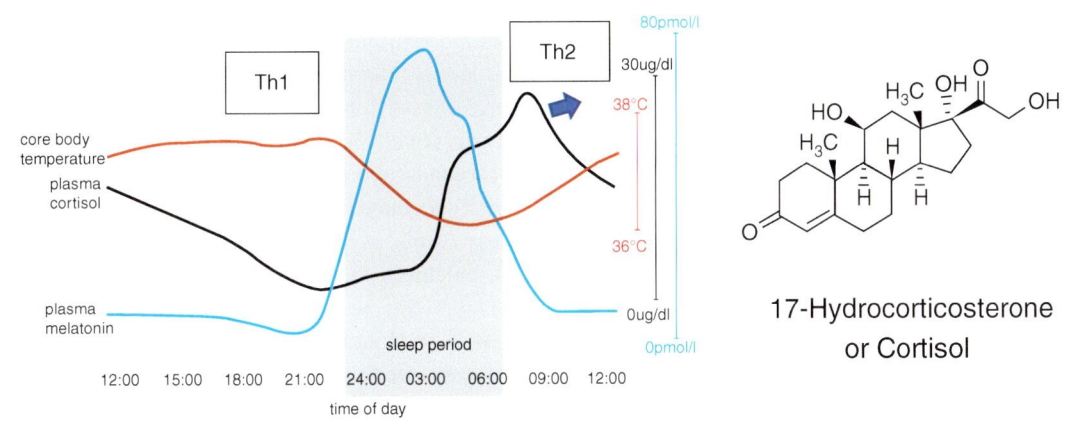

Fig. 5.4 Remember that cortisol follows a circadian rhythm. If the environmental stress becomes chronic, cortisol tends to increase leading to the dominance of Th2 immune circuit (blue arrow) (from Hickil et al. (2013) modified)

ics. As shown in Fig. 3.9, principles, and methods, of landscape bionomics can evaluate the different health states of our environment using systemic parameters and functions. Thresholds of environmental stress alarm can be measured.

As displayed in Fig. 3.15 (Berne and Levy 1990) the sympathetic nervous system and the hypothalamus-pituitary-adrenal axis mediate the integrated responses of the human organism to stress. The main factors that cause stress (stressors) simultaneously activate (a) neurons in the hypothalamus, which secrete CRH (Corticotropin-releasing hormone), and (b) adrenergic neurons.

Many of these stressors are inevitably due to landscape structural dysfunctions, even in the absence of pollution. Remember that cortisol follows a circadian rhythm.

These responses potentiate each other, both at the central and the peripheral level. The final effect of the activation of neurons that secrete CRH is the increase in cortisol levels, while the net effect of adrenergic stimulation is to increase plasma levels of catecholamine (Dopamine, nor-epinephrine, and epinephrine).

The negative feedback exerted by 17-hydroxicorticosterone (cortisol) has the function to limit an excessive reaction, which is dangerous for the organism. However, when the stress became chronicle, the circadian rhythm melatonin/cortisol (Fig. 5.4) is altered. Consequently, plasma levels of cortisol bring to a dominance of the Th2 immune circuit, with production of typical catecholamine (e.g., IL-4, IL-5, and IL-13) and the circuit Th17 (Bottaccioli and Bottaccioli 2017).

Remember that the Th2 immune response is not available to counteract viral infections, neoplastic cells, auto-immune syndromes, which need a Th1 response. We can deduce that the majority of illness is linked with chronicle stress conditions, confirming the intuitions of Selye (Selye and Timiras 1949; Selye 1956, 1979). A sharp increase in morbidity inevitably brings to an increase in mortality rate (MR).

5.1.2.3 Lack of Defense Agents

The Gut Microbiome. As it is known, the populations of host bacteria and other microorganisms in the human body are 10^{14}, therefore 10 times more numerous than our cells. The main functions of the intestinal microbiota (GM) can be summarized as follows.

1. Barrier against the proliferation of pathogens, with a mechanism known as "colonization resistance" or "barrier effect" or "competitive exclusion." It is the mechanism used by bacteria already present in the intestine to maintain their presence in this environment, avoiding the colonization of the same intestinal sites by other microorganisms, ingested or already present.
2. Regulation of the maturation of the immune system and its modulation—Production of vitamins (folic acid, vitamins K and of group B).
3. Regulation of intestinal motility, therefore of evacuation.
4. Partial recovery of energy from dietary fiber.

The intestinal microbiota can change and can be influenced in diversity and composition by the diet, by the supply of food supplements, by the *geographical-ecological area* or, better, by the *bionomic-ecological habitat* with its vegetation and fauna, in which the organism/individual lives, by drugs and by the action of agents coming from the environment such as pesticides, fertilizers, and antibiotics that are administered to the animals one eat. Here, we must underline the crucial importance of direct and indirect contact of men with natural vegetation systems, to maintain and enrich their GM, even independently from nutrition. Environment plays a much greater role than host genetics in determining the composition of the human gut microbiome, according to a study published today (Rotschield et al. 2018) in Nature.

The Phytoncides. It should be added that the influence of forests also highlights direct health-friendly contributions, as shown by the studies of the group coordinated by Quing Li of the University of Tokyo (Li et al. 2009). In Japan, a Shinrin-yoku (forest bath, i.e. a path in a natural forest landscape of a couple of days) was traditionally recommended as a therapy. Note that the forest bath's psycho-physical relaxation condition enhances the beneficial action of terpenes and monoterpenes (α-pinene, β-pinene, γ-pinene, linalool, and limonene) which have anti-bacterial, analgesic effects, and increase the activity of the immune system.

Quing Li has shown that phytoncides (forest essential oils) have beneficial effects on natural killers (Nk) because they increase the number and effectiveness of the cytotoxic enzymes granulysin, perforin, and A and B granzymes expressed by peripheral blood lymphocytes cells (PBLs), i.e. CD8 lymphocytes (T effectors), strengthening the immune system (Fig. 5.5).

5.1.2.4 Lack of Hierarchical Relationships

The argument is summarized from Chap. 3, Fig. 3.2. Levels in the hierarchy are isolated from each other because they operate at distinctly different rates. So, the inferior level components

Fig. 5.5 Effect derived from a 2-day "forest bath" on the number of cytotoxic enzymes capable of destroying dangerous cells (e.g., neoplastic) (Li et al. 2009). This effect lasts for almost a month

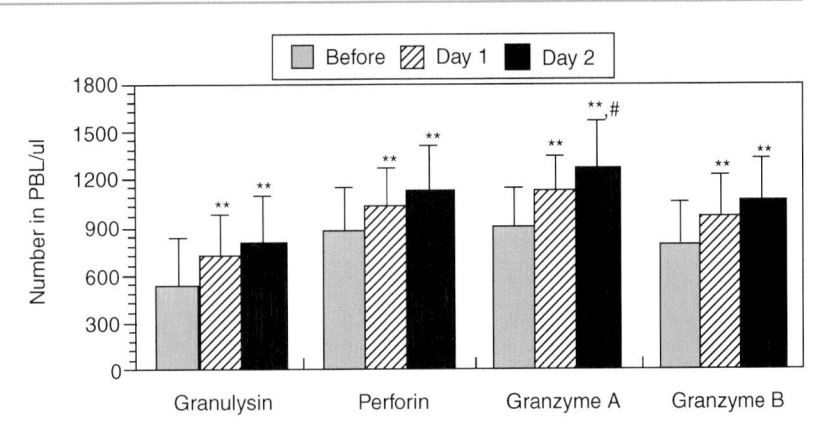

explain the origin of the level of interest. In contrast, a superior level system *explains the significance* of the level of interest (i.e., allow the characters derived from transferred conditions to be understood) (Allen and Hoekstra 1992). The concept of *constraint* shows the behavior of an ecological system as limited: (a) by the behaviors of its components and (b) by environmental bonds imposed by superior levels of the organization. Thus, there is a linkage between constraint and information.

Genome and Cell Scales: **Epigenetics** and the environment. The information written on the chromosome bases is insufficient; we also need the modulation of information given by the *environment* and our behavior. So, the gene–environment interaction is crucial. This fact implies a direct linkage between cell scale and environmental scale, e.g., the health state of the landscape unit alteration and the genome of an organism living there. Signaling molecules and transcription factors modulate genetic information producing anti-inflammatory cytokines anti-oxidant, anti-mutation, a decrease of cortisol, etc. We have known for some time that the DNA-methyl tags are copied too, so that both daughter cells have the same pattern of DNA methylation. The pattern of histone tags is also mostly duplicated as cells divide, although this is currently less well understood (Bird 2013; Carey 2012).

Organism and Ecocoenotope Scales: **Population** and the environment. An example: the complete homogeneity of industrialized crops, due to cultivars with the same genetic characters, is dangerous for our immune system and mental health, disrupting gut microbiota. Therefore, ecology suggests to plant *populations* of cereals. These populations evolve under different managements and in the face to specific combinations of *landscape unit characters* (LU) as microclimate, insects, weeds, trees, and shrubs, BTC, HS/HS*, TD, etc. In this way, the frequency of genotypes that have adapted to local conditions gradually increases. Local environmental characters guide DNA expressions and make them transmitted to the following generations (Bird 2013; Carey 2012).

Landscape Unit and Region Scales: **Bionomics** and the environment. For instance, we report the main processes demonstrating the influence of broader scales on detailed ones: (a) LU characters are indispensable to the evolution of populations; (b) the zoological reconversions, due to LU structure changes, lead to an *exacerbation of virus* and parasites influencing man and animals; (c) the interactions "landscape-apparatuses/vital-space per capita" express other meaningful linkages among scales, permitting the measure of heterotrophy of a LU and also enlighten if man's life in a specific place should or not be sustainable.

5.2 Correlation between Bionomic Dysfunctions and Mortality Rate

5.2.1 Bionomic Dysfunctions of the Territory Milan-Monza-Brianza

The choice of a land area of experimentation presenting an available gradient of landscape types (from dense urban to agricultural-forested) was facilitated by living in Milan, one of Europe's largest metropolitan areas (five million inhabitants). As plotted in Fig. 5.6 (left), the blue-gray lines indicate a territory covered by the province of Monza-Brianza (North) and a portion of the province of Milan (South): in summary 655.8 km^2 and 2.5×10^6 inhabitants with a gradient of six landscape types (Table 5.3). Note that not only Milan but also Monza and Brianza present one of the most air-polluted areas of Europe. Therefore, pollution could be considered as homogeneous in our sample land area. We have to note that the city of Milan was analyzed by dividing into its nine administrative sections, to have data comparable with the other municipalities' ones in surface and population. Few municipalities reach this goal in our territory.

The biological territorial capacity of vegetation (BTC) was estimated using field surveys and the registered statistical data on the main types of vegetation. The evaluation of CBSt better allowed the assessment of the forest vegetation in the entire Monza-Brianza Province. This function, named "concise bionomic state of vegetation" (Ingegnoli 2013), is available to be applied as the vegetation's bionomics efficiency.[1] Figure 5.6 exposes, in the right part, the most significant set of forest assessment surveyed on the field. The modest value of the mean BTC = 5.84 Mcal/m^2/year is confirmed by 57.14% of altered and weak forests vs. only 19.05% of good ones.

Considering our 72 municipalities as Landscape Units (LU), we can register these

---

[1]Efficiency relates the maturity level (MtL) of a vegetation coenosis and its bionomic quality (bQ). It is fundamental for the comparison among different types of vegetation in providing ecological services.

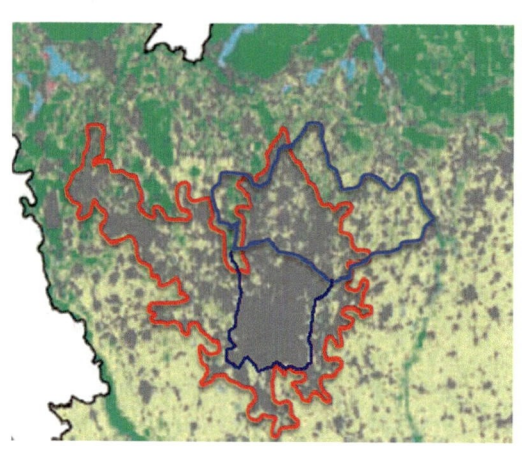

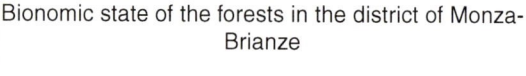

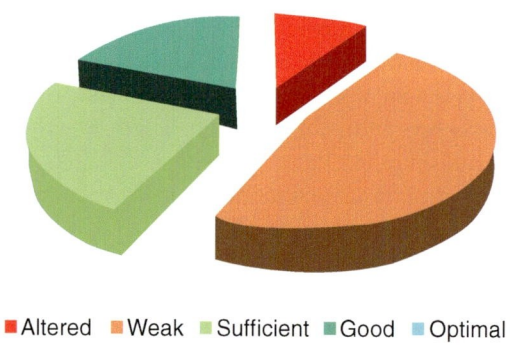

Fig. 5.6 (Left). The blue-gray lines indicate the land area of experimentation: Monza-Brianza (North) and Milan (South). The two pictures to the left represent the extreme differences in the landscape gradient. This territory covers 655.9 km^2 with a population of 2.5×10^6 inhabitants and with a gradient of 6 landscape types (base map from DUSAF, Milan). (Right) The bionomic state of the forest formation on the Province of Monza-Brianza shows only 19.05% of them in right conditions while no one is optimal

Table 5.3 Six landscape types and their bionomic characters in the study area

km²	Mun.	Landscape Type	%for	%urb	%agr	HH	BTC	MR	PA	BF
96.2	14	Agricultural	21.73	36.10	41.83	72.00	1.68	7.11	42.38	1.08
89.9	14	Rural	13.03	40.19	45.99	78.16	1.25	8.33	42.66	0.98
75.5	13	Suburban-Rural	6.10	44.05	48.65	83.62	0.89	7.62	42.57	0.84
133.4	14	Suburban-Tech.	2.76	60.54	33.83	88.41	0.61	8.66	43.47	0.67
69.1	8	Urban	0.95	72.30	24.24	92.17	0.44	9.45	44.75	0.54
181.8	9	Dense urban	1.35	86.25	11.08	93.65	0.39	9.57	45.08	0.41
655.90	**72.00**	**Mean**	**7.65**	**56.57**	**34.27**	**84.67**	**0.88**	**8.46**	**43.49**	**0.75**

Mun. Municipality, *for* forest, *urb* urban, *agr* agricultural, *HH* human habitat (%LU), *BTC* Mcal/m²/year, *MR* mortality rate, *PA* population age, *BF* bionomic functionality

range of values: human habitat HH = 72.0–93.7 (% of LU); biological territorial capacity of vegetation BTC = 0.39–1.68 (Mcal/m²/year); mortality rate MR = 4.10–11.35 (×1000 inhabitants); population age PA = 38.6–47.1 (years); level of alteration or degradation of functional bionomic state BF = 1.10–0.34.

Only 22/72 = 30.56% of the examined municipalities were found in a healthy condition (BF = 1.15–0.85), while 25/72 = 34.72% were in a dysfunctional or degraded condition; the others are altered. These levels of BF compare the mean BTC of each LU with the value given by the BTC model (see Chap. 3, Fig. 3.9) corresponding to the LU value of HH.

5.2.2 The Correlation Bionomic Functionality/Mortality Rate (72 LU)

Plotting the results of the previous Table 5.3, related to the six types of landscape units, we reach the first evidence of the increase of mortality rate (MR) when the bionomic functionality (BF) is decreasing from normality (BF = 1.0). Traditional ecology states that the increase in the mean age of a population (PA) indicates a very developed country and less negative environmental factors. On the contrary (Fig. 5.7) in this consumer society, the increase in population age grows with the increase of landscape degradation. However, the contrast between traditional and bionomic ecology interpretation is not so strong as it may appear. We may observe that PA grows proportionally to MR.

Nevertheless, the increase of $PA = f(BF)$ cannot explain the similar increase in MR. If we examine our two comparing functions $MR = f(BF)$ and $PA = f(BF)$ (Fig. 5.7, right) we note a sharp difference: e.g., going from the point of full normality (BF = 1.0) to the point of very high degradation (BF = 0.4), PA increases only 7.3% (gray line) while the MR increases 30.3%. If only 24% of the MR increase depends on PA, 76% is due to other causes. Being very contained, the increase of PA seems to imply limited negative environmental factors, as stated by general ecology (Ingegnoli and Giglio 2017).

Detailing the 72 LU, the correlation MR/BF (mortality rate/bionomic functionality) emerges (Fig. 5.8). The result of $MR = f(BF)$ presents the $R^2 = 0.252$, but the Pearson's correlation coefficient is −0.438 that is twice the minimum value of significance, therefore, MR increases with BF. At full normality BF = 1.00 and MR = 7.64, becoming MR = 7.95 at BF = 0.85 but 10.27 at BF = 0.35, representing a deep dysfunction.

A first base to estimate a Risk Factor depends on the MI-MB Model. From Table 5.3 the mean BF = 0.75, therefore, only considering the 76% of MR due to other causes,

$$\Delta MR_{BF} = MR_{measuredBF} - MR_{BF=1} = (8.253 - 7.64) \times 0.76 = 0.466 \text{ every } 1000 \text{ inhabitants}$$

Consequently, the estimation of premature death (PD) probability due to the altered Bionomic Functionality (PD_{BF}) can be measured, remembering that the total population in the study area is 2.524 million:

$$PD_{BF} = 2.524 \times 10^6 \times 0.466 \times 10^{-3}$$
$$= 1176.2 / \text{year}$$

This huge number of premature deaths/years can be compared with the total death number

LANDSCAPE GRADIENT	N°
AGRICULTURAL - Monza 1	1
RURAL - Monza 2	2
SUBURBAN – RURAL Monza 3	3
SUBURBAN - TECHNOLOGIC Milan 1	4
URBAN - Monza 4	5
DENSEURBAN – Milan 2	6

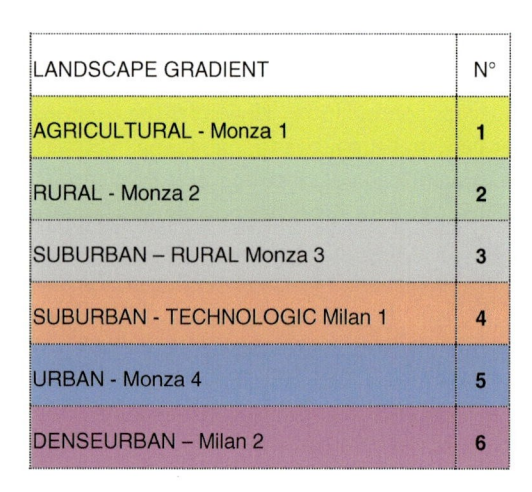

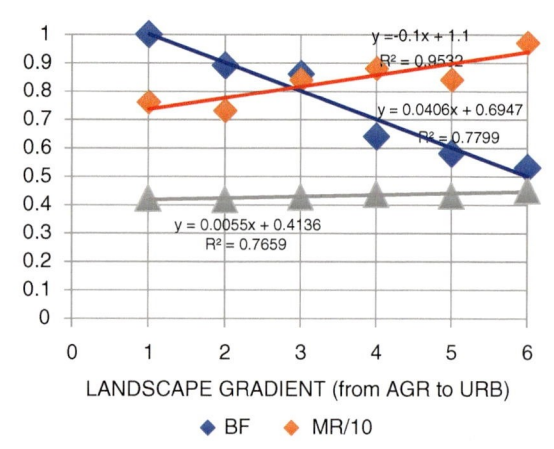

Fig. 5.7 Change of the parameters: MR (mortality rate), BF (bionomic functionality), and PA (population age) related to the landscape gradient from agricultural to dense urban

Fig. 5.8 Correlation between Mortality Rate and Bionomic dysfunction in 72 Landscape Units of the Milano-Monza (MI-MB) Area. The Pearson coefficient is twice the minimum values of significance. The mean BF is 0.75

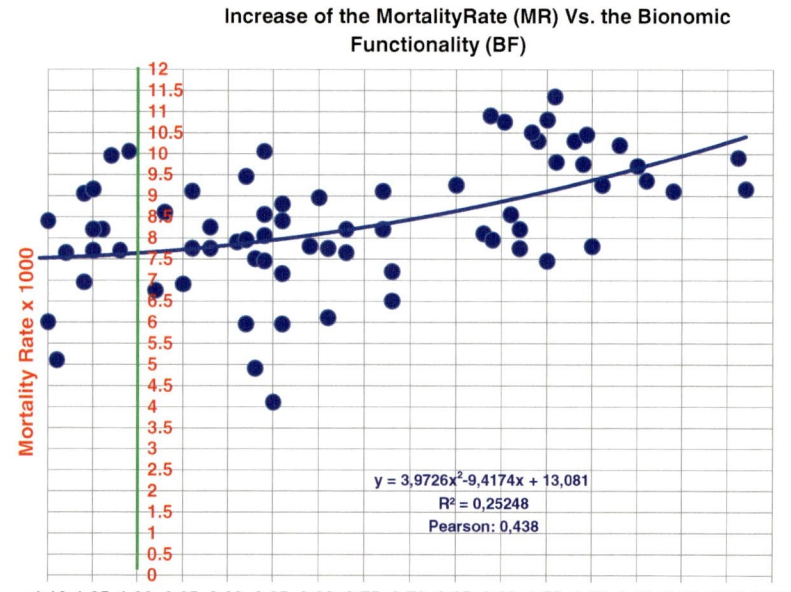

(TD) in this area, pair to 21,050 people/year. The rate: $PD_{BF}/TD = 5.6\%$ Note that the high MR result is quite similar to the fine dust premature death $(PD_{FD}) = 7.0\%$ related to the Milan Metropolitan Area, but it must be added to it!

5.2.3 Agrarian Landscape Destruction and Increase of Mortality Rate

To investigate the Health of an agricultural landscape, let us consider the case study of Albairate. This is chosen because of its participation in the Regional Agricultural Park of South Milan, thus verifying the bionomics state of this Landscape Unit (LU) in a protected regimen. The territory of Albairate (see forward Fig. 5.10a), forming the LU in the examination, has a surface of 1538.2 ha, 4700 inhabitants, and is located about 20 km West to Milan, confining (West) with the Natural Regional Park of Ticino River. This landscape is mainly agrarian.

The study concerns regional data of land use and land cover (Dusaf-ESRSAF 2015), but the vegetation analysis was made directly on the fields. The historical data were based on the maps of IGM (Military Geographic Institute) for the year 1900, while for sequent years 1954 and 1989 again, the bases were from Dusaf-ERSAF (Regional Agency for Environmental Protection). The most significant changes emerge in the process of urbanization, where the residential apparatus[2] (RSD) passed from 2.06% in 1900 to 7.05% in 2016, and the subsidiary apparatus (SBS) from 0.6% to 6.65%. So, the total PRD reduced from 83.11% to 72.67% (−12.56% of the agrarian surface). In add, cropland with trees disappeared, while grasslands and rice fields had an exceptional increase: from 0.32% to 6.76% (grass) and from 0.97% to 9.94% (rice).

The protective apparatus (PRT), on the contrary, decreased from 8.11% in 1900 to 3.94% in 2016. The resilient apparatus (RSL), composed by shrubs and permanent prairies, is today small, but larger than before. The resistant apparatus (RNT), woods and forests, remained the same 4.9% (apparently) but had a violent reduction after the war (1954). Even rivers and canal (HGL) remained near the same. The main bionomics parameters HH and BTC present different changes: the human habitat (HH) is nearly constant, from 78.8 in 1900 to 79.4 in 2016, after a small grow in the medium period; the vegetation capacity (BTC) presents an evident decrease, from 1.24 to 1.02 Mcal/m^2/year (−17.7%): this is not a good sign.

About 1/5 of this territory uses Organic farming, more than the regional average. We investigated two Organic farms (62.5 ha) and compared with two Conventional farms, with the assistance of two graduating students of the Department of Agrarian Science DiSAA-UNIMI. Focusing the two bionomics functions BTC and CBSt, the result indicates a superior ecological condition of Organic Farming (Vaglia and Zamprogno 2017).

However, the forest has regrown and a right presence of Organic farming is not sufficient to maintain the examined LU in a healthy state, mainly for the disruption of the tree-corridor network and for the scarce bionomics' efficiency of the local forests (Table 5.4). Note that, even if the soil humus efficiency (RxN) is the same, the mean BTC of Albairate forests is 5.27 vs. 6.33 Mcal/m^2/year of the Milan hinterland, lower than the regular temperate forests (BTC = 7.38).

---

[2]A Landscape Apparatus (L-Ap) is a functional system of ecocoenotopes (even not connected), performing a specific physiologic function within the landscape unit: it forms a specific configuration within the ecotissue. Remember that the **ecocoenotope** is a *multifunctional* entity in a definite geographic locality: it is the "tessera" of the basic mosaic of an ecotissue. See also Sect. 3.3.1 and (Ingegnoli 2002, 2011, 2015).

Table 5.4 Comparison between the forests of Albairate and vs. the Hinterland of Milan

Forests of MI hinterland		HC	PB	BTC	CBSt	RxN
Albairate	Albairate mean Forest	19.38	268.0	5.27	14.36	34.44
Milan hinterland Forests	Milan Hinterland Forests	23.50	433.7	6.33	24.70	34.14

HC canopy height, *PB* plant biomass volume, *RxN* soil humus condition

Applying the local diagnostic model HH/BTC to the case study of Albairate, the results confirm the mentioned alterations but, primarily, they indicate the recent drastic change of direction of the movement of this complex system (Fig. 5.9).

Note that all the movements of the system during at least 90 years were done strictly near the curve of normality (green). Only since 1990, the direction has been going out of the tolerance (dotted), while BF has been resulting altered (BF = 0.83).

The disruption of the tree-corridor network needs an explanation. In Fig. 5.10b and c, we can see and compare the maps of this network in 1954 (left) and 2016 (right). The differences are immediately visible.

- In 1954 the network was formed by 440 Nodes and 669 Links. It covered 114.99 ha, and the BTC was higher than today, due to the frequency of autochthone species (*Quercus robur* and *Alnus glutinosa*). So, BTC was pair to 3.20 Mcal/m^2/year and the energy flux of maintenance was Mcal/year = 3.68×10^6.
- In 2016 the same network presented 133 Nodes and 405 Links. It covered 50.68 ha, and the BTC decreases to BTC = 2.90 Mcal/m^2/year. Consequently, the energy flux resulted Mcal/year = 1.47×10^6.

We have underlined that the decrease in land cover of tree corridors (115 ha vs. 51) means a loss of 55.7%, but the decrease of the bionomics

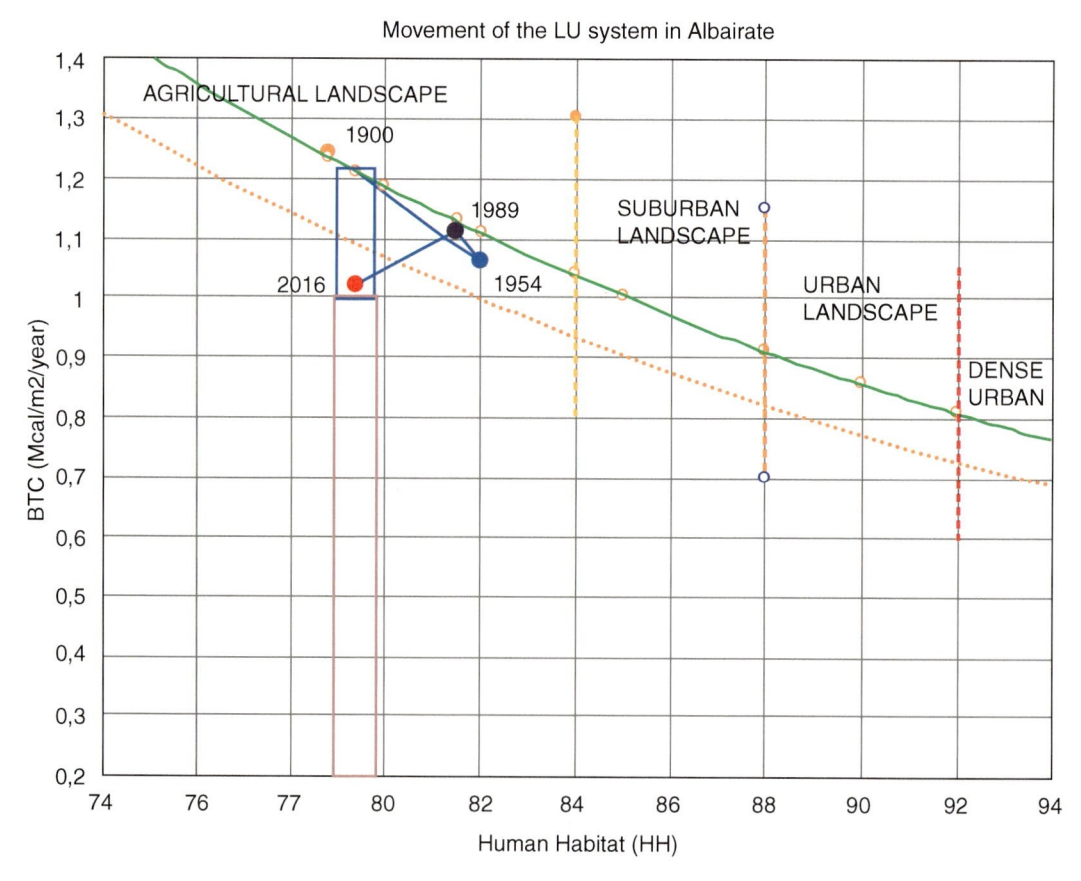

Fig. 5.9 The transformation of the LU of Albairate since 1900. Note that the movements of the complex system during at least 90 years were done strictly near the normality (green). Only recently, the direction went over the tolerance (dotted), and BF = 0.83 resulted altered

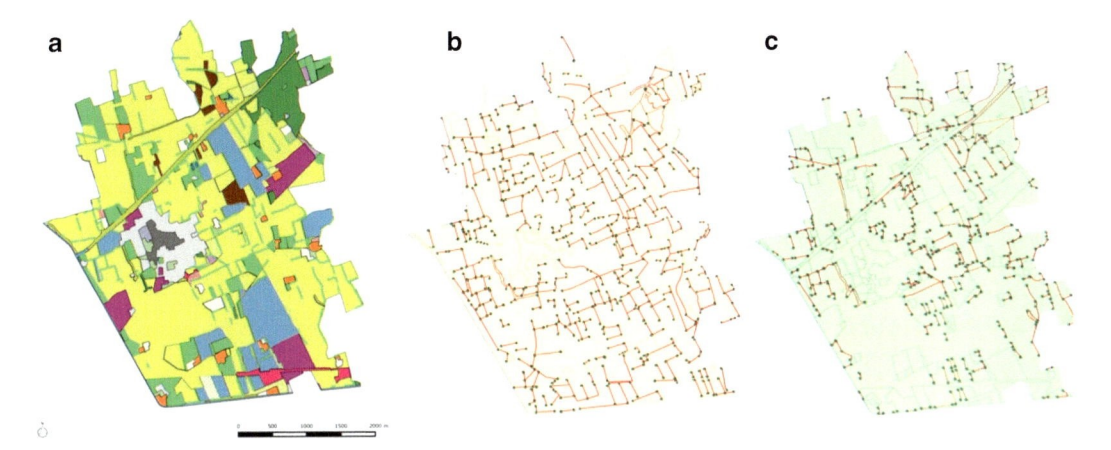

Fig. 5.10 (a) Map of Albairate, 20 km West of Milan. Urban areas in gray, industrial areas in violet, dark green the forest cover, light green poplar cultures, blue-gray the rice fields. This territory forming the landscape unit (LU) in examination has a surface of 1538.2 ha and 4700 inhabitants. (b, c) Evident significant changes in the ecological network of the trees corridors in Albairate from 1954 (b) to 2016 (c)

flux of maintenance of the network (3.68 vs. 1.47 × 10^6 Mcal/m²/year) means a loss of 60%! This is due to the reduction of the traditional semi-natural tree corridors characterized by *Quercus robur, Alnus glutinosa,* and *Ulmus campestris*, the corridor having a width > 12 m, the threshold to host plant and animal of interior species as underlined by Forman & Godron (Forman and Godron 1986). Today, these autochthonous plant species have been frequently substituted by *Populus nigra x canadensis* and corridors have been reduced in their width < 12 m.

The destruction of the tree-corridor network leads to out-of-scale dimensions of the crop fields, one of the basilar aspects of Agriculture Industrialization. Today, too many agrarian landscapes cannot be adequately defined as landscape, due to their absurd homogeneity (remember Fig. 5.3, right). These environmental changes follow Neoclassical Economy, which reduces the economic values of these fields only to the amount of labor. Therefore: (a) the "ecological services" are only perceived toward human society and are very limited; consequently (b) the market values of agrarian lots are shallow, and their conversion in urbanized lots appears acceptable.

One of the most important therapeutic action to rehabilitate the Agricultural Landscapes depends on the ecological and bionomic impellent need to restructure the crop fields through this crucial green infrastructure: the tree-corridor network. The bionomic influence of tree corridors on field dimensions depends mainly on two aspects: (1) the protection from climatic disturbances and (2) the sufficient bionomic balance of BTC. In the first case, Forman and Godron (Forman and Godron 1986) shown that a dense windbreak may protect to nearly 5–6 times the trees' height, generally in Europe, from 75 to 120 m. These results confirm and enhance what already underlined in Fig. 5.9: the examined LU of Albairate, despite its participation in the Regional Agricultural Park of South Milan, reveals a clear disease (Fig. 5.10).

If we plot the relationships between the LU bionomic dysfunction (BF = 0.83) and the mortality rate (MR = 9.9/1000 inhabitant/year) and compare this situation with BF/MR correlation (remember Fig. 5.9), the result is alarming (Fig. 5.11).

Note that a tolerance of ±5% has been considered, so the distance is reduced (ΔMR = 9.80–7.41 = 2.39/1000). Let us evaluate the *Risk Factor* from the BTC/HH Model [*BF = 0.83*]:

Fig. 5.11 Alteration of an agricultural landscape. An evident dysfunction appears in the BF/MR model through the distance of the present mortality rate (red) from the condition of normality (blue). Even considering a tolerance BF = 0.95, the gap remains too high

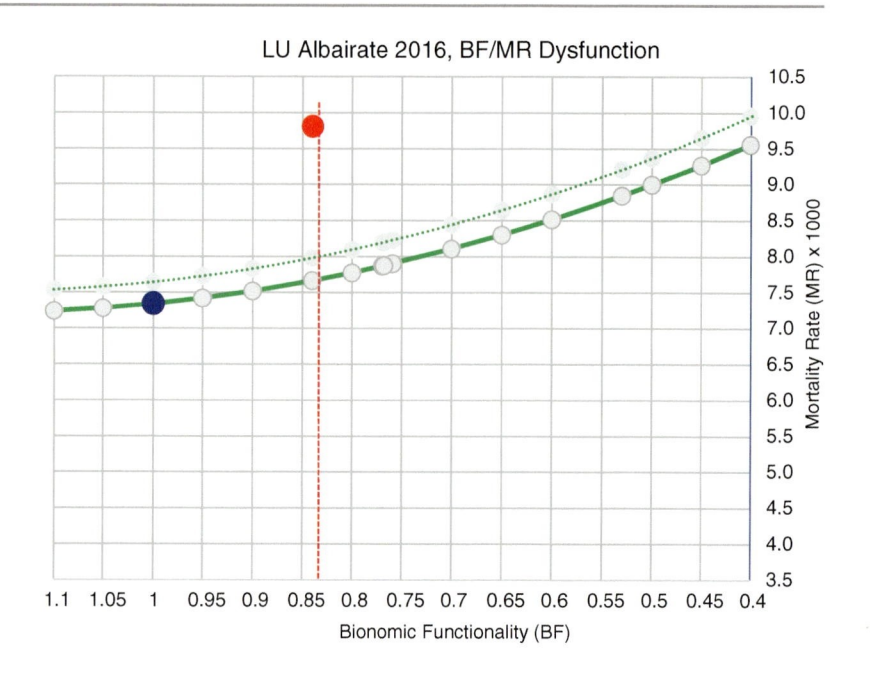

$$\Delta MR_{BF} = (MR_{BF} - MR_{BF=0.95}) \times 76\%$$
$$= (9.80 - 7.41) \times 0.76$$
$$= 1.8164 \times 10^{-3}$$

The population of Albairate is 4708 inhabitants, so it is possible to estimate the premature death risk related to Bionomic Dysfunction per annum:

$$4.71 \times 1.8164 / 1000$$
$$= 8.53 \text{ premature deaths / year}$$

Note that an MR = 9.80 in a municipality counting 4700 inhabitants brings to 46 premature deaths/year, and the ratio 8.5/46 = 0.191 is very high. In the Hinterland of Milan, this ratio is generally ½ of those values. So, probably the premature death risk is due to two components: (a) the environmental stress, dependent from the bionomic alteration of the LU, (b) the pollution due to agrarian chemical (herbicides, pesticides, fertilizers, etc.).

The economic costs of premature deaths can be estimated, taking inspiration from similar calculations regarding the risks for pollution (Lattarulo and Plechero 2005) (L: the cost estimate considers both direct costs (hospitalization, medicines, ambulances, etc.) and indirect ones (e.g., life expectancy) and the sum of these costs is equal to 75.000–100.000 € + 450.000 € = 550.000 €/dead.

5.2.4 The Correlation Bionomic Functionality/COVID-19 Insurgence

The northern blue delimitation above Milan city (go back to Fig. 5.6, left) is the M-B province, counting 55 municipalities or operative Landscape Units (LU). Their landscape types can be grouped in six, as we have shown in Table 5.3. Their BF values range from 1.11 (agrarian PRT) to 0.44 (dense urban of Monza. In M-B, the COVID-19 pandemic started on March first, 2020; this research refers to April 19th, when its curve was going to the plateau. The percentage of COVID-19 infected people appears growing inversely to the bionomic functionality (BF). This impression seems not so clear, because the COVID-19 percentage is not always strictly

5 Landscape Bionomics Dysfunctions and Human Health

Fig. 5.12 Infections by COVID-19 in Monza-Brianza Province (MB) near Milan, data of 19/04/2020 (top of the plateau). The correlation with bionomic functionality (BF) of the 55 landscape units (municipalities) is evident: at BF = 1.0, people infected by COVID-19 was pair to 0.36%, while at BF = 0.5, COVID-19 concerned 0.54% of people

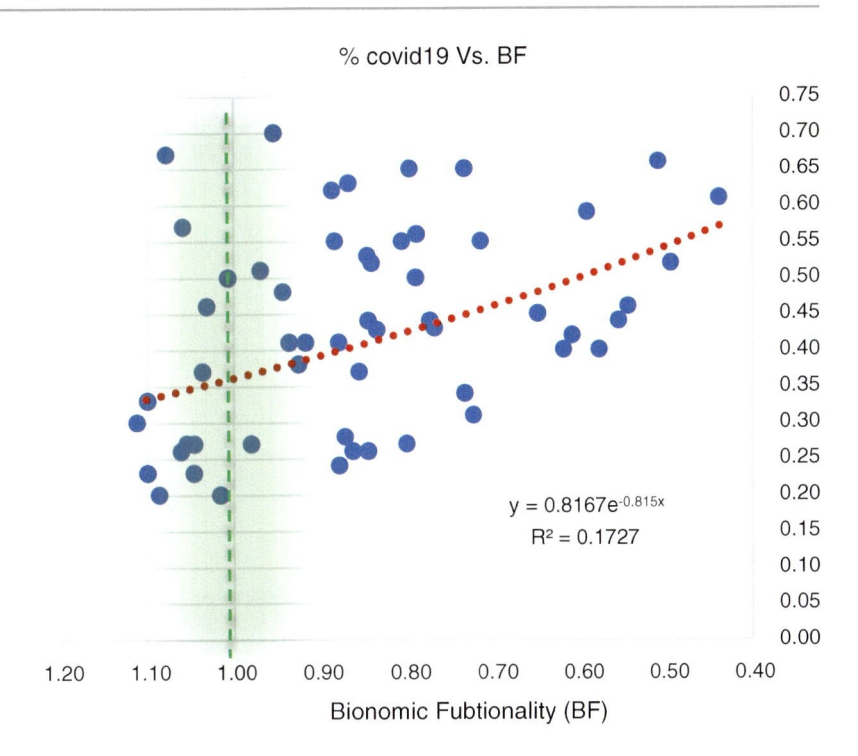

growing vs. BF decrease. Therefore, it is necessary to test the correlation of all the 55 LU, concerning their BF vs. %COVID-19, as shown in Fig. 5.12. The correlation with bionomic functionality (BF) of the 55 landscape units (municipalities) is evident: at BF = 1.0, people infected by COVID-19 was pair to 0.36%, while at BF = 0.5, people with COVID-19 was 0.54%. As already underlined, the BF index derives from the diagnostic model HH/BTC (see Sect. 3.3.3), so it is a very synthetic control linked with the basilar state function of BTC, the biological territorial capacity of vegetation correlated with the level of the human habitat (HH).

It could be interesting to test even other ecological parameters supposed to be drivers for the COVID-19 epidemy: (a) mean population age (years), (b) urbanization (%), (c) forest cover (%), (d) ecological people density (ab/ha), (e) human habitat (%), (f) agricultural cover (%). For instance, see Fig. 5.13.

The best-known control for the significance of the possible correlation between two variables (X, Y) is the statistical population Pearson's correlation coefficient (r X, Y), with the equation:

$$r\,X,Y = \frac{\mathrm{cov}(X,Y)}{\mathrm{sd}x\,\mathrm{sd}y}$$

(where : cov = covariance, sd = standard deviation)

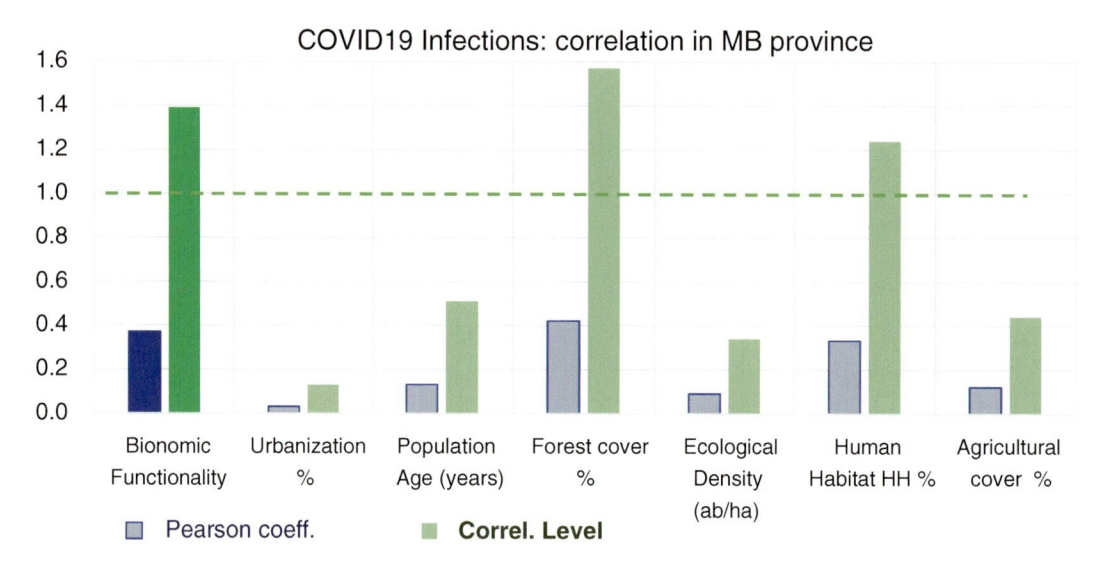

Fig. 5.13 Correlations of the main ecological "driver parameters" of COVID-19 infections and comparison with BF (systemic index of Bionomic Functionality). Note that BF can synthesize the integration of the six plotted parameters and more. We can also see that forest cover and human habitat present a significantly higher value than Population Age or People's ecological density

Pearson correlation coefficient and its value of significance related to the sample sizes. In this case (55 LU), the value of significance is equal to 0.266. The statistical population of 55 LU of M-B province presents a threshold of significance for the Pearson coefficient equal to 0.266; the ratio with r X, Y measures the correlation level (CL):

$$CL = \frac{r X,Y}{0.266} > 1.0$$

As exposed in Fig. 5.12, only forest cover and human habitat have CL > 1.0, while all the other possible causes show weak-low correlations. The bionomic functionality (BF) can be considered a mediating parameter of each LU's bionomic state, so it seems reasonable, even compared to the other drivers: note that the BF correlation level (CL = 1.39) shows an intermediate value between two parameters, Forest cover (CL = 1.57), favorable for a natural environment, and Human habitat (CL = 1.24) favorable for an anthropized environment.

This research is in progress because the dead number/LU is not fully available, nor the disease course and/or aggressiveness. So, in Fig. 5.14, the correlation between BF and the dead number, elaborated with 35 LU on 55, seems to give a result similar to the BF/COVID-19 incidence and better for the moment: Pearson Coefficient = 0.46, $R^2 = 0,301$, CL = 1.73.

Fig. 5.14 No. of dead people during the Infections by COVID-19 in Monza-Brianza Province (MB) near Milan, 04/04/2020 (near the top of the plateau). The correlation is evident

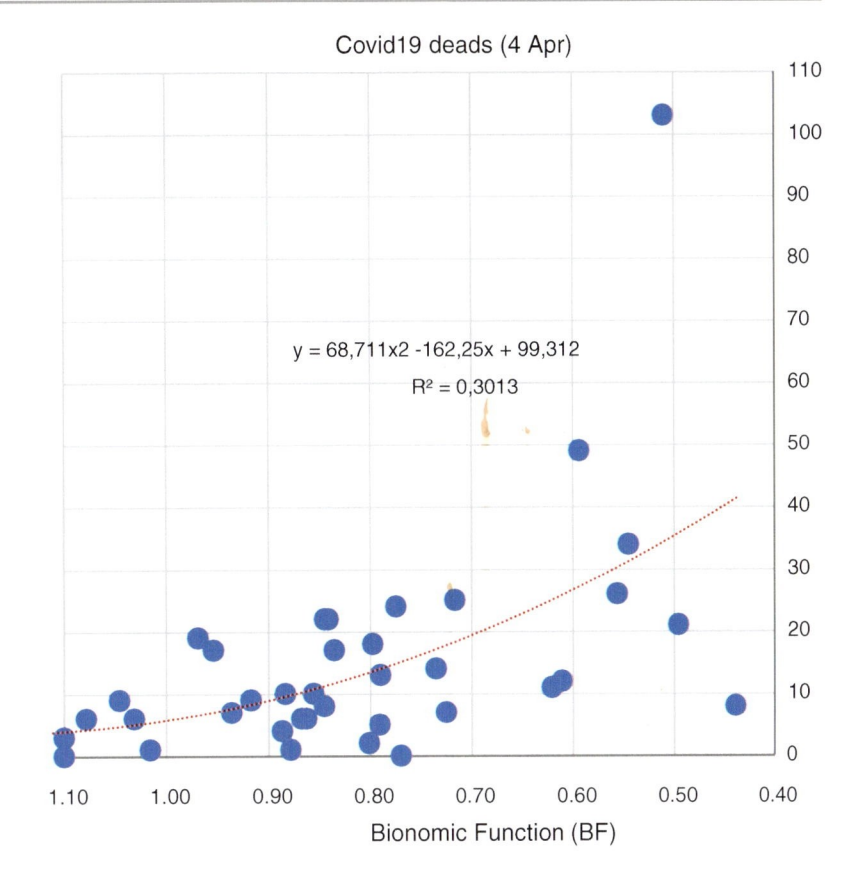

Covid19 deads (4 Apr)

$y = 68{,}711x^2 - 162{,}25x + 99{,}312$
$R^2 = 0{,}3013$

Bionomic Function (BF)

5.3 Correlation between Bionomic Dysfunctions and Cancer Incidence in Europe

5.3.1 Forest Types and BTC Evaluation in European Countries

In Chap. 3, we affirmed that the bionomics-ecological diagnosis of territorial complex systems is impossible without an advanced analysis of vegetation, following the Landscape Bionomic Survey of Vegetation (LaBiSV) methodology (Ingegnoli 2002; Ingegnoli and Giglio 2005; Ingegnoli and Pignatti 2007). In this research, we need to start from a rework of data from European Ecological Agency (EEA) (Barbati et al. 2018; EEA 2012) on European Forests (Table 5.5). In this table, the forest distribution, expresses as its composition in 14 forest types, and the high/low % of the presence of forest by country shows 6 countries with presence >45% vs. 8 < 25% (average 33.78%).

As we can see in Chap. 3, Sect. 3.4.1, the BTC of vegetation is a bio-systemic function able to measure the level of order and metastability reached by a phytocoenosis (natural or anthropic), as the negentropic flows of energy-related to its metabolic functions [Mcal/m^2/year]. BTC levels differ very much: values for grass and fields range between 0.50 and 1.40, while vines, shrubs, plantations, and gardens from 1.50 to 3.50, adult/mature

Table 5.5 Distribution of forest types in European Countries (data: EEA statistics, 2008–10, reworked)

	EU Nations	Forest km²	Forest %	F % EU	1	2	3	4	5	6	7	8	9	10	11	12	13	14
1	Sweden	307.875	68.42	16.81	50	39		1	1	1							6	2
2	Spain	283.007	56.07	15.46			3	2	2		2	10	26	43				12
3	France	246.640	44.84	13.47		4	9	6	24	7	5	14	4	10		1	2	15
4	Finland	233.320	69.03	12.74	88	3								4			6	
5	Germany	113.176	31.71	6.18		51	4	1	8	12	6					1	3	14
6	Italy	106.736	35.43	5.83		1	23	1	2		16	40	4	4			3	6
7	Poland	90.000	28.78	4.92		75	5	1	7	2	2					1	5	2
8	Romania	63.700	26.82	3.48		1	16		21	22	21	10					2	7
9	Austria	39.600	47.22	2.16		24	65		3	1	6					1		
10	Greece	37.520	28.43	2.05					9	2	10	19	16	43				1
11	Bulgaria	36.250	32.39	1.98		8	26		5	12	7	17		17				8
12	Portugal	32.400	35.19	1.77					1			4	48	29			1	17
13	Latvia	28.807	44.32	1.57	19	59											22	
14	United Kingdom	28.650	11.70	1.56		4		4	16	14						1		61
15	Czech Rep.	26.000	32.97	1.42		69	1	1	9	4	4					1	3	9
16	Croatia	24.901	44.00	1.36		1	6	2	15	21	11	14	2	6		20		2
17	Estonia	23.066	51.26	1.26	7	77			1						12		3	
18	Lithuania	21.223	32.65	1.16	5	76			2								17	
19	Slovak Rep.	20.006	40.96	1.09		5	39		16	25	10	1						4
20	Hungary	18.513	19.91	1.01		5			21	7		19				5	7	36
21	Slovenia	12.574	62.02	0.69		12	19		2	21	29	5				2	2	8
22	Switzerland	12.425	30.10	0.68		15	50		8	6	13	6						2
23	Ireland	6.690	9.56	0.37														100
24	Belgium	6.607	21.64	0.36				10	10									80
25	Denmark	5.172	12.00	0.28					10	30								60
26	Netherlands	3.650	8.79	0.20				9	27									64
27	Cyprus	1.740	18.81	0.10										100				
28	Luxemburg	870	33.64	0.05		25			25	25	25							
29	Malta	3	0.95	0.00											100			
	EU	1.831.121	33.78		20.1	19.2	6.98	1.62	6.96	4.2	4.19	7.38	5.99	10.1	0.67	0.63	3.44	8.78

forest from 6.50 to 9.50 within Temperate Ecoregions. Remember the positive correspondence with metastability, reducing the arboriculture to low BTC (e.g., 3.0–4.0) because of its reduced capacity to increase and maintain the order condition of the bionomic system, not being a type of wood or forest from a bionomic point of view.

Each one of the EEA forest typologies (Barbati et al. 2018; EEA 2006) (Table 5.7) has been estimated with the most frequent range of BTC (Mcal/m²/year) resulted from surveys by Ingegnoli and Ingegnoli & Giglio (Ingegnoli 2002; Ingegnoli 2015; Ingegnoli and Giglio 2005). So, the distribution of the diverse typologies of forest permits to evaluate the examined European countries.

Here is reported the example of France, the most representative of European nations, with 12 forest types. The land cover quantification based on Corine Land Cover Inventory (EEA 2012) initiated in 1985 (the reference year 1990); updates have been produced in 2000, 2006, 2012, and 2018. It consists of an inventory of land cover in 44 classes. CLC uses a Minimum Mapping Unit (MMU) of 25 hectares (ha) for areal phenomena and a minimum width of 100 m for linear phenomena.

The eight main components of France's territory were detailed in other 2–4 classes, then elaborated in Table 5.6. It shows the 20 elements for which it is possible to synthesize both the human habitat and the BTC per element (HH' and BTC'), whose weighted mean gives the values of HH = 52.37 (land %) and BTC = 2.39 Mcal/m²/years. We will see that these values are very similar to the HH and BTC of Italy, but decidedly different from Belgium or Denmark for example.

Table 5.6 Distribution of forest types in France (2012) with related values of HH and BTC

Forest types	Land use	%	HH'	HH	BTC'	BTC
2	Hemiboreal conifers and mixed	0.88	3	0.0264	8	0.0704
3	Alpine-boreal	1.99	2	0.0398	7	0.1393
4	Acidophilous oak and oak-birch	1.33	4	0.0532	6.8	0.09044
5	Mesophytic deciduous	5.31	4	0.2124	7.5	0.39,825
6	Beech forest	1.55	5	0.0775	7	0.1085
7	Montanious beech	1.10	4	0.044	6.7	0.0737
8	Thermopilous deciduous	3.01	5	0.1505	6.6	0.19,866
9	Broadleaved evergreen	0.88	5	0.044	7	0.0616
10	Mediterranean coniferous	2.20	6	0.132	5.9	0.1298
12	Floodplain forest	0.22	6	0.0132	7.5	0.0165
13	Non riverin alder, aspen, birch	0.44	5	0.022	7	0.0308
14	Plantations woods	3.32	15	0.498	3.2	0.10,624
	Other shrubs/wooded	7.08	9	0.6372	3.7	0.26,196
	Arable lands	14.20	88	12.496	0.78	0.11,076
	Orchard, vine, hedgerow, etc.	15.90	90	14.31	2.3	0.3657
	Grasslands and pastures	29.00	60	17.4	0.65	0.1885
	Wetlands and waters	3.00	25	0.75	0.6	0.018
	Barren soils, rocks, and glaciers	3.00	2	0.06	0.01	0.0003
	Urban residential (+ urb. Green)	4.00	98	3.92	0.47	0.0188
	Industrial, traffic, mines, etc.	1.60	99	1.485	0.01	0.00015
	Tot	100.0		**52.37**		**2.39**

5.3.2 Cancer Incidences in Europe and Correlation with Carcinogens Parameters

Data on Cancer incidence derived from a study published by the European Journal of Cancer, titled "The European Cancer Observatory: A New Data Resource" (Steliarova-Foucher et al. 2014). The European nations present a wide variation in Cancer incidence, from 192/100.000 of Greece to 454/100.000 of Denmark.

The main classes of Cancer Incidence distribution in Europe, elaborated on the European Journal of Cancer's data, 2012, resulted in a gradient W-E and partially N-S, except for the Iberian Peninsula. The comparison among cancer incidence in European countries was made correlating the three primary state bionomic functions of a territorial complex system, and the main three conventional parameters supposed to influence Cancer incidence, to which the other 5 were added. In facts, the bionomic principles underline the necessity to evaluate at least vegetation BTC, Human habitat, and carrying capacity (Table 5.7).

The leading traditional causes of cancer insurgence, exposed in Table 5.7, are as follows:

(i) Economic power (GDP) (€/capita), which leads to the growth of stress and artificialization, beef diet and fat body.
(ii) Air pollution PM_{10} (micro-grams/m^3), which leads to increase toxic/chemical inputs and inflammatory cytokines.
(iii) Pesticides in agriculture (kg/ha), which leads to Food pollution and direct toxicological agents.

Twenty-one European countries occur in Table 5.7, (15 nations and six *regions*): Austria, Bayern, Belgium, Denmark, Finland, France, Germany, Greece, Hungary, Ireland, Italy, *Latium, Lombardy, Nord-Tyrol, NR-Westphalia*, Poland, Portugal, Romania, Spain, *Trentino-Sud Tyrol*, and the United Kingdom. For each of them the values of the cited 11 parameters were calculated, but only the six most meaningful are exposed in Table 5.7. So, cancer incidence appears with them. Remember that about 4/5 of the European population, 452 million in 2012, is represented in this study.

Table 5.7 Preliminary data on environmental diseases vs. cancer incidence in some European Countries 2012

Regions and countries	Surface km² × 10³	Population P/10⁶	*Cancer* C/10⁵	Pesticides kg/ha	Carrying Cap. SH/SH*	Air pollution PM10	GDP € x capita	HH (land %)	BTC Mcal/m²/yr
Denmark	44.02	6.0	**454**	2.2	3.75	34	43.933	80.93	1.54
NR-Westphalia	34.08	17.93	**395**	3.3	0.65	32	37.500	56.70	2.16
Belgium	30.50	10.75	**388**	9.6	1.09	46	41.647	63.22	1.89
Ireland	84.4	6.1	**382**	0.6	4.98	28	46.058	52.22	0.87
United Kingdom	242.50	65.0	**371**	3.1	1.25	33	37.994	55.38	1.60
France	543.90	66.5	**370**	3.9	2.95	41	39.251	52.37	2.39
Bayern	70.36	13.4	**355**	3.5	1.84	30	44.100	57.72	2.64
Germany	357.03	81.5	**345**	3.4	1.48	42	44.266	55.74	2.74
Italy	302.07	59.5	**342**	4.7	1.82	49	35.051	51.84	2.69
Lombardy	23.88	9.8	**328**	3	0.97	50	36.500	59.08	2.02
Finland	338.4	15.79	**324**	1.9	1.11	27	40.340	15.79	3.71
Latium	17.20	5.9	**323**	4.1	1.89	30	31.500	55.40	2.77
Hungary	93.0	9.2	**319**	1.95	4.44	57	23.086	63.75	1.98
Trentino-Sud Tyrol	13.62	1.25	**317**	10.5	1.72	20	38.700	28.68	3.91
Austria	83.9	8.6	**304**	1.2	2.30	45	45.466	34.13	3.26
Nord-Tyrol	12.64	0.95	**296**	3.5	3.35	18	43.600	21.38	3.11
Poland	312.7	37.5	**284**	1.6	2.81	73	23.395	55.53	2.97
Spain	504.6	46.5	**265**	3.7	2.83	36	32.076	44.74	2.81
Portugal	90.6	10	**263**	6.1	4.54	41	26.093	42.20	3.09
Romania	238.4	18.5	**254**	1.5	5.48	47	17.859	51.14	2.57
Greece	131.9	10.4	**192**	0.85	3.41	56	25.433	39.03	2.84
Mean			*327.2*	*3.53*	*2.59*	*42.66*	*35.897*	*48.08*	*2.64*

Data from: EEA, Land Cover Country Fact Scheets 2012; Steliarova-Foucher E (2015) The EU Cancer Observatory: A new data resource. EU J of Cancer, Elsevier; elaborations from Ingegnoli (2015). HH and BTC elaborated following Bionomics methods (Ingegnoli 2015); Other data from EEA statistics

We register three groups: (1) High cancer incidence (454–365) Denmark, *NR-Westphalia*, Belgium, Ireland, United Kingdom, France; (2) Medium incidence (355–300) Bayern, Germany, Italy, *Lombardy*, Finland, *Latium*, Hungary, *Trentino-Sud Tyrol*, Austria; and (3) Low incidence (296–192) *Nord-Tyrol*, Poland, Spain, Portugal, Romania, and Greece.

If we compare the values of Denmark, which has the max cancer incidence, with other parameters, the result is surprising: pesticides are inferior to the mean value (kg/ha = 2.20 < 3.53), Carrying capacity SH/SH* = 3.75 > 2.59, air pollution PM_{10} = 34 micro-g < 42.6. Only economic/technologic power (GDP = 43,933 > 35,897) and still more human habitat and BTC result sharply different from their averages (HH = 80.9 > 48.08 and BTC = 1.54 < 2.64 Mcal/m²/year). Anyway, Germany presents an economic power bigger than Denmark (+ 333€/capita), but a cancer incidence equal to 345 vs. 454!

5.3.3 Cancer Incidences in Europe and Correlation with Carcinogens Attributes

The primary *diagnostic model* at the local (LU) scale is the BTC/HH one (Ingegnoli 2015), available from 10 to 10^4 km²: a similar model (but with a different trend) refers to broader scales, from 10^4 to 10^6 km² (regional scale; see Fig. 3.9): at the country scale, we have to use the last one. Applied to Europe (Fig. 5.15), this model presents four sections, in which four types of countries are listed as follows:

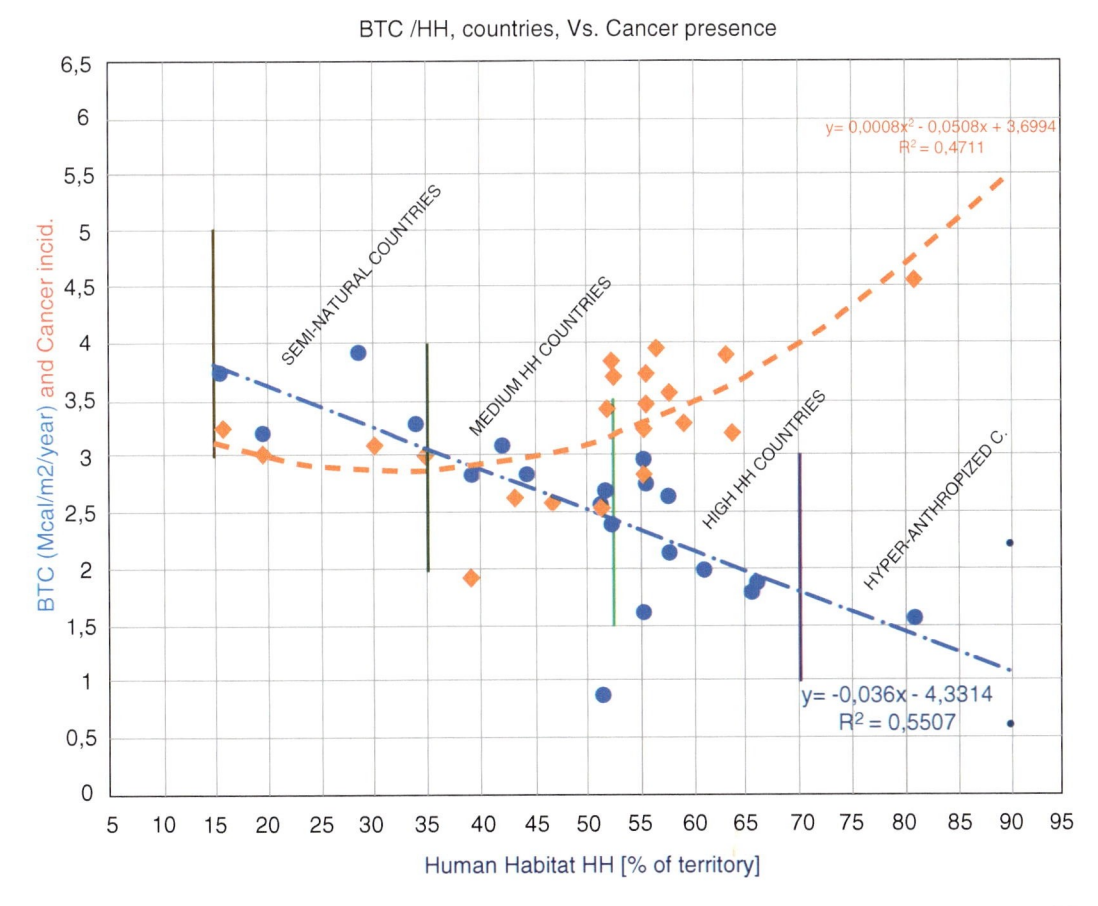

Fig. 5.15 Countries and related data of Table 5.7 are distributed in four sections, depending on their HH level. The orange line represents the trend line of the incidence of cancer vs. the HH; the blue line represents the trend line and the situation of each country in the diagnostic model HH/BTC. The decrease of BTC and the increase of HH present an inverse correlation with cancer incidence/100.000

1. semi-natural (e.g., Finland),
2. with medium HH values (e.g., Spain),
3. with high HH values (e.g., Germany),
4. Hyper-anthropized (e.g., Denmark).

The orange curve represents cancer incidence, which shows a robust correlation with the increase of HH and BTC's decrease.

Note that at a regional scale, HH has severe limits, because over HH = 70% the HH values become typical of local scale (see Fig. 3.9 left). So, hyper-anthropization is an environmental syndrome, as values like Denmark' one, because the level of HH is decidedly out-of-scale for a country of 44,000 km^2: for example, the presence of Danish forests is pair to 0.28% at European scale (Table 5.6), but only 40% are real forest! Consequently, at a regional scale, the most functional-systemic parameters are not BF but BTC + BF.

The exposed results indicate a real correlation between cancer incidence and BTC/HH bionomics functions, while the other parameters seem to be more marginal. The problem is that people have been induced by doctors to correlate cancer incidence with (a) nutrition factors (e.g. too much butter and or meat); (b) pesticides and or chemical additives in agriculture (so in water, etc.); (c) air pollution, especially PM$_{10}$ (but also PM$_{2.5}$, etc.); (d) geographical density, too high people density, etc.; (e) economic power per capita, so too much technology, energy use, artificial behavior, etc.; very few people (and few doctors too) may imagine the reasons by which BTC and HH have to do with cancer: the conflict between *the flux of energy needed by a living system to reach and maintain a proper level of organization and structure* (BTC) and *the measure of the human control and limitation exerted on the self-regulation of the same system (HH)*.

Figure 5.16 shows the correlation significance of BTC, HH, and Carrying Capacity (SH/SH*) with four traditional cancer incidence parameters: GDP/capita, Air pollution PM10, Physical Activities, and Pesticides. It is possible to see that only BTC and HH reach and bypass the full level of Pearson correlation significance; so, if we evaluate the capacity to influence cancer incidence putting BTC 1.27 equal to 100% of incidence, we should have these secondary influencing values: HH = 90%, GDP = 67%, PM$_{10}$ = 50%, SH/SH* = 49%, Physical Activity = 43%, pesticides = 8%. Therefore, it could be interesting to test other conventional parameters.

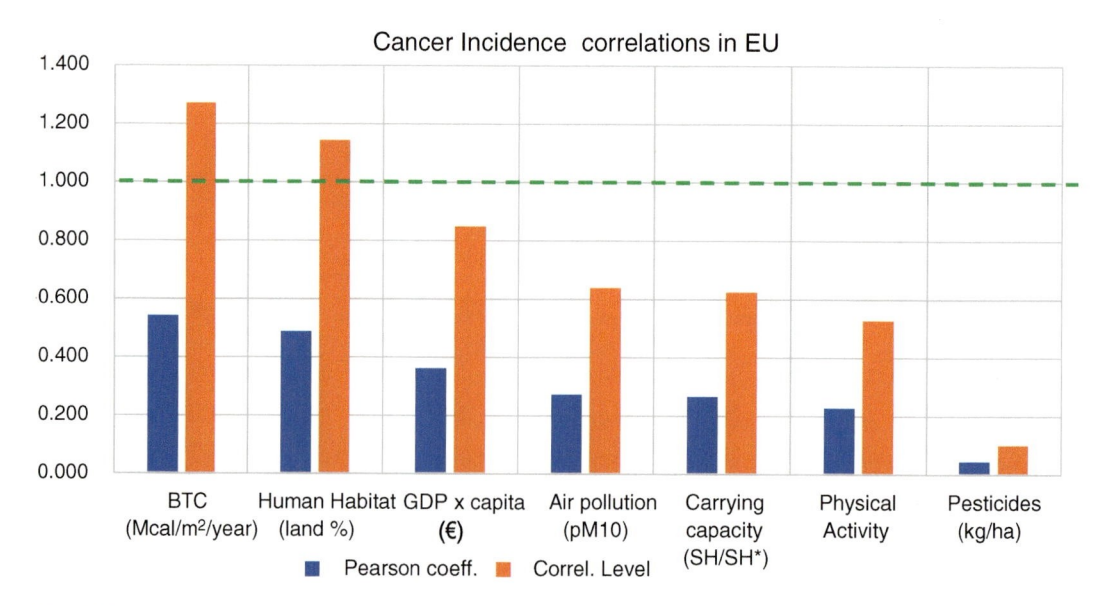

Fig. 5.16 Cancer incidence of six parameters. Minimum values ($P = 0.425$ in a set of 21 countries) of significance by the Pearson coefficient emerge when they are below 1. BTC reached 1.27 and HH 1.18

5.4 Relations among Bionomics Parameters and Brain/Environment

5.4.1 The Importance of Bionomic Parameters in Cancer Incidence

Table 5.8 summarizes the leading 12 causes of cancer insurgence, divided into (a) traditional and (b) bionomics. The first are (1) economic power, €/capita, (2) air pollution, PM_{10}, (3) smoking, cigarettes n°/capita, (4) lack of physical activity of the population, (5) median population age, (6) alcohol consumption/capita, (7) pesticides in agriculture, kg/ha, (8) lack of fruit and vegetables consumption /capita. The second are: (i) too low Vegetation BTC [Mcal/m²/year], leading to the destruction of forest and agrarian landscapes, lack phytoncides' and GMB (Gut Microbiome); (ii) too much Human Habitat HH [land %] leading to the destruction of natural habitat and vegetation infrastructures; (iii) Carrying capacity SH/SH* [autotrophy >1.5], leading to too much urbanization and more import and artificialization of food, (iv) Complex landscape unit dysfunctions (BTC + BF): we have to add psychosomatic alterations and the insurgence of stress alarm, as affirmed and underlined in previous pages.

Each parameter is presented with the correlation significance, following the Pearson correlation coefficient and its value of significance related to the sample size. Note that the butter consumes per capita in Europe, typical of a non-Mediterranean diet, was not considered among the other parameters, because the primary consumer is France 8.2 kg/capita vs. 6.4 in Denmark, 4.5 in Poland, 2.0 in Italy: all values not congruent with cancer incidences. Meat consumption per capita (from about 40 to 80 kg/year) brought to similar incongruence. Smoking cigarettes is a

Table 5.8 Main causes of cancer insurgence: conventional and bionomics

Corr %	Conventional parameters	Main causes of cancer insurgence	Corr %	Bionomic parameters	Main causes of cancer insurgence
85	Economic power (GDP) (€/capita)	Growth of stress and artificialization and beef diet	127	Too low Vegetation BTC [Mcal/m²/year]	Destruction of forest & agrarian landscapes, lack of phytoncides' and GMB
65	Air pollution PM_{10} (micro-grams/m³)	Pollution and inflammatory citokines	114	Complex landscape unit dysfunctions (BTC + BF)	Heavy and chronic environmental stress, Th2 dominance…
65 (?)	Smoking cigarettes	Lung pollution and chemicals (e.g. benzo(a) pyrene damages DNA	112	Too much Human Habitat HH [land %]	Destruction of natural habitat and vegetation infrastructures
53	Lack of physical activities of population	Maintaining a healthy weight and decrease stress and depression	62	Carrying capacity SH/SH* [autotrophy] >1.5]	Too much urbanization and more import and artificialization of food
27	Median population age	Cell degeneration and immune system deficiency growing with time			
24	Alcohol consumption	Acetaldehyde damages DNA and prevents body from repairing it			
16	Pesticides in agriculture (kg/ha);	Food pollution and direct toxicological agents			
8	Lack of fruit and vegetable consumption	Lack of polyphenols and vitamins			

Corr.% = correlation significance. (?) Smoking is dangerous, especially for lung cancer; but its correlation with cancer incidence is ambiguous: e.g. Denmark has 12.3% of smoking people Vs. 27.0% of Greece, when their cancer incidence is 454 vs. 192/100.000… [data Eurostat]

dangerous cancerogenic parameter, as all air pollution, which correlation significance results in about 64–66%, especially for lung cancer, but the direct cigarette correlation with cancer incidence is ambiguous: e.g., Denmark has 12.3% of smoking people vs. 27.0% of Greece, when their cancer incidence is, respectively, 454 vs. 192/100.000… (Eurostat 2018).

We underline that only 8/12 parameters result in >50% of significance, and the most important are the three bionomics: BTC, BTC + BF, HH, all >100%. To understand the reasons of the importance of bionomics, we must remember what sustained on etiology in Sect. 5.1.2 about environmental stress, gut microbiome, phytoncides, and in Chap. 3, Sect. 3.4 on Hierarchical relationships and epigenetics.

As already discussed in Sect. 5.1.2, etiological processes due mainly to *"Too Much,"* that is an excess of abundancy in a human body are the most studied: (a) *Pollution agents* and (b) *Infective agents*, (c) *fat/sweet foods*. More recently, also etiology due to (d) *Stress agents* begins to be studied, especially after the foundation of the new medical discipline of PNEI (Psycho-Neuro-Endocrine-Immunology), while etiological processes mainly due to *"Not Enough,"* i.e. (e) *Lack of Defense Agents* and (f) *Lack of Hierarchical Relationships* in a human body are the less studied. Only recently, in section (e) a steady increase of researches is made on "Gut Microbiome," but still few studies have been done on the "Phytoncides" (VOC from vegetation related to the immune system), as in the entire section (f), which premises are presented in Chap. 3, Sect. 3.3.4 and Sect. 3.3.5.

We underlined that life organization can be described in *hierarchic levels*: this implies the necessity to interrelate all the levels of biological organization, thus confirming the bionomic principles. However, only after the acquisition of the importance of *epigenetics,* these interrelations began to be possible. Differences between the old view (DNA, RNA, and Protein) and the *modulation* of information linked to the environment (Signaling Molecules, Transcription Factors, DNA, RNA, and Protein) are crucial. The new concept of life, at the base of bionomics principles, not limited to a single organism or a group of species, brings to the *linkage* due to epigenetic processes to be very significant. Moreover, epigenetics allows a braiding of interrelations with other new disciplines, particularly in environmental fields (Fig. 3.14). Food, health, and environment are strictly linked with gene expression, but so it is also for the bionomic state of landscape units, especially where man lives.

Boundaries, which are not only the physical ones, separate the set of processes from components in the rest of the system. One of the most important consequences of the hierarchical structure of a system is the concept of *constraint*. It is correct more than the non-systemic concept of "limiting factor," and it shows the behavior of an ecological system as limited: (a) by the behaviors of its components, which *explain the origin* of the level of interest (i.e., allow an inner description) and (b) by environmental bonds imposed by superior levels of the organization, which *explains the significance* of the level of interest (i.e., allowing the characters to be derived from transferred conditions to be understood). Remember: there is a linkage between constraint and information.

As we will see forward ahead, our brains are designed to enhance their fitting with the environment and social behaviors (Fig. 5.17); a direct relationship between psychic stress and inflammation is shown (Fig. 5.2). Figure 5.18 shows the main three Feedback Interactions of landscape pathologies and human health. Figure 5.19 shows biochemical paths following them chronic stress increases the risk of cancer onset and Fig. 5.21 shows (in synthesis) the complex but real interactions between the environmental conditions and

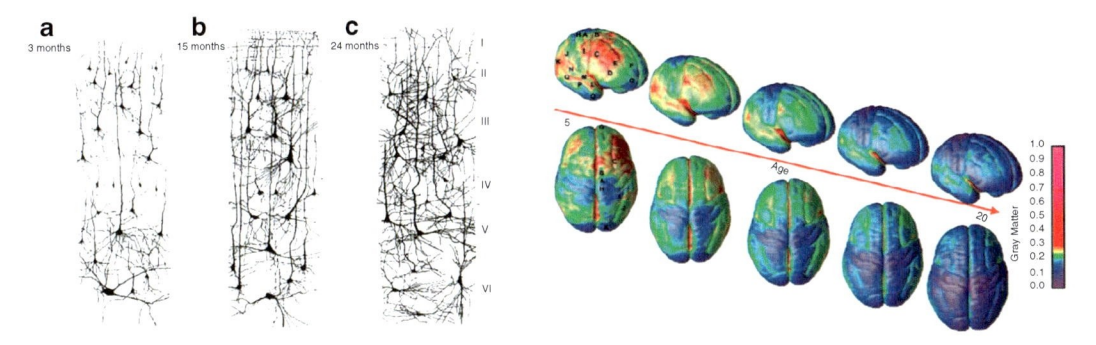

Fig. 5.17 (left) Cerebral cortex maturation in a baby is the period of exceptional plasticity of the brain structure due to necessary fitting with the environment (Ramon y Cajal 1909). (right) At right, the well-known representation of the "pruning effect" in the human brain (Gogtay et al. 2009)

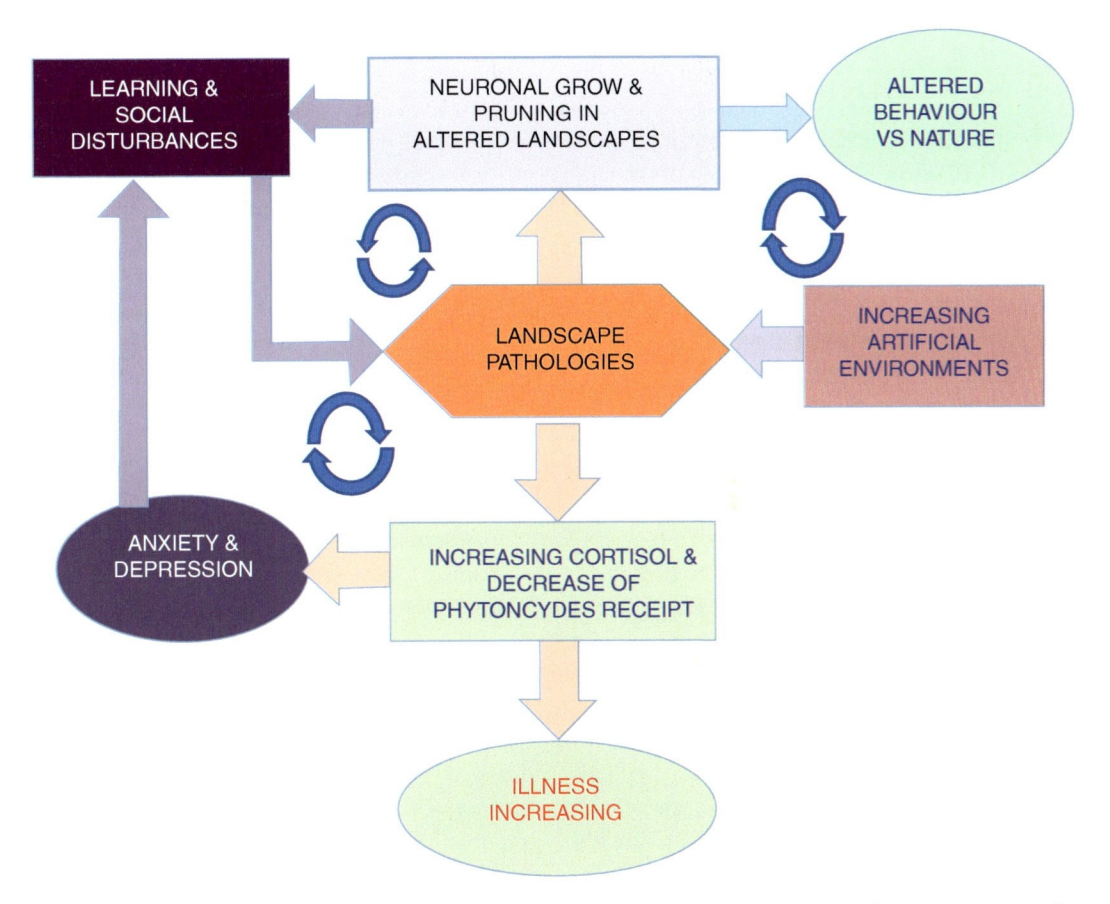

Fig. 5.18 The landscape pathologies, influencing the behavior vs. Nature, increase artificial environment and produce exalted feedback, which in turn generates other exalted feedbacks. Note the formation of at least three self-exerting feedback loops, which can aggravate the landscape syndromes, therefore the consequences on illness

Cancer incidence. Therefore, we can affirm that body reaction to bionomic-ecological parameters:

(a) are not substantially different from what happens in traditional etiology, e.g., psychic stress/inflammation, pollution, and agrofood problems and,
(b) can be more important than other parameters for the reliable survival power given by the just mentioned processes and cumulative impacts.

5.4.2 Complex Relationships Brain-Environment

The bionomic alteration of a Landscape Unit (LU), even independently of pollution, is a severe environmental degradation. Therefore, it becomes necessary to deepen the investigation of the etiopathogenesis of environmental stress and its consequences again.

We must premise that during the first growth of superior animals and still more of man's, cerebral cortex's components mature at different times (Fig. 5.17, left). Medicine doctors knew the possible existence of a process like this since 1909 (Ramon y Cajal 1909) when Ramon j Cajal compared the development of the neuron of the brain cortex of the babies from 3 to 24 months. For decades neuroscientists have been believing that neural pruning ended few years after birth. However, in 1979 the late Huttenlocher (Huttenlocher 1979), a neurologist at the University of Chicago, demonstrated that this excess production and pruning strategy have been continuing for synapses long after birth.

The consequences of this process could be very notable. In fact, since 1999, Ingegnoli had been speculating a possible link between cerebral maturation in an altered environment and the increase of artificialization and environmental degradation (Gogtay et al. 2009). A confirm of this hypothesis arrived in the study "Function and Dysfunction of Microglia during Brain Development: Consequences for Synapses and Neural Circuits" (Paolicelli and Ferretti 2017). Environmental factors, such as infections, stress, and dietary intake, are associated with synaptic dysfunction and are linked to increased risk for neurodevelopmental/psychiatric disorders. Importantly, all of the above factors activated the immune system during early brain development.

As well expressed by Gogtay et al. (Gogtay et al. 2009) between 5 and 20 years of age there is a decrease in the volume of gray matter, due to the so-called pruning effect, i.e., a selective cut of neuronal synapses to achieve the best "fitting" with the environment (Fig. 5.17, right).

Pay attention: if a young person grows up in a degraded environment, his adaptation to the altered environment can lead him to *aggravate* the stress conditions repeatedly mentioned. There are three negative self-enhancing feedbacks (Fig. 5.18):

1. the pathologies of the landscape units (LU), or environmental syndromes, can influence the "pruning" process, leading to incorrect behavior toward nature; this allows you to pay no attention to the artificiality of the environment, which—by continually increasing—worsen the LU syndromes;
2. these syndromes also lead to increased "learning and social behavior disorders" (see industrial suburbs) which are aggravated by the "pruning" effect, which in turn leads to worsening pathologies of LU;
3. the worsening of LU diseases and their persistence, the alteration of the bionomic and ecological state, increase stress and its consequences (cortisol), therefore increases morbidity and the risk factor for premature mortality.

Growing in an altered environment from birth leads man to lose sensitivity toward nature. As expressed in Fig. 5.1, an emotional activation indicated by an increase in signal in dACC, *Anterior Cingulate Cortex,* correlates in the graph with an increase in the plasma levels of the soluble TNF-alpha type II receptor, a basic inflammatory cytokine. This process shows a direct relationship between psychic stress and inflammation (Ramon y Cajal 1909).

Moreover, epigenetics allowed a braiding of interrelations with other new disciplines, particularly in environmental fields (Fig. 3.13). Food, health, and environment, are strictly linked with gene expression, and the bionomic state of landscape units, especially where man lives. The consequences are evident following what happens in our vast metropolitan areas, where artificial environments seem to attract more and more young people and submit urban green to technology in a sort of new Babel (see also Fig. 5.22).

We affirmed in Sect. 3.2 that landscapes present a modality of transformation led by environmental laws, which may cause a change in the *culture* and the *ethology* of man to maintain a stable equilibrium, inducing a buffer effect, when landscapes suffer a heavy changing pressure. The cultural change of the shape of gardens during the Industrial Revolution followed this unconscious modality. In Sect. 5.1, we underlined that the *psychic* inputs are recently considered in a more systemic vision, passing from an agent of diseases due to unconscious conflicts (sensu Freud) to the psychosomatic medicine, mostly depending on physiopathology. All these observations are convergent; however, we need to refer also to Environmental Ethics.

5.5 Etiology of Cancer Onset Due to Environmental Alterations

5.5.1 Cancer, as a Systemic Multifactorial Disease, and Stress

As affirmed in Chap. 7, by Caterina La Porta, most cancers are due to sporadic mutations happening during the lifetime. The etiology of the disease can involve specific predispositions such as a family history of cancer but appears mainly related in a complex way with many environmental factors from smoking to sun and radiation exposure to chemicals, viruses, bacteria, alcohol, overweight, lack of physical activity, and poor diet. All these factors work in a complex way alone or in combination with our genes and affect epigenetic regulation, leading at the end to a heterogeneous phenotypic tumor.

Recent research made by Pelicci, Dellino et al. (Dellino et al. 2019), gave to sporadic or random mutations less importance.

Bertolaso (Chap. 1) notes that a typical epistemic operation in life sciences coincides with an ontological *reductionist assumption* for which activities belong to fundamental unities or entities. For instance, in this sense, tumor cells have and retain the properties to proliferate abnormally, to avoid apoptosis, etc. *independently* of other (micro)environmental factors. This automy is well represented also by the paradox that emerges in *cancer research* when accounting for carcinogenesis based on DNA mutations, where quite often the *same genes* are involved in oncogenetic or tumor *reversion* processes or, even, seem to *follow* more than anticipate, the neoplastic transformation of the cells.

Bottaccioli (Bottaccioli and Bottaccioli 2017) converge with these colleagues remembering that the metilation of onco-suppressor genes impede the production of proteins having an anti-proliferative role (e.g., p53 or p16) in colon cancer or leukemia. Cancer is a pathology presenting multiple factors and crossing various phases. So, genes and epigenetics, environment, diet, and behavior play essential roles in carcinogenesis and, as underlined, chronic stress has to be enhanced.

Figure 5.19 (elaborated from Bottaccioli and Bottaccioli 2017) shows biochemical paths following which chronic environmental stress increases the risk of cancer onset. Noradrenaline, Adrenaline induce the formation of VEGF (growth factor of vascular endothelium) and metalloproteinases (MMP), which increase angiogenesis, helping tumor proliferation, facilitated by the immune deregulation due to cortisol, which leads to the weakening of TH1. This last condition brings to surveillance failure and an increase of inflammation. Stress is also linked with telomere length reduction, increasing the enzyme telomerase, which can stop the telomer reduction blocking the apoptosis.

This logic flow chart on the relations stress/cancer is significant, but if you will try to deepen our systemic linkages, it is necessary to remember the last section of Chap. 4, Fig. 4.18. Note that in this figure are exposed two processes, one following the neurological path and the other the hormonal one, with ulterior subdivisions and interactions: each section may have a more profound etiological development. Each section may present a cumulative impact, and a complex combined, but a complex, combined and cumulative impact, together with interferences, are possible. This situation indicates that in many cases, it becomes challenging to arrive at an understandable etiology.

However, we have to clarify these processes as far as possible if the damage to human health is to be addressed and contained; therefore, we present a fuller frame (Fig. 5.20), in which the previous Fig. 5.19 can be inserted.

Figure 5.20 shows that the relation *man/environment* via compared ethology and landscape bionomics presents two different aspects enhanced by the value judgment (order vs. disor-

Fig. 5.19 Biochemical paths with whom chronic stress increases the cancer risk. Noradrenaline, Adrenaline, and Cortisol induce the formation of VEGF (growth factor of vascular endothelium), and metalloproteinases (MMP), which help the tumor proliferation, facilitated by the immune deregulation with the weakening of TH1 (from Bottaccioli & Bottaccioli (Bottaccioli and Bottaccioli 2017); modified and re-plotted)

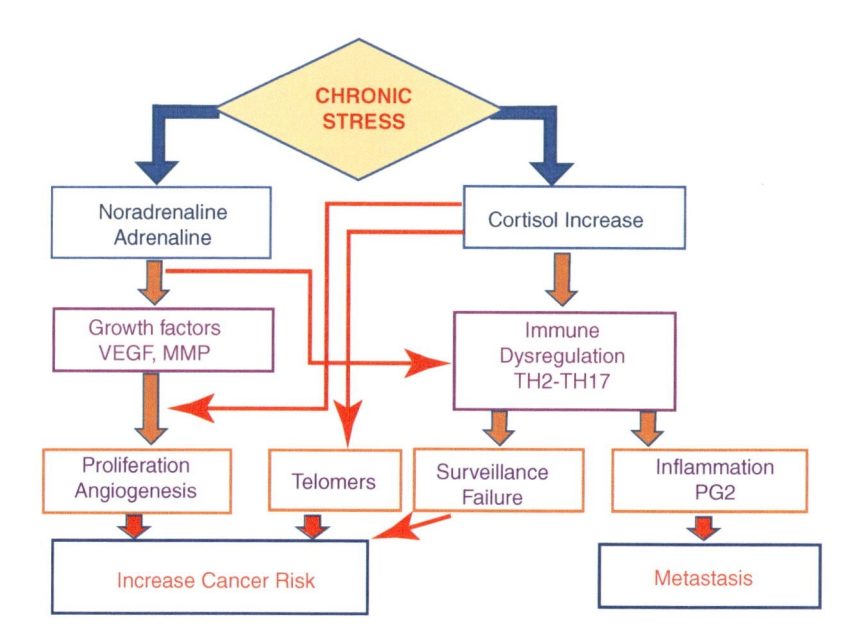

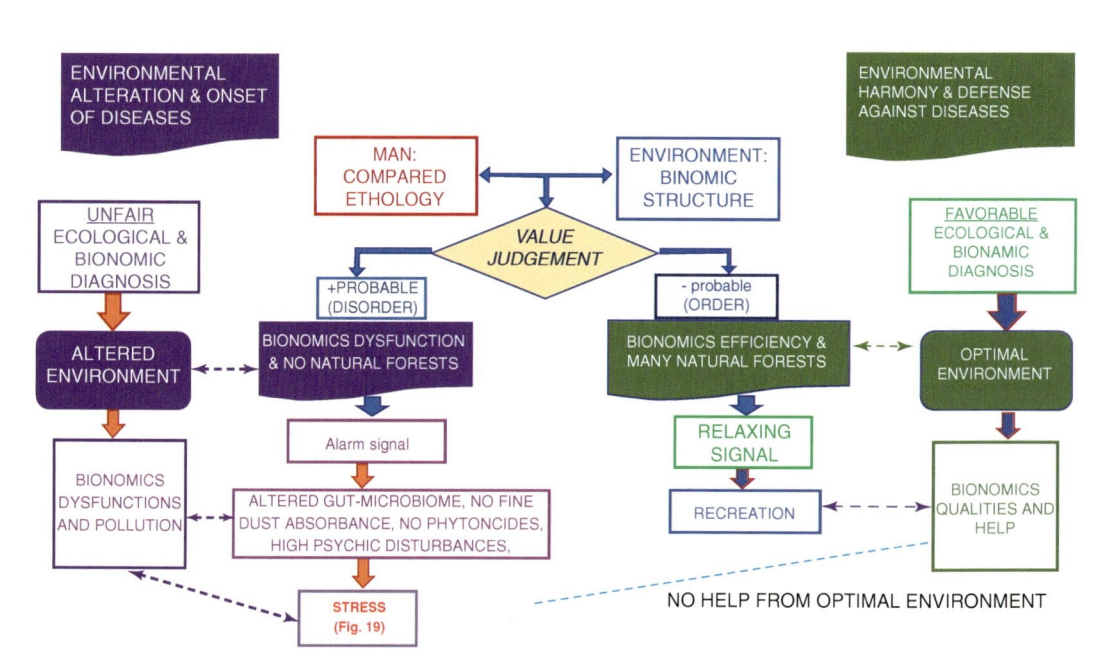

Fig. 5.20 A logic flow chart for a full-frame of processes like Stress/Cancer. Note that we have to consider two different situations related to the environment: (violet) nature alteration and cancer onset, (green) harmony with nature and defense against cancer. Even if the second is becoming rare, we must consider it, because prevention is concerned with the rehabilitation (green)

der) and these two are linked with two more extensive fields: (a) environmental alteration and cancer onset, violet, (b) harmony with nature and defense against cancer, green.

Even if (b) is becoming rare, we must consider it, because prevention is concerned with rehabilitation and therapy. For instance, the increase of polyphenols in wine depends on the entire bionomic condition of the environment. Both altered environment and optimal one lead to sequences of processes concerned with our health interacting with stress and recreation and linked to landscape pathologies and the pruning effect previously mentioned. It is necessary to enlarge these fields.

5.5.2 Trying to Deepen the Logic Flow Chart of Cancer Onset and Prevention Again

Beginning with the environmental alteration and the emergence of cancer (left part of the flow chart, Fig. 5.21), we find a violet frame containing four agents of modification:

(a) Stressing roads, labor, and home environment,
(b) Industrial agriculture and poor feeding,
(c) No natural forests and poor phytoncides,
(d) Presence of pollution.

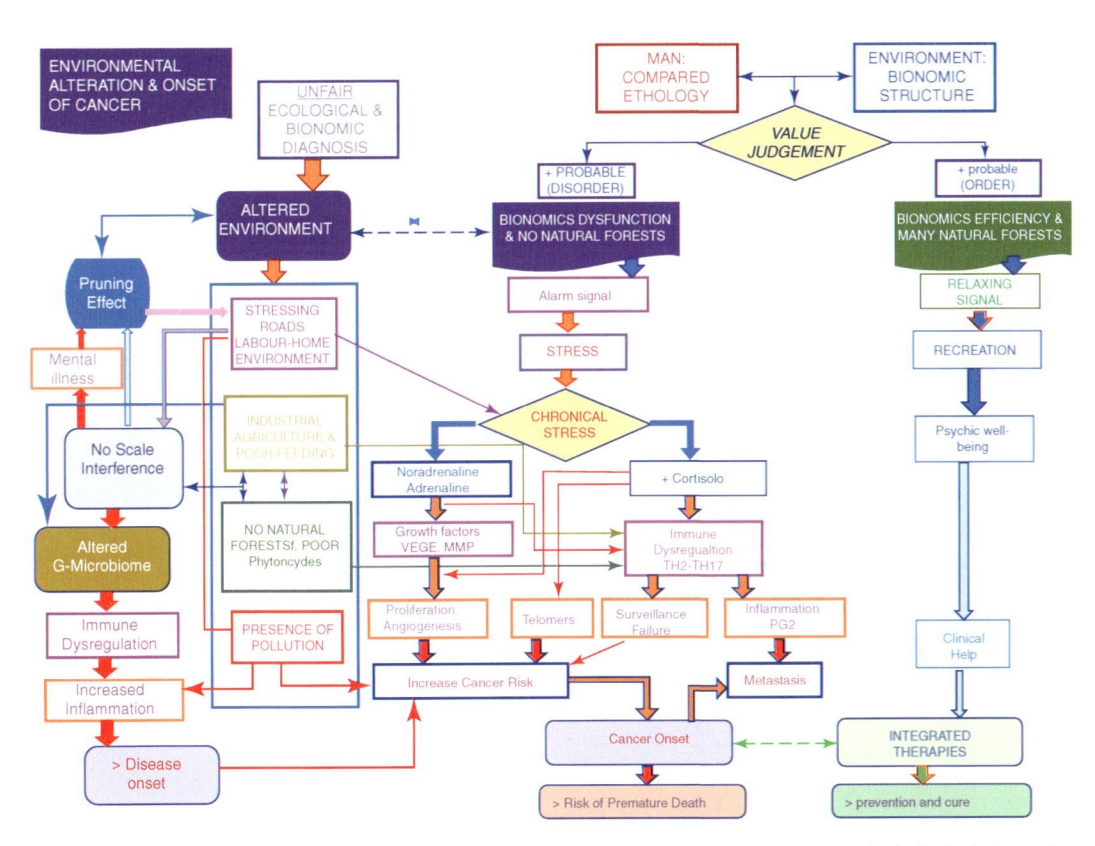

Fig. 5.21 An inauspicious diagnosis of the syndromes of a landscape unit (LU) shows various possible pathways of morbidity: e.g., (1) Chronic stress (yellow), (2) presence of pollution (red), (3) lack of defensive contributions (green), (4) lack of scale interference (blue-gray). Complex interrelationships and convergences exacerbate health changes. The emergence of ethological alarm signals, capable of activating the adrenaline and hormonal processes, can be reinforced by feedback due to the neurocerebral "pruning" effect, which leads to an increase in technology in anthropized environments

These agents present many linkages with surrounding other agents and processes, derived from environmental stress (to the right). For instance, left: (e) the Pruning effect, strictly linked with mental fitness, (f) no scale interference, related to the destruction of agrarian landscapes and inadequate food and also to (g) altered gut microbiome, leading to immune deregulation and increased inflammation, so to major diseases onset and cancer risk (Fig. 5.21).

Complex linkages may characterize some processes of carcinogenesis due to environmental alterations. Let us consider the agrarian landscape syndromes: the dominance of industrialization in agriculture destroys the landscapes and produces mediocre food, and the lack of phytoncides reinforces these alterations. This process influences the chronic stress sequences, but also it blocks or alters the scale interrelations, which influence mental illness and the pruning effect, cutting the sensibility toward nature and enhancing the artificialization of the environment. Simultaneously, the alteration of the gut microbiome leads to the depression of the immune regulation, which can worsen the analogous immune deregulation due to chronic stress (cortisol increase, Th2 dominance, etc.). Moreover, the presence of pollution depending on pesticides, which itself is not among the most important agents of carcinogenesis, could reinforce the other mentioned factors. A similar effect may be done by the presence in the agrarian landscape of an extraordinary road and its disturbances and pollutions. Finally, the risk of premature death is increasing.

On the other side, the possibility to evaluate the bionomic state of the landscape units and consequently to correlate its bionomic functionality (BF) with the mortality rate give the possibility to control the environmental syndromes and to reduce the impacts of transformation and to advise the local Authorities for the necessity to environmental rehabilitation. These bionomics consulting acquire a specific function of health protection and prevention. We can also add clinical functions brought by harmonic natural LU, related to integrated therapies, as proposed by Psycho-Neuro-Endocrine-Immunology (Bottaccioli and Bottaccioli 2017).

That is why a complete frame of the primary relations on altered environment/reduced health has to be presented (Fig. 5.22), with the addition of a healthy natural environment. The right part of this broad logic flow chart shows the interrelations in a semi-specular way with the alarming signals. Thus, the relaxing signals, leading to psychic well-being, may develop favorable biochemical processes. For instance:

(a) decreasing inflammation enhancing IL-10, an immunomodulatory cytokine with a critical role in limiting inflammation in immune-mediated pathologies, or,
(b) increasing IFN-γ, which activates macrophages and makes them better able to mount an effective *immune* response,
(c) enhancing antigen processing and presentation through upregulation of class II MHC, increased ROS and NOS production,
(d) inducing autophagy for clearance of intracellular pathogens etc. (remember Fig. 5.2).

Going on with the environmental rehabilitation and the prevention of cancer (right part of the flow chart, Fig. 5.22), we find again a frame containing four agents of healthy state:

(a) Right roads, labor, and home environment,
(b) Agroecology and well-feeding,
(c) Natural mature forests and good phytoncides,
(d) *Absence of pollution.*

The healthy agents are three because of the absence of pollution. Agroecology is particularly essential: plant populations evolve under different agronomic management types and in the face of specific combinations of LU characters (from diseases to insects, weeds, drought, etc.) In this way, the frequency of genotypes that have adapted to local conditions gradually increases (Ceccarelli 2013). Epigenetic processes lead this EPB; local environment characters guide DNA expressions and make them transmitted. These populations do not need to be treated with pesticides and herbicides. They are deeply linked with natural forest patches, of which we have to

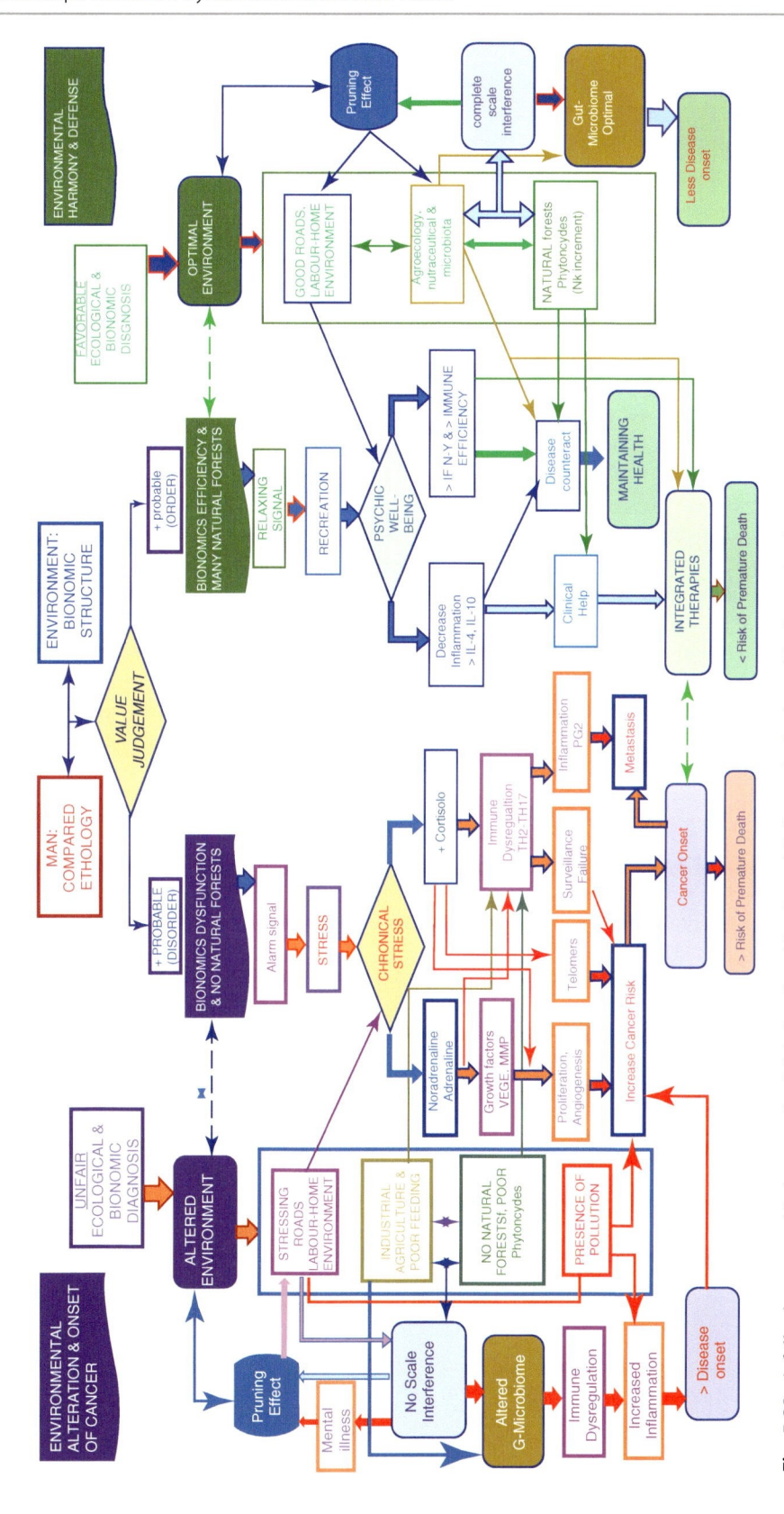

5 Landscape Bionomics Dysfunctions and Human Health

Fig. 5.22 A full scheme, considering negative (violet) and positive (green) linkages. In the favorable diagnosis of LU, the morbidity and cancer onset present lower risk, and their defensive contributions may help to develop integrated therapies

remember Fig. 5.5: Phytoncides (forest essential oils) have beneficial effects on natural killers (NK).

In this positive case, the role of the neural Pruning Effect stabilizes, leading to developing man's sensibility toward nature in the correct direction.

5.6 Conclusion: Environmental Ethic Implications

Globalization is a changing culture; it is evident to everyone, but today, two opposing cultural positions are facing each other: the era of the *Anthropocene* and that of the *Ecozoic*. A vision, Anthropocene, based on utilitarianism, individualism, competition, driven anthropocentrism, artificialization, little or no respect for Mother Earth, vs. Ecozoic, a term coined by Thomas Berry of Columbia University (Berne and Levy 1990), on the other hand, attempts to align human activities with other forces and with the laws that govern the Earth to achieve a creative and protective balance.

The first vision, still decidedly dominant today, might seem rational: it is not. The race for enrichment ignores spirituality and considers ecology as dependent on the economy, adoring the *Progress*. Gandhi's disciple Lanza del Vasto had questioned man precisely on this idolatry's domain for progress. He replied: Progress or haste makes it necessary more haste, overabundance more overabundance, vanity more vanity, violence more violence: everything accelerates, enlarging like an avalanche. The avalanche must meet a rock to shatter, a village to take away, and, at its peak, it stifles all life (Lanza del Vasto 1978). Progress underlines the mentioned exalting feedbacks. The second vision is not that of the Greens; instead, the *Ecozoic* vision has as its progenitor Saint Francis, who said: "Every day we use creatures and without them, we cannot live, and in them, the human race greatly offends the Creator" (*Legenda Antiqua Sancti Francisci*) and then He left to nature a part of the vegetable gardens (Augusta and Perugia 1244).

Let us underline that the cultural change of the shape of gardens during the Industrial Revolution followed this unconscious modality led by Mother Earth when the majority of Western People maintained a great sensibility toward nature. However, today too many people live in an altered environment. Under the Anthropocene culture and the neural "pruning effect," no doubt may influence their behavior to technology. This fact is so relevant today that many bioethics scientists and epistemologists affirm that the Technological World is the authentic environment of man (Agazzi 2004, 2014).

Most parts of people prefer to live in broad metropolitan areas and megalopolis instead of in the rural countryside or semi-natural landscape (Fig. 5.23, Dx): on a global scale, from 30% in 1950, the urbanization increased to 55% in 2017.

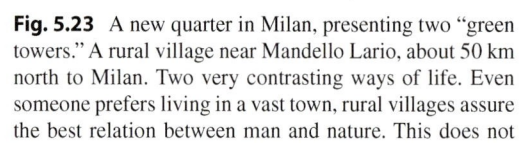

Fig. 5.23 A new quarter in Milan, presenting two "green towers." A rural village near Mandello Lario, about 50 km north to Milan. Two very contrasting ways of life. Even someone prefers living in a vast town, rural villages assure the best relation between man and nature. This does not mean that all the inhabitants of the towns must relocate, turning small villages and the country into a chaotic rural/sub-urban region! On the contrary, towns and cities must be rehabilitated following landscape bionomic principles

Most parts of people assert that agrarian industrialization and biotechnology represent the best way to produce our food. However, even many people who may admit the ecological crisis continue to follow technology! Many architects, for instance, propose "vertical woods" in urban landscapes, covering the prospects of tower-buildings with plants or even "green walls" (Fig. 5.23, left), employing sophisticated technologies and much energy; being convinced to act as ecologists!

This chapter aims to deepen the impact of human alteration of Earth's ecological systems on the health of humanity. Human activities drive fundamental biophysical change at the steepest rates in our specie's history, leading to dangerous health effects. With the help of advanced ecological principles capable of dialog with systemic medicine, this chapter tries to demonstrate the importance of systemic medicine in studying the effects of ecological alterations on human health.

Pay attention: it must be clear that the Technological World is not the real environment of man. The main reasons are as follows:

(i) Technology is not able to solve all the problems, as the present pandemic of COVID-19 is demonstrating;
(ii) Landscape bionomics has demonstrated the complementarity of human and natural habitats;
(iii) It is possible to prove the effective relations between environmental syndromes and human pathologies.

All these reasons, and other more, can be shown following Ecozoic vision and bionomics principles.

References

Agazzi E (2004) Right, wrong and science. The ethical dimension of the techno-scientific enterprise. Rodopi, Amsterdam-New York
Agazzi E (2014) Science, metaphysics, religion. F. Angeli, Milano
Allen TFH, Hoekstra TW (1992) Toward a unified ecology. Columbia University Press, New York
Augusta B, Perugia (1244) Collezione di testimonianze scritte su san Francesco di Assisi fatte nel 1244 dal Generale dell'Ordine Francescano. Co, pilatio Perusina (Latin)
Baccarelli, Zanobetti, Martinelli et al (2008) Exposure to particulate air pèollution and risk of deep vein thrombosis. Arch Intern Med:920–927
Barbati et al (2018) European forest types: toward an automated classification. Ann For Sci:75. https://doi.org/10.1007/s13595-017-0674-6
Berne RM, Levy MN (1990) Principles of physiology. The CV Mosby Company, USA
Bird A (2013) Epigenetics. Instant Expert No.29 in: New Scientist, No.2898
Bottaccioli F, Bottaccioli AG (2017) Psico-Neuro-Endocrino-Immunologia e Scienza della Cura. Edra Ed, Milano
Carey N (2012) The epigenetics revolution: how modern biology is rewriting our understanding of genetics, disease and inheritance. Iron Books Publ
Ceccarelli S (2013) Produrre i propri semi. Manuale per accrescere la biodiversità e l'autonomia nella coltivazione delle piante alimentari. Libreria Editrice Fiorentina, Firenze
Dellino GI, Palluzzi F, Piccioni R, Chiariello AM, Bianco S, Furia L, De Conti G, Bouwman B, Melloni G, Guido D, Giaco L, Luzi L, Cittaro D, Faretta M, Nicodemi M, Crosetto N, Pelicci PG (2019) Release of stalled RNA-Polymerase II at specific loci and chromatin domains favors spontaneous DNA double strand breaks formation and predicts cancer translocations. Nat Genet 51:1011
Dusaf-ESRSAF (2015) Banca dati dell'uso e copertura del suolo. Regione Lombardia, Milano
EEA (2006) Environmental statement. ISBN: 92–9167-878-3
EEA, Corine Land Cover (2012) Copernicus Land Monitoring Services
Eibl-Eibesfeldt I (2007) Human ethology. Routledge
Eurostat (2018) Are you looking for statistics on cancer? European Commission, Brussel
Forman RTT, Godron M (1986) Landscape ecology. John Wiley &Sons, New York, p XIX+619
Freud S (1923) The Ego and the Id. (the Standard Edition of the complete Psychological Works of S.F.), 1990. W.W. Norton Comp
Gilmour PS, Morrison ER, Vickers MA et al (2005) The procoagulant potential of environmental particles (PM_{10}). Occup Environ Med 62: 164–171
Gogtay N, Giedd JN, Lusk L et al (2009) Dynamic mapping of human cortical development during childhood through early adulthood. Proc Natl Acad Sci U S A 101(21):8174–8179
Hickil et al (2013) Manipulating the sleep-wake cycle and circadian rhythms to improve clinical management of major depression. BMC Med 11(1):79
Huttenlocher P (1979) Synaptic density in human frontal cortex: developmental changes and effects of aging. Brain Res 163:195–205

Ingegnoli V (1991) Human influences in landscape change: thresholds of metastability. In: Ravera O (ed.) Terrestrial and aquatic ecosystems: perturbation and recovery. Ellis Horwood Ltd. Chichester, England, pp. 303–309

Ingegnoli V (1999) Definition and evaluation of the BTC (Biological Territorial Capacity) as an indicator for landscape ecological studies on vegetation. In Sustainable Landuse Management: The Challenge of Ecosystem Protection. EcoSys: Beitrage zur Oekosystemforschung, Suppl Bd 28:109–118

Ingegnoli V (2001) Landscape Ecology. In: Baltimore D, Dulbecco R, Jacob F, Levi-Montalcini R (eds) Frontiers of Life, vol IV. Academic Press, New York, pp 489–508

Ingegnoli V (2002) Landscape ecology: a widening foundation. Springer, Berlin, New York, p XXIII+357

Ingegnoli V (2011a) Bionomia del paesaggio. L'ecologia del paesaggio biologico-integrata per la formazione di un "medico" dei sistemi ecologici. Springer-Verlag, Milano, pp. XX+340. Foreward by Giacomo Elias

Ingegnoli V (2011b) Non-equilibrium Thermodynamics, Landscape Ecology and Vegetation Science. In: Moreno-Pirajan JC Thermodynamics. Systems in equilibrium and non-equilibrium. InTechWeb.org, Croazia. pp. 139–172

Ingegnoli V (2013) Concise evaluation of the bionomic state of natural and human vegetation elements in a landscape. Rend Fis Acc Lincei. https://doi.org/10.1007/s12210-013-0252-2

Ingegnoli V (2015) Landscape bionomics. Biological-integrated landscape ecology. Springer, Heidelberg, Milan, New York, p XXIV + 431

Ingegnoli V, Giglio E (2005) Ecologia del Paesaggio: manuale per conservare, gestire e pianificare l'ambiente. Sistemi editoriali SE, Napoli, pp. 685+XVI

Ingegnoli V, Giglio E (2017) Complex environmental alterations damages human body defence system: a new biosystemic way of investigation. WSEAS Transactions on Environment and Development, Pp. 170–180

Ingegnoli V & Pignatti S (2007) The impact of the widened Landscape Ecology on Vegetation Science: towards the new paradigm. Springer, Rendiconti *Lincei Scienze Fisiche e Naturali*, s.IX, vol.XVIII, pp. 89–122

Lanza del Vasto GG (1978) Pellegrinaggio alle sorgenti. Jaca Book, Milano

Lattarulo P, Plechero M (2005) Traffico e Inquinamento: i danni per la salute dell'uomo e i Costi Sociali. Irpet, Firenze

Li Q, Kobayashi M, Park BJ et al (2009) Effect of Phytoncide from Trees on Human Natural Killer Cell Function. Int J Immunol Pharmacol 22(4): 951–959

Lorenz K (1973) Die Ruckseite der Spiegels. Piper & co, Munchen

Odum EP (1971) Fundamentals of ecology. Saunders, Philadelphia, PA

Odum EP (1983) Principles of ecology. Saunders, Philadelphia, PA

Paolicelli RC, Ferretti MT (2017) Function and dysfunction of microglia during brain development: consequences for synapses and neural circuits. Front Synaptic Neurosci 9:9

Ramon y Cajal S (1909–1911) Histologie du Systeme Nerveux de l'Homme et des Vertebres

Rotschield D et al (2018) Environment dominates over host genetics in shaping human gut microbiota. Nature. https://doi.org/10.1038/nature25973

Selye H (1956) The stress of life. McGrow-Hill, New York

Selye H (1979) Stress of my life: a scientist's memoris. Van Nostrand Reinold Company, New York

Selye H, Timiras PS (1949) Participation of brown fat tissue in the alarm reaction. Nature 164:745

Slavich GM, Thornton T, Torres LD, Monroe SM, Gotlib IH (2009) Targeted rejection predicts hastened onset of major depression. J Soc Clin Psychol 28(2):2 23–243

Steliarova-Foucher E, O'Callaghan M, Ferlay J, Masuyer E, Rosso S, Forman D, Bray F, Comber H (2014) The European Cancer Observatory: a new data resource. Eur J Cancer 2015(51):1131–1143

Sukopp H, Hejny S (eds) (1990) Urban ecology. SPB Academic Pub, The Hague

Vaglia V, Zamprogno L (2017) *Landscape Bionomics:* Valutazione Diagnostica di un'unità di Paesaggio Agrario con un Metodo Sperimentale Innovativo. Applicazione ad un Contesto Agricolo-Forestale. Tesi di laurea Corso di Laurea Magistrale in Scienze Agroambientali

6. Agrofood System and Human Health

Stefano Bocchi, Simone Villa, Francesca Orlando, Ludovico Grimoldi, and Mario Raviglione

Abstract

Food is the foundation of human development and prosperity, but its production chain can also jeopardize Earth's biosphere and, consequently, human health if its footprint on ecosystems is not timely and adequately contained. The relationship between humans and environment should be assessed considering the complex and convoluted structure of the current food system. Its establishment dates back several millennia to the latest post-glacial period, but its complexity had risen in recent decades starting with the Green Revolution in the 1940s and through the subsequent agroecological transition. The redesigned food production ecosystem has a deep impact on human health with the opposite effects of reducing malnutrition and related (mainly infectious) diseases, and increasing overweight and obesity that are coupled with non-communicable diseases. Furthermore, new evidence shows that environmental pollution is also associated to food production. For instance, antimicrobial resistance is strongly associated with the abuse of these medicines for food-producing plants and animals aiming at maximizing production. The United Nations' Sustainable Development Goals (SDG) adopted in 2015 by all Member States seek to achieve sustainability also with regard of food production, supply, and consumption. This global, cross-sectoral, and multidisciplinary framework provides the opportunity to pursue an economy at societal disposal and in keeping with the Earth's safe operating space.

Keywords

Agroecology · Sustainability · Agenda 2030 Interdisciplinary · Multidisciplinary

S. Bocchi (✉)
Department of Environmental Science and Policy, University of Milan, Milan, Italy
e-mail: Stefano.bocchi@unimi.it

S. Villa · M. Raviglione
Centre for Multidisciplinary Research in Health Science (MACH), University of Milan, Milan, Italy
e-mail: simone.villa@unimi.it

F. Orlando
Department of Environmental Science and Policy, University of Milan, University of Milan, Milan, Italy

L. Grimoldi
SCIBIS, University of Milan, Milan, Italy

6.1 Introduction

6.1.1 The Human–Environment Relationship: The Case of Agrofood System

Planet Earth is currently going through a relatively stable geological era, called Holocene, that has been lasting since the last glacial period starting about 11,700 years ago, the mild climate of which has allowed agriculture and complex human societies to develop and flourish. Consequently, humankind has become inherently dependent upon the current "warm" interglacial period for its nutrition and overall survival. However, the impact of humanity on the Planet have become century-by-century progressively pronounced as the population grew in dimension, knowledge, and wealth.

Nowadays, in order to provide nourishment to about 8 billion people worldwide, food production has become so complex to be deeply interconnected with different sectors of human development and planet's environmental sustainability. Such complexity has resulted in the existing agrofood system which can be profoundly faulty and negatively affect human and environmental wealth. In general, a food system is the set of all inputs, activities, and actors associated with the production, processing, distribution, packaging, and consumption of food. Such systems also encompass outputs ranging from food production and consumption shaped in health- and economy-driven terms to social and environmental well-being of individuals and communities.

The processes underlying the food system can strongly contribute to the erosion of the Earth. The latter is described in the environmental assessments as a complex of nine main planetary impacts: (a) climate change; (b) biodiversity; (c) biogeochemical flow, including the nitrogen and phosphorus cycles; (d) stratospheric ozone depletion; (e) ocean acidification; (f) global freshwater use; (g) change in land use; (h) atmospheric aerosol loading; (i) chemical pollution. Even if the ecosystem has some resilience capacity, each of these mechanisms has its own intrinsic thresholds which are especially endangered by the current food system.

Humankind seems to be finally acknowledging the depth of its print and insult on the planet and its own kind. Hence, the global community attempts to mitigate the impact of unhealthy ecological policies of countries and communities by building multisectoral, comprehensive, and integrated global frameworks. The uppermost of these is the United Nations (UN) Sustainable Development Goals (SDGs) framework launched in 2015 and endorsed by all UN Member States that includes a set of 17 broad goals and 169 specific targets underpinning those goals in the most critical development sectors. Reaching the goals is a sine qua non for sustainable development. Not by chance, in the perspective of the SDG framework, all sectors of development, including ecosystems, biodiversity on lands and oceans, climate change, food, health, education, clean energy, society and institutions, and economy, are closely interconnected and promoted as integrated and indivisible.

6.2 The Current Scenario and its Aftermaths

6.2.1 Structure and Functions of the Current Industrial Agrofood System: From Green Revolution to Agroecological Transition

Green revolution (GR) has been an innovation strategy firstly created and implemented in USA starting from '40s of the last century, spreading out all over the World after the Second War World. It was an extremely focused process on the increase in crop production obtainable thanks to a technological modernization of the agricultural production sector. It is possible to distinguish three phases: the first one, related to the creation and first application of the model prototype; the second period of wider adoption in 1966–1995, during which the model was adopted by other countries in the world with increases in production of the main crops, followed by the third maintained along the two following decades, during which several critical issues emerge.

The first phase was developed thanks to three pre-conditions existing in North America: a defined, focused, and shared goal, that is to increase the yield of some principal crops; the availability of resources (funds, research centres, trained researchers, and technicians) for designing and feeding the new gene revolution, crop intensification, and technology dissemination (i.e. extension service); a prototype of an innovative agrofood policy useful at large scale. The innovation was defined "top down" both from research centres to farmers (Cochrane 1993; Perkins 1997), and from USA to the entire world. The model was tested with the introduction of hybrid corn in the U.S. Corn Belt area, from the 1940s (Ryan and Gross 1943; Pingali 2012) up to the 60s. New cultivation and breeding technologies available in high-income countries were transferred to low- and middle-income settings, aiming at providing better growth and including irrigation capacity and the use of fertilizers and agrochemicals (or pesticides). These, in turn, provided the basis for production of high-yielding varieties (HYV) of cereals, including maize (*Zea mays* L.), wheat (*Triticum* spp.) and, in Asia, novel varieties of rice (*Oryza sativa* L.). The production capacity in many farms significantly increased thanks to the increasing specialization, simplification, use of agrochemicals and standardized agro-techniques. This was considered a valid test for optimizing the conceptual framework and for launching the intensification-industrial model outside of North America. The important research centres created from '40s to '70s thanks to the funding of Private Foundations and National funds subsequently constituted the widespread network worldwide, named Consortium of International Agricultural Research Centers (CGIAR). The GR model was implemented after the Second World War also to innovate European agriculture and to recover from the War consequences. Many farms intensified production through processes of technological convergence and specialization, functional both for industry and international markets that required standardized agricultural products (commodities).

GR achieved its goal of increasing worldwide the yields of some important crops that represented the staple food in numerous areas of the planet. During the second phase the total, irrigated plus rainfed, cultivated area of our world increased from 1.4 billion ha (1961) to 1.5 billion ha (2010). Irrigation spread, especially in the plains, increasing from 139 million ha to more than 300 million ha, so that the global productivity of the agrofood system significantly increased by 150–200% and the land required for feeding a single person decreased from 0.45 ha (1961) to 0.22 ha (2010). The agri-food system that came so transformed at the beginning of the twentieth century was considered, for these results, one of the most important achievements of humanity ever.

The revolution was economic and technological, but above all cultural: the farm was no longer considered as a complex system, dynamic component of the territory, in constant evolution, but as an industry. For, this innovation system designed and developed with industrial framework and values began to show the negative sides that had not been foreseen in the first two phases. The latest reports published by Food and Agriculture Organization (FAO) annually denounce and document the negative consequences that the model of industrial specialized agriculture can have for environmental and human resources.

The cultivated soil undergone to degradation processes are 25% of the total area. The desertification processes, affecting soil quality of more than 50% of the cultivated area, are a particular concern to sub-Saharan Africa, south America, south and south-east Asia, and northern Europe. Industrial agriculture contributed significantly also to the degradation of natural ecosystems and losses of biodiversity (Scherr and McNeely 2008). In just a few decades 75% of the agrobiodiversity was lost. There are today more than 50.000 plants potentially suitable for producing food, but only 1500 are utilized, and only 120 crops have national importance; 15–20 of them are important at global scale. The three main cereals are cultivated in more than half of the agricultural areas. About 85% food requirement worldwide depends on only 8 crops. The food system is responsible for 19–29% of the global anthropogenic greenhouse gas (GHG) emissions (Vermeulen et al. 2012) due by the fossil fuel-intensive production of chemical fertilizers and pesticides and by the industrial livestock breeding. Agriculture is responsible for nitrate, phosphorus, pesticide, soil sediment, and pathogen pollution in soil and water (Parris, 2011). In the post-harvest phases, i.e. transport, food processing, storage, packaging, conservation, and retail, other impacts and emissions are added. Some consumers have become accustomed and require highly processed and unseasonal products so that in the richer markets (North of the Planet) around 80% of the food is not local, whereas in the "South" of the world the food is 80% local. The unsustainable use of resources water, soil, biodiversity, labour can cause a global loss of productivity of the cultivated soil by 0.2% every year (UNEP, 2019) vanishing the terrific achievements of genetics, mechanics, and chemistry. Agri-food system is showing its weaknesses in relation to at least five aspects: (a) unsustainable use of resources; (b) population growth and food insecurity; (c) uncertainty due to climate change; (d) inequality of the richness distribution; and (e) distance between the production/consumption systems or farmer/consumer.

For these reasons, it is generally accepted that agri-food systems require innovation both at local and global level, but most of research and innovation projects are still based on focused, specific issues (innovation of inputs, products, and processes), without a broad picture perspective, still considering only the direct effects of innovation, seldom the indirect ones. Rarely, the transformation of agrofood systems is analysed and proposed, anyway still at the scale of specific agricultural systems and for specific supply chains (Wezel et al. 2009). The agroecological vision of innovative agrofood system at territorial scale with all the dimensions and meanings involved is still neglected. An innovative vision based on a systemic approach, on complex thinking, capable of providing solutions at different territorial scales and which can represent a new language of integration is therefore needed.

In this context, the agroecology represents the innovation front. Agroecology, whose main characteristics are presented in Table 6.1, is defined generally as "A new research and development paradigm for world agriculture" (Altieri 1989) or "the science of applying ecological concepts and principles to the design and management of sustainable food systems" (Gliessmann 2016) but also as a "scientific discipline, practices, policy and social movement" (Wezel et al. 2009) or "the integrative study of the ecology of the entire food system, encompassing ecological, economic and social dimensions" (Francis et al. 2015).

6.2.2 Human Health: How the Environment Shapes Human Health

Diseases, in general, can flourish when a set of individual-level characteristics (i.e., genetics) and environmental factors act together and deflect health of organisms from its natural course. For instance, besides the intra-individual features, the main contributor to disease burden in humans is the environment (Prùss-Ustùn et al. 2016).

Globally, in 2016, the top ten causes of death were: ischaemic heart diseases; stroke; chronic obstructive pulmonary disease; lower respiratory infections; Alzheimer's disease; and other dementias; trachea, bronchus, lung cancer; diabetes mellitus; road injury; diarrhoeal disease; and tuberculosis (World Health Organization 2020). Among those, six are non-communicable diseases (NCDs) and three are communicable, maternal, neonatal, and nutritional diseases (CMNNDs) (Naghavi et al. 2017). Most of NCDs are attributable to obesity and overweight, therefore to unhealthy diets and sedentary lifestyle, while most of CMNNDs can be related to poverty and undernutrition. Consequently, diseases are disproportionally distributed across the globe with low- and lower-middle-income countries (LICs and LMICs, respectively) bearing most of the burden from CMNNDs while high-income countries with well-performing health care systems have to cope mostly with NCDs (Fig. 6.1) (World Health Organization 2020). However, many LICs and LMICs are overwhelmed by a double burden of diseases as NCDs and CMNNDs diseases nowadays coexist (Maher et al. 2010), especially in those settings transitioning to a better socioeconomic level, a phenomenon that carries sometimes harmful consequences linked to profit aims (Moodie et al. 2013). Noteworthy, the association between income and proportion of NCDs and CMNNDs can be derived from these

Table 6.1 From green revolution/industrial model to agroecological innovative approach (Modified Bocchi 2019)

Specialized Industrial agriculture	Agroecology-based agri-food systems
Reductionism approach	System approach
Specialization: The socioeconomic paradigm pushed actors of the agrofood systems to specialize. Industrial agriculture as the model analogous to industrial processes	Diversification: To maintain multiple sources of production, and varying what is produced and offered to the market in terms of ecosystem services
Disconnections from natural cycles	Connection with natural cycles
No limiting factors in agriculture production	Considering local limiting factors
Sectorialization	Local integration among sectors
Focus on commodities for global market	Products and ecosystem services locally markets
Focus on technologies (genetics, chemical, mechanical, and informatics)	Technologies based on local knowledges and capabilities
Intensification of favourable areas, marginalization of many territories	Integration of areas. Landscape design
Upgrading of dimension as dominant trend (big farm is better than little farm)	Cooperation, creation of association, local districts
Intensification as technological function (more factors form/for global market)	Intensification based in quantity and quality of labour. Labour-intensive systems
Specialization (farm, research, and policy)	Multifunctionality, multisector, interdisciplinarity, transdisciplinary. Participated citizen science
Crop monocultures; concentrated animal feeding operations (CAFOs).	Temporal diversification (e.g. crop rotation) and spatial diversification (e.g. intercropping, mixed farming, agroforestry)
Intensive use of external inputs, e.g. fossil fuel, chemical fertilizer, pesticides, and antibiotics	Low external inputs

(continued)

Table 6.1 (continued)

Specialized Industrial agriculture	Agroecology-based agri-food systems
Reductionism approach	System approach
Use of genetically uniform varieties or breeds selected mainly for high productivity.	Use of wide range of species and less uniform, locally-adapted varieties/breeds, based on multiple uses (including traditional uses), cultural preferences
Disconnecting past-present-future	Connecting past, present, future
Vertical and horizontal segregation of product chains, e.g. animal feed production and animal rearing in separate farms, value chains and regions	Natural synergies emphasized and production types integrated (e.g. mixed crop-livestock-tree farming systems and landscapes)
Neutral, ascetic, no values and ethics	Priority of local values, material and nonmaterial
Only one global model. Uniform system	Comparison among models, locally analysed. Diversified system for diverse outputs
No linkages between farm and territory	Place-based strategy
External, and specialized research	Interdisciplinarity, transdisciplinarity, participation
Global market	Global Health
Productive and intensive agriculture for maximizing yield/economic returns	Maximization of multiple outputs. Ecosystem functions/services, nutrition-sensitive agriculture.
Privatization of the resources and unequal distribution of richness	Increase of the social and territorial richness
Ecosystem services not considered	Evaluation of ecosystem services

figures: as the income increases, the proportion of deaths caused by CMNNDs decreases, while the proportion caused by NCDs increases.

6.2.2.1 The Main Diseases Related to Diets

Adverse effects on the health of populations appear in large part linked with better socioeconomic conditions induced by the "Green revolution" (GR): increased consumption of high-calorie foods, often heavily processed, and foods of animal source linked to new dietary pat-

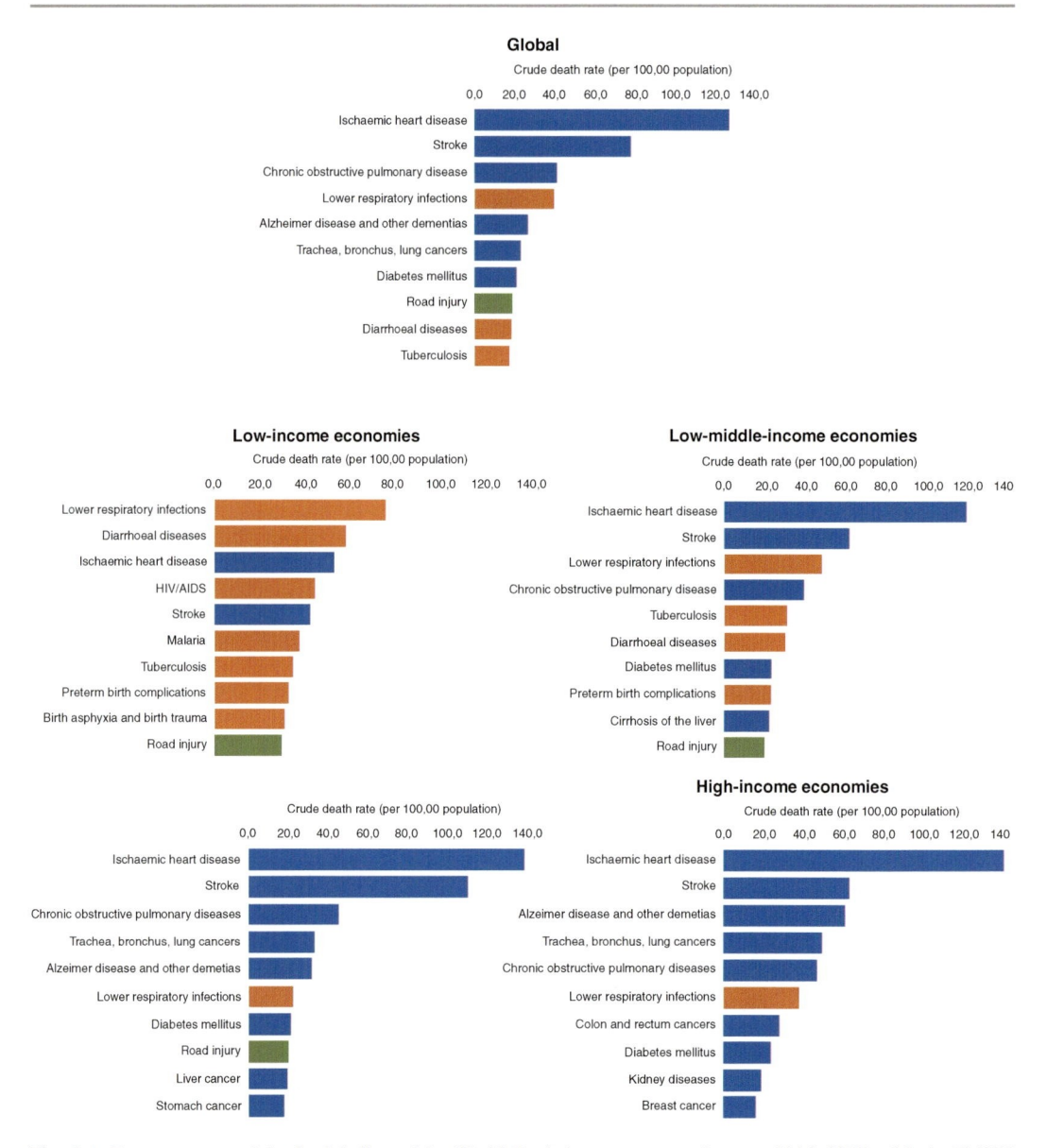

Fig. 6.1 Top ten cause of death globally and by World Bank income groups for year 2016: NCDs (blue), CMNN (orange), and the injuries (green). (World Health Organization 2020)

terns and behavioural changes driven by rapid urbanization and increasing income. In England, for example, estimates indicate that consumption of sugar has increased 5–10 fold in the past two centuries, while consumption of complex carbohydrates has drastically declined (Shammas 1994). Evidence regarding the potential negative consequences of unhealthy diets on human health have been collected in the last century, especially the ones regarding the effects of obesity and its link to some chronic diseases such as diabetes mellitus and cancer (Mozaffarian 2016; Quail and Dannenberg 2019). Another serious consequence of the agricultural advances for the health of people is due to the massive use of pesticides that may have carcinogenic effects and pollute drinking water from dwells.

While the GR has undoubtfully brought prosperity and availability of nutrient-rich and high-calorie foods and changes in dietary behaviours, current diets in many parts of the world still struggle in meeting the appropriate quantity of macro- and micronutrients and, in general, the recommended calorie intake. Malnutrition is shaped in many forms and disproportionally affects certain human populations across the globe leading to the development of diseases which require broader actions to be prevented, reverted, and managed.

6.2.2.2 Malnutrition and Health

Malnutrition is a wide-spectrum term covering deficiencies, excesses, and imbalances in energetic intake and nutrients that can be classified in three groups: (a) overweight and obesity; (b) undernutrition; and (c) micronutrient-related malnutrition.

While undernutrition is well-recognized as a condition of poverty affecting mostly population living in LICs, the common sense could make us assume that overweight and obesity are typical of high-income countries. However, as mentioned above, many LICs and LMICs are now facing a "double burden" of malnutrition consisting in the increasing co-existence of CMNNDs associated with undernutrition and NCDs linked to obesity and overweight (World Health Organization 2017). Globally, the number of overweight people now exceeds the number of those underweight in every region except for sub-Saharan Africa and parts of Asia. Finally, today the vast majority of overweight or obese children live in LICs and LMICs, where the rate of increase has been much faster than that of high-income countries (Blüher 2019).

Overweight and Obesity

Obesity is one of the main causes of NCDs (e.g. type 2 diabetes mellitus) and, as such, a major contributor to the decline in life expectancy. In 2016, worldwide more than 1.9 billion adults were overweight (body mass index [BMI] between 25 and 30 kg m^{-1}) and, among those, 650 million were obese (BMI beyond 30 kg·m^{-1}), which is three times as much as compared to 1975 (World Health Organization 2017). Furthermore, 38 million children below 5 years of age and over 340 million children and adolescent aged 5–19 were overweight or obese.

At individual level, obesity is driven by long-term energy imbalance between high level of calories consumed and low calories expended. Increased consumption of nutrient-poor and calorie-dense (rich in fat and sugar) foods coupled with increased physical inactivity are the driving factors of the phenomenon of "globesity" (i.e., escalation of the global epidemic of overweight and obesity).

Noteworthy, in history, to survive periods of undernutrition caused by famine, humans have undergone a selection pressure to genotypes at risk of low energy expenditure characterized by overeating and highly efficient storage of excessed calories into adipose tissue leading, eventually, to overweight and obesity. This slow selection process in the population genetics is supported by evidence that obesity may also be genetically inherited as mutation in genes encoding for some hormones (i.e., leptin, a hormone produced by adipose tissue) and hormonal receptors (i.e., leptin receptor, melanocortin-4 receptor) that are documented to be associated with severe obesity (Blüher 2019).

Worldwide, "globesity" is disproportionally affecting populations of central and south America (e.g. Colombia, Guatemala, Mexico), whereas almost no changes in BMI have occurred in eastern European countries such as Belarus and the Russian Federation) (Blüher 2019). Many countries especially those in east Asia, Latin America, and the Caribbean have witnessed a rapid shift from underweight to overweight and obesity in the past few decades. In Pacific island nations, where a large body size is considered a symbol of beauty, overweight, and obesity are widespread in the population (Di Cesare et al. 2016). For example, in 2007, in Nauru—an island of Micronesia in the Central Pacific area—the estimated prevalence for overweight and obesity were 95% and 72%, respectively. Conversely, in sub-Saharan Africa overweight and obesity are less common, as the region is still struggling from undernutrition, although there is a rapidly increasing prevalence of diabetes (Di Cesare et al. 2016; World Health Organization 2020).

Children and adolescents, among other, are the most at risk of overweight and obesity and their prevalence in this population subset is alarmingly growing. This phenomenon is particularly important in some populations like those of the small Pacific islands (e.g. Cook Islands, Nauru, and Palau) where the estimated prevalence exceeds 30%. These trends will likely generate a future rise in both overweight and obesity, and ultimately the associated NCDs (Blüher 2019). In fact, an increase in weight gain in children between age 2 and 6 years will lead in about 90% of cases to overweight and obesity in adolescence. In high-income countries, thanks to health promotion efforts, since 2000 excess weight in children and adolescents has plateaued, while the increase in LICs and LMICs continues.

So far, health policies have tried, and failed, to tackle overweight and obesity with interventions only focused to modify single individual-level factors and relying on individual responsibility to reduce food intake and enhance physical activity. Bariatric surgery has spread in economic wealthier countries to revert the weight gain and its detrimental effects (Blüher 2019). However, this cannot be the solution at the global level. Other measures have been fashioned to reduce unhealthy dietary behaviours (e.g. promoting diets rich in fruit and vegetables and poor in fat) and sustain physical activity. Nevertheless, other indirect upstream determinants, including those of cultural and structural nature, undoubtfully exist and may hinder actions taken at individual level. Therefore, a broader, multidisciplinary, and cross-sectoral approach is required to address upstream determinants that act upon dietary behaviours and patterns, and sedentary lifestyle.

Undernutrition

In 2018, worldwide about 795 million people, in particular children under age 5, were undernourished (World Health Organization 2017). This sheer figure has been however in regular decline over recent years, especially in LICs and LMICs, resulting in a reduction by 167 million in the last decade. Nevertheless, in some settings, food intake in children and adolescents is so low that it fails to meet essential needs for physiological functions and growth.

Since 1970s, undernutrition has been grouped in two major categories: wasted (low weight-for-height) and stunted (low height-for-age); in general, children underweight (low weigh-for-age) (BMI below 18.5 kg m^{-1}) may be stunted, wasted, or both. This classification, however, is a simplification fashioned to indicate either an acute (i.e., wasting) or chronic (i.e., stunting) undernutrition with the purpose of designing adequate nutritional therapies.

Globally, in 2018, 151 and 51 million children still suffered from, respectively, stunting and wasting especially in LICs and LMICs of Asia and Africa where access to health care, safe water, and sanitation is generally poor. Indeed, wasting and stunting can be the consequence of low food intake as well as infectious diseases occurring during childhood. Stunting, in particular, is mainly driven by diets lacking a range of nutrient-rich foods and by poor hygiene which leads to diarrhoeal disease generating environmental gut enteropathy associated with malabsorption of nutrients. This produces diversion of energy to promote the immune response against enteric pathogens causing diarrhoea rather than supporting physical and neurological growth.

Childhood undernutrition and infectious diseases are connected by a positive reinforcement loop as both can be cause and consequence of one another: undernutrition can impair the immune system thus predisposing children to other infections. Diarrhoea, for example, is the eighth leading cause of deaths among all ages worldwide and the fifth among children under 5 years of age (Troeger et al. 2017). In general, diarrhoea is well known to disproportionally affect settings with poor access to health care, safe water, and sanitation especially in Africa. Direct interventions to reduce the proportion of wasted children have proven the most effective measures to reduce diarrhoea deaths globally, while direct actions to prevent stunting and underweight are estimated to be ineffective compared to others (i.e., access to safe water and oral rehydration solution). The relationship between undernutrition and other infectious diseases such as tuberculosis and human immunodeficiency virus (HIV) infection/acquired immunodeficiency syndrome (AIDS) exemplifies another vicious cycle linked to immunological dysfunctions.

Undernutrition in children can have irreversible long-lasting aftermaths such as psychophysical growth retardation thus hampering school performance and reducing intellectual capacity. Furthermore, women of short stature due to stunting are at high risk of obstetric complications because of a smaller pelvis and the risk of delivering an infant with low birth weight, who, in turn, will probably remain smaller in size as an adult originating the intergenerational cycle of malnutrition.

Micronutrient-Related Malnutrition
Vitamins and minerals required and consumed in small quantities (micronutrients) are essential for physical and mental development and to orchestrate a range of physiological functions, including the immunological ones, to maintain health throughout life. In 2000s, more than 2 billion people are estimated by the World Health Organization [WHO] to suffer from this "hidden hunger"—the so-called because often micronutrient deficiencies are not immediately visible—in particular in the case of vitamin A, iodine, iron, and zinc (FAO 2014). Examples of naturally micronutrient-rich foods are fruits, vegetables, and animal products, the access to which is poor in most resource-limited settings where people living in poverty cannot afford these products. In general, micronutrient deficiency results in increased susceptibility to infection diseases such as diarrhoea, measles, malaria, tuberculosis, and pneumonia.

Children, adolescents, pregnant women, and lactating women are at the highest risk of consuming diets deprived of micronutrients. Children and adolescents fed with micronutrient-poor diets, especially in sub-Saharan Africa, can experience a high rate of infectious diseases. Pregnant women can frequently be subjected to iron deficiency anaemia increasing their risk of dying during childbirth. In lactating women micronutrient deficiency can also influence the health and development of breast-fed children (i.e., high risk of infections).

Micronutrient deficiencies can be effectively prevented by meeting the recommended daily intake ensuring consumption of a balanced nutrient-rich diet. One way is to enrich, or fortify, foods with essential vitamins and minerals (i.e., vegetable oil enriched with vitamin A and iodized salt), especially in settings where daily income is so low to hinder access to balanced diets. Food fortification policies may be addressed by public health authorities for the general population (mass fortification) or for specific subsets such as young children (targeted fortification) or it may be driven by food industry (market-driven fortification). However, imprudent food fortification may result in excessive increase in micronutrient intake thus leading to health consequences in some individuals. For example, in individuals suffering from chronic iodine deficiency, excessive iodine intake can cause iodine toxicity due to iodine-induced hyperthyroidism.

6.2.2.3 On the Plate: Diseases Attributable to Food and Nutrition

Non-communicable Diseases
Alongside the increasing prevalence of overweight and obesity, the rise of NCDs is widely documented worldwide. In general, the burden of disease substantially changed over time as NCDs surpassed infectious diseases as the main cause of death and disability worldwide, a phenomenon known as "epidemiological transition". While the development of NCDs can also partly be attributable to undernutrition, especially due to chronic stunting causing neural and psychological impairment, unhealthy diets leading to obesity are the most prominent risk factors. The main dietary risks and associated NCDs are shown in Table 6.2.

To date, each year, NCDs are responsible for 42 million deaths (71% of all deaths globally), especially in people under age 70 years (>40%) and in LICs and LMICs (~80%). Globally, diet and dietary risks are the second leading cause of death (19% of all deaths) and DALYs (Disability

Table 6.2 Dietary risks and associated non-communicable diseases (Afshin et al. 2019)

Dietary risk[a]	Associated non-communicable diseases
Low healthy food intake	
Diet low in fruits (<200 g/day)	Cancer (tracheal, bronchus, and lung and oesophageal), ischemic heart disease, stroke, and diabetes mellitus
Diet low in vegetables (<290 g/day)	Ischemic heart disease and stroke
Diet low in legumes (<50 g/day)	Ischemic heart disease
Diet low in whole grains (<100 g/day)	Ischemic heart disease, stroke, and diabetes mellitus
Diet low in nuts and seeds (<16 g/day)	Ischemic heart disease and diabetes mellitus
Diet low in milk (<350 g/day)	Cancer (colon and rectum)
Diet low in fibre (<19 g/day)	Cancer (colon and rectum) and ischemic heart disease
Diet low in calcium (<1 g/day)	Cancer (colon and rectum)
Diet low in seafood omega 3 fatty acid (<200 g/day)	Ischemic heart disease
Diet low in polyunsaturated fatty acids (<9% of total daily energy)	Ischemic heart disease
High unhealthy food intake	
Diet high in red meat (>27 g/day)	Cancer (colon and rectum) and diabetes mellitus
Diet high in processed meat (>4 g/day)	Cancer (colon and rectum), ischemic heart disease, and diabetes mellitus
Diet high in sugar-sweetened beverages (>5 g/day)	Cancer (colon and rectum, liver, pancreatic, breast, uterine, and ovarian, kidney, thyroid, non-Hodgkin lymphoma, multiple myeloma, acute and chronic lymphoid leukaemia, acute and chronic myeloid leukaemia), ischemic heart disease, stroke, hypertensive heart disease, atrial fibrillation and flutter, asthma, gallbladder and biliary diseases, Alzheimer's disease and other dementias, diabetes mellitus, chronic kidney disease (i.e., due to diabetes mellitus, hypertension, and glomerulonephritis) osteoarthritis, low back pain, gout, and cataract
Diet high trans fatty acids (>1% of total daily energy)	Ischemic heart disease
Diet high in sodium (>2 g/day)[b]	Cancer (stomach), rheumatic heart disease, ischemic heart disease, stroke, atrial fibrillation and flutter, aortic aneurysm, peripheral vascular disease, endocarditis, and chronic kidney disease (i.e., due to diabetes mellitus, hypertension, and glomerulonephritis)

[a]Per each dietary risk is reported the lower or upper threshold based on the high or lower consumption attributed to the development of NCDs
[b]The threshold indicated is the one recommended for adults by WHO and need to be adjusted downward for children based on their energy requirements (World Health Organization 2007)

Adjusted Life Years) lost (10% of all DALYs lost). DALYs are a measure of the burden of a disease that sums the number of years lost due to ill health or disability or due of early death compared to life expectancy (Afshin et al. 2019). In recent years, epidemiological evidence supports the role of specific dietary factors in triggering NCDs development. Worldwide, the consumption of healthy foods and nutrients was suboptimal in 2017, especially for nuts and seeds (global average intake of 3 g/day), milk (global average intake of 71 g per day), and whole grain (global average intake of 29 per day). The vast majority of deaths and DALYs lost were accounted for by diets low in whole grains (5% of all deaths and 3% of all DALYs) and diets low in fruits (4% of all deaths and 3% of all DALYs). Conversely, unhealthy foods and nutrients intake like sugar-sweetened beverages [SSBs] (global average intake of 49 g/day) especially in young adults aged 25–49 years, processed meat (global average intake of 4 g/day), and sodium (global average intake of 6 g/day) were exceeding the maximum recommended level. SSBs consumption greatly increased in recent years. SSBs are major contributors to the increase of obesity and related NCDs with obesity and dental caries in children and adolescents, and type 2 diabetes mellitus in adults (Afshin et al. 2019).

Finally, in 2015, the International Agency for Research on Cancer classified red meat (defined as fresh unprocessed mammalian muscle meat) as "probably carcinogenic to humans" (Group 2A) and processed meat (any meat processed e.g. through salting, curing, fermentation, smoking) as "carcinogenic to humans" (Group 1), especially for cancer of the digestive tract e.g. colorectal and stomach cancers, when frequently consumed (International Agency for Research on Cancer 2018).

Communicable Diseases: Foodborne Diseases

Globally each year 600 million people (~10% of the global population) are estimated to develop foodborne diseases with 420,000 among them dying, especially children below age 5 (30% of foodborne diseases). The vast majority of foodborne diseases are diarrhoeas (550 million cases yearly) caused especially by noroviruses (120 million) and *Campylobacter* spp. (96 million), although the main cause of diarrhoeal deaths (230,000 deaths) is due infection with non-typhoidal *Salmonella enterica* (59,000 deaths) and enteropathogenic *Escherichia coli* (37,000 deaths) (World Health Organization 2015). Pathogens causing human foodborne diseases have frequently developed resistance to antimicrobial agents given their use in agriculture and livestock husbandry together with other non-antimicrobial compounds (i.e., heavy metals) (FAO and WHO 2019).

In general, to prevent foodborne diseases, strict rules of hygiene such as the hazard analysis and critical control points and the "cold chain", and a public veterinary service devoted to monitor animal production in the food system are needed (FAO 2003). Food processing techniques, such as pasteurization, have been developed with the purpose of eliminating pathogens and their enzymes in food and consequently prevent foodborne diseases.

6.2.2.4 Determinants of Dietary Patterns

The impact of diet on human health has been described to have a spatial pattern which reflects the level of socioeconomic development. For example, high BMI, red meat consumption, SSBs, and alcohol have been documented to have a strong relationship with increased socioeconomic development (Afshin et al. 2019; Blüher 2019). Unhealthy foods, in particular, are socially patterned in high-income countries where individuals of low socioeconomic status are more exposed to their use.

The role of economy in shaping environments and communities and in driving individual's behaviours and choices has been widely recognised. Nevertheless, unregulated multinational food and beverage companies exploit this knowledge for profit by increasingly promoting cheap palatable, tasty, and unhealthy food production. Food and beverage industries have developed strategies and approaches to introduce into the market unhealthy products thus establishing the so-called "commercial determinants of health" (Kickbusch et al. 2016; Allen et al. 2019). Alarmingly, in LICs and LMICs, an inexorable penetration of food sales and corporations is underway. Food and beverage companies act mainly through four channels: (a) marketing and (b) supply chain focused on promoting the production and consumption of unhealthy products (Allen et al. 2019); (c) lobbying and (d) corporate citizenship (that is the responsibility of companies towards society) to deflect public attention and prevent policy barriers. A clear example is the soda industry Coca-Cola which was discovered in 2015 to fund researchers to withhold evidence of the detrimental risks (e.g. diabetes mellitus) caused by soda consumption (Kickbusch et al. 2016).

This situation notwithstanding, many countries have moved forward in implementing policies to reduce unhealthy food and beverage consumption. Taxing sugary drinks, especially when done proportionally to sugar content rather than volumetrically, has indirectly promoted (a) beverage reformulation with soft drink companies reducing sugar content in their products to decrease or remove tax liability (e.g. beverages with sugar level <5 g/100 mL are not taxed in United Kingdom); and, in general, (b) price change in SSBs and, consequently, drop in sales (Briggs et al. 2017).

In conclusion, to tackle malnutrition and NCDs steps need to be taken cross-sectorally through measures in various fields devoted to improvement of socioeconomic status of individuals and prevention of for-profit interferences. The United Nations Agenda for Sustainable Development, agreed upon and adopted by Governments and Heads of State of countries across the globe, constitutes the perfect foundation upon which to renovate food production, processing, and marketing jointly with changes in dietary behaviours of people.

6.2.2.5 Environmental Pollutants in Food and Human Health

Environmental pollution is estimated to be responsible for nine million premature deaths in 2016 (Landrigan et al. 2018). The food system is one of the main contributors to human ill health in many other parallel ways besides the duality, and sometimes dual burden, of undernutrition and overnutrition. Evidence of the detrimental effects on human health from agrochemical (i.e., pesticides and fertilizers) exposure of food and groundwater is constantly emerging (Tóth et al. 2016; Landrigan et al. 2018; FAO, 2020a).

Through food, humans can be exposed to many environmental pollutants especially in two ways and at different concentrations. Farmer and, in general, workers in the agricultural and livestock sectors can be heavily exposed to high-level pesticides, fertilizers, and other agrochemicals compounds which can lead to a professional exposure that may be neglected in some settings, especially in LICs. Food consumption, however, can expose consumers to these residues and to harmful products used in food processing and as food contact materials (World Health Organization 2018b).

Heavy Metals

Many pesticides and fertilizers widely used in agriculture can contain traces of heavy metals [HMs] like arsenic (As), cadmium (Cd), mercury (Hg), lead (Pb), and selenium (Se). These elements can be adsorbed and accumulated in plants entering the animal and human food chain or, anyway, contributing to the environmental pollution as HMs can accumulate in soil, freshwater used for irrigation, or in groundwaters (Tóth et al. 2016; Landrigan et al. 2018).

Plants accumulate HMs disproportionally. Rice, for example, can contain high quantities of As and Cd that can threaten human health especially for those with micronutrient deficiency (i.e., Fe, Zn, and Ca) (Wu et al. 2016). The biological explanation is that people with micronutrient deficiencies up-regulate genes encoding for non-specific metal-transporters along the gut able to adsorb As and Cd. Less worrying concerns for human health seem to exist for absorption and accumulation from food of Pb, Hg, and Se either because these elements are less retained in edible plants or accumulate in parts of plants usually not eaten (e.g. roots). In humans, in general, the main source of As are contaminated groundwaters rather than food as this element is commonly present in soil and rocks, while Cd is usually absorbed through water and rice. Long-term health effects likely attributable to chronic exposure to As are skin manifestations (e.g. pigmentation changes), cancer (e.g. skin, bladder, lung cancer), diabetes, pulmonary and cardiovascular diseases (e.g. myocardial infarction), especially if exposure happened early in life; whereas exposure to Cd can cause kidney dysfunction, type 2 diabetes mellitus, and cancer (Peralta-Videa et al. 2009).

Endocrine-Disrupting Chemicals

Endocrine-disrupting chemicals [EDCs] (or endocrine disruptors) are compounds and products used in pesticides, food packaging, and food contact materials (Gore 2016). The name comes from their properties to alter the (human) endocrine function of exposed subjects, especially when exposure is chronic. EDCs consumption can have detrimental consequences, especially if eaten by pregnant women where they can alter embryonic development of the foetus, in animal models, leading, for example, to congenital defects (Gore et al. 2015). Other outcomes of exposure to EDCs through food can be cancer (i.e., breast, prostate, and reproductive cancers), endocrine dysfunction, immune impairment, neurological and cognitive problems, and respiratory distress.

Among the EDCs, Bisphenol A is used in food contact materials, especially in beverages cans (Geens et al. 2010) and therefore it is now banned in European Union in products used for children under 3 years of age. Bisphenol A is known to interact with a large set of nuclear receptors (e.g. oestrogen receptors) and several studies have suggested its role in overweight, type 2 diabetes mellitus, altering neurodevelopment and behaviour (e.g. hyperactivity) of children, and increasing the risk of developing breast and prostate cancer in adults (Nagel and Bromfield 2013). Another EDCs are perfluoroalkyl substances [PFASs] a group of molecules contained in in food contact materials, from which they can migrate to food. High level of PFASs intake are documented in those who consume high quantities of fast foods (Begley et al. 2005). The biological effect of PFASs seems to be associated to low level of circulating thyroid hormones (Kim et al. 2018).

In general, emerging evidence shows how diets with high levels of organic foods—that are produced without synthetic pesticides, fertilizers, or any other synthetic growth promoting agent which are, instead, widely used in non-organic foods—are associated with low frequency of cancer development, thus further supporting the supposed role of pesticides in affecting human health (Smith-Spangler et al. 2012). In general, the role and impact of pesticides and fertilizers in NCDs need to be further investigated to adequately design and develop national and international policies and best practices to reduce their impact on human health.

Indoor Air Pollution

In many low-resource settings food cooking as well as heating and lighting are based on household fuel combustion, women and children are seriously threatened by indoor air pollution which includes particulate matter [PM] (World Health Organization 2018a). In Africa, for example, the main contributor of fine particulate pollution ($PM_{2.5}$), both indoor and outdoor, is food cooking (Landrigan et al. 2018).

This type of pollutants is source of non-negligible morbidity and mortality as, overall, 1.6 million people are estimated each year to prematurely die, especially in LICs, with a 1000-fold gap between the mortality estimated in low- and high-income countries (World Health Organization 2002). Moreover, household pollution is one of the main risk factors for most of the lower respiratory infections (64% and 57% of the global burden), chronic obstructive pulmonary disease (51% and 52% of the global burden), and tracheal, bronchial, lung cancer (48% and 38% of the global burden) in LICs and LMICs, respectively. In general, low-level exposure to household air pollution has mostly chronic effects thus its consequences are largely perceived after many years and cause death especially in those aged 65 years and above (Landrigan et al. 2018).

Ambient Air Pollution

The food system is one of the main contributors of greenhouse gas [GHG] [e.g. methane and carbon dioxide (CO_2)] emissions as it produces more than a quarter of the overall GHG emissions (26%) (Poore and Nemecek 2018). Livestock and fisheries, in particular, are the principal sources of GHGs produced in the food sector (52%), which includes crop cultivation for animal use (6%) and land use (e.g. deforestation for livestock production) (16%). The so-called enteric fermentation, which consists in methane produced from ruminants' digestion is the most potent GHG as it has 23 times more global warming potential compared to CO_2 (Landrigan et al. 2018). Nevertheless, the main contributor of GHG emissions is food loss and waste (4.4 gigatonnes of CO_2; 8% of total human-made GHG emissions), which represent one-third of the food produced annually (Food and Agriculture Organization of the United Nations 2015).

The effects of GHG emissions on human health, however, are not direct but through global warming and climate change which undoubtfully jeopardize human health and the sustainability of food systems (Landrigan et al. 2018). The level of GHG emissions from agriculture is highly dependent on diet composition and if the current dietary pattern is maintained until 2050, when the global population in projected to reach 10 billion people (Springmann et al. 2018a), there will be an increase in GHG emissions with food systems expected to impact half of total GHG emissions. Diets with associated with lower level of GHGs are usually the one labelled as "unhealthy" as high in sugar, fats, or carbohydrates (Springmann et al. 2018b).

Noteworthy, agriculture can be also be the source of particulate pollution, especially $PM_{2.5}$ matter, generally in large population centres surrounded by intensive agriculture fields.

6.2.3 Emerging Risks: Antimicrobial Resistance

The current agrofood system requires protection from infectious diseases carried by food-producing animal and plants to maintain, given its complexity and dimension, the capacity to feed nearly 8 billion people without originating health hazards to them. Protection from infectious diseases is required to sustain, at least in the short period, the large-scale productivity and avoid massive costs to the food industry and its employees. Antibiotics are used for this purpose, besides the additional aim of accelerating weight gain and promoting rapid growth in food-producing animals.

The use of antibiotics in agriculture and husbandry of livestock is deeply rooted in the common practice within our society as their use began in 1938, almost a century ago, soon after the first antimicrobial molecule—sulfamidochrysoïdine (Prontosil)—was developed by Bayer. Since then, an endless race started, opposing the introduction of new antimicrobials to the emergence of antimicrobial resistance (AMR) among strains of viruses, bacteria, fungi, and parasites. The onset of this phenomenon had not only consequences for the food industry but its rapid spread has come to the point of affecting also the health and wealth of humans.

The genotype and phenotype of all organisms are shaped by the environment where they live. The same process happens in viruses, bacteria, fungi, and parasites. It is often so fast—due to the inherent cellular simplicity of microorganisms—that in certain settings new strains end up replacing the previous populations. Antimicrobials fashion the environment where microorganisms grow and adapt by developing or reducing a different range of features (e.g. increasing the synthesis of efflux pumps or alteration of binding sites). Once the resistance is acquired and encoded in the genome (either in the DNA or RNA), it can be transferred to other microorganisms within and across species. Whatever type of AMR is developed, it provides an evolutionary advantage compared to other competitors in the same ecological niche. Hence, drug-resistant strains thrive and replicate.

Antibiotics are not only used to control and treat diseases in food-producing animals and plants but are often employed to improve the profit of the food industry by promoting the growth of animals at-scale by reducing reproductive hurdles and augmenting their weight by increasing the food intake and nutrient absorption. The underlying mechanism which enables growth promotion through antibiotics is, however, unclear. Their use as (antibiotics) growth promoters (AGPs) goes back to 1940 when streptomycin—a powerful antibiotic and the first anti-tuberculosis medicine—was found to improve the growth rate in chickens. Since then, penicillin and tetracycline were used as AGPs not only in broiler chickens and poultry but in general with the aim to increase their body weight (up to +12%), but also in pigs and cattle because they could optimize intestinal health and improve reproductive capacity. To date, more than ten classes of antibiotics, ranging from beta-lactams to macrolides, are used in animal food industry for these purposes coupled with market strategies developed to ensure more and more food-producing animals and sell them rapidly.

The use of antimicrobial compounds in the food industry has, however, adverse effects and consequences for human health and microorganisms that have a wide range of hosts can transfer acquired drug resistance to other species (e.g. through plasmid transfer). This phenomenon is important as many of the antimicrobials used in the food industry are the same or are closely related to the ones used in human medicine. For instance, the use of antibiotics against *Escherichia coli* in animals and humans is associated with an increased occurrence of resistance, bidirectionally. The use of antibiotics such as the macrolides and tetracyclines in food-producing animals has been associated with an increased occurrence of resistant strains of bacteria such as *Campylobacter* spp. and *Salmonella* spp. affecting humans. Furthermore, food is not completely free from chemical contaminants and residues including hormones (i.e., progesterone) especially in meat and meat products. These chemical compounds can accumulate in fatty tissues of animals exposing humans who consume their meat.

An additional complication linked to the abuse of antibiotics in the animal sector is the excretion in waste through manure and urine of the majority of the drug volume which is not absorbed by the animals. This produces environmental pollution. Manure converted into fertilizers further complicates matters eventually polluting water and soil. Millions of kilogrammes of antibiotics, some of them poorly biodegradable under certain conditions, can be detected in the soil months after their release and can accumulate in plant tissues eventually entering the human food chain. Furthermore, emerging evidence now suggests the role played by heavy metals, contained in fertilizers, in increasing the environmental pressure upon bacteria to co-select ARGs and metal resistant genes (MRGs) especially in Enterobacteriaceae such as *S. enterica*, *E. coli*, and *Klebsiella pneumoniae*.

To prevent disastrous consequences of AMR in May 2015 Ministries of Health worldwide convened at the Sixty-eight World Health Assembly (WHA) drafted and adopted a global action plan to respond to the threat of AMR. A successful global plan must be addressing challenges that go beyond the field of human health recognizing that numerous microorganisms cross the line between species and kingdoms (e.g. *Pseudomonas aeruginosa* can infect both plants and animals). Therefore, to fulfil its goal, the WHO's Global Action Plan (GAP) pursues a cross-sectoral approach, named "One Health", through other multilateral and multinational organizations such as FAO and the Office International des Epizooties (OIE) that is the international organization for animal health.

In 2016, the AMR debate was taken at the highest level possible as Heads-of-States and governmental delegations from all countries were convened at the Seventy-first United Nations General Assembly (UNGA) to address such threat. This resulted in the UNGA Political Declaration A/RES/71/3 where the GAP was reaffirmed and a new ad-hoc Inter-Agency Coordination Group (IACG) was established to effectively address AMR through a comprehensive multi-sectorial approach.

While in 2006 the European Union was the first to ban the use of AGPs for animal husbandry, in 2016, after the 68th WHA decision to promote the GAP, the United States Food and Drug Administration (FDA) also restricted their use. Since then, the food-producing animal industry has faced difficulties in sustaining profitable production. As a result, the food industry redesigned its strategies and began using other compounds and approaches. Different policies have been developed to sustain the animal food industry without increasing the risk of AMR, including: vaccination; phytochemicals; immune-related products; and other innovative drugs, chemicals and enzymes.

Vaccination strategies for livestock and warmwater aquaculture have been implemented in the last few years and shown to be the most cost-effective strategies to prevent infectious disease outbreaks. This can ensure the sustainability of the food-producing animal marketing and the better taste and texture of food produced as animals have not been exposed to AGPs. Furthermore, the employment of vaccination policies in food production can shelve the use of antimicrobial agents and ultimately avoid the emergence of AMR. One exemplary story is that of Norwegian salmons the productivity of which has hugely increased through a seven-antigen vaccine. The new vaccination strategies were developed through a collaboration between the Norwegian government and the salmonid industry.

Phytochemicals are bioactive derivates from plants with antiviral, antibacterial, antifungal properties that were firstly implemented starting in 2006 to replace AGPs. The first types of phytochemicals to be used in the food-producing animal industry were essential oils derived from seeds, roots, barks, herbs, leaves, flowers, or fruits. Essential oils display a positive growth effect by increasing broiler animals' body weight, feed intake, and nutrient absorption with beneficial antioxidant and antimicrobial effects. Other phytochemical products such as capsaicin and others capsaicinoids, substances derived from Solanaceae plants of the genus *Capsicum* that includes more than 200 varieties of red peppers, increase feed intake and body weight and also offer antibacterial properties; for instance, they can inhibit colonization in the colon of food-producing animals by *Clostridium perfringens* and *E. coli*.

In agriculture, these policies are not always easy to implement at all levels, especially for small-scale producer in low- and middle-income countries. Hence, the need, beyond establishing regulations of pesticides at national and international levels, to intensify education of farmers in the prevention of over- and mis-use of agrochemicals. FAO has been among the first agencies to support an integrated approach in implementing biosecurity and infection control practices without resorting to antibiotics and pesticides with the aim of preventing both AMR and environmental pollution.

Within all these promising developments, there are still formidable challenges. In the current food system, for instance, there is a need to acknowledge the role played by the genetic diversity of crops and livestock resources even if in the short run it may mean reduced market profits. In the long term, even if the trend of AMR emergence is reversed, losses in genetic diversity and human-made clonal selection of food-producing plants and animals can lead to mass extinction of species if environmental conditions or pathogens change. This threat is extremely pronounced for bananas belonging to a single clonal variety, the Cavendish, that dominates the global production and exportation. Each year banana plantations face major threats from fungal infections against which wild bananas, the production of which is rapidly declining, are the source of resistant genes.

Moreover, the current food system at global scale and the continuous increase in beef production is leading to an intensification in nutrient flows, resulting in an anthropogenic distortion of the delicate phosphorus and nitrogen cycles. Nitrogen extracted from the atmosphere and phosphorus obtained from fossil rocks are used to manufacture agricultural chemicals. They are eventually poured into freshwater polluting lakes and oceans and contributing to multiply and extend dead zones, i.e., areas where marine life is not possible due to the low oxygen concentration. In conclusion, AMR is undoubtfully the next challenge humankind needs to prepare to confront assertively, but other threats which defy the planetary boundaries (e.g. climate change) lie ahead and urgently need to be addressed.

6.3 Building a Sustainable Circular Agrofood System in the SDGs Framework

6.3.1 The Sustainable Development Framework

Food is the keystone of human, societal, and economic development. Dietary behaviours and food can both sustain and oppose health and wealth of individuals. In recent years, food systems to sustain large-scale food production and marketing able to feed ~8 billion people worldwide have become so convoluted to turn into an interacting and multidimensional complex of ecosystems, agricultural lands, pastures, inlands fisheries, labour, infrastructure, technology, policies, culture, traditions, institutions, and markets involved in producing, processing, distributing, and consuming food: the "eco-agri-food systems" (TEEB 2018). Their structures are now populated by several multinational shareholders whose actions cross national boundaries and legislations. Interactions between these actors have become so complex that food systems are now exceedingly intertwined in many non-food domains to require trans-national, cross-sectoral, and multidisciplinary interventions. In such multifaced landscape, the current agrofood system is irremediably deteriorating Earth resources driving the planet beyond Holocene range of variability of sub-ecosystems and challenging its planetary boundary thresholds. Food chain production, for example, largely relies on nitrogen- and phosphorus-based fertilizers which alter the biochemical flow cycles. Approximately half of these elements remains trapped in the soil and freshwater and can be transported towards seas and oceans where they contribute to engender oceanic dead zones—areas of oceans with levels of oxygen so low that life cannot be sustained. This detrimental mechanism is further worsened because of animal breeding that generates a surge in nutrient flows across continents through the international trade of animal food.

To reverse this route and preserve humans as a species, humankind must "meet the needs of the present without compromising the ability of future generations to meet their own needs" (United Nations 1987). To pursue this objective, the global community has gathered in 2015 at its highest level possible, at UNGA, and adopted the SDGs (United Nations 2015). The new and comprehensive 2030 UN Agenda (Fig. 6.2, Table 6.3), shaped to "leave no one behind" in 17 goals, 169 target, and 232 indicators, aims to take humankind closer to its sustainability in this planet. All the "5P" dimensions of human development and its sustainability—People, Planet, Prosperity, Peace, and Partnership—are, therefore, covered within the framework. By design, SDGs symbolize the holistic nature of sustainable interventions needed to act upon plenty of upstream determinants that influence health, society, and economy. Among the goals SDG2, the specific targets and indicators of which are displayed in Table 6.4, is the one devoted to food and its directly related challenges as well as the promotion of a "sustainable agriculture". To renovate the eco-agri-food system, policies and strategies need to act in concert and across sectors and disciplines. The SDGs framework, internationally agreed upon by

Fig. 6.2 The United Nations' sustainable development goals

Table 6.3 The United Nations' sustainable development goals

Goal 1	End poverty in all its forms everywhere
Goal 2	End hunger, achieve food security and improved nutrition, and promote sustainable agriculture
Goal 3	Ensure healthy lives and promote Well-being for all at all ages
Goal 4	Ensure inclusive and equitable quality education and promote lifelong learning opportunities for all
Goal 5	Achieve gender equality and empower all women and girls
Goal 6	Ensure availability and sustainable management of water and sanitation for all
Goal 7	Ensure access to affordable, reliable, sustainable, and modern energy for all
Goal 8	Promote sustained, inclusive and sustainable economic growth, full and productive employment, and decent work for all
Goal 9	Build resilient infrastructure, promote inclusive and sustainable industrialization, and foster innovation
Goal 10	Reduce inequality within and among countries
Goal 11	Make cities and human settlements inclusive, safe, resilient, and sustainable
Goal 12	Ensure sustainable consumption and production patterns
Goal 13	Take urgent action to combat climate change and its impacts
Goal 14	Conserve and sustainably use the oceans, seas, and marine resources for sustainable development
Goal 15	Protect, restore, and promote sustainable use of terrestrial ecosystems, sustainably manage forests, combat desertification, and halt and reverse land degradation and halt biodiversity loss
Goal 16	Promote peaceful and inclusive societies for sustainable development, provide access to justice for all and build effective, accountable and inclusive institutions at all levels
Goal 17	Strengthen the means of implementation and revitalize the global Partnership for Sustainable Development

governments, non-governmental organizations (NGOs) and civil society, plays a key role in this purpose. Indeed, renewing the food system needs to go beyond the targets of SDG2 and requires interaction with other goals within the framework through the adoption of multidisciplinary and cross-sectoral actions. Likewise, transforming the food system will in turn contribute to other global objectives such the ones concerning health, biodiversity, and climate change.

6.3.1.1 Wedding Cake Model: The Role of Food in the Sustainability

All the aspirational goals set in the SDGs to contribute to societal (SDGs 1–5, 7, 11, 16) and economic (SDG 8–10, 12) development needs to occur within planetary boundaries. To achieve that, the first priority is to acknowledge the importance of the biosphere (SDGs 6, 13–15) and that what preserves the Earth is not negotiable. These elements, however, are several times erroneously perceived as limitations to human prosperity, transformation, and success rather than an opportunity to sustainably promote human development. The so-called wedding cake model (Fig. 6.3) clearly represents this transformation and inversion of the world logic: the economy needs to be at society's disposal in keeping humans within the safe operating space of planet Earth, and not the contrary. In this perspective, both economy and society need, therefore, to be perceived as embedded parts of the biosphere, the worldwide sum of all Earth ecosystems. Food becomes then a prerequisite for human success and prosperity and the single strongest element to optimize human health, societies and economies, and our environmental sustainability on Earth (EAT-Lancet Commission 2019).

Food and the Biosphere

Oceans and seas cover more than 70% of the surface of the planet and support life by providing half of the world's oxygen, sequestering carbon dioxide, and hosting 80% of life on Earth. "Life below water" nowadays feeds more than 3.1 billion people providing them with ~20% of their daily animal protein intake. Oceans and seas have the potential to feed the ~10 billion people projected to live on Earth in 2050 (FAO 2016). Nevertheless, fishing practices exceed the market demand and consumption, as ~30% of the current fish stocks are estimated as overfished, thus threatening marine biodiversity. Moreover, the

Table 6.4 The Sustainable Development Goals number 2: "End hunger, achieve food security and improved nutrition and promote sustainable agriculture"

Targets		Indicators	
2.1	By 2030, end hunger and ensure access by all people, in particular the poor and people in vulnerable situations, including infants, to safe, nutritious, and sufficient food all year round	2.1.1	Prevalence of undernourishment.
		2.1.2	Prevalence of moderate or severe food insecurity in the population, based on the food insecurity experience scale (FIES)
2.2	By 2030, end all forms of malnutrition, including achieving, by 2025, the internationally agreed targets on stunting and wasting in children under 5 years of age, and address the nutritional needs of adolescent girls, pregnant, and lactating women and older persons	2.2.1	Prevalence of stunting (height-for-age < −2 standard deviation from the median of the WHO child growth standards) among children under 5 years of age
		2.2.2	Prevalence of malnutrition (weight-for-height > +2 or < −2 standard deviation from the median of the WHO child growth standards) among children under 5 years of age, by type (wasting and overweight)
2.3	By 2030, double the agricultural productivity and incomes of small-scale food producers, in particular women, indigenous peoples, family farmers, pastoralists, and fishers, including through secure and equal access to land, other productive resources and inputs, knowledge, financial services, markets, and opportunities for value addition and non-farm employment	2.3.1	Volume of production per labour unit by classes of farming/pastoral/forestry enterprise size
		2.3.2	Average income of small-scale food producers, by sex and indigenous status
2.4	By 2030, ensure sustainable food production systems and implement resilient agricultural practices that increase productivity and production, that help maintain ecosystems, that strengthen capacity for adaptation to climate change, extreme weather, drought, flooding, and other disasters and that progressively improve land and soil quality	2.4.1	Proportion of agricultural area under productive and sustainable agriculture
2.5	By 2020, maintain the genetic diversity of seeds, cultivated plants and farmed and domesticated animals and their related wild species, including through soundly managed and diversified seed and plant banks at the national, regional, and international levels, and promote access to and fair and equitable sharing of benefits arising from the utilization of genetic resources and associated traditional knowledge, as internationally agreed	2.5.1	Number of plant and animal genetic resources for food and agriculture secured in either medium or long-term conservation facilities
		2.5.2	Proportion of local breeds classified as being at risk, not-at-risk or at unknown level of risk of extinction
2.A	Increase investment, including through enhanced international cooperation, in rural infrastructure, agricultural research and extension services, technology development and plant and livestock gene banks in order to enhance agricultural productive capacity in developing countries, in particular least developed countries	2.A.1	The agriculture orientation index for government expenditures
		2.A.2	Total official flows (official development assistance plus other official flows) to the agriculture sector
2.B	Correct and prevent trade restrictions and distortions in world agricultural markets, including through the parallel elimination of all forms of agricultural export subsidies and all export measures with equivalent effect, in accordance with the mandate of the Doha development round	2.B.1	Producer support estimate
		2.B.2	Agricultural export subsidies
2.C	Adopt measures to ensure the proper functioning of food commodity markets and their derivatives and facilitate timely access to market information, including on food reserves, in order to help limit extreme food price volatility	2.C.1	Indicator of food price anomalies

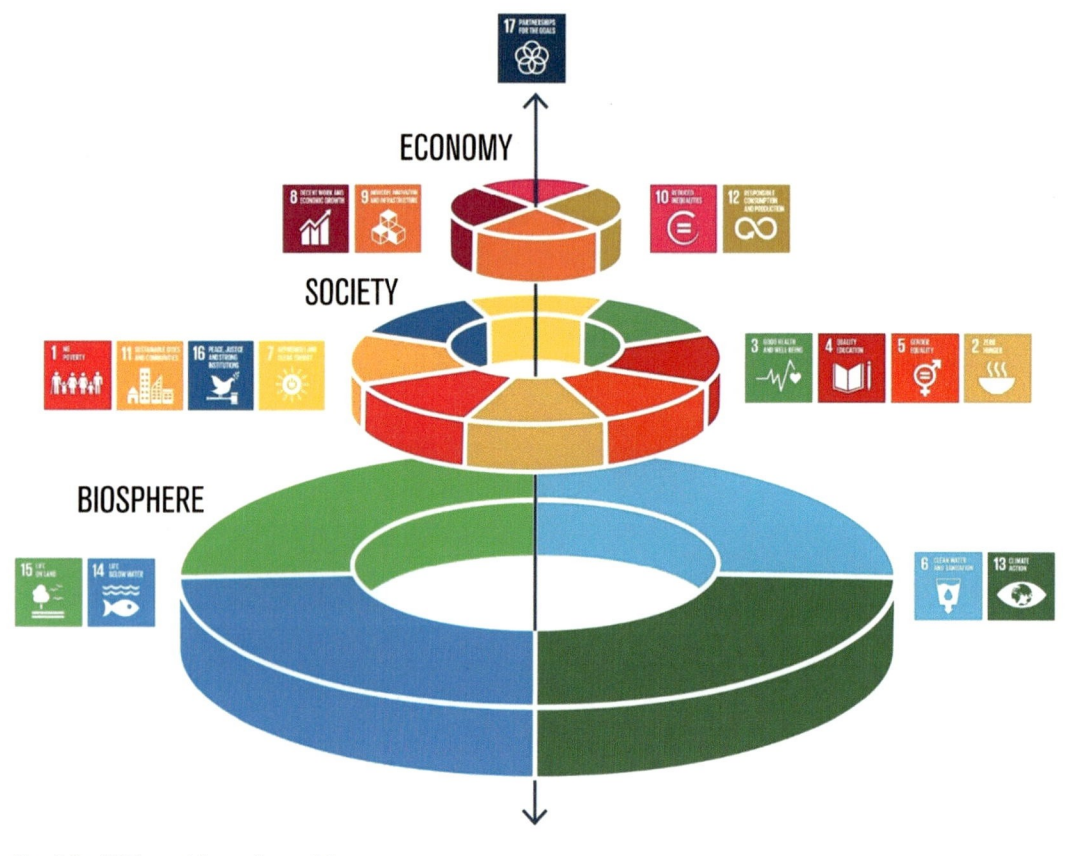

Fig. 6.3 SDGs wedding cake model

marine ecosystem is largely responsible for the planet resilience to climate change as it accumulates greenhouse gas (GHG) and water pollutants derived from agricultural and animal breeding wastes. This property is, regrettably, challenged and consequences can now clearly be observed in the form of changes in oceans currents, El Niño Southern Oscillation, oceans acidification, sea level rise, and meteorological events such as severe storms.

Land use for agriculture and animal breeding, to date, accounts for 40% of the global land surface and if these practices remain in the "business-as-usual" approach in feeding ~10 billion population projected for 2050 the additional calories required will lead to extension of land use up to 70% of Earth's land. To be able to sustain this model in a projected scenario, the amount of water required will match the total freshwater availability worldwide as already 70% of all freshwater withdrawal is currently used in agriculture (World Water Assessment Programme 2009). Water is the key element for maintaining forest and freshwater ecosystems and their biodiversity and its use need to be rationalized to safeguard the biosphere. Moreover, if dietary patterns do not rapidly shift from meat-based animal proteins towards a fish-based animal protein intake, the large freshwater use will undoubtfully jeopardize Earth's biosphere. Moreover, the use of agrochemical compounds such as fertilizers and pesticides used in agriculture is already high and can further increase in the next decades contributing to the environmental pollution especially of water reserves. Water will become an even bigger issue in the near future and access of quality water for human consumption and sanitation may be endangered. Furthermore, human-made direct and indirect events linked to deforestation and desertification because of climate change, can further accelerate biodiversity loss as most of "life on land" is concentrated in forests (United Nations 2015; FAO 2017). In general, ecosystems rely on biodiversity to survive to environmental changes and diseases, and current practices in agriculture, livestock, and fishery are hampering this genetic diversity for marketing profits.

To safeguard planet ecosystems, the food system needs to be transformed in a sustainable way thus avoiding unnecessary and hazardous routines. Fishing practices must be controlled by regulating harvesting, ending overfishing, prohibiting certain forms of fishery subsidies, and engage small-scale producers (SDG2 target 2.3 and SDG14) (United Nations 2015). Agriculture and animal breeding need to be regulated to become sustainable in the long time by ensuring a more efficient and productive water use with low impact on the environment (SDG2 target 2.4 and SDG6).

Beyond the food system, cross-sectoral interventions must be implemented to tackle climate change. For example, governments and the global community ought to further commit and strengthen the United Nations Framework Convention on Climate Change (UNFCCC)—a treaty adopted in 1992 during the Earth Summit in Rio de Janeiro (Brazil)—and the Paris Climate Change Agreement. Furthermore, "climate-smart" approaches are necessary, including adaptation and building resilience to climate change by diversifying production through exploitation of genetic diversity of different species, and by reducing GHG emissions (SDG2 target 2.4 and SDG13). Indeed, food production can become resilient and survive to environmental changes and natural disasters if the food system maintains and protects the genetic diversity of seeds, plants, and animals (SDG15 and SDG2 targets 2.4 and 2.5) (TEEB 2018).

Food and Society

Currently agriculture, livestock, and fishery practices provide food disproportionally across the globe. The world is halved in economic wealthier communities enjoying food availability, affordability, and its overproduction while low-resource countries struggle to ensure access to food leading to undernutrition and eventually to morbidity (communicable and non-communicable diseases) and premature death especially among women, children, and adolescents. Overnutrition and overconsumption, with diets high in calories, lead to overweight and obesity especially if coupled with a sedentary lifestyle. Nowadays, current practices to maintain the large-scale food production are backed by wide reliance on fertilizers, pesticides, and agrochemicals. However, such practice is leading to detrimental effects on human health and environment by polluting soil and water.

Low- and middle-income countries are the one most affected by the vicious cycle of food insecurity and poverty. This scenario together with high rate of unemployment, low-salary jobs, and social distress can result in conflicts and migration outflows. Another form of social attrition is gender inequality that still endures in many settings especially in economically poorer countries where women, despite being key actors in the society and at workplace (for instance, in agriculture they comprise 43% of workforce), still do not entirely benefit from adequate wage and the social position derived from agrofood system.

Moreover, modern food systems are currently consuming, directly (e.g. electricity and fossil fuels) and indirectly (e.g. to manufacture farm equipment), about 30% of the world's available energy, producing >20% of world's GHG emissions. Most of the energy consumed by the food industry (~70%) is required for food transport and distribution and in food preparation. In the past 50 years, the surge in energy consumption has resulted in the increase in food price that further hinders food access for people living in poverty (FAO 2012). Nevertheless, even if food price will overall decrease by addressing energy issues alone, market-driven fluctuation in food prices may still prevent vulnerable groups in their purchasing quality food, thus leading to food insecurity.

The transition required in the agrofood system depends on countries' decisions to move forward in reducing poverty and sustaining affordable access to quality food (SDG2 targets 2.1 and 2.2). This is especially important among women, children, and adolescents as it will prevent undernutrition consequences, including premature mortality. Health programmes can assist in this task and can also contribute in promoting balanced and "healthy" diets to prevent development of NCDs and in monitoring the food chain to prevent and mitigate foodborne disease outbreaks (SDG3).

Ending poverty can safeguard the right of people to access quality food and improve the general level of education in the population, hence alleviating inequalities. As agriculture provides work to large part of the global population, the most effective tools in tackling poverty and improving people income through the food sector, are the adoption and implementation of policies that facilitate connections of small-scale food producers and smallholders to markets (SDG2 target 2.3). In addition, social protection measures to prevent financial shocks among farmers and assist vulnerable groups in accessing financial services (SDG1) are necessary. Improvement in daily income can enable families to realize the importance of quality education for their children and the role played by education and professional training in improving their skills and capacities throughout life (SDG4). More importantly, strategies to promote food security and availability together with policies focused on ending poverty is expected to contribute in alleviating social difficulties and eventually related crimes, violence, and conflicts (SDG16). Furthermore, measures to offset gender inequality in society and at workplace need to be stressed and implemented (SDG2 target 2.3 and SDG5).

To reach sustainability, agri-food systems need to reduce their overall energy consumption (i.e., avoiding mechanical soil tillage), decouple from fossil fuels, and promote renewable energy sources employment (SDG2 target 2.4 and SDG7). Finally, food wasting management is already challenging in many parts of the world and it will become even more complicated in the near future with the global population projected to further increase and be concentrated in highly populated and polluted cities. Therefore, sustainable cities and national policies to handling food wastes need to be simultaneously and urgently implemented (SDG11).

Food and Economy

Inequalities are widespread across sectors and within the food system. One source of injustice is food lost or wasted alongside the food chain. Annually, indeed, one-third of overall food produced for human consumption (1.3 billion tonnes) is lost or wasted along the food chain (FAO 2019) and these wastes hamper the food system sustainability further deepening the human footprint on planet Earth. Moreover, the food industry generates also a large amount of social and economic inequalities that hamper some individuals and groups in accessing income, food, health, and education Agriculture, the world largest employer (~1.3 billion workers worldwide) is, regrettably, also responsible for the largest pool of working children (108 of 152 million working children) (FAO 2020a, b). They consequently are exposed to physical and psychological consequences on their health status that may perdure throughout life. Indeed, agriculture is a hazardous job as exposes workers to agrochemicals and injuries and considering that fewer than 20% of them are covered by social protection measures. This exposes workers and their households to financial insecurity if they become ill, injured, or die (FAO-ILO-IUF 2007).

Food industries, countries, and the international community need to revert the current trends of food lost and wasted during its production, processing, transport, sale, and conservation once purchased. To achieve this goal, policies and strategies are needed to reduce and, when feasible, eliminate food waste including by educating the society (SDG12). To prevent further income and social gaps, countries need to ensure decent working conditions and salaries for all in compliance with labour rights based on recommendations by the International Labour Organization (ILO) and rapidly put an end to child labour (SDG2 target 2.2 and SDG8). Agricultural infrastructure and technical innovations ought to be properly financed by governments and investors to support small-scale producers in accessing to (food) markets (SDG2 target 2.3 and SDG9) and financial and social protection measures should be developed to safeguard vulnerable groups and promote their social, economic, and political inclusion in policies (SDG10).

In conclusion, due to the complexity of the current eco-agri-food system there is a desperate need to forge and revitalize partnerships focused on global food sustainability and, in general, human development gathering stakeholders across sectors beyond that specifically involved in food (SDG 17).

6.3.2 From Sustainable Food Production to Sustainable Food Supply and Consumption: Healthy Diets for Sustainable Agriculture

6.3.2.1 What Is a Healthy Diet?

The guidelines of the World Health Organization summarize in the following points the essential features of a healthy diet that meets the individual needs determined by different factors (e.g. age, gender, bodyweight, physical activity, etc.).

Concerning the nutrients composition:

– intake adequate amounts of good quality protein, taking into account that the quality depends on the composition of amino-acids and how easily the protein is digested and then used by the body (WHO 2007).
– intake enough amounts of fibre every day, from fruits, vegetables, and whole grains (Mann et al. 2007).
– intake carbohydrates, preferring the unrefined complex carbohydrates, supplied by food rich in fibre, vitamins, and minerals (Mann et al. 2007), while free sugars (i.e. simple carbohydrates contained in industrial foods or drinks, honey, fruit juices, etc.) should not exceed more than 10% of total calories (WHO 2015a, b).
– consume moderate quantities of fats, that should not provide more than 30% of total calories. Priority should be given to unsaturated fats (e.g. olive, fish, avocado, nuts, sunflower, and rapeseed oils), while saturated fats (e.g. butter, palm and coconut oils, fatty meat, cheese, and cream) should not exceed more than 10% of total calories and the partially hydrogenated vegetable oils (i.e. industrial trans-fats) should be completely avoided (WHO 2003; FAO 2010).
– consume moderate quantities of salt, less than 5 g/day, including that contained into the processed foods (e.g. bread, snacks, cheese, processed meats, and ready meals), preferring iodized salt (WHO 2012).
– finally, consume the adequate daily calories, avoiding the overeating and the undereating.

Concerning the foods composition:

– eat a wide variety of foods, from different food groups,
– promote the plant-based foods, eating more than 400 g/day of vegetable and fruits (WHO 2003). In particular, eat legumes regularly (e.g. beans, peas, lentils), as low-fat food sources, rich in protein and fibre, and as possible alternative to meat.
– cereals, potatoes and other starchy roots do not count as portions of vegetables, but are an important source of carbohydrates that need to be included in a balanced diet, preferring the no-processed or minimal processed forms (e.g. whole grain), rather than the refined forms (e.g. bread, pasta, white rice).
– do not exceed in the consumption of meet, eating not more than 500 g of cooked meat per week, avoiding, if possible, the processed meat and limiting the red meat, associated with an increased risk of certain cancers (WHO 2003, AICR 2007). On the other hand, assure to cover the needs of protein and minerals with animal source foods, given their important role during pregnancy and children's growth.
– consume moderate amounts of milk and dairy food, choosing the products low in fat, salt, and sugar.
– eat fish or shellfish regularly (e.g. twice a week).

- consume moderate amounts of fats and oils, preferring the vegetable sources with unsaturated fats.
- promote foods from recognized or certified sustainable sources, locally produced, fresh and home-prepared.

Moreover, it is now important to add the concept of health or functional food to that of healthy diet: this is gaining more and more attention as it recognizes the "medical" role of food. This concept is used for all the foods that, beside correctly satisfying the nutritional needs of humans, are able to prevent or reduce the risk of nutrition-related diseases (e.g. celiac disease), interacting with a variety of target body functions (Roberfroid 2000; Jnawali et al. 2016).

6.3.2.2 What Is Meant by a Sustainable Diet?

The assessment of food environmental sustainability involves a wide range of aspects that can be hardly addressed through a single assessment and require an integrated and systemic approach. The evaluations should be able to analyse the food system, considering the several levels in which it is articulated (i.e. field, farm, landscape, territorial, and foodshed region level), and then its specificity compared to other production processes. The sustainability of a food product or a pattern of food consumption (i.e. a diet) are often evaluated through the Life Cycle Assessment (LCA) methodology, with the support of the related inventory and modelling software (Tilman and Clark 2014; Clark et al. 2019). The LCA is focused on single products, field or farm level, and takes into account ten main impact categories: climate change, ozone depletion, photo-oxidant formation, acidification, eutrophication, land use, water use, non-renewable energy use, human toxicity, and ecotoxicity.

However, the LCA, developed at first for industrial systems, shows weaknesses when applied to complex and live agro-ecosystems characterized instead by interactions with exogenous variables, long-term processes, cumulative effects and many sources of variability in time and space. This led to misinterpretations of LCA outcomes, neglecting the information at landscape level and many key factors relevant for the long-term sustainability (Notarnicola et al. 2017). Then, in order to overcome these criticalities, the sustainability assessment should include further indicators (Notarnicola et al. 2017; Tuomisto et al. 2015; Kremen and Miles 2012), evaluating the following aspects:

- the soil quality (e.g. organic matter, structure, biological fertility, and edaphic biodiversity);
- the biodiversity (i.e. natural fauna, flora, and production system genetic sources) and the natural habitats;
- the ecosystems services generated as benefits (e.g. with respect to soil erosion, water retention, easiness of tillage, maintenance of soil long-term productivity, pest natural enemies, pollinators, etc.);
- the risk, due to the use of pesticides or other chemicals, for the ecosystem or human health, depending on the volume of the receiving compartment, the concentration and the time of exposure;
- the weighting of post-harvest steps on sustainability (transport, storage, packaging, and food preparation);
- the dependency of the environmental impacts from the cultivation and geographical context (e.g. climate, soil, seasonality, local availability of natural resources, transport distances for inputs, the spectrum of practice existing within the same category of farming system based on farmer' expertise and preferences, etc.)

Finally, an environmentally sustainable diet, in addition to being healthy, should be acceptable from a social, economic and ethical point of view with respect to: food need and food security issue, culinary tradition, the farming system resilience (e.g. risks due to yield or price variability), the employment, the workers and farmers conditions and pay, etc.

6.3.2.3 The Change of Diet over Time and Consequences

Slightly more than 50% of the world population live in urban areas and an increase up to 60% is expected by 2030. Urbanization seems to be a necessary step that characterizes the transition from poverty and low-income status to middle-income status worldwide (Elkins et al. 2019). The related increase of income and of other linked socioeconomic phenomena have determined changes in feeding behaviour, food consumption basket, and food supply system, towards unhealthy diets favoured by a high-environmentally impacting agriculture and food industry:

- since the urban food demand overcomes what could be produced in the surroundings rural areas, the cities exploit a complex and large global supply chains that replaces the short and local supply chains and breaks the consumer-producer connection, with many intermediaries and long-distance transport (Satterthwaite et al. 2010),
- the urbanization is often accompanied by an increasing consumption of non-traditional food, fast-food, processed products or prepared meals, and a decline in fresh or raw products, home-prepared and home-cooked food, also in relationship with more opportunities for women to work outside home (Regmi and Dyck 2001).
- in the global food system, the ultra-processed products (e.g. frozen products, snacks, soft drinks) are becoming dominant in high- and middle-income countries. These hyper-palatable, cheap, ready-to-consume foods are part of energy-dense, fatty, sugary or salty, and obesogenic diets. They are connected with the worldwide rapid increase of obesity and related diseases (e.g. hypertension, type 2 diabetes) (Monteiro et al. 2013).
- the shift towards moderate or high income is in general accompanied by an increasing demand of meat (i.e. ruminant meat, pork, poultry, and processed meat), dairy and egg, fish and seafood processed meat, and "empty calories" foods (i.e. refined sugars, refined animal fats, oils, and alcohol) (Tilman and Clark 2014; Godfray et al. 2018), while a relative decrease of fruits, vegetables, and plant protein (i.e. legumes). This diet is associated with an increase of non-communicable diseases (i.e. chronic diseases: cardiovascular diseases—coronary artery diseases, heart attacks -, diabetes and cancers - colorectal cancer).
- the structural change in food consumption follows also the trend drawn by the green revolution, with a collective decline in the consumption of traditional coarse grain cereals and minor cereals (e.g. barley, oats, rye, millet, and sorghum), in favour of high-yielding cereals, relatively poor in micronutrient, such as rice and wheat bread cultivars (DeFries 2018; Regmi and Dyck 2001).

These changes negatively affect not only human health but also drive the agricultural sector towards farming systems and productions that impact heavily on the environment. One just needs to consider the higher environmental and energy costs involved by the long-distance transports and the cold chain, the industrial food processing and packing. Moreover, the abandonment of minor and traditional crops, usually characterized by low needs in terms of agronomic inputs (e.g. agrochemicals, water, tillage), and the decline of leguminous plants with a key role in crop rotation and soil fertility, foster the shift towards monocropping and high-input farming systems. Finally, the demand for meat-protein in replace of plant protein involves higher resources used per unit of supplied energy—due to the energy loss at each trophic level-up—and higher greenhouse gases (GHG) emissions with usually in the lead the ruminant production, followed by the non-ruminant mammals and the poultry (Godfray et al. 2018).

The policy should address the nexus between diet, health, and environment, and identify the trade-off in this trilemma, minimizing the negative externalities on both people's health and natural resources. History suggests that change in dietary behaviours is slow, but targeted norms and interventions, coordinated with civil society, can redraw the food consumption trends and redesign the food system, within the unavoidable framework of development and modernization (Godfray et al. 2018).

6.3.2.4 Alternative Diets

Since food production is driven by consumer's demand, targeted changes in food behaviour, with widely adoption of alternative diets, can foster the environmental performance of agriculture.

LCA-based studies (Tilman and Clark 2014; Clark et al. 2018; Clark et al. 2019) evaluated the benefits of alternative diets, such as: "vegan" (i.e. only plant-based foods), "vegetarian" (i.e. mainly plant-based products, and inclusion of eggs and dairy), "pescetarian" (i.e. vegetarian diet but with the inclusion of fish and seafood), and "Mediterranean" (i.e. diet rich in vegetables, fruit, fish and seafood, inclusive of grains, eggs, dairy, and moderate amounts of meat), comparing them with the typical omnivorous diet (i.e. the global average and income-dependent diet). The latter includes all food groups with a higher consumption of "empty calories", dairy, eggs, meat and then an exceeding intake of saturated fat and red meat.

In scenario analyses, the alternative diets were able to reduce the risk of overweight, obesity, and non-communicable diseases (e.g. incidence reduction up to minus 41% for diabetes, −13% for cancer, −26% for coronary artery disease, −18% for overall mortality rates for all causes combined; Tilman and Clark 2014). In parallel, basically, the environmental impact decreases with the replacing of animal-based foods with plant-based products (i.e. fruit, legumes, nuts, whole grains, and vegetables) in the diet (Clark et al. 2018). For nearly all impact categories considered, the ruminant meat and processed red meat are the "very high impacting" foods, followed by the "intermediate-high" impacting foods, such as pork and poultry meats, dairy, and eggs, while the fish impact varies between the two groups depending on the production systems: the fish from trawling fishery (where nets are dragged across the seabed) and recirculating aquaculture (where water is consistently cycled and filtered) have a higher impact than non-trawling fishery and non-recirculating aquaculture, because of their higher energy inputs. Therefore, the alternative diets result in major environmental benefits with a substantial reduction of impacts per unit of supplied kcal. This concerns: the GHG emissions (e.g. GHG emissions reduction: −30% with Mediterranean, −45% with pescetarian, and − 55% with vegetarian diets), the land clearing due to the cropland needs (Tilman and Clark 2014), the acidification and eutrophication potential, and the release of nitrogen pollutants (i.e. Clark et al. 2018; Clark et al. 2019). Reduction of livestock products leads to improved food system nitrogen use efficiency, better air and water quality, with less nitrogen leaching and gasses emissions, and major land availability, previously used for animal feed production, for cereal or perennial energy crops (Westhoek et al. 2014). In particular, the ecologically-inefficient ruminants (e.g. cattle, goats, and sheep) show bigger carbon and nitrogen footprints and pressure on ecosystems and biodiversity due to the higher land use and land clearing, than the monogastrics (e.g. poultry, pigs) (Westhoek et al. 2014; Machovina et al. 2015).

Despite the general correlation between a healthy diet and a low environmental impact, some considerations about these results are needed, with the aim to correctly identify the sustainable diets on the basis of a trade-off between different key aspects.

First, the environmental impacts are referred to coarse units, such as total protein or calories supplied, neglecting the nutritional and functional values of foods and raising up the need of a definition of more sophisticated nutritional quality indices, through collaboration between environmental and nutritional sciences (Heller et al. 2013). For instance, foods that are not calorie-dense or protein-dense, such as fruits and vegetables, are important sources of micronutrients, antioxidants, and fibre; the high impacting meat and fish are sources of essential fatty acids, micronutrients, minerals and vitamins; on the other hand, unhealthy foods such as butter, sugar, oils, and salt result in lower GHG emissions (Tilman and Clark 2014; Godfray et al. 2018).

Secondly, most of time the reference unit for the environmental impact assessments is derived from the unit of product per hectare. This often leads to penalizing the extensive and low-input systems in favour of high-yielding and intensive agriculture, neglecting the impact determined by the pressure of the agricultural activities per unit of area, even if it directly affects the agro-ecosystem' biodiversity and habitats, the ecosystem services, the soil and water quality, the toxicity of pesticide, and other chemicals (Notarnicola et al. 2017). These aspects are described by non-LCA indicators that should be taken into account in the evaluation of the long-term sustainability of a diet.

Third, the assessments pay little attention to the wide variability in the impact, derived from the same food, depending on the adopted production system (Notarnicola et al. 2017). The sustainability evaluation of a diet should include this aspect, especially considering that farming systems can affect not only the environmental performance but also the food nutritional value: e.g. the grass-fed beef and dairy have nutritionally superior fatty acid and vitamin content than grain-fed cattle (Daley et al. 2010); the organic foods may reduce the risk of allergic disease and pesticides exposure, and they show a lower cadmium content and the milk has higher content of omega-3 fatty acids (Mie et al. 2017).

Finally, the sustainability of a diet is a site-specific concept that needs to take into account the geographical and social variables affecting the natural resource availability, the food system beyond the field level (i.e. most of evaluations are "from cradle to farm gate") and the compatibility with food security issues: e.g. livestock is crucial for small-scale farms placed in marginal lands unsuitable for crops cultivation, because it uses resources otherwise not available as food, converting grass into a resilient and precious reserve of concentrated protein (Laurance et al. 2014).

Aubert et al. (2015), starting from IPCC investigations and a recent EU document, compared the impacts of a European's current diet compared to a possible 2050 diet improved both in human and planet health (Fig. 6.4).

Starting from a significant reduction in the consumption of meat from intensive livestock

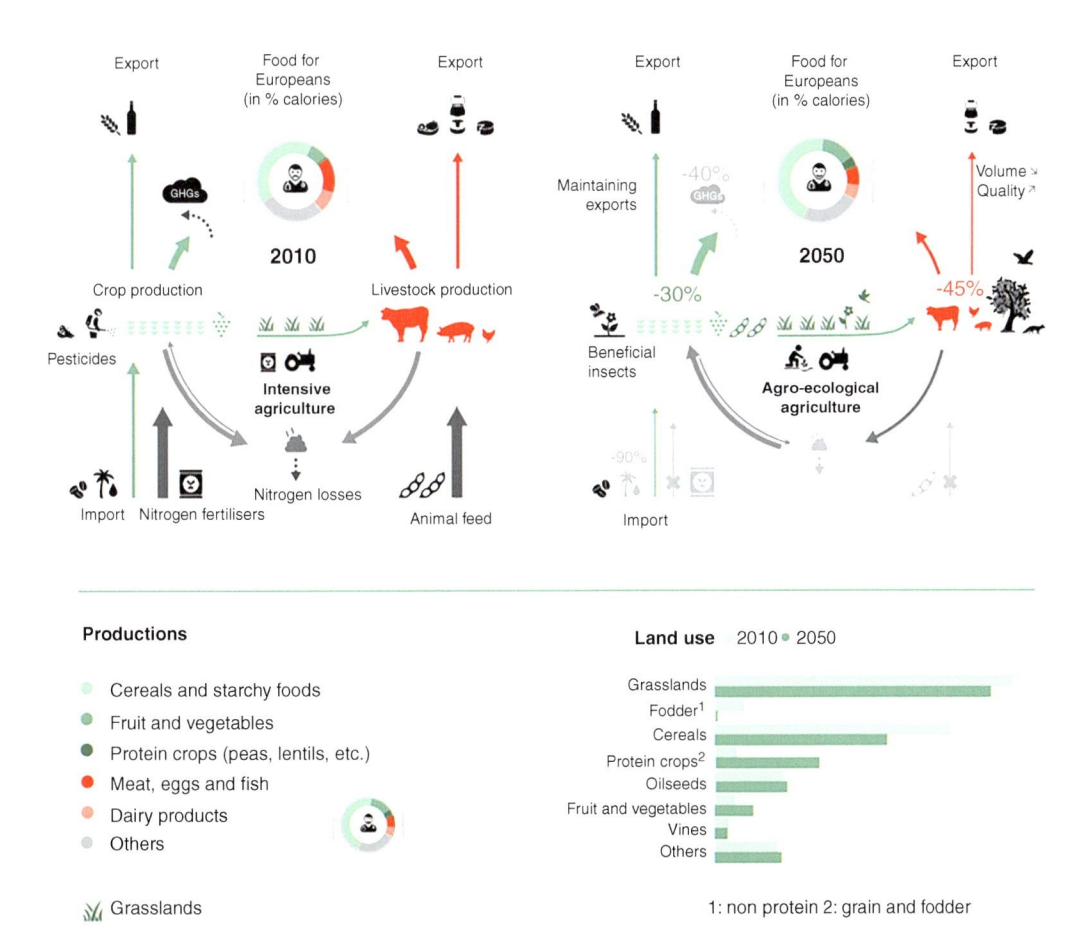

Fig. 6.4 New diets for human and planetary health with permissions from Poux, X., Aubert, P.-M. (2018). Une Europe agroécologique en 2050: une agriculture multifonctionnelle pour une alimentation saine. Enseignements d'une modélisation du système alimentaire européen, Iddri-AScA, *Study* N°09/18, Paris, France, 78 p

and from an increase in forage produced locally by adopting agroecological criteria, greenhouse gas emissions, impacts from pesticides, and soy imports would be reduced.

6.3.3 Education for Sustainable Development and FAO Principles of Agrofood Sustainability

"With a world population of 7 billion people and limited natural resources, we, as individuals and societies need to learn to live together sustainably" (UNESCO). In this context, the reorienting of education, improving the access to quality education and including sustainable development issues into teaching and learning, empower people to change the way they think and work towards a sustainable future. The education of sustainable development fosters new thinking approaches, knowledge, skills, values, and behaviours, that are needed to be responsible actors and to plan changes at all levels and in all social contexts.

Since 1972, the United Nations plans International Earth Summits, grouping the world leaders with the aim to cooperate and address the global challenges of development. The International Earth Summits retraces the main stages that led to the Global Agenda of sustainability (Hammoud and Tarabay 2019). During the first summit in Stockholm (1972), the environmental issues were institutionally recognized at international level, highlighting the need of international environmental laws. After that, other four summits followed: in Nairobi (1982), Rio de Janeiro (1992), Johannesburg (2002), and in Rio de Janeiro (2012, also known as Rio +20). However, the environmental safeguard has been linked to development only in the 1992 Rio Earth Summit with a declaration that sustainable development involves social and economic development as well as environmental protection (Meakin 1992). Before that, it was the Brundtland Commission (United Nations 1987) that defined the sustainable development concept as we understand it today. In 1987, the Brundtland Commission in an international initiative aiming to face together the global environmental challenges, produced a report ("Our Common Future") where sustainable development was defined as a paradigm of resource use that "meets the needs of the present without compromising the ability of future generations to meet their own needs".

In parallel, other initiatives led by high-level educational institutions (i.e., universities) drew the path of integration between the sustainable development issues and the educational aspects, defining the Higher Education Agenda (Hammoud and Tarabay 2019), starting from the Magna Charta of European Universities (Magna Charta Universitatum 1988).

One of the principles of the Magna Charta Universitatum, adopted by 388 universities in 1988, stated that "universities must give future generations education and training that will teach them, and through them others, to respect the great harmonies of their natural environment and of life itself". After that, in 1990, the Talloires Declaration, signed by over 500 universities, defined the "sustainable university", and the action plan aimed to integrate the sustainability and environmental literacy in teaching, research, and outreach. In 1993, the International Association of Universities (IAU) adopted the Kyoto Declaration on Sustainable Development, in which universities are called to support the dissemination and the understanding of sustainable development (International Association of Universities 1993). In 1994, as answer to the Rio de Janeiro Earth Summit of 1992, hundreds of European universities (currently more than 320) signed the COPERNICUS University Charter for Sustainable Development that declared the relevant role covered by the sustainable development issue in the university' priorities (COPERNICUS 1994).

In 2001, the Declaration on Higher Education for Sustainable Development (COPERNICUS 2001) was the result of the concepts expressed in the International COPERNICUS Conference "Higher Education for Sustainability Towards the World Summit on Sustainable Development (Rio + 10)". The document defined "the ultimate goal of education for sustainable development, to impart the knowledge, values, attitudes and skills needs to empower people to bring about the changes required to achieve sustainability", calling for specific actions in this direction.

In March 2005, the United Nations officially opened the "Decade of Education for Sustainable Development" (2005–2014). The "Education for Sustainable Development" section of UNESCO offered oversight and advice, coordinating the efforts of member states. The Decade aimed "to integrate the values inherent in sustainable development into all aspects of learning to encourage changes in behaviour that allow for a more sustainable and just society for all" since "education for sustainable development strengthens the capacity of individuals, groups, communities, organizations and countries to make judgments and choices in favour of sustainable development (United Nations 2005).

Sustainable development itself is a constantly evolving concept that identifies different priorities depending on the historical or geographical contexts, and that can be linked to quantitative concepts of economic growth, as well as to the qualitative sphere underlying the social aspects (e.g. reduction of inequalities, marginalization, and war conflicts) or to environmental issues (e.g. mitigation of global warming and safeguard of natural heritage for future generations). Sustainable development can be understood as the learning process required to improve the humanity conditions, or as a target to reach, or as an alternative system of thinking and acting, and a different lifestyle for individuals and society. Whatever the interpretation is, sustainable development is intrinsically linked to new educational processes.

The education for sustainable development is a dynamic process in continuous expansion over different levels, that envisions "a world where everybody has the opportunity to benefit from education and learn the values, behaviour and lifestyles required for a sustainable future and for positive societal transformation" (UNESCO 2014). The United Nations Economic Commission for Europe stated that "education, in addition to be a human right, is a prerequisite for achieving sustainable development and an essential tool for good governance, informed decision-making and promotion of democracy" UNECE (2009).

The education for sustainable development attempts to bring three educational traditions together: environmental education, development education, and citizenship education (Mannion et al. 2011). During the first decade of the twenty-first century, the concept of Education for Sustainable Development progressively replaced that of Environmental Education. There is a close relationship between these two terms, since the environmental educators were the first to endorse education for sustainable development. On the other hand, the Education for Sustainable Development focuses on all the spheres of sustainability, while the Environmental Education is more reductive and referred to school curricula, contributing, like other educational fields (e.g. human rights education, ecological economics education), to Education for Sustainable Development, in terms of content and pedagogy (UNESCO 2012).

Finally, the education for sustainable development encompasses a global vision because, as UNESCO states, "while the world may be increasingly interconnected, human rights violations, inequality and poverty still threaten peace and sustainability" and the Global Citizenship Education is the response to these challenges. Therefore, education for sustainable development should work by empowering learners to understand the global dimension of sustainability, promoting their active role for more peaceful, tolerant, inclusive, secure and sustainable societies" (UNESCO 2015).

6.4 Conclusions

The widespread innovation model of the so-called Green Revolution, after making improvements to the quantitative and qualitative levels of the diets of a large part of the world population, has recently highlighted significant problems. The negative impacts of an agri-food model driven by mainly industrial international interests and based on a growing supply of standardized food goods are being highlighted both on natural resources (soil, air, biodiversity) and on human health. The new frontiers of innovation, which require a new systemic approach to the complex problems of sustainability, push us to find systemic solutions in the definition of sustainable diets supported by production processes that respect the limited natural resources. Agenda 2030 offers an integrated framework for sustainable development and invites us to reflect on the importance of the quality of citizen education, required to achieve and maintain a new attitude of respect and care for our health and that of the planet.

References

Afshin A et al (2019) Health effects of dietary risks in 195 countries, 1990–2017: a systematic analysis for the Global Burden of Disease Study 2017. Lancet. https://doi.org/10.1016/S0140-6736(19)30041-8

AICR (2007) 2007. World Cancer Research Fund/American Institute for Cancer Research. Food, nutrition, physical activity and prevention of cancer: a global perspective. AICR, Washington DC

Allen LN, Hatefi A, Feigl AB (2019) Corporate profits versus spending on non-communicable disease prevention: an unhealthy balance. Lancet Glob Health. https://doi.org/10.1016/S2214-109X(19)30399-7

Altieri M (1989) Agroecology: a new research and development paradigm for world agriculture. Agric Ecosyst Environ 27:37–46

Aubert PM, Schwoob MH, Poux X (2015) Agroecology and carbon neutrality in Europe by 2050: what are the issues? Findings from the TYFA modelling exercise. Findings from the TYFA modelling exercise. IDDRI, Study N°02/18

Begley TH et al (2005) Perfluorochemicals: potential sources of and migration from food packaging. Food Addit Contam. https://doi.org/10.1080/02652030500183474

Blüher M (2019) Obesity: global epidemiology and pathogenesis. Nat Rev Endocrinol. https://doi.org/10.1038/s41574-019-0176-8

Bocchi S (2019) Agroecology: relocalizing agriculture accordingly to places. In: Fanfani D, Mataran A (eds) 2019, bioregional planning and design: volume II. Issues and practices for a bioregional regeneration. Springer, Cham

Briggs ADM et al (2017) Health impact assessment of the UK soft drinks industry levy: a comparative risk assessment modelling study. Lancet Public Health. https://doi.org/10.1016/S2468-2667(16)30037-8

Clark M, Hill J, Tilman D (2018) The diet, health, and environment trilemma. Annu Rev Environ Resour 43:109–134

Clark MA, Springmann M, Hill J, Tilman D (2019) Multiple health and environmental impacts of foods. Proc Natl Acad Sci 116(46):23357–23362

Cochrane WW (1993) Farm prices, myths and reality. Univ. Minnesota Press, Minneapolis, MN

COPERNICUS (1994) The university charter for sustainable development

COPERNICUS (2001) The Luneburg Declaration on higher education for sustainable development. https://www.iau-hesd.net/sites/default/files/documents/2001_-_the_luneburg_declaration_fr.pdf

Daley CA, Abbott A, Doyle PS, Nader G, Larson S (2010) A review of fatty acid profiles and antioxidant content in grass-fed and grain-fed beef. Nutr J 9:10

DeFries R (2018) Trade-offs and synergies among climate resilience, human nutrition, and agricultural productivity of cereals–what are the implications for the agricultural research agenda. In Background paper for the CGIAR ISPC Science Forum (pp. 2018-09)

Di Cesare M et al (2016) Trends in adult body-mass index in 200 countries from 1975 to 2014: a pooled analysis of 1698 population-based measurement studies with 19.2 million participants. Lancet. https://doi.org/10.1016/S0140-6736(16)30054-X

EAT-Lancet Commission (2019) Healthy diets from planet: food, planet, health. Lancet. https://eatforum.org/content/uploads/2019/07/EAT-Lancet_Commission_Summary_Report.pdf [Internet]

Elkins P, Gupta J, Boileau P (eds) (2019) Global environment outlook: GEO-6: healthy planet, healthy people. Cambridge University Press, p 31

FAO (2003) General principles of food hygiene. Rome

FAO (2010) Fats and fatty acids in human nutrition: report of an expert consultation. FAO Food and Nutrition Paper 91. Food and Agriculture Organization of the United Nations, Rome

FAO (2012) Energy-Smart Food at FAO—An Overview. Food and Agriculture Organization of United States

FAO (2014) Global hunger index: the challenge of hidden hunger. Rome. https://doi.org/10.2499/9780896299269GHI2010

FAO (2016) Fisheries, aquaculture and climate change. Rome

FAO (2017) Keeping on SDG 15: working with countries to measure indicators for forests and mountains [Internet]. KEEPING ON SDG 15 AN EYE Working. Rome. www.fao.org

FAO. The state of food and agriculture [internet]. 2019 [cited 2020 Jun 26]. http://www.fao.org/state-of-food-agriculture/2019/en/

FAO (2020a) Polluting our soils is polluting our future. http://www.fao.org/fao-stories/article/en/c/1126974/ (Accessed: 15 July 2020)

FAO (2020b) FAO framework on ending child labour in agriculture. FAO framework on ending child labour in agriculture

FAO and WHO (2019) Joint FAO/WHO Expert Meeting in collaboration with OIE on Foodborne Antimicrobial Resistance: Role of the Environment, Crops and Biocides—Meeting report. Microbiological Risk Assessment Series no. 34. Rome

FAO-ILO-IUF (2007) Agricultural workers and their contribution to sustainable agriculture and rural development. ILO. Organization, Geneva

Francis CA, Wezel A, (2015) Agroecology and Agricultural Change. In: James D. Wright (editor-in-chief), International Encyclopedia of the Social & Behavioral Sciences, 2nd edition, Vol 1. Oxford: Elsevier. pp. 484–487

Geens T et al (2010) Intake of bisphenol a from canned beverages and foods on the Belgian market. Food Addit Contam Part A Chem Anal Control Expo Risk Assess. https://doi.org/10.1080/19440049.2010.508183

Gliessmann S (2016) Agroecology: the ecology of sustainable food systems, 3rd edn. CRC Press, Boca Raton, p 405

Godfray HCJ, Aveyard P, Garnett T, Hall JW, Key TJ, Lorimer J, Pierrehumbert RT, Scarborough P, Springmann M, Jebb SA (2018) Meat consumption, health, and the environment. Science 361(6399):eaam5324

Gore AC (2016) Endocrine-disrupting chemicals. JAMA Intern Med. https://doi.org/10.1001/jamainternmed.2016.5766

Gore AC et al (2015) EDC-2: the Endocrine Society's second scientific statement on endocrine-disrupting chemicals. Endocr Rev. https://doi.org/10.1210/er.2015-1010

Hammoud J, Tarabay M (2019) Higher education for sustainability in the developing world: a case study of Rafik Hariri University in Lebanon. Eur J Sustain Dev 8(2):379–379

Heller MC, Keoleian GA, Willett WC (2013) Toward a life cycle-based, diet-level framework for food environmental impact and nutritional quality assessment: a critical review. Environ Sci Technol 47(22):12632–12647

International Agency for Research on Cancer (2018) Red meat and processed meat: IARC monographs on the evaluation of carcinogenic risks to humans Volume 114. https://doi.org/10.1103/PhysRevA.86.012307

International Association of Universities (1993). Kyoto Declaration on sustainable development. https://www.iau-aiu.net/IMG/pdf/sustainable_development_policy_statement.pdf

Jnawali P, Kumar V, Tanwar B (2016) Celiac disease: overview and considerations for development of gluten-free foods. Food Sci Human Wellness 5(4):169–176

Kickbusch I, Allen L, Franz C (2016) The commercial determinants of health. Lancet Glob Health. https://doi.org/10.1016/S2214-109X(16)30217-0

Kim MJ et al (2018) Association between perfluoroalkyl substances exposure and thyroid function in adults: a meta-analysis. PLoS One. https://doi.org/10.1371/journal.pone.0197244

Kremen C, Miles A (2012) Ecosystem services in biologically diversified versus conventional farming systems: benefits, externalities, and trade-offs. Ecol Soc 17(4):40

Landrigan PJ et al (2018) The Lancet Commission on pollution and health. Lancet. https://doi.org/10.1016/S0140-6736(17)32345-0

Laurance WF, Sayer J, Cassman KG (2014) The impact of meat consumption on the tropics: reply to Machovina and Feeley. Trends Ecol Evol 29(8):432

Machovina B, Feeley KJ, Ripple WJ (2015) Biodiversity conservation: the key is reducing meat consumption. Sci Total Environ 536:419–431

Magna Charta Universitatum (1988). http://www.magna-charta.org/resources/files/the-magna-charta/english

Maher D, Smeeth L, Sekajugo J (2010) Health transition in Africa: practical policy proposals for primary care. Bull World Health Organ. https://doi.org/10.2471/BLT.10.077891

Mann J, Cummings JH, Englyst HN, Key T, Liu S, Riccardi G et al (2007) FAO/WHO scientific update on carbohydrates in human nutrition: conclusions. Eur J Clin Nutr 61(Suppl 1):s132–s137

Mannion G, Biesta G, Priestley M, Ross H (2011) The global dimension in education and education for global citizenship: genealogy and critique. Glob Soc Educ 9(3-4):443–456

Meakin S (1992) The Rio Earth Summit: Summary of the United Nations Conference on Environment and Development. http://publications.gc.ca/Collection-R/LoPBdP/BP/bp317-e.htm

Mie A, Andersen HR, Gunnarsson S, Kahl J, Kesse-Guyot E, Rembiałkowska E et al (2017) Human health implications of organic food and organic agriculture: a comprehensive review. Environ Health 16(1):111

Monteiro CA, Moubarac JC, Cannon G, Ng SW, Popkin B (2013) Ultra-processed products are becoming dominant in the global food system. Obes Rev 14:21–28

Moodie R et al (2013) Non-communicable diseases 4 profits and pandemics : prevention of harmful effects of tobacco, alcohol, and ultra-processed food and drink. Lancet. https://doi.org/10.1016/S0140-6736(12)62089-3

Mozaffarian D (2016) Dietary and policy priorities for cardiovascular disease, diabetes, and obesity. Circulation. https://doi.org/10.1161/CIRCULATIONAHA.115.018585

Nagel SC, Bromfield JJ (2013) Bisphenol A: a model endocrine disrupting chemical with a new potential mechanism of action. Endocrinology. https://doi.org/10.1210/en.2013-1370

Naghavi M et al (2017) Global, regional, and national age-sex specifc mortality for 264 causes of death, 1980-2016: a systematic analysis for the Global Burden of Disease Study 2016. Lancet. https://doi.org/10.1016/S0140-6736(17)32152-9

Notarnicola B, Sala S, Anton A, McLaren SJ, Saouter E, Sonesson U (2017) The role of life cycle assessment in supporting sustainable agri-food systems: a review of the challenges. J Clean Prod 140:399–409

Parris K. 2011. Impact of agriculture on water pollution in oecdcountries: recent trends and future prospects. Int J Water Res Dev, 27:1, 33–52, https://doi.org/10.1080/07900627.2010.531898

Peralta-Videa JR et al (2009) The biochemistry of environmental heavy metal uptake by plants: implications for the food chain. Int J Biochem Cell Biol. https://doi.org/10.1016/j.biocel.2009.03.005

Perkins JH (1997) Geopolitics and the green revolution: wheat, genes, and the cold war. Oxford University Press, p 352

Pingali PL (2012) Green revolution: impacts, limits, and the path ahead. PNAS 109(31):12302–12308

Poore J, Nemecek T (2018) Reducing food's environmental impacts through producers and consumers. Science. https://doi.org/10.1126/science.aaq0216

Prùss-Ustùn A, Wolf J, Corvalàn C, Bos RNM (2016) Preventing disease through healthy environments: a global assessment of the burden of disease from environmental risks. Geneva

Quail DF, Dannenberg AJ (2019) The obese adipose tissue microenvironment in cancer development and progression. Nat Rev Endocrinol. https://doi.org/10.1038/s41574-018-0126-x

Regmi A, & Dyck J (2001) Effects of urbanization on global food demand. Book chapter in Changing structure of global food consumption and trade, 23-30

Roberfroid MB (2000) A European consensus of scientific concepts of functional foods. Nutrition (Burbank, Los Angeles County, Calif) 16(7-8):689–691

Ryan, Gross (1943) The diffusion of hybrid seed corn in two Iowa Communities. Rural Sociol 8:15

Satterthwaite D, McGranahan G, Tacoli C (2010) Urbanization and its implications for food and farming. Philos Trans R Soc B Biol Sci 365(1554):2809–2820

Scherr SJ, McNeely JA (2008) Biodiversity conservation and agricultural sustainability: towards a new paradigm of 'ecoagriculture' landscapes. Philos Trans R Soc. B.477-494363

Shammas C (1994) Changes in English and Anglo-American consumption from 1550 to 1800. in Consumption and the World of Goods

Smith-Spangler C et al (2012) Are organic foods safer or healthier than conventional alternatives?: a systematic review. Ann Intern Med. https://doi.org/10.7326/0003-4819-157-5-201209040-00007

Springmann M, Clark M et al (2018a) Options for keeping the food system within environmental limits. Nature. https://doi.org/10.1038/s41586-018-0594-0

Springmann M, Wiebe K et al (2018b) Health and nutritional aspects of sustainable diet strategies and their association with environmental impacts: a global modelling analysis with country-level detail. Lancet Planet Health. https://doi.org/10.1016/S2542-5196(18)30206-7

TEEB (2018) The economics of ecosystems and biodiversity. For agriculture and food: scientific and economic foundations. Geneva

Tilman D, Clark M (2014) Global diets link environmental sustainability and human health. Nature 515(7528):518–522

Tóth G et al (2016) Heavy metals in agricultural soils of the European Union with implications for food safety. Environ Int. https://doi.org/10.1016/j.envint.2015.12.017

Troeger C et al (2017) Estimates of global, regional, and national morbidity, mortality, and aetiologies of diarrhoeal diseases: a systematic analysis for the Global Burden of Disease Study 2015. Lancet Infect Dis. https://doi.org/10.1016/S1473-3099(17)30276-1

Tuomisto HL, Hodge ID, Riordan P, Macdonald DW (2015) Farming for the future Optimizing farming systems for society and the environment. Chapter in book Wildlife Conservation on Farmland Volume 1: Managing for nature on lowland farms Ed. by Macdonald DW and Feber RE. ISBN-13: 9780198745488

UNECE (2009) Learning from each other: the UNECE strategy for education for sustainable development. https://www.unece.org/fileadmin/DAM/env/esd/01_Typo3site/LearningFromEachOther.pdf

UNESCO (2012) Education for sustainable development sourcebook. Learn Training Tools N 4. https://sustainabledevelopment.un.org/content/documents/926unesco9.pdf

UNESCO (2014) UNESCO roadmap for implementing the global action programme on education for sustainable development. https://unesdoc.unesco.org/ark:/48223/pf0000230514

UNESCO (2015) Education 2030: Incheon declaration and framework for action: for the implementation of sustainable development goal 4. World Educators Forum. https://unesdoc.unesco.org/ark:/48223/pf0000245656

UNEP (2019) Global environment outlook - geo-6: healthy planet, healthy people. Nairobi. https://doi.org/10.1017/9781108627146

United Nations (1987). Our common future: report of the World Commission on Environment and Development. http://www.exteriores.gob.es/Portal/es/PoliticaExteriorCooperacion/Desarrollosostenible/Documents/Informe%20Brundtland%20(En%20ingl%C3%A9s).pdf

United Nations (2005) UN decade of education for sustainable development 2005–2014. The DESD at a glance. ED/2005/PEQ/ESD/3. New York: 2005

United Nations (2015) 2015. Transforming Our World: the 2030 Agenda for Sustainable Development United Nations United Nations Transforming Our World: the 2030 Agenda for Sustainable Development. A/RES/70/1. New York

Vermeulen SJ, Campbell BM, Ingram JSI (2012) Climate change and food systems. Annu Rev Environ Resour 2012(37):195–222

Westhoek H, Lesschen JP, Rood T, Wagner S, De Marco A, Murphy-Bokern D, Leip A, van Grinsven H, Sutton MA, Oenema O (2014) Food choices, health and environment: effects of cutting Europe's meat and dairy intake. Glob Environ Chang 26:196–205

Wezel A, Bellon S, Doré T (2009) Agroecology as a science, a movement and a practice. A review. Agron Sustain Dev 29:503–515. https://doi.org/10.1051/agro/2009004

WHO (2003) Diet, nutrition and the prevention of chronic diseases: report of a joint WHO/FAO expert consultation. WHO Technical Report Series, No. 916. Geneva: World Health Organization

WHO (2007) Protein and amino acid requirements in human nutrition: report of a joint FAO/WHO/UNU expert consultation. WHO technical report series; no. 935. World Health Organization, Geneva

WHO (2012) Guideline: sodium intake for adults and children. World Health Organization, Geneva

WHO (2015a). Connecting global priorities: biodiversity and human health: a state of knowledge review. World Health Organization and Secretariat of the Convention on Biological Diversity

WHO (2015b) Guideline: sugars intake for adults and children. World Health Organization, Geneva

World Health Organization (2002) World Health Report 2002 - Reducing Risks, Promoting Healthy Life, World Health Report

World Health Organization (2007) Prevention of cardiovascular disease: Guidelines for assessment and management of cardiovascular risk. Geneva

World Health Organization (2015) WHO estimates of the global burden of foodborne diseases: foodborne disease burden epidemiology reference group 2007-2015. Geneva. https://doi.org/10.1007/978-3-642-27769-6_3884-1

World Health Organization (2017) The double burden of malnutrition—Policy Brief, Groundwater. Geneva. https://doi.org/10.1111/j.1745-6584.1983.tb00740.x

World Health Organization (2018a) Household air pollution and health. https://www.who.int/news-room/fact-sheets/detail/household-air-pollution-and-health (Accessed: 15 July 2020)

World Health Organization (2018b) Pesticide residues in food. https://www.who.int/news-room/fact-sheets/detail/pesticide-residues-in-food (Accessed: 15 July 2020)

World Health Organization (2020) The global health observatory. https://www.who.int/data/gho/data/indicators (Accessed: 29 June 2020)

World Water Assessment Programme (2009) The United Nations World Water Development Report 3: Water in a Changing World. The United Nations World Water Development Report 3. Paris: UNESCO, and LONDON: Earthscan. 2009

Wu C et al (2016) The effect of silicon on iron plaque formation and arsenic accumulation in rice genotypes with different radial oxygen loss (ROL). Environ Pollut. https://doi.org/10.1016/j.envpol.2016.01.004

Environmental Alterations and Oncological Diseases: The Contribution of Network Medicine

Caterina A.M La Porta

Abstract

The majority of cancers are due to sporadic mutations happening during the lifetime. In the present review I discuss the involvement of environmental alterations, such as climate change or chemicals factors in the water and food, in the development of diseases like cancer considering particular cross-reactions.

The view put forward this review is the use of network medicine to study the complex interactions between many pollutants. In the era of "BigData" and thanks to the high performance of computers and their capacity to store and analyze a large amount of data, we are able to identify emergent properties of a complex system and to build robust models capable to predicting and exploring the dynamics of a pathological system or the development from a physiological system to a pathological one.

Keywords

Network medicine · Pollution · Cancer Climate change · Heavy metal · Food and water contamination

7.1 Cancer and the Environment

Hereditary cancers represent about 5–10% of all cancers. The majority of cancers are due to sporadic mutations happening during the lifetime. The etiology of cancer can involve specific predisposition such as family history of cancer but appears mainly related in a complex way with many environmental factors from smoking to sun and radiation exposure, to chemicals, virus, bacteria, alcohol, overweight, lack of physical activity, and poor diet (Fig. 7.1).

All these factors work in a complex way alone or in combination with our genes and affect epigenetic regulation, leading at the end to a phenotypic heterogeneous tumor. The phenotypic heterogeneity can derive from a combination of factors, in which the genetic make-up, the interaction with the environment and the ability of the cells to adapt to the evolving microenvironments and mechanical forces play a major role. Epigenetic mechanisms most often involve changes that affect gene activity and expression, without becoming heritable. Waddington in 1942 defined epigenetic as "the causal interactions

C. A.M La Porta (✉)
Department of Environmental Science and Policy UniMi, University of Milan, Milan, Italy

Center for Complexity & Biosystems (CC&B) UniMi, Milan, Italy

Innovation For Well-Being And Environment (CRC-I-WE) UniMi, Milan, Italy
e-mail: Caterinla.laporta@unimi.it

Fig. 7.1 Interaction between genes and environmental factors on human

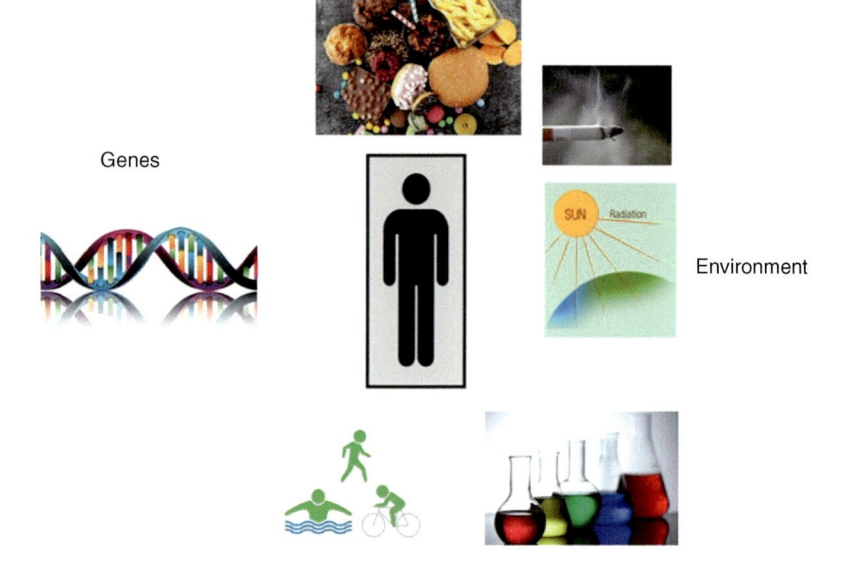

between genes and their products, which bring the phenotype into being" (Waddington 1942). Since the crucial role of epigenetic in controlling gene expression, they are involved during both physiological and pathological processes including cancer. Epigenetic regulation activities include DNA methylation, histone modifications, and RNA-associated silencing (i.e. noncoding RNAs). The critical role of epigenetic regulation was, for the first time, shown in colorectal cancer where less DNA methylation than normal tissue from the same patients occurs (Feinberg and Vogelstein 1983). Because methylated genes are typically turned off, loss of DNA methylation causes abnormally high gene activation by altering the arrangement of chromatin (Feinberg and Vogelstein 1983). Moreover, DNA methylation occurs at CpG sites and there are stretches of DNA near promoter regions that have higher concentrations of CpG sites (known as CpG islands) that are usually free of methylation in normal cells. These CpG islands become excessively methylated in cancer cells, thereby affecting genes that should not be silenced to be turned off (Egger et al. 2004; Jones and Baylin 2002).

Although epigenetic changes do not alter the sequence of DNA, they can cause mutations. For example, hypermethylation of the promoter of MGMT causes the number of G-to-A mutations to increase (Egger et al. 2004). On the other hand, hypermethylation can also lead to instability of microsatellites, which are repeated sequences of DNA. Microsatellites are common in normal individuals, and they usually consist of repeats of the dinucleotide CA.

Microsatellite instability has been shown to be linked to many cancers, including colorectal, endometrial, ovarian, and gastric cancers (Jones and Baylin 2002).

The factors linked to the environment involved in cancer development can be summarized into three categories for simplicity as follows:

1. Climate change.
2. Chemical factors in the water.
3. Chemical factors in the food.

An important issue is to understand how these factors not only can affect biological behavior but also how they impact through a cross-reaction with others factors. In the next paragraphs, I will discuss in depth the main factors related to cancer belonging to the three categories described above. In the last paragraph, I will show how using a network medicine approach can be useful to understand and build up a strategy that is able to predict the possible impact of environmental changes on human physiology.

7.2 Climate Changes and Oncological Diseases

The connection between climate change and tumors is not an obvious one. We know that climate change is linked to temperature changes, precipitation, sea level rise, and extreme weather. Climate changes can alter cancer risk, increasing the exposure to particulates and carcinogenic chemicals, to ultraviolet (UV) radiation and the quantity and quality of key staples, resulting in micronutrient deficiencies. If we focus on lung cancer, the most important tumor causing mortality in the world (Turner et al. 2011), fine particulate matter has been shown to be involved extensively (Turner et al. 2011; Raaschou-Nielsen et al. 2013). Climate changes lead to an indirect exposure to particulate matter, increasing the risk of wildfires and inducing anti cyclonic conditions that increase local anthropogenic air pollution and wild fire smokes (Wilbanks et al. 2007; He et al. 2017a; Caserini et al. 2017a). For example, fine particulate matter (PM2.5: particles <2.5 microns in diameter) represent the most dangerous matter for cancer lungs (Shwartz 1993). These small particles can contain carcinogens, such as polycyclic aromatic hydrocarbons (PAHs), acrolein, benzene, and formaldehyde, that can cause acute and/or chronic inflammation, thus worsening not only existing cardiovascular and respiratory conditions but increasing the risk of lung cancer too (Abdel-Shafy and Mansour 2016; Chen and Liao 2006a; Li et al. 2016). It has been shown that anticyclonic conditions can exacerbate human exposure to harmful PM8 (Caserini et al. 2017b; He et al. 2017b).

Wildfires are also harmful for human health since they produce large quantities of carbon dioxide, PM2.5, carcinogenic volatile organic compounds, including benzene, formaldehyde, acrolein, and PAHs (Ramesh et al. 2012). The relationship between human exposure to volatile organic compounds and cancer risk is well reported in the literature (Chen and Liao 2006b; Schoeny and Poirier 1993; Stec et al. 2018). In this connection, climate change significantly alters environmental factors that exacerbates wildfires such as shifting precipitation patterns, increasing temperatures, strengthening winds, and straining ecosystems (Flannigan et al. 2000; Jin et al. 2015; Matthews-Trigg et al. 2009; Westerling et al. 2006; Flannigan et al. 2009; IPCC 2007).

The possible impact of climate change on UV radiation and the possible connection with the increased risk of skin cancer is still a debate. In fact, the correlation between these factors appears to be complex and not always completely clear. Actually, it is likely that behavior associated with climate change mitigation, rather than ozone depletion, seems to be an important factor in UV exposure and consequently skin cancer (Diffey 2003). People living in areas expected to experience warmer, drier conditions have an increased tendency to spend more time outdoors (Bharath and Turner 2009), yet when temperatures reach uncomfortable levels, time outside may start to decline. Hence, increasing temperatures and heat waves may cause sun avoidance and act as a protective factor in southern regions for some individuals. There is also another and not obvious impact of the climate change on human health and is related to food security (Porter et al. 2014). In fact, Increasing temperatures there are changes in the hydrologic cycle, and in the frequency and intensity of extreme weather and climate events for example heat waves. All these factors alter the yields of cereal crops, the world's most important sources of food (Challinor et al. 2014). In addition, climate change affects the nutritional quality of wheat, rice, and other staples enhancing malnutrition in human population (Loladze 2014). In particular, the nitrogen decline with higher CO_2 concentrations leads to a decrease in B vitamins (Zhu et al. 2018). Therefore, climate change will result in tropical and some temperate regions producing lower yields staple crops, with those crops of lower nutritional quality with important consequences on human health.

7.3 Heavy Metals and Cancer

Heavy metals are naturally present in the earth's crust. However, human activities from technological-based to research-related ones increase them dramatically. Heavy metals are metallic elements with a relative high density with an atomic density greater than 4 g/cm^3 or at least 5 times greater than the density of water. Within this group are included lead (Pb), cadmium (Cd), zinc (Zn), mercury (Hg), arsenic (As), silver (Ag), chromium (Cr), copper (Cu), iron (Fe), and platinum (Pt) are included. They are not degradable and they enter the body through food, air, and water and therefore they can bioaccumulate over a period of time. Anthropogenic sources of heavy metal contamination include agricultural activities such as the use of pesticide and herbicide (Alloway and Jackson 1999; Gray et al. 1999). Additional anthropogenic sources of heavy metals include traffic emissions, cigarette smoking, metallurgy and smelting, aerosol cans, sewage discharge, and building materials, such as paints (Nriagu 1990). The atmosphere can be loaded with heavy metals through the breakdown of applied waste materials, which gradually release the heavy metals in the air. Heavy metal contamination is dangerous because of, for example, bioaccumulation through the food chain (Aycicek et al. 2008) mainly by ingestion of food crops contaminated by these harmful elements (Zukowska and Biziuk 2008). Uptake of heavy metals by plants through absorption and subsequent accumulation along the food chain is a potential threat to animal and human health (Sprynskyy et al. 2007; Jordao et al. 2006).

There are heavy metals such as zinc, copper, iron, manganese, and cobalt which are required by the human body, and, therefore, they may be toxic if ingested at higher concentrations. On the other hand, there are other heavy metals such as mercury which do not play any physiological role and are deleterious to human health when they accumulate over time in the body. An example of release of mercury is dental fillings (read for more information ref. (Guzzi and La Porta 2008). The mechanisms by which heavy metals alter body's metabolic functions are multiple. First of all, they may accumulate in vital body organs such as liver, heart, kidney, and brain disturbing normal biological functioning. A tragic example has been Minamata disease due to the release of high quantity of methylmercury by a chemical factory in Japan in 1956. After this ecological disaster there was a bioaccumulation of heavy metals in fish and shellfish that were eaten by local population causing a severe mercury poisoning (Guzzi and La Porta 2008). Considering more likely conditions, heavy metals can enter at lower level into the human body in a number of ways everyday: through consumption of contaminated food, drinking water, and/or air. Since we cannot degrade them, they accumulate into the body and during the time they can reach harmful levels. Various carcinogenic pathways are induced by heavy metals exposure leading to oxidative stress (Beyersmann and Hartwig 2008; Genestra 2007). In fact, metal ions such as chromium, nickel, vanadium are involved in biological system redox reactions and lead to the production of free radicals that can damage proteins and DNA, leading to the activation of transcription factors which are redox sensitive and interfere with DNA repair and therefore helping tumor development. Numerous studies have recognized the ability of Cd for contributing in the induction and propagation of different kind of cancers either in animals model (Waalkes 2003) and in vitro mammalian cell studies (Stohs et al. 2001; Valko et al. 2006). For example, it has been shown that both glutathione reductase and superoxide dismutase enzymes are inhibited by Cd exposure (Stohs et al. 2001; Valko et al. 2006). In humans, Cd is shown to be one of the main causes of prostate cancer (Waalkes et al. 1997). Arsenic is also a well-known carcinogenic factor which facilitates various type of cancers (Yoshida et al. 2004) due to its capability to react with thiol groups which are present in various zinc binding structures in many transcription factors and in proteins that control the cell cycle (Schoen et al. 2004). Bladder, stomach, and lung cancers are known to be caused by Pb exposed to humans (IARC 2004). Ni was shown to be toxic through the modulation of the oxidative stress damaging DNA (Kasprzak 1995) it is shown to be involved in the development of oral (Su et al. 2007), skin (Su et al. 2007), and lung cancers (Silvera and Rohan 2007).

7.4 Food and Water Contamination and Risk of Cancer Development

An open issue is the impact of food and water contamination on human health including cancer development by chemicals. There is always a contamination both in the processed food due to packaging, transportation, and storage processes and the law should regulate strictly the allowed limit of concentration of each single chemical element that science shows not to be toxic. On the other hand, fresh products can be contaminated by chemicals due to the contamination of the soil where they grow, by disinfection by-products, air, water, cooking procedures and, at the end, packaging material too. One of the main issues that the scientific community studied intensely is the impact of multiple chemicals on human health. The answer to this question is not easy to address because we are everyday simultaneously exposed to many factors at low level for long time. We can understand the impact of a single factor at a high dose when a contamination or ecological disaster is going on but it is more complicated to address scientifically the impact of multiple factors chronically present in the environment each at low dose. In developing countries, where legislation is not in place or is poorly enacted there are many papers showing the disaster of contamination by pesticides, heavy metals, plastic, or other pollutants. Related to cancer, it has been reported that pesticide residues, such as diazinon and fenpropimorph, are involved in leukemia and other neurological and immunological cancers (Mekonen et al. 2016; Shi et al. 2018), i.e. radon for lung cancer (Gunnarsdottir et al. 2016; Jobbágy et al. 2017).

7.5 Network Medicine to Study the Complex Interaction between Many Pollutants on Human Diseases

In a simplified world, a cell is a complex system with proteins, transcription factors, DNA, RNA, miRNA, etc. connected together in a network, but many cells cooperate with each other at the higher level of organization of organs, creating another level of the network and, on the other hand, organs are also coordinated together by the nervous system, giving rise to a superior level of network of organization and giving rise to a human body. In the last two decades, biological researchers focused on a specific part of this complex system, inside the cell studying metabolic pathways or specific transcriptomes and detangling the mechanism of function of these proteins and some little interactions. Nowadays, we have a huge literature regarding each single factor/protein/epigenetic mechanism but we still miss the complexity of the system. Thanks to the high performance of computers and their capacity to store and analyze a lot of data, network biology using the convenient representation of the biological factors in graphs is able to study the relationship between various biological components combining graph theory, system biology, and statistical analyses. The use of quantitative biology allows also the possibility to understand the cellular organization and to capture the impact of perturbations on these complex intracellular networks. Network medicine is an extension of network biology with a set of focused goals related to disease biology, including understanding disease eziology, identifying potential biomarkers, and designing therapeutic interventions. In this framework, I have worked with other colleagues to build the stepping stones of these new disciplines and to detangle the complexity of cancer (La Porta and Zapperi 2018). The possibility to build a robust model able to predict and assessing the future is mainly depended by understanding how perturbations propagate in the system by identifying the pathways and highlight key components in the networks that can be targeted in clinical interventions. This approach allows a personalized medicine (La Porta and Zapperi 2018).

The use of network medicine to understand how a pollutant and, even better, how many pollutants cross-react together and how they could cooperate to give rise a cancer phenotype appears crucial to build up a model that is able to predict and demonstrate the combined impact of pollutants and also to understand how we could repristinate a physiological network at the level of population (Fig. 7.2).

In the near future, we need to build up a medicine of aggregation collecting together many disciplines and creating a new expert figure to handle the new challenge posed by diseases due to pollutants in the environment and in particular cancer that is the leading disease in the world, mainly in developing countries (Shrestha et al. 2019).

7.6 Some Shorts Note on Network Science and how it Can Help in Understanding the Connection between Pollution and Cancer

In the last 10 years, an important shift happened in biomedical research. In fact, while during the era of the sequencing of the human genome it was believed that knowing the complete sequence of the bases was enough to understand the extraordinary complexity of our organism, now we now know that the sequence alone can't be enough to understand the complex biological functions which regulate our organism. During this period, I started to study biomedical topics using a network medicine approach combining complex systems and systems biology to identify the emergent properties and understand the deep causes of human diseases. This strategy is the only one that can be successful since pathologies are driven by complex interactions among a variety of molecular mediators linked in a network. Therefore, the capability to study the complex interactions between these factors and their relation with the environment can help in the identification of the critical networks involved. In particular, for understanding the impact of pollution on the human body. This approach appears very useful, since it tries to look at the whole system at once. The first thing to do is to break down

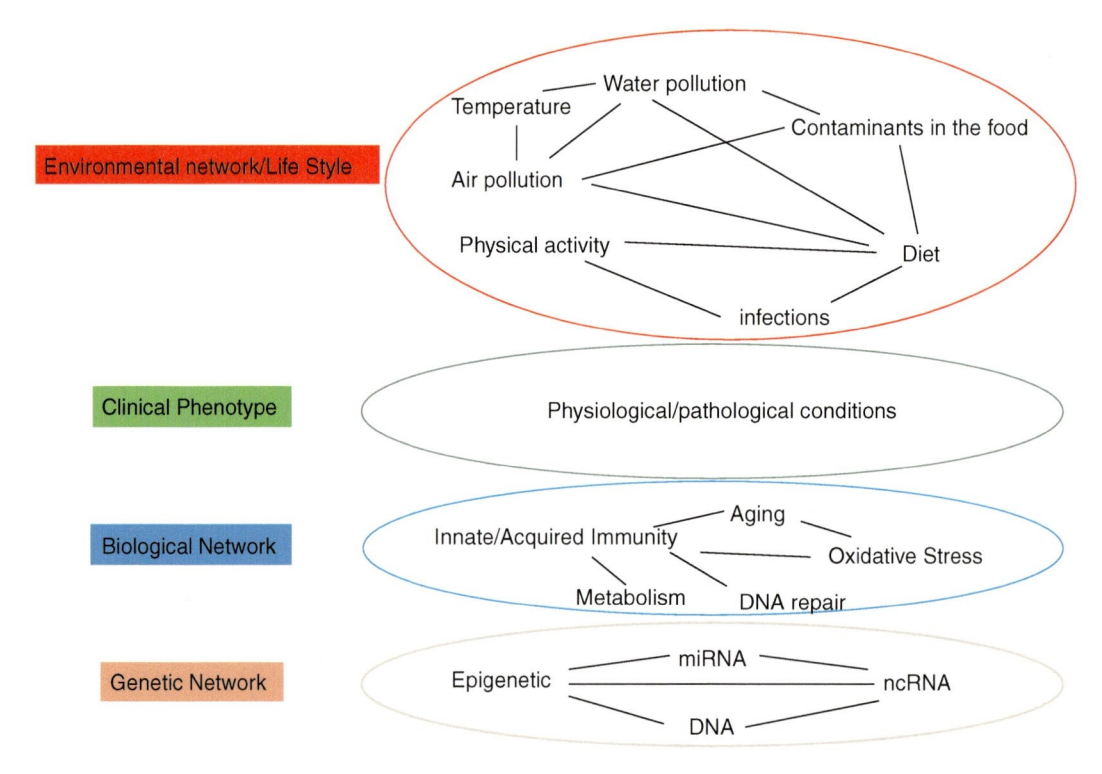

Fig. 7.2 Environmental factors and lifestyle leads to a clinical phenotype such as tumor affecting biological and genetic networks

a network into smaller components, focusing on individual pathways and modules. Then, we can compute global statistics describing the network as a whole. We can compare networks within the same context (e.g., between two gene regulatory networks) or cross disciplines(e.g., between regulatory networks and governmental hierarchies). In this context, the availability of large amounts of biomolecular data, the integration of in silico quantitative methodologies, and the "big data" analysis tools typical of "data science," provides the potential to build a predictive model. The interdisciplinary approach is the main ingredient to reach this goal.

References

Abdel-Shafy HI, Mansour MSM (2016) A review on polycyclic aromatic hydrocarbons: source, environmental impact, effect on human health and remediation. Egypt J Pet 25:107–123

Alloway BJ, Jackson AP (1999) Behaviour of trace metals in sludge-amended soils. Sci Total Environ 100:151–176

Aycicek M, Kaplan O, Yaman M (2008) Effect of cadmium on germination seeding growth and metal contents of sunflower (*Helianthus annus* L). Asian J Chem 20:2663–2672

Beyersmann D, Hartwig A (2008) Carcinogenic metal compounds: recent insight into molecular and cellular mechanisms. Arch Toxicol 82:493–512

Bharath AK, Turner RJ (2009) Impact of climate change on skin cancer. J R Soc Med 102:215–218

Caserini S, Giani P, Cacciamani C, Ozgen S, Lonati G (2017b) Influence of climate change on the frequency of daytime temperature inversions and stagnation events in the Po Valley: historical trend and future projections. Atmos Res 184:15–23

Caserini S, Giani P, Cacciamani C et al (2017a) Influence of climate change on the frequency of daytime temperature inversions and stagnation events in the Po Valley: historical trend and future projections. Atmos Res 184:15–23

Challinor AJ, Watson J, Lobell DB, Howden SM, Smith DR, Chhetri NA (2014) Meta-analysis of crop yield under climate change and adaptation. Nat Clim Chang 4:287–291

Chen S-C, Liao C-M (2006a) Health risk assessment on human exposed to environmental polycyclic aromatic hydrocarbons pollution sources. Sci Total Environ 366:112–123

Chen S-C, Liao C-M (2006b) Health risk assessment on human exposed to environmental polycyclic aromatic hydrocarbons pollution sources. Sci Total Environ 366:112–123

Diffey B (2003) Climate change, ozone depletion and the impact on ultraviolet exposure of human skin. Phys Med Biol 49(1):R1

Egger G, Liang G, Aparicio A, Jones PA (2004) Epigenetics in human disease and prospects for epigenetic therapy. Nature 429:257–263

Feinberg AP, Vogelstein B (1983) Hypomethylation distinguishes genes of some human cancers from their normal counterparts. Nature 301:89–92

Flannigan M, Stocks B, Turetsky M, Wotton M (2009) Impacts of climate change on fire activity and fire management in the circumboreal forest. Glob Chang Biol 15:549–560

Flannigan M, Stocks B, Wotton B (2000) Climate change and forest fires. Sci Total Environ 262:221–229

Genestra M (2007) Oxyl radicals, redox-sensitive signaling cascades and antioxidants. Cell Signal 19:1807–1819

Gray CW, Mclaren RG, Roberts AHC, Condron LM (1999) The effect of long-time phosphatic fertilizer applications on the amounts and forms of cadmium in soils under pasture in New Zealand. Nutr Cycling Agroecosyst 154:267–277

Gunnarsdottir MJ, Gardarsson SM, Jonsson GS, Bartram J (2016) Chemical quality and regulatory compliance of drinking water in Iceland. Int J Hyg Environ Health 219:724–733

Guzzi GP, La Porta CAM (2008) Molecular mechanisms triggered by mercury. Toxicology 244:1–12

He C, Wu B, Zou L, Zhou T (2017a) Responses of the summertime subtropical anticyclones to global warming. J Clim 30:6465–6479

He C, Wu B, Zou L, Zhou T (2017b) Responses of the summertime subtropical anticyclones to global warming. J Clim 30:6465–6479

IARC (2004) IARC monographs on the evaluation of carcinogenic risks to humans. Inorganic and organic lead compounds. International Agency for Research on Cancer

IPCC. Climate change 2007: impacts, adaptation and vulnerability. Contribution of Working Group II to the Fourth Assessment Report of the Intergovernmental Panel on Climate Change. Parry ML, Canziani OF, Palutikof JP, van der Linden P, Hanson CE eds. Cambridge University Press. 2007

Jin Y, Goulden M, Faivre S, Veraverbeke S, Sun F, Hall A, Hand MS, Hook S, Randerson JT (2015) Identification of two distinct fire regimes in southern California: implications for economic impact and future change. Environ Res Lett 10:094005

Jobbágy V, Altzitzoglou T, Malo P, Tanner V, Hult M (2017) A brief overview on radon measurements in drinking water. J Environ Radioact 173:18–24

Jones PA, Baylin SB (2002) The fundamental role of epigenetic events in cancer. Nat Rev Genet 3:415–428

Jordao CP, Nascentes CC, Cecon PR, Fontes RLF, Pereira JL (2006) Heavy metal availability in soil amended with composted urban solid wastes. Environ Monit Assess 112:309–326

Kasprzak KS (1995) Possible role of oxidative damage in metal-induce carcinogenesis. Cancer Investig 13:411–430

La Porta CAM, Zapperi S (2018) Explaining the dynamics of tumor aggressiveness: at the crossroads between biology, artificial intelligence and complex systems. Semin Cancer Biol 53:42–47

Li X, Yang Y, Xu X, Changqin X, Hong J (2016) Air pollution from polycyclic aromatic hydrocarbons generated by human activities and their health effects in China. J Clean Prod 112:1360–1637

Loladze I (2014) Hidden shift of the ionome of plants exposed to elevated CO_2 depletes minerals at the base of human nutrition. elife 3:02245

Matthews-Trigg NT, Mike F, Brian S, Merritt T, Mike W (2009) Impacts of climate change on fire activity and fire management in the circumboreal forest. Glob Chang Biol 15:549–560

Mekonen S, Argaw R, Simanesew A, Houbraken M, Senaeve D, Ambelu A, Spanoghe P (2016) Pesticide residues in drinking water and associated risk to consumers in Ethiopia. Chemosphere 162:252–260

Nriagu JO (1990) The rise and fall of leading gasoline. Sci Total Environ 92:13–28

Porter JR, Xie L, Challinor AJ, Cochrane K, Howden SM, Iqbal MM, Lobell DB, Travasso MI (2014) Food security and food production systems. In: Field CB, Barros VR, Dokken DJ, Mach KJ, Mastrandrea MD, Bilir TE, Chatterjee M, Ebi KL, Estrada YO, Genova RC, Girma B, Kissel ES, Levy AN, Mac Cracken S, Mastrandrea PR, White LL (eds) Climate change 2014: impacts, adaptation, and vulnerability. Part a: global and sectoral aspects. Contribution of Working Group II to the Fifth Assessment Report of the Intergovernmental Panel on Climate Change. Cambridge University Press, Cambridge/New York, pp 485–533

Raaschou-Nielsen O, Andersen ZJ, Beelen R et al (2013) Air pollution and lung cancer incidence in 17 European cohorts: prospective analyses from the European study of cohorts for air pollution effects (ESCAPE). Lancet Oncol 14:813–822

Ramesh A, Hood D, Guo Z, Loganathan B (2012) Global environmental distribution and human health effects of polycyclic aromatic hydrocarbons. Global contamination trends of persistent organic chemicals, vol 1. CRC Press, Taylor & Francis Group, pp 97–128

Schoen A, Beck B, Sharma R, Dube E (2004) Arsenic toxicity at low doses: epidemiological and mode of action considerations. Toxicol Appl Pharmacol 198:253–267

Schoeny R, Poirier K (1993) Provisional guidance for quantitative risk assessment of poly cyclic aromatic hydrocarbons. U.S. Environmental Protection Agency, Office of Research and Development, Office of Health and Environmental Assessment, Washington, DC, EPA/600/R-93/089 (NTIS PB94116571)

Shi P, Zhou S, Xiao H, Qiu J, Li A, Zhou Q, Pan Y, Hollert H (2018) Toxicological and chemical insights into representative source and drinking water in eastern China. Environ Pollut 233:35–44

Shrestha AD, Vedsted P, Kallestrup P, Neupane D (2019) Prevalence and incidence of oral cancer in low- and middle-income countries: a scoping review. Eur J Cancer Care:e13207. https://doi.org/10.1111/ecc.13207

Shwartz J (1993) Particulate air pollutants and respiratory diseases. Environ Res 62:7–13

Silvera SAN, Rohan TE (2007) Trace elements and cancer risk: a review of the epidemiologic evidence. Cancer Causes Control 18:7–27

Sprynskyy M, Kosobucki P, Kowalkowski T, Buszewsk B (2007) Influence of clinoptilolite rock on chemical speciation of selected heavy metals in sewage sludge. J Hazard Mater 149:310–316

Stec AA, Dickens KE, Salden M, Hewitt FE, Watt DP, Houldsworth PE, Martin FL (2018) Occupational exposure to polycyclic aromatic hydrocarbons and elevated cancer incidence in firefighters. Sci Rep 8:2476

Stohs SJ, Bagchi D, Hassoun E, Bagchi M (2001) Oxidative mechanisms in the toxicity of chromium and cadmium ions. J Environ Pathol Toxicol Oncol 20:77–88

Su C-C, Yang H-F, Huang S-J, Lian I-B (2007) Distinctive features of oral cancer in Changhua County: high incidence, buccal mucosa preponderance, and a close relation to betel quid chewing habit. J Formos Med Assoc 106:225–233

Turner MC, Krewski D, Pope CA, Chen Y, Gapstur SM, Thun MJ (2011) Long-term ambient fine particulate matter air pollution and lung cancer in a large cohort of never-smokers. Am J Respir Crit Care Med 184:1374–1381

Valko M, Rhodes C, Moncol J, Izakovic M, Mazur M (2006) Free radicals, metals and antioxidants in oxidative stress-induced cancer. Chem Biol Interact 160:1–40

Waalkes M, Rehm S, Coogan T, Ward J (1997) Role of cadmium in the etiology of cancer of the prostate. Endocrine toxicology, 2nd edn. Taylor and Francis, Washington DC, pp 227–243

Waalkes MP (2003) Cadmium carcinogenesis. Mutat Res 533:107–120

Waddington CH (1942) The epigenotype. Endeavour 1:18–20

Westerling AL, Hidalgo HG, Cayan DR, Swetnam TW (2006) Warming and earlier spring increase Western U.S. forest wildfire activity. Science 313:940–943

Wilbanks TJ, Romero P, Lankao M et al (2007) In: Parry ML, Canziani OF, Palutikof PJ et al (eds) Industry, settlement and society. Climate change 2007: impacts, adaptation and vulnerability. Contribution of Working Group II to the Fourth Assessment Report of the Intergovernmental Panel on Climate Change. Cambridge University Press, Cambridge, pp 357–390

Yoshida T, Yamauchi H, Fan Sun G (2004) Chronic health effects in people exposed to arsenic via the drinking water: dose-response relationships in review. Toxicol Appl Pharmacol 198:243–252

Zhu C, Kobayashi K, Loladze I, Zhu J, Jiang Q, Xu X, Liu G, Seneweera S, Ebi KL, Drewnowski A, Fukagawa NK, Ziska LH (2018) Carbon dioxide (CO2) levels this century will alter the protein, micronutrients, and vitamin content of rice grains with potential health consequences for the poorest rice-dependent countries. Sci Adv 4:eaaq1012

Zukowska J, Biziuk M (2008) Methological evolution of method for dietary heavy metal intake. J Food Sci 73:R21–R29

Zootechnical Systems, Ecological Dysfunctions and Human Health

Luigi Bonizzi, Francesco Campana, and Alessio Soggiu

Abstract

Global health, as a collective issue, must be addressed by looking not only at humans but at the entire ecosystem. In this perspective, research aimed at pursuing public health protection requires a multidisciplinary approach. In this chapter, the assessment of environmental quality and in particular the assessment of the environmental footprint of livestock and food supply chains and the related environmental impacts are described and the technological innovations to implement a sustainable livestock and bioeconomy for a better protection of public health are exposed. In this sense, preventive health systems are also defined both in human and veterinary field and their integration with environmental and socio-economic data in a one-health perspective.

Keywords

One-health · Environmental impact Livestock production · Food chain Bioeconomy · Technological innovations Prevention systems

L. Bonizzi (✉) · F. Campana · A. Soggiu
Department of Biomedical, Surgical and Dental Sciences, Section One Health, University of Milan, Milan, Italy
e-mail: luigi.bonizzi@unimi.it

8.1 Introduction

EU environmental policies are guided by three political priorities: (1) protecting, preserving and improving the EU's natural capital; (2) transforming the EU into a low-carbon, resource-efficient, green and competitive economy; (3) protecting EU citizens from environmental pressures and risks to health and well-being.

The Seventh Environmental Action Programme (seventh EAP) states that in order to live well within the ecological limits of our planet, by 2050 prosperity a healthy environment will be based on an innovative circular economy without waste, where natural resources are managed sustainably and biodiversity is protected, valued and restored in a way that strengthens the resilience of our society. Our growth will be characterized by low-carbon emissions and will be long decoupled from resource use, thus setting the pace for a safe and sustainable global society. Health, as a collective issue, must be tackled by looking not only at people but at the whole ecosystem. In this perspective, research aimed at pursuing the protection of public health requires a multidisciplinary approach; this principle must be reflected in the public institutions that are called upon to protect the health of the community.

The term "Public Health" refers to the discipline that aims to prevent disease, prolong life and promote health. Health, defined by the World

Health Organization as "a state of complete physical, mental and social well-being and not simply the absence of disease or infirmity" is an inalienable human right, constitutionally guaranteed by the fundamental law of all modern states. Hence the transition to a model of medicine that lends itself to the idea of prevention and health promotion. A model of clearly systemic approach, which takes into account different levels of interpretation of health (biological aspects, psychological aspects, social aspects) and which integrates them in an increasingly multidisciplinary perspective.

The quality of the environment, i.e. the set of conditions and factors that surround the individual organism in a defined space, is the main determinant of the health status of a population (plant or animal). This is the context in which species of zootechnical interest that both in the intensive and extensive farming module are influenced by the physical, chemical and biological factors that make up the habitat in which they perform their vital functions. The "continuous" monitoring of these factors constitutes the main pre-occupation of breeders, since the preservation of[1] animal welfare, as well as a legal obligation, represents a powerful lever of management efficiency. The "holistic" management of farms is at the heart of many research projects, in particular two strands of study that unite academies around the world: the protection of biodiversity[2] and the protection[3] of water resources. The preservation of genetic variety is essential to ensure the adaptability of species of zootechnical interest to the environment; likewise, water is indispensable for the proper functioning of the vital functions of living organisms and in the various processes of the agri-food chains that extend from the fields to the table. Animal husbandry reproduces, in fact, a complex system[4] made up

---

[1] Animal welfare is an important area of EFSA's (European Food Safety Authority) mandate. The safety of the food chain is indirectly affected by the welfare of animals, in particular those bred for food production, due to the close link between animal welfare, [1]animal health and food poisoning. Stress factors and poor welfare conditions may result in animals becoming more susceptible to disease. This may lead to a risk to consumers, as for. Example. in the case of common food poisoning caused by the bacteria Salmonella, Campylobacter and E. Coli. The welfare of food-producing animals depends to a large extent on human management practices. A number of factors may affect their welfare, such as the type of housing and resting areas, space allowances and stocking densities, transport conditions, stunning and slaughter methods, castration of males and cutting of the tail. The European Union has some of the highest animal welfare standards in the world. The general framework for EU action on animal welfare is set out in the EU Strategy for the Protection and Welfare of Animals 2012–2015. Harmonized standards at EU level are currently in place for many animal species and for various issues affecting animal welfare. Council Directive 98/58/EC lays down minimum standards for the protection of all animals kept for farming purposes, while other EU rules define the welfare standards for farm animals during transport and at the time of stunning and slaughter. Specific Directives cover the protection of individual animal categories such as calves, pigs and laying hens. In addition to farm animals, animals used in laboratory experiments and wild animals kept in zoos are also protected by EU-wide harmonized standards. Other international organizations have also issued recommendations and guidelines on animal welfare, such as the World Organisation for Animal Health (OIE) and the Council of Europe. The EU is a signatory to the European Convention for the protection of animals kept for farming purposes adopted by the Council of Europe (Source: EFSA).

[2] Biodiversity has been defined by the Convention on Biological Diversity (CBD) as the variability of all living organisms included in aquatic, terrestrial and marine ecosystems and the ecological complexes to which they belong. Interactions between living organisms and the physical environment give rise to functional relationships that characterize different ecosystems ensuring their resilience, their maintenance in a good state of conservation and the provision of the so-called ecosystem services (MATTM) (Peterson et al. 1998).

[3] The expression water resources indicates, in a strictly general sense, all the various forms of water availability: a substance indispensable to man and the ecosystem in which he lives, which at the same time can be abundant and available or scarce and unavailable (Wikipedia).

[4] Complex adaptive systems have been studied by Holland, who defines them as: groups of bound agents in a co-adaptive process, in which each one's adaptation moves have consequences for the whole group of individuals. Holland showed that, under certain conditions, simple models have surprising self-organizing capabilities. Complex systems are closely related to non-linear systems. Holland's definition of a non-linear system states that: a non-linear system is a system whose behaviour is not equal to the sum of its individual parts. If, therefore, in order to study linear systems, they are broken down and each of their parts is studied analytically, this cannot be done for the study of non-linear systems. The behavior of

of different sub-sets typical of "natural habitats" in which different elements interact to ensure a systemic function indispensable to humanity: the availability of food.

A distorted vision of environmental functions, which coincided with the process of industrialization of the economy, determined a linear development model with harmful consequences for the places inhabited by living organisms (animals and plants). Pollution, consequent to the breaking of natural balances, which degrades habitats (natural and/or artificial), and the excessive consumption of natural resources, due to the increase in the "carrying capacity" of ecosystems with the consequent recurrent explosion of pandemics, are only some elements of reflection that require a cultural rethinking in the way of thinking and planning human activities. Sustainability[5] is the paradigm that allows us to keep together the development of economic activities, the protection of the environment and the well-being of citizens.

8.2 The Environmental Footprint of Livestock Production Chains

The environmental footprint measures how much land and water the human population needs to produce, with available technology, the resources it consumes and to absorb the waste produced. It is possible to measure the environmental footprint of an individual, a city, a population, but also of a farm, a livestock or a product (Mekonnen et al. 2010).

To better explain the concept of environmental footprint, we quote researchers Mathis Wackernagel[6] and William Rees[7] who say: *"A typical example to explain the theory of footprint is that of a city enclosed in a glass dome, which lets in*

---

such systems, in fact, depends more on the interaction of the parts than on the behavior of the parts themselves. It is therefore necessary to consider the non-linear system as a whole not equal to the sum of the parts and, therefore, it is necessary to focus on the dynamics of interaction between the elements that make up the system. Complex phenomena cannot be studied with traditional mathematical tools, but they can be analyzed by observing the interaction of the elements of the system, in an attempt to see some coherence. This type of coherence, typical of complex systems, is called "emerging phenomena". The term "complex" is not synonymous with "complicated", in the sense of difficult. By complexity we mean a mathematically definable phenomenon or an organic and structured aggregate of interacting parts, which assumes properties not deriving from the simple sum of the parts that compose it. As an example, imagine the engine of a car, composed of many mechanisms, even sophisticated ones. The engine is defined "banal machine" because its functioning, however difficult and complicated, is the result of the sum of the parts that compose it and can be studied by breaking it down into these parts. In contrast to the engine, let us imagine the anthill, a set of ants that interacting with each other are able to maintain, for example, the temperature inside the anthill on constant values and with minimum variations between summer and winter. The anthill is a complex system because every single ant does not know the mechanism of thermal regulation of the surrounding environment but, simply from the interaction of many ants, a "complex" or "emerging" phenomenon manifests itself. In this second case it is not possible to study the anthill by studying the behavior of the single ants. Let us think of a flock of starlings, for example. Those who have observed them will have been amazed by the synchronism of the movements with which they move, in search of food or the most comfortable tree to rest. The flock has a completely different behaviour from that of individual isolated birds. The rules that are thought to be followed by the birds in the flock are very simple: a) imitate the behaviour of the nearest bird, b) keep its direction, the same speed and c) try not to bump into it. The results of this group cooperation are surprising. Without a leader or a central government, a complex structure emerges, endowed with a kind of distributed intelligence, capable of determining unpredictable and original evolutions, with the aim of finding better and better solutions for the survival of the group (John H. Holland. Adaptation in Natural and Artificial Systems: An Introductory Analysis with Applications to Biology, Control and Artificial Intelligence. MIT Press, Cambridge, MA, USA, 1992).

[5] According to the definition proposed in the report "Our Common Future" published in 1987 by the World Commission on Environment and Development (Brundtland Commission) of the United Nations Environment Programme, sustainable development is defined as development that "meets the needs of the present generation without compromising the ability of future generations to meet their own needs". The concept of sustainability, in this sense, is linked to the compatibility between the development of economic activities and environmental protection (TRECCANI).

[6] Mathis Wackernagel, founder and current president of the Global Footprint Network.

[7] William Rees, Professor Emeritus at the University of British Columbia...

light but prevents material things of any kind from coming in and out. Let us suppose that this city is surrounded by a diversified landscape, in which cultivated land and pastures, forests and reservoirs, i.e. all types of ecologically productive land, are represented in proportion to their current presence on Earth and that the city has an adequate amount of fossil fuel energy available to sustain current levels of consumption and its prevailing technologies. Let us also assume that the glass dome is elastically expandable. The question, at this point, is: how big must the dome become so that the city at its centre can sustain itself indefinitely only thanks to the terrestrial and aquatic ecosystems and the energy resources contained within the dome itself? In other words: what is the total area of terrestrial ecosystems needed to sustain all the social and economic activities of the inhabitants of that city on an ongoing basis? This surface area, necessary for the continuous existence of the city, is in fact its Ecological Footprint on Earth. It is clear that the ecological footprint of a city will be proportional to both its population and material consumption per capita".

The European Commission has taken this approach and, with regard to the environmental footprint of products (PEF) and organizations (OEF), has issued Recommendation 2013/179/EU on the use of common methodologies to measure and communicate environmental performance throughout the life cycle of products and organizations. Annex 1 of the above-mentioned "Recommendation sets out the scope of application for OEF and PEF methodologies".

In particular, for the *PEF methodology,* they are:

- process optimization during the life cycle of a product;
- support for product design that minimizes environmental impacts over the life cycle;
- provision of information on environmental performance throughout the life cycle of products (e.g. through product documentation, websites and apps) by individual companies or through voluntary schemes;
- programs relating to environmental statements, in particular ensuring sufficient reliability and completeness of the statements;
- programs that create reputation by giving visibility to products that calculate their environmental performance over their life cycle;
- identification of significant environmental impacts in order to establish eco-label criteria;
- incentives based on environmental performance over the life cycle, where appropriate.

The fields of application of the OEF methodology are:

- optimization of processes along the entire supply chain of an organization's product range;
- communicating life-cycle environmental performance to stakeholders (e.g. through annual reports, in sustainability reports, in response to investor or stakeholder questionnaires);
- programs that build reputation by giving visibility to organizations that calculate their environmental performance over their life cycle or to organizations that improve it over time (e.g. from year to year);
- programs that require reporting of environmental life-cycle performance;
- a means of providing information on environmental life-cycle performance and on the achievement of objectives under an environmental management system;
- incentives based on the improvement of environmental performance over the life cycle, calculated according to the OEF methodology, where appropriate.

8.2.1 Environmental Footprint Assessment Systems

Recommendation 2013/179/EU *on the use of common methodologies for measuring and communicating environmental performance throughout the life cycle of products and organizations* explicitly states that when a PEF (Environmental Product Footprint) study is carried out, certain steps should be completed such as: definition of objectives and scope, resource use and emissions profile, impact assessment of the environmental footprint and interpretation and communication of the footprint. In this perspective, the European Commission has launched an experimental phase, which ended in 2016, with the aim of developing specific methods for the calculation of PEF and OEF (Environmental Footprint of the Organization), for specific product categories. One of the most relevant objectives of this research was to develop a unique indicator that can make the environmental performance of different types of products comparable; this in order to guide the customer/consumer towards the choice of products that can be defined as "green" on the basis of methodologies approved by the EU and increase the competitiveness of companies in a green economy perspective. The experimental phase pursued three main objectives:

- test the process of developing product- and/or industry-specific rules;
- test the different approaches to verification;
- test the means to communicate environmental performance over the entire life cycle to all stakeholders.

This experimentation was carried out on groups of organizations that voluntarily joined and offered to develop the rules for their products and/or sectors. With regard to food products, the pilot studies covered the following products:

- beer;
- coffee;
- dairy products;
- animal feed;
- fish products;
- meat (beef, pork, sheep);
- olive oil;
- bottled water;
- noodles;
- dog and cat food;
- wine.

Environmental footprint certification interfaces with other environmental and product certification systems. In this regard, mention should be made of the European Commission Decision no. 2017/1508/EU, which approved the reference document on best environmental management practice, sectoral environmental performance indicators and examples of excellence that are taken into account by organizations in the food and beverage production sector that adhere to the EMAS registration system (Regulation no. 2009/1221/EC).

8.2.2 Measuring the Environmental Footprint in the Food Sector

Environmental impacts related to food products can be assessed using environmental footprints and related indicators:

- the carbon footprint;
- the water footprint;
- the product environmental footprint.

As regard the carbon footprint, the European Commission launched a study on food products which showed that about 20–30% of global warming is attributable to the food sector. The different food sectors contribute in different ways according to both the quantities consumed and the way they are grown/farmed and processed, as shown in the table below.

Product	Contribution % to global warming
Meat and meat products Meat	12
Dairy products	5
Cereal products	1
Fruits and vegetables	2

Data source A. Tukker, B. Jansen, Journal of Industrial Ecology (2006)

It is therefore clear that the meat industry is the one with the highest environmental footprint and, consequently, the one with the highest greenhouse gas emissions.

The table below provides an estimate of greenhouse gas emissions emitted for the production of certain foods.

Food	Quantities of greenhouse gases issued
1 burger	2.5 kg of CO_2
1 veggie burger	1 kg of CO_2
1 orange	1 kg of CO_2
6 eggs	1.8 kg of CO_2
1 litre of milk	720 g of CO_2
1 cheese wheel	12 kg of CO_2
1 bottle of beer	900 g of CO_2

Data source A. Tukker, B. Jansen, Journal of Industrial Ecology (2006)

As far as the water footprint is concerned, first of all, it should be specified that water consumption is not only for drinking or cooking food. In fact, a large amount of water is used in the production phases of the food itself from its origin to its arrival on our tables (from the fields to the table). The following table shows the quantities of water used for the production of some foods.

Food	Litres of water for the production
1 egg	200
1 kilo of potatoes	900
1 litre of milk	1000
1 burger	2400
1 kilo of chicken meat	3000
1 kilo of rice	3400

Data source A. Tukker, B. Jansen, Journal of Industrial Ecology (2006)

When calculating the water footprint, both the water used for production and the water used for consumption must be taken into account, also taking into account the water withdrawal point, considering the availability (or scarcity) of the water resource in the specific geographical area of production or origin of the food.

As far as PEFs are concerned, pilot projects promoted by the European Commission are currently being finalized. Of particular note is the LIFE+ PREFER Project—Product Environmental Footprint Enhanced by Regions, aimed at testing a new European methodology to assess the environmental footprint of products and services.

Within the PREFER project, the environmental impacts of the following food products are analysed:

- canned tomatoes from the Emilia-Romagna and Lombardy industrial tomato district,
- the sparkling wine of the Asti Wine District,
- pasta and tomato from the Nocera Gragnano agri-food district,
- the cheese of the Lombard milk district.

On the basis of the above, it is obvious that the problem of the environmental footprint in the food sector involves everyone at various levels: raw material suppliers, processing, preparation and trade in food. The problem, however, also concerns consumers who, through their food choices and behaviour, can direct the organization of supply towards more sustainable models.

8.3 Mitigation of Environmental Impacts

In general, strategies aimed at mitigating the impact of economic activities on the environment must take into account:

- of the entire production process: interaction between direct and indirect emissions (LCA),
- of the animal species and productive aptitudes considered,
- of "local" environmental conditions.

For some types of emissions there are international regulations or agreements in force with specific interest in animal production:

- Kyoto Protocol to reduce greenhouse gases,
- Gothenburg Protocol for the abatement of acidification, eutrophication and combating ozone depletion,
- NEC (National Emission Ceiling Directive, Directive 2001/81/EC) for emissions,
- IPPC (Integrated Pollution Prevention and Control, Directive 96/61/EC) for intensive pig and poultry farming,
- Nitrates Directive 91/676/EEC.

8.3.1 The Bioeconomy Paradigm

The bioeconomy,[8] with its huge innovative potential, can contribute to solving most of the challenges of the new millennium, from environmental restoration to climate change, the invention of new medicines and the need to feed a world where food needs will increase by 70% between now and 2050.

However, in order to ensure the environmental and economic sustainability of our societies, there is a need, in an increasingly shared form, for a profound change in both the political and research framework.

On the first point, global challenges require a shift from sectoral policies and governance mechanisms to a much more integrated approach,[9] promoting the integration of the bioeconomy with other policies, including the Common Agricultural Policy (CAP). This approach is clearly present in the new European Bioeconomy Strategy,[10] as agriculture has great potential to promote a sustainable bioeconomy through different instruments under the CAP, in particular in rural development, contributing to diversification of activities and ensuring decent living, working and economic conditions for operators. The same approach can be found in the UN Agenda 2030, which has become an integral part of European policies.[11] The common objective of all these strategic programming documents is the reorientation of the European development model, promoting the bioeconomy as a tool for growth and job creation.

The process of transforming the economy towards sustainability requires reorienting the system of research and university education. This is already happening at different levels, as shown, for example, by the launch of the new EU Framework Program for Research and Technological Development "Horizon Europe" and the emergence of new training pathways for the bioeconomy.[12] All these initiatives have one vision in common: the world of research and higher education must be reoriented towards a more inclusive model based on the "convergence" of different disciplines, recognizing that, while more in-depth knowledge at sectoral level plays a crucial role, the attempt to make research and higher education more interdisciplinary is essential to address complex problems, such as those posed by current challenges.

8.3.2 Technological Innovations for Sustainable Animal Husbandry

Consumer interest in the quality of agricultural products is growing alongside the ethical aspects of agri-food production, both because of the increased attention paid to animal welfare and the impact of agricultural and livestock activities on the environment.

Within this framework, Italian agriculture is looking with growing interest at the targeted adoption of Information and Communication Technologies (ICT), in order to improve resource management and reduce the environmental foot-

---

[8] The bioeconomy can be defined as an economy based on the sustainable use of renewable natural resources and their transformation into final or intermediate goods and services (European Commission, 2012b). Therefore, the bioeconomy includes not only traditional sectors such as agriculture, fisheries, aquaculture and forestry, but also more modern economic sectors such as biotechnology and bioenergy. Overall, in 2009 the bioeconomy in Europe had an added value of more than €1 trillion, a turnover of more than €2 trillion and about 21.5 million employees (Clever Consult, 2010). The prospects for further growth are even more promising: according to an OECD study (Oecd, 2009) it is estimated that in 2030 biotechnology will account for 35% of chemicals and industrial products, 80% of pharmaceuticals and diagnostics and 50% of agricultural products in developed countries (Source: European Commission).

[9] see, The Government of the Parks, AA.VV., Ed. ARACNE, Year 2019.

[10] Commission Communication of 11 October 2018, "A sustainable bioeconomy for Europe: strengthening the link between economy, society and the environment".

[11] Council conclusions "Towards an increasingly sustainable Union by 2030"; Commission discussion paper "Towards a sustainable Europe by 2030".

[12] The Bioeconomy Science Center at the University of Aachen, the Bioeconomy Institute at Iowa State University, the Master in Bioeconomy Management, Innovation and Governance at the University of Edinburgh.

print of the sector, to present itself to the consumer not as an enemy of the environment, but as a full and consciously (Tagliavini et al. 2019). In particular, a sector is developing in animal husbandry that is called "precision animal husbandry", i.e. a set of control techniques based on an intensive and targeted use of ICT, which in turn can be identified as "digital animal husbandry".

Digital zootechny is the set of ICT technologies applied to the livestock sector to manage data and information produced also by sensors or other devices able to measure variables of interest, acquired from different data sources, to translate them into useful information to make decisions.

Precision animal husbandry is based on the integrated management of all the information available in animal husbandry (derived from sensors as well as from traditional sources, however translated into numerical data that can be processed to obtain information) to examine it in the light of a model and, following its greater or lesser compliance, to make decisions and, where possible, trigger corrective feedback.

A significant contribution to the development of certain aspects of digital animal husbandry coincided with the introduction of automated milking systems, the so-called milking robots. These systems have seen the necessary development of technologies for the correct detection of teats, of sensors to detect the suitability of milk to be collected (colour and electrical conductivity sensors as indicators of hygiene or health problems), as well as for the detection of milk flow and production from each quarter of the udder.

Today a large part of the digital revolution in animal husbandry is played on the one hand on the development of sensors to be placed "on board" the animal[13] and on the other hand on the integration of information available through various devices (farm and not) that put the operator in a position to increase the ability to first diagnose problems in the herd and, at the same time, to activate almost automatic feedback to remedy disadvantageous or potentially dangerous conditions for the herd or for milk production.

The innovative potential related to the adoption of ICT in agricultural and livestock management processes is well represented by some examples. Precision animal husbandry contributes to the control of the individual herd and, as a result, significantly improves the possibility of managing the herd as a whole.[14] Sensors aimed at the early detection of reductions in motor activity, ingestion of food and rumination, as well as milk production, allow an important reduction in the time it takes to identify and diagnose a disease or poor welfare problems that can compromise the health and performance of the animal.[15] This early detection of the welfare status of the individual animal is directly translated into these

---

[13] A striking example is that of ruminometry, i.e. the precise and punctual H 24 measurement of ruminal activity, a valuable indicator that detects any disease at an early stage in subclinical form and therefore not detectable by the human eye (CREA).

[14] The monitoring levels (single leader vs. herd) are composed, decomposed, intersected and completed in a virtuous way. In fact, it is intuitive that if the herd or group has always been a reference also in the past, it is the observation on the individual that detects the acute problem and, on the other hand, it is always the group and the deviation from it of an individual measurement that allows to judge the extent of the anomaly on the individual (decomposition and recomposition effect). On the other hand, for example, the precision observation of the ingestion of dry matter in the distributed diet makes it possible to detect whether, beyond the individual, there is a latent problem in terms of the quality of the food used or, very important, whether the generalized decrease in ingestion can be ascribable to problems of appetite reduction as the first sign of widespread inflammatory or subclinical infectious states (Daniel M. Weary, University of British Columbia—Vancouver).

[15] Respect for animal welfare is certainly one of the aspects that makes it possible to make the most of the use of precision animal husbandry, making the image of this sector more "animal friendly". In fact, in addition to what is available for a prompt milking diagnostics especially for the udder, the availability of motion and rumination sensors supports the preparation of algorithms that allow you to understand if there are limb problems or poor rumination problems that can often be an early signal of environmental discomfort, thermal in particular. In fact, we have had the opportunity to highlight how, above all, summer heat stress can determine an immediate and significant reduction in the time dedicated to rumination (Abeni and Galli, 2016), with a preferential shift of this activity towards the night hours (less stressful).

advantages: an increase in the conversion index of food into milk[16]; a reduction in the use of drugs and various therapeutic devices; a better prediction of estri; a[17] greater probability of success in the recovery of the animal's state of health; a better control of the food ration; a[18] reduction in the recovery time of the animal; a lower amount of lost production.

The process of transformation of the food production ecosystem is driven by the need to respond to the growing challenges faced by entrepreneurs as a result of the changes brought about by globalization, both in terms of economic competitiveness and in terms of increasing consumer demands and in terms of the sustainability of agricultural products. Moreover, for some time it has been evident the need to have an increasingly easy (and inexpensive) traceability of products according to the recognizability of their peculiarities (geographical origin, raw materials used in the production cycle, compliance with specifications). The amount of information necessary to testify all these aspects in a punctiform way and in a reasonable time (if not in real time) can only pass from digital zootechny and the possible applications with precision zootechny.

The implementation of these technological systems has revolutionized the data management process, with the growing importance of Big Data and the transfer of important shares of "field" decisions from the operator to expert systems. A particularly innovative computer system for the ability to integrate multiple types of data and obtain information from them, which in turn can determine corrective feedback, is the "precision feeding" system based on real-time analysis of food during the preparation of the daily ration. This type of system includes a Nir[19] sensor system able to perform a quantification of the main components of the food, generally silage: dry matter, fibrous fractions, protein, lipids, starch. The inclusion of this technology in a computer system that regulates the management of the wagon with which the ration is prepared allows to correct in real time the amount of food to be loaded to respect the actual amount of dry matter

---

[16] In dairy farming, the item "food" accounts for more than 30% of the cost of producing cow's milk. One of the efficiency parameters generally used is the conversion efficiency of feed into milk. The relative ease with which this index can be measured on a daily basis today (thanks to the so-called precision feeding systems, integrated with automated detection systems of individual milk production) allows a real-time monitoring of the food-animal system able to warn through the detection of any sudden changes in efficiency (Crpa, 2015).

[17] Different systems of sensors "on board" of the cow are spreading, mainly aimed at detecting the moment of oestrus through algorithms that interpret the variations in the motor activity of the animal and, in some cases, are also based on the detection of rumination activity which, with its decrease in conjunction with the oestrus itself, helps to improve the predictive capabilities of the system.

[18] In agricultural areas where large quantities of wet fodder silage and especially whole maize silage (silomais) and grass silage are used, it is very important to ensure that the actual intake of dry matter itself is in line with the feeding programme. Today there are systems capable of integrating a rapid analysis (using a method based on reflectance in the near infrared, Nir) of the feed that is loaded into the mixing wagon for ration preparation, with an automated system that corrects the amount of feed to be loaded according to the dry matter actually analysed during loading. Equally important is the possibility of using methods able to highlight in real time the risk of a potential deterioration in the quality of the stored food. An example that well represents the important economic impact of a silomais quality decline is reported by Tabacco et al. (2011), which correlated the processing capacity of a tonne of dry matter of silomais in milk according to the presence of possible moulds in the food or an increase in the temperature of the silage mass (index of yeast activity that can deteriorate the silomais). This study shows that an efficient monitoring of silage front temperature can prevent efficiency drops from values close to 1.6 to less than 1.2. For this reason, thermal imaging cameras are now available that, through the quantification of the infrared emission of the bodies, allow to detect critical points in the silage mass that are not visible to the naked eye and that could represent hotbeds of yeast activity that trigger phenomena of aerobic deterioration on which moulds are grafted (with risks of mycotoxin production) and sporogenic clostridia activity (with great risks of contamination of the chain of hard and long matured cheeses).

[19] The spectrometric analysis of the Nir (Near InfraRed) system detects, in a variable way depending on the food considered, parameters such as moisture, crude protein, Adf and Ndf fiber, starch, ash and ethereal extract. Since the nutrient content of a food is typically expressed as a percentage of dry matter, by measuring the actual amounts of dry matter loaded for each ingredient in the ration it is possible to monitor the actual amounts of nutrients that are forming the mixture and make adjustments in real time to approach the parameters defined by the nutritionist.

in the daily diet. In this way, waste and inefficiencies deriving both from the variability induced by atmospheric events and the climate on the material to be loaded, and from the variability that naturally tends to occur in different points of the large trenches where the silage is stored, are avoided. Another type of computer system is the one that relates the information derived from two types of sensors seen before: motion detector and rumination detector. This type of system is able to produce important synergies between two sensors that are already interesting in themselves: the drop in rumination can serve to confirm the presence of an estrus already indicated by a variation in body activity; at the same time, variations (generally drops) in rumination not justified by the suspicion of an estrus are important signals on the inadequacy of the ration or on the state of well-being, even thermal, of the cow.[20]

With regard to work, the main changes introduced by the automation of production processes concern the skills and quality of life of operators.[21] With reference to the first point, digital zootechnics and, with it, precision zootechnics, require professional figures who are adequately and specifically prepared. The improvement in the quality of life of the operators has become evident with the introduction of automated milking with voluntary access of the cows in the milking parlour, going beyond the traditional fixed working scheme that provided for milking at night or at restricted hours due to the need to deliver at predetermined times one or two milks to the processor. Today, this begins to apply also to other routine operations, first of all the detection of estri, which in traditional conditions provides for the careful observation of the herd by the operator at different times of the day, while thanks to the new detection systems is replaced by the examination of the company computer or mobile phone that receives the alert message for the presence of animals probably in estrus. In a Dutch study (W. Steeneveld and others—Wageningen University, 2015), it became clear that the number of working hours per cow per week is lower in farms where there are sensor systems to support the detection of animals and the breeding environment.

In conclusion, we can say that a large part of the economic assessment of the adoption of new technologies does not only involve a direct increase in production, but also progress in the efficiency of the management of human and material resources, while ensuring the improvement of animal welfare and the sustainability of farming systems.

8.3.3 The Cross-Compliance Criteria of the CAP (Common Agricultural Policy)

The 2003 CAP reform introduced "cross-compliance", which requires farmers to respect certain standards relating to good agricultural land management, environmental protection, public health, animal health and animal welfare in order to be eligible for EU payments. Cross-compliance requirements therefore have as fundamental objectives consumer food safety, environmental protection and animal welfare. Those concerned by the respect of cross-compliance commitments and prohibitions are farmers and stockbreeders, i.e. those who provide food to the community and contribute to the protection of the environment, the landscape and biodiversity. In contrast, society calls for greater attention to food safety, environmental issues and animal welfare (Luise et al. 2016).

Regulation (EC) No 1782/2003 established the principle that farmers who do not respect certain public, animal and plant health, environmen-

---

[20] Improve food efficiency for cattle in heat stress, F. Abeni and A. Galli, Supplement of L'Informatore Agrario n.21/2016.

[21] These aspects of improving the quality of work are confirmed by the study by Hostiou et al. (2017), which highlights how other factors than purely economic ones can motivate milk producers to adopt the new technologies. In fact, as it is pointed out, if the reduction of work is in itself an objective to which the entrepreneur generally tends, in these cases it goes hand in hand with the appreciation of a better quality of life in the workplace, less stressful, carried out on the farm. Nathalie Hostiou, Jocelyn Fagon, Sophie Chauvat, Amélie Turlot, Florence Kling, et al. Impact of precision livestock farming on work and human–animal interactions on dairy farms. A review. Bioscience, Biotechnology and Biochemistry, Taylor & Francis, 2017, 21, pp.1–8.

tal and animal welfare requirements are subject to reductions in payments or exclusion from direct support. "Cross-compliance" requirements are an integral part of Community support under direct payments. Indeed, the decoupled single payment is now essential to cross-compliance, which makes all payments to agricultural holdings conditional on the respect of mandatory management requirements (SMRs) and the maintenance of land in good agricultural and environmental condition (GAEC). In case of non-compliance with the standards imposed by cross-compliance, a reduction of direct aids is applied until they are completely withdrawn.

The new Council Regulation (EC) No 73/2009 of 19 January 2009 establishing common rules for direct support schemes for farmers under the common agricultural policy repeals Regulation (EC) No 1782/2003.

The general principle of cross-compliance is laid down in Article 4, according to which "every farmer receiving direct payments shall comply with the statutory management requirements listed in Annex II and the good agricultural and environmental condition referred to in Article 6".

Non-compliance with these rules results in the total or partial reduction of certain EU payments to farmers. The reductions shall be proportionate to the severity, extent, permanence, frequency and intentionality of the non-compliance.

The EU Regulation n.1306/2013 has defined the new cross-compliance regime for the period 2015–2020. At national level, cross-compliance is regulated by DM n.2490 of 25 January 2017 which repealed DM 3536/2016.

8.3.3.1 Goals
Cross-compliance has the double objective of increasing the environmental sustainability of agricultural activities and meeting the expectations and interests of consumers.

8.3.3.2 Stakeholders
Conditionality applies to those who benefit from:

– Direct payments (first pillar).
– Payments for the restructuring and conversion of vineyards (OCM Wine) and for green harvesting.
– Silvo-climatic-environmental payments.
– Payments for organic farming.
– Natura 2000 payments and Water Framework Directive.
– Compensatory allowances for mountain areas or areas subject to natural or other specific constraints.

8.3.3.3 Application
The discipline relating to cross-compliance consists of two types of commitments: the Mandatory Management Criteria (SMRs) deriving from the application of Community regulations and directives and the Good Agricultural and Environmental Conditions (GAEC), also called standards, defined at national level and relating to the proper maintenance of land. The SMRs are represented by 13 EU directives and regulations, most of which have been in force for many years. The GAEC are 7 standards and represent the minimum agronomic and environmental conditions under which agricultural land should be maintained. The GAEC to be respected are agronomic (erosion, surface water regulation, soil structure and fertility) and environmental (set-aside management, minimum soil cover, maintenance of landscape features) to avoid the risks of soil and habitat deterioration.

SMRs and GAECs are classified into three areas:

1. Environment, climate change and good agronomic condition of the soil.
2. Public health, animal and plant health.
3. Animal welfare.

These three areas are in turn divided into 9 themes:

- Waters.
- Soil and carbon stock.
- Biodiversity.
- Minimum maintenance level of landscapes.
- Food safety.
- Identification and registration of animals.
- Animal diseases.
- Plant protection products.
- Animal welfare.

8.3.3.4 Duration of Commitments

The duration of cross-compliance commitments varies depending on the application submitted. For the Single Payment Application and/or Rural Development Program (RDP) payment applications, commitments must be fulfilled for the duration of the entire calendar year in which the application is submitted. If, at any time in a given calendar year, the cross-compliance rules are not complied with and such non-compliance is directly attributable to the beneficiary, reductions shall be applied to the aid.

Beneficiaries of the payment of support for restructuring and conversion of vineyards must, on the other hand, respect their commitments for the 3 years following the receipt of those payments, while beneficiaries of the payment of green harvesting support programs must respect their cross-compliance commitments in the year following the receipt of that payment. Therefore, for those beneficiaries, the reduction of the Community aid shall apply in case of non-compliance with cross-compliance rules at any time in the 3 years following the calendar year in which the first payment for restructuring and conversion of vineyards was granted or at any time in the year following the calendar year in which the green harvesting payment was granted. From 2017, reductions and exclusions shall also apply where the amount is equal to or less than EUR 100 per beneficiary and per calendar year. The obligation to implement the corrective actions notified to the beneficiary by the competent authority remains unchanged.

8.3.3.5 Non-compliance with cross-compliance rules.

In case of non-compliance with commitments and cross-compliance rules (SMR and GAEC), as a result of actions or omissions directly attributable to the individual farmer, payments due to the farmer in the calendar year in which the non-compliance was found shall be reduced or cancelled depending on the severity, extent and duration of the breach. In case of repetition of the breach, the level of penalties applicable shall be multiplied by three.

8.4 The Protection of Public Health

In order to achieve its objectives, "Public Health" must implement a holistic and multidisciplinary approach, making use of all the necessary expertise. Relevant disciplines include public hygiene, occupational medicine and hygiene, epidemiology, biostatistics, infectious disease clinics, microbiology, toxicology, nutrition sciences, but other disciplines may also be part of the group. These include, for example, odontostomatology and geriatrics. Moreover, in recent times, the awareness that Public Health is not an exclusively medical discipline has begun to spread: for example, given the close link between nutrition, food quality and the health of the population, present both on the purely nutritional side and on the toxicological and microbiological quality of food, agronomy, veterinary medicine and food technology have become fundamental components of modern "Public Health". The starting point of public health is the awareness that all diseases attributable to a causal agent that is detectable and external to the individual can be prevented, simply by removing exposure or neutralizing the agent: this is the case of infectious diseases, which modern society has been able to deal with, with the consequent dramatic reduction in the "burden of disease" from biological agents highlighted in particular in countries with higher socio-economic levels (Prüss-Ustün et al. 2016). Even diseases that are not communicable by environmental etiopathogenesis recognize a clear and external cause, but their impact on public health and the overall cost of disease has not been satisfactorily reduced to the point that the World Health Organization has identified the fight against non-communicable diseases ("non-communicable diseases") as one of the health priorities of our time, without of course underestimating infectious diseases, and in particular zoonoses, which not infrequently originate from the rearing of farm animals, in particular from incorrect practices of intensive animal husbandry. The link between animal husbandry and the spread of antibiotic resistance is also a priority.

On this basis the theory of the "One-Health Approach" has developed and taken root (Zinsstag et al. 2015). It recalls that the health of the general population is promoted by the intake of adequate quantities of healthy food, of high nutritional value, free from chemical and biological contamination and that in livestock farming the health of animals and that of workers and consumers are strongly correlated; the concept of One-Health was born from the observation that human health, animal health and the ecosystem are inextricably linked and that a holistic approach is necessary to understand, protect and promote the health of all species.

8.4.1 The Health, Human and Veterinary Prevention System

When reference is made to the activity of doctors and veterinarians, the exercise of the profession is automatically associated with the care of the patient (human or animal), aimed at maintaining the state of well-being and reducing the damage caused by any disease, until recovery. However, when dealing with particularly dangerous infectious diseases, as is the case with many zoonoses, the role of the professional expands to assume the functions (and responsibilities) typical of public health protection. In these situations, the doctor and the veterinarian are called to play a more comprehensive role, which requires—in addition to the ability to diagnose and treat pathologies (individual and collective)—multidisciplinary knowledge and timely exchange of information between the two professionals. Such an approach is not improvised, but requires a process of preparation and training, of knowledge of each other's skills that is still to be achieved. There is no other way to effectively prevent zoonoses and successfully manage possible emergencies (Rabozzi et al. 2012). The problem exists in all countries of the world (King et al. 2004).

The lack of communication and integration between human and animal health has long been a leitmotiv in the literature on zoonoses and public health problems in general (Cipolla et al. 2015).

Already in the 1960s, the American epidemiologist Calvin Schwabe, to whom we owe the concept of One Medicine, advocated the need for a unified view of health and disease without distinction between humans and other animal species, opposing the growing compartmentalization between different disciplines and specializations. This approach has been taken up both in scientific research and clinical practice. P.J. Cripps, Director of the Livestock Health and Welfare Division of the Faculty of Veterinary Medicine, University of Liverpool, stressed that many veterinarians are unaware of the importance of zoonoses. The problem of underdiagnosis and misdiagnosis of zoonoses is more dramatic in developing countries: for example, in Africa, human brucellosis, characterized by recurrent fever, is almost always mistaken for drug-resistant malaria. Moreover, many of the less common zoonoses are easily misdiagnosed even in developed countries: suffice it to mention the dramatic case at the end of 2004 in Germany (but with precedent in the USA) of a patient suffering from unrecognized rabies, whose organs, transplanted after death, led to the death of three other people.

At the Conference on emerging zoonoses held in The Hague in September 2004, European Health Commissioner Byrne reiterated the importance of a new culture of collaboration between doctors and veterinarians, both in the European institutions and in the Member States. Appropriate mechanisms are therefore needed to ensure a common approach between the two professions at national level. The joint initiative of the British Medical Journal and the Veterinary Record—which in November 2005 devoted an issue to zoonoses and the need for close cooperation between doctors and veterinarians—could be a sign of a real breakthrough towards convergence between human and animal medicines. It is time to reflect on how doctors and veterinarians can collaborate more closely for the benefit of patients of all species, promoting in university and postgraduate courses a greater opportunity for comparison and integration of skills.

On September 24, 2004, during a symposium organized by the Wildlife Conservation Society at Rockfeller University, a group of scientists denounced the danger posed by increasingly common phenomena, such as the disappearance of some species and the invasion of alien species, environmental degradation, pollution and climate change that can alter life on our planet, from the deepest oceans to the most populous cities. In that context the document "The Manhattan Principles on One World, One Health" was presented, consisting of 12 principles considered indispensable for the protection of the integrity of the planet (Cook et al. 2004):

1. recognize the essential link between the health of humans, domestic animals and wildlife and the threat that disease poses to people, food and economic security, and the biodiversity needed to maintain a healthy environment and a well-functioning ecosystem that we all need;
2. recognize that all decisions concerning land and water use have significant health implications. Whenever we ignore this relationship, changes in the ecosystem and the emergence of new diseases occur;
3. include the study of the health of wild species as an essential component of global disease prevention, surveillance and control;
4. recognize that public health programs can make an important contribution to the conservation of different species;
5. to promote innovative, holistic and forward-looking approaches to the prevention, surveillance, monitoring and control of emerging and re-emerging diseases, taking into account the complex interconnection between species;
6. seek opportunities for full integration between a biodiversity conservation perspective and human needs when taking measures to control infectious diseases;
7. reduce trade and regulate the conservation and hunting of wild species, not only to protect these species, but also to reduce the risk of disease transmission, including between species, and the development of new hosts for pathogens;
8. reduce the planned killing of wild species free for disease control only to specific situations based on a scientific, multidisciplinary and international consensus that such a population does indeed pose a significant threat to public health, food safety, or other wild species;
9. increase investment in global health infrastructure, both human and animal, appropriate to the severity of emerging and re-emerging threats to humans and animals, by strengthening health surveillance on animals and humans and improving coordination between governmental and non-governmental agencies, vaccine and drug companies, and all possible partners;
10. to create collaboration between governments, populations, public, private and non-profit sectors to address the challenges of global health and biodiversity conservation.
11. provide resources and support for the development of global wildlife health surveillance networks able to exchange information with the public and veterinary health system as part of an alert system for emergency and re-emergence of diseases.
12. invest in education and awareness raising for the world's population to influence the policy process to improve the awareness that we need to better understand the relationship between health and ecosystem integrity to successfully improve a planet's health prospects.

These principles represent a prodromal decalogue to the formulation of "holistic policies" capable of guaranteeing the biological integrity of ecosystems, collective health, animal health and, ultimately, the conservation of our planet's environment for future generations.

8.4.2 The "Risk Assessment" of Agricultural Operators

Agriculture is one of the sectors at greatest risk, both in terms of the extent and frequency of accidents reported, and therefore the legislator has drawn up a series of rules to protect the health and safety of agricultural operators.

The Consolidation Act on health and safety at work (Legislative Decree no. 81 of 9 April 2008) introduced key concepts for the protection of agricultural workers, provisions for employers, technical, procedural and organizational preventive measures and the use of protective equipment. Article 21 of Legislative Decree no. 81/2008 also extends to self-employed workers, including direct farmers and members of simple companies operating in the agricultural sector, two obligations previously incumbent only on employers with employees or similar:

- Using machines and equipment that comply with standards;
- Equip personal protective equipment.

In addition, in the last two years, some important ministerial decrees have also been published, which also have a significant impact on safety at work in agriculture.

The first (Ministerial Decree 30/11/12) imposes the obligation to draw up the Risk Assessment Document on all companies, even with less than 10 workers, replacing the previous possibility of self-certification. Therefore also the farmer who makes use of the work of seasonal and occasional workers.

The second one (D.M. 27/3/2013) introduces simplifying measures with regard to information, training and health surveillance for companies that employ seasonal workers for less than 50 days/year.

The measures defined in the Consolidation Act and subsequent ministerial decrees are applied in various ministerial circulars and especially in the issuing of regional guidelines, which explicitly define how to adopt Community provisions at a territorial level and go into detail on the prevention of the various possible risks.

Among these, priority attention is given to the risk of manual handling of loads, the risk associated with the use of agricultural vehicles, the risk associated with the use of hazardous substances and contact with biological agents, the risk due to the presence of overhead power lines and some aspects related to the risk of exposure to noise.

8.4.2.1 Risk Associated with Manual Handling of Loads

In all the activities of the agricultural sector there are numerous operations in which there is the risk of Manual Handling of Loads: for lifting, transporting, towing or pushing loads, even heavy loads. This risk determines the possibility of injury to the spinal column and disturbance of the skeletal muscle system. It is therefore essential to protect workers by introducing procedural measures that require controlled and limited efforts, organizational measures such as Health Surveillance, and technical measures that provide, for example, the use of automatic lifting equipment where possible.

8.4.2.2 Risk Related to the Use of Agricultural Means of Transport

Very statistically significant is the impact of accidents related to the risk inherent in the use and handling of means of transport and agricultural vehicles: without prejudice to the prescriptions regarding the legislative conformity of the vehicles, it is important to adopt strict maintenance protocols, exercise periodic controls and use procedures that are appropriate, both in terms of applicability and in terms of training, dissemination and control.

8.4.2.3 Risk of Exposure to Hazardous Substances

In addition, the risk from exposure to dangerous substances, especially from the use of plant protection products, is not negligible. The term plant protection product means a substance or mixture of substances used to prevent, destroy or control any parasite, which may cause damage or interfere with the production, processing, storage of food, agricultural raw materials, wood. The use of chemicals in the agricultural sector must be regulated by internal procedures with appropriate provisions for specific risk assessment, storage, use and disposal. Also for this particular risk exposure, the health surveillance protocol and specific training is mandatory.

8.4.2.4 Biohazard

There are a number of working situations in the agricultural sector that can potentially expose people to the biological risk of contracting infectious diseases, when these are transmitted from animals to humans they are called zoonoses. Disease transmission can occur during cleaning of shelters, milking, grooming of animals, handling of excrement or by insects and parasites.

8.4.2.5 Risks and Effects of Noise Exposure

Furthermore, as mentioned, it is important not to underestimate certain risks with a possibly less significant impact at epidemiological level, but no less important from an operational point of view. For example, activities in which equipment producing high noise levels is used, which are generally periodic and discontinuous activities, against which operators must be protected by means of appropriate Personal Protective Equipment.

8.4.2.6 Risk Related to the Presence of Overhead Power Lines

The risk assessment should take into account the possibility of accidents due to overhead power lines, which should be well reported and identified in order to avoid interference with even life-threatening consequences.

The adoption of safety systems and behaviours, which affect the organization of work, is not only a regulatory obligation, but implies a real change in the behaviour of the people involved, changes that can only be achieved through a cultural and educational path.

8.5 Environment and Health

Over the last hundred years, our concept of health has been expanded from the limited view of health as the absence of disease or other physical problems to a broader, more complex concept that includes social and personal components, as well as physical abilities and the environment. In this regard, health must include the ability to realize hope, to meet needs and to change or confront others and the environment. Never as in this historical context is the relationship with the environment one of the fundamental determinants that condition the state of health of the human population. From the polluted city to the uncontaminated forest, the relationship between the individual and different environmental factors can result in different states of well-being or illness. Understanding what elements must be taken into account, from an epidemiological point of view, to assess the impact of different factors on health status is a very complex task. It is only by crossing environmental, territorial and urban, epidemiological, mortality and other health, demographic, cultural and social indicators that possible welfare scenarios for a given population can be drawn. The environment can directly or indirectly affect health. It can promote the circulation of pathogens and other biological factors, such as toxins, pollens and other allergens, it can carry chemical and physical environmental contaminants that can interact with the animal and plant populations present.

Epidemics are created when there are great imbalances "between man and the environment"; we remember the Aids virus that has infected man from monkeys or the Ebola virus that has reached man because of deforestation that has forced wild animals to adapt to urbanized areas.

Finally, the environment can be the source of accidents and disabilities when, at work as well as on the road, adequate safety and protection measures for people and the containment of animal and plant species are not observed. The most studied and best known impacts on health are associated with air pollution, poor water quality and poor sanitary conditions, together with the knowledge of dangerous chemical substances, the so-called contaminants. Other emerging problems are noise and climate change, stratospheric ozone depletion, loss of biodiversity and soil degradation.

8.5.1 Health and Ecosystem Services

Recently in the 1980s the term "ecosystem services" was introduced (Ehrlich & Ehrlich 1981) focusing on the importance of the environment for human survival. These are not simply "green areas", but a structural and functional network of natural and semi-natural systems capable with their "services" of improving the quality of life also in social and economic terms. They are therefore systems of green areas essential to improve the resilience of habitats, ecological efficiency, full functionality of ecosystems, ecological connectivity and, at the same time, the aesthetic-perceptual perception linked to the presence of natural systems and nature-based solutions provided for in the municipal urban green plan. Therefore, in line with the European Commission's requests, a new model of urban planning and design more attentive to climate change mitigation and adaptation, but also to the removal of pollutants from urban green areas is proposed. In short, more attentive to the well-being of citizens, the protection of biodiversity and the reduction of the artificialization of urban spaces and therefore more attentive to the reduction of soil consumption. Finally, more recently, and in my opinion at last, the environment and nature are recognized as a global system, not relegated only to constrained areas and the survival of man, but linked to the whole territory, indispensable for the well-being of the citizen and as productive added value, able therefore not only to perform a passive function of existing, mitigating and furnishing a landscape, but as an active element of the socio-political and economic context of the country.

Therefore, it is essential to increase the effectiveness of the environmental policies underpinning the UN Convention on Biological Diversity by fostering a better understanding of the true economic value of ecosystem services and promoting organizational and economic tools that enhance it.

8.5.2 Environment and Social and Economic Development

The demand for health and the demand for environmental quality, far from being an economic brake, are a factor in the development and growth of the economy. Because they trigger a virtuous spiral: to achieve greater health welfare and environmental quality, in fact, more innovation and more organization and social integration are needed. It is no coincidence that the European countries with the best economic performance: Germany, the Scandinavian countries and northern Europe are also the countries that best meet the demand for health, environmental and welfare quality.

Well-being, which means to be well, to exist well, is the term that specifies the aspects, the characteristics, the quality of life of each individual and of the environment. In the report of the Health Commission of the European Observatory on Health Systems and Policies the definition of well-being as "the emotional, mental, physical, social and spiritual state of well-being that enables people to reach and maintain their personal potential in society" has been proposed. Human well-being is determined by individual well-being (to which attributes such as health, educational level, etc. are linked) and social well-being (to which correspond attributes shared with other people, i.e. family, friends, society as a whole). Human well-being therefore needs some "supporting pillars" (environment, culture, economy). Now in its sixth edition, the Bes Report published by the Institute, which annually presents the statistical framework and the innovations introduced by the project on fair and sustainable well-being indicators in Italy, offers an integrated picture of the main economic, social and environmental phenomena that characterize our country, through the analysis of a wide set of indicators divided into 12 domains. In this framework, the exchanges between the sphere of human well-being and the well-being of the ecosystem are relevant. The domains that, through an adequate set of indicators identified by the BES ("Fair and Sustainable Welfare", ISTAT 2018), are taken into consideration are twelve: (1) Health; (2) Education and training; (3) Work and life time balance; (4) Economic well-being; (5) Social relations; (6) Politics and institutions; (7) Security; (8) Subjective well-being; (9) Landscape and cultural heritage; (10) Environment; (11) Research and innovation; (12) Quality of services. In the analysis of equitable and sustainable well-being, it is important to consider not only the levels of well-being and their trend over time, but also the differences in their distribution and the articulation of well-being profiles. In this perspective, the territorial dimension is a fundamental key to understanding because it brings out more precisely the areas of advantage or deprivation. The Bes 2017 report confirmed that 2015 and 2016 mark an improvement in many domains of well-being, even if territorial differences remain both in levels and dynamics. The data indicate that territorial inequalities in welfare levels affect all domains, albeit with varying intensity, and that they are rather persistent over time, but at the same time highlight more articulated territorial gradients than the usual North/South contraposition. In various domains, coexistence emerges, in the same region or distribution of areas with very different, sometimes opposite welfare profiles and trends. Read in the geographical space, the differences between neighbouring territories or between provinces in the same region sometimes draw different boundaries between the Centre-North and the South. In particular, the inequalities observed in the domain, landscape and cultural heritage essentially at the territorial level, show how the constitutional principle of landscape and cultural heritage protection is not, in substance, implemented equally throughout the national territory, with significant consequences for collective well-being. The expenditure of municipalities for the management of cultural assets and activities is unevenly distributed over the territory: the administrations of the Centre-North spend, on average, almost three times as much as those of Southern Italy (23.8 euros per capita against 8.9 in 2016). The expenditure is 49.6 euros per capita in Trentino-Alto Adige (55.1 in the province of Bolzano) and exceeds 30 euros in Friuli-Venezia Giulia and Emilia-Romagna, while in all regions of Southern Italy, except Sardinia, the municipalities spend less than 10 euros per capita on culture, and in Campania less than 5. In the period 2010–2016, the current expenditure of municipalities for culture has decreased in all regions, but to different degrees and in such a way as to accentuate the inequality between the allocations: −21.9% in the South, −16.6% in the Centre and − 10.3% in the North. In 2016, only in the provinces of Trentino-Alto Adige and Veneto, Abruzzo and Sicily this item of expenditure increased compared to the previous year. Therefore, as already mentioned, the perception of well-being has a subjective and personal, objective and social value in relation to times, places, circumstances and people. Subjective well-being is influenced by objective well-being conditions. If one has money but lives in a highly polluted environment or physical health is compromised, the quality of life is poor and there is no well-being. Well-being does not only lie in the comforts in which and with which one lives and works, but in the satisfaction, one obtains by acting. Well-being is also psychological, relational. It draws on the emotions of the individual, his anxieties and hopes, his fears and all that is deep. Well-being is perceived when there is an authentic human relationship, when one is welcomed and recognized, when one is called by name and is a person, not just a "client" or "user", with one's own uniqueness and potential. Well-being is then everything that concerns security, tranquility, the absence of difficulties. Who does not, after all, want a life without problems, without risks and unforeseen events? But, sometimes, well-being is right there, the result of what difficulties teach us to grow inwardly. Because it is precisely crises that illustrate how to "feel good", how to build new life paths, how to draw on one's personal resources, how to develop the ability to activate oneself, how to discover emotional and managerial potential to overcome obstacles. Resilience means learning something more about oneself, about one's own abilities to face all kinds of problems and build new balances. Physical well-being is a dynamic condition for seeking balance, based on the individual's ability to interact with himself and the environment in a positive way, while changing the surrounding reality. Talking about physical well-being means taking responsibility and attention for caring for oneself, for feeling well in the best possible way. It means taking on the responsibility of doing physical activity with constancy, of thinking about one's own diet, going from eating indiscriminately to feeding oneself in order to have vital energy, of drinking water to satisfy the need for hydration of a body made up of about 70% water and to purify oneself from waste, of breathing oxygen in ways that are not only vital, automatically taking care of the quality of the air and the green environment. In a word, wellness, i.e. well-being deriving from the practice of movement and physical exercise from a correct diet, from a positive and proactive attitude, from the search for one's own psychophysical balance. The person can acquire direct control and management of his/her own condition of well-being (individual habits, mental attitude). The mind is directly connected to the body and can become a tool and a resource to increase well-being. The term well-being in this context indicates a philosophy that sees the individual as responsible, actively involved in the process of improving and increasing one's own health. Well-being also means empowerment, i.e. the ability to take control of one's life, to master it, to acquire an active role towards one's existence and the environment by facing difficulties with a positive and constructive attitude. There is a well-being that is acquired through educational and cultural processes that serve to give fullness to life, to form free and thinking personalities; that well-being, true, is found in pursuing the truth in continuous formation, in sincere and honest human relationships. Well-being is also the capacity of the individual to realize himself with satisfaction and gratification, with awareness and autonomy, having at his disposal all the accessible, personal and community resources. In the future, human existence will depend greatly on the search for one's own condition of well-being, the attention paid to oneself, the subjective sense of well-being and the quality of life. Once we reach our own well-being, we will be able to cope with the changing living conditions of a technological reality and its continuous change.

References

Abeni F, Galli A. Monitoring cow activity and rumination time for an early detection of heat stress in dairy cow. Int J Biometeorol 2017 Mar;61(3):417–425. https://doi.org/10.1007/s00484-016-1222-z. Epub 2016 Aug 8

Bes Report 2018: Equitable and sustainable well-being in italy. ISBN 978-88-458-1967-4 2018 Istituto nazionale di statistica (ISTAT). https://www.istat.it/en/archivio/225140

Cipolla M, Bonizzi L, Zecconi A (2015) From "one health" to "one communication": the contribution of communication in veterinary medicine to public health. Veterinary Sciences. Open Access 2(3):135–149

Clever Consult (2010). Albrecht J & Carrez D & Cunningham, P. & Daroda, Lorenza & Mancia, R. & Máthé, L. & Raschka, Achim & Carus, Michael & Piotrowski, Stephan. (2010). The Knowledge Based Bio-Economy (KBBE) in Europe: Achievements and Challenges. https://doi.org/10.13140/RG.2.2.36049.94560

Crpa (2015) Periodico C.R.P.A. NOTIZIE n. 1/2015-2.74 –Febbraio 2015 costi di produzione e di trasformazione del latte in Emilia-Romagna

Ehrlich PR, Ehrlich AH (1981) Extinction: the causes and consequences of the disappearance of species. Random House, New York

European Commission (2012b) Communication from the commission to the european parliament, the council, the european economic and social committee and the committee of the regions innovating for sustainable growth: a bioeconomy for europe. website: https://op.europa.eu/en/publicationdetail/-/publication/84e7a360-6970-4cb8-939d-8acbf-33f0ae8/language-en

Hostiou N, Fagon J, Chauvat S, Turlot A, Kling-Eveillard F, Boivin X, Allain C (2017) Impact of precision livestock farming on work and human-animal interactions on dairy farms. A review, Biotechnologie, Agronomie, Société et Environnement, 21(4), 268–275

King LJ (2004) "Emerging and re-emerging zoonotic diseases: challenges and opportunities." Compendium of technical items presented to the International Committees or to Regional Commissions of the OIE: 21–29

Luise A, Postiglione A, Cordini G. Convegno Nazionale "Clima, biodiversità e territorio italiano"; Atti Abbazia di Montecassino 23 aprile 2016. https://www.isprambiente.gov.it/files2017/pubblicazioni/atti/ATTI_2016_ICEF.pdf

Massimo Tagliavini, Bruno Ronchi, Carlo Grignani, Piermaria Corona, Roberto Tognetti, Marco Dalla Rosa, Paolo Sambo, Vincenzo Gerbi, Mario Pezzotti, Francesco Marangon e Marco Marchetti) 2019. Intensificazione Sostenibile. Strumento per lo sviluppo dell'Agricoltura italiana, La posizione dell'Associazione Italiana delle Società Scientifiche Agrarie (AISSA), Casa editrice: Società di Ortofrutticoltura Italiana (SOI) ISBN: 978-88-32054-01-9, pp. 74

Mekonnen MM, Hoekstra AY (2010) The green, blue and grey water footprint of farm animals and animal products. Value of Water Research Report Series no.48, UNESCO-IHE, Delft, the Netherlands

Oecd (2009) The Bioeconomy to 2030: designing a policy agenda. 15 Apr 2009 ISBN: 9789264056886 (PDF). https://doi.org/10.1787/9789264056886-en

Peterson G, Allen CR, Holling CS (1998) Ecological resilience, biodiversity, and scale. Ecosystems 1:6–18

Prüss-Ustün A, Wolf J, Corvalán C, Bos R and Neira M (2016) Preventing disease through healthy environments: A global assessment of the burden of disease from environmental risks. World Health Organization. https://www.who.int/publications/i/item/9789241565196

Rabozzi G, Bonizzi L et al (2012) Emerging zoonoses: the "one health approach". Saf Health Work Open Access 3(1):77–83

Robert A. Cook, William B. Karesh, and Steven A. Osofsky. The manhattan principles on "one world, one health" September 29, 2004. Wildlife Conservation Society, Bronx, New York, USA. http://www.oneworldonehealth.org/

Tabacco E, Righi F, Quarantelli A, Borreani G (2011) Dry matter and nutritional losses during aerobic deterioration of corn and sorghum silages as influenced by different lactic acid bacteria inocula, J Dairy Sci, 94, 1409–1419

Zinsstag J, Schelling E, Waltner-Toews D, Whittaker M, Tanner M (Eds) (2015) One health: the theory and practice of integrated health approaches. Publisher: CABI International https://doi.org/10.1079/9781780643410.0000

Environmental Pollution and Cardiorespiratory Diseases

9

Cristina Sestili, Domenico Barbato, Rosario A. Cocchiara, Angela Del Cimmuto, and Giuseppe La Torre

Abstract

There is sufficient evidence that associate the exposure to a variety of environmental air pollutants, such as nitrogen oxides, ozone and particulate matter, to respiratory diseases, such as asthma, chronic obstructive pulmonary disease, lung cancer and respiratory infections, as well as cardiovascular diseases.

The air in the urban environment is made of a complex mixture of chemicals and carcinogens, derived mainly from combustion sources. However, quantifying the magnitude of pollutants on health presents considerable challenges due to the limited availability of information on exposures to air pollution. The association between air pollution, mainly particulate matter (PM) and ozone exposure, and cardiopulmonary diseases has long been recognized. There is evidence coming from time series studies of hospital admissions for respiratory diseases, which indicates that admissions for chronic obstructive lung diseases, asthma and pneumonia are more frequent on days with high air pollution concentrations. These associations are usually observed in association with PM, O_3 and NO_2.

Moreover, in last decades, several studies have showed that hospital admissions for cardiovascular diseases were more frequent on days with high concentrations of PM and ozone. Different studies found associations between ambient air pollution and hospital admissions for various cardiovascular diseases, such as ischaemic heart disease, congestive heart failure and dysrhythmia including congestive heart failure.

The aim of this chapter is to give an update of these relationship with the latest evidence from the scientific literature.

Keywords

Climate change · Environment · Pollution
Cardiovascular · Respiratory diseases
Environmental tobacco smoke

C. Sestili · D. Barbato · R. A. Cocchiara
A. Del Cimmuto · G. La Torre (✉)
Department of Public Health and Infectious Diseases, Sapienza University of Rome, Rome, Italy
e-mail: cristina.sestili@uniroma1.it;
domenico.barbato@uniroma1.it;
rosario.cocchiara@uniroma1.it;
angela.delcimmuto@uniroma1.it;
giuseppe.latorre@uniroma1.it

9.1 Introduction

The relationship between the exposure to air pollution and respiratory as well as cardiovascular diseases has been deeply studied.

There is sufficient evidence that associate the exposure to a variety of environmental air pollutants, such as nitrogen oxides, ozone and particulate matter, to respiratory diseases, such as asthma, chronic obstructive pulmonary disease, lung cancer and respiratory infections, as well as cardiovascular diseases (Ab Manan et al. 2018; Doiron et al. 2019; GBD 2015 Collaborators for Chronic Respiratory Disease 2017; Peel et al. 2005).

The air in the urban environment is made of a complex mixture of chemicals and carcinogens, derived mainly from combustion sources. However, quantifying the magnitude of pollutants on health presents considerable challenges due to the limited availability of information on exposures to air pollution. The association between air pollution, mainly particulate matter (PM) and ozone exposure, and cardiopulmonary diseases has long been recognized (Tsai et al. 2013; Nuvolone et al. 2013). On the other hand, there is clear evidence that a mitigation of these effects can occur, through urban green infrastructures that can play an important role in improving air quality, because of not only their well-known aesthetic and recreational benefit but also their capability to reduce air temperature and to remove air pollutants (Manes et al. 2008).

It is well known that the surface of the respiratory tract represents the largest interface between human body and the environment. This surface is constantly exposed to a spectrum of gaseous contaminants and particulates dispersed in the respired air.

There is evidence coming from time series studies of hospital admissions for respiratory diseases, which indicates that admissions for chronic obstructive lung diseases, asthma and pneumonia are more frequent on days with high air pollution concentrations. These associations are usually observed in association with PM, O_3 and NO_2 (La Torre et al. 2018).

Moreover, in last decades, several studies have showed that hospital admissions for cardiovascular diseases were more frequent on days with high concentrations of PM and ozone (Zanobetti et al. 2000; Schwartz et al. 2001). Different studies found associations between ambient air pollution and hospital admissions for various cardiovascular diseases, such as ischaemic heart disease, congestive heart failure and dysrhythmia including congestive heart failure.

The aim of this chapter is to give an update of these relationship with the latest evidence from the scientific literature.

9.2 Respiratory Diseases

9.2.1 Respiratory Infections

The link between the exposure to air pollution and respiratory infections has been evident since the mid of last century. The work published by Logan in the Lancet (Logan 1953) found that there was a link between the London smog and the increase of death due to pneumonia.

More recently, the association between air pollution and respiratory infections has been studied by several observational studies and reviews. One of the main focus of this evidence is represented by children. There is evidence that traffic-derived particulate matter is strongly associated with an increased risk of respiratory infections in early childhood. Moreover, there is indirect evidence that suggests that a cleaner air can be able to rapidly decrease these risks.

A meta-analysis coming from the ESCAPE project demonstrated that the chronic exposure to traffic-derived air pollution is strongly associated with an increased risk of respiratory infection in childhood. Particularly, the following pollutants were studied: nitrogen dioxide (NO_2), nitrogen oxide (NO_x), particulate matter ≤ 2.5 μm ($PM_{2.5}$), $PM_{2.5}$ absorbance, PM_{10}, $PM_{2.5-10}$ (coarse PM). All the pollutants, except $PM_{2.5}$, were significantly associated with pneumonia, with the highest effects for PM_{10} (OR = 1.76; 95% CI: 1.00–3.09). Moreover, they found a significant association between the exposure to NO_2 and otitis media and croup (OR = 1.09; 95% CI: 1.02–1.16 per 10-μg/m^3) (MacIntyre et al. 2014).

Peel et al. particularly studied the relation between ambient air pollution and respiratory emergency department visits in the general population in Atlanta, USA. They reported a 1–3% increases in upper respiratory infection visits in relation to a standard deviation increases of ozone, NO_2, CO and PM_{10}. Moreover, they found a 3% increase in pneumonia visits due to a 2 μg/m increase of $PM_{2.5}$ organic carbon.

Vanker et al. (2017) posed their attention to the relationship between the exposure to maternal smoking and environmental tobacco smoking (ETS) and children respiratory infections. They found that maternal smoking and ETS have a great influence on infant lung development and consequently are associated not only with wheezing or asthma but also with childhood upper and lower respiratory tract infections.

As far as upper respiratory tract infection is concerned, they demonstrated that ETS exposure is associated with recurrent otitis media, and the need of undergoing tonsillectomy for recurrent tonsillitis. Moreover, maternal smoking is associated with an increase (almost double) of the risk of middle ear disease. In relation to lower respiratory tract infections, the exposure to ETS is associated with an 22% and 62% increase of the odds of having an infection if there are one or two parents who smoke. And the main risk is related to the development of bronchiolitis.

The main pathologies that are associated with an increased risk are tuberculosis with a pooled effect estimate of OR 1.9 (95% CI: 1.4–2.9) for TB infection and OR of 2.8 (95% CI: 0.9–4.8) for TB disease for those exposed to ETS (Jafta et al. 2015). Other well-recognized infections associated are those sustained by respiratory syncytial virus and the influenza virus.

And concerning influenza virus, a time-series analysis carried out in China shows that air pollution may be associated with the risk of influenza (Meng et al. 2021). The increase of the relative risk (RR) of influenza was associated with a 10 μg/m^3 increase in SO_2 (RR: 1.099; 95% CI: 1.011–1.195), NO_2 (RR: 1.039; 95% CI: 1.013–1.065) and O_3 (RR: 1.005; 95% CI: 0.994–1.016).

In relation to indoor air pollution, Enyew et al. (2021) estimate the magnitude of the association between biomass fuel used for domestic purposes and acute respiratory infection (ARI) among children younger than 5 years in Ethiopia. These authors found a pooled prevalence of ARI of 22% (95% CI: 17–29), and significant association between biomass fuel use and ARI with an OR = 2.6 (95% CI: 2.05–3.30).

In the adult population, this association is weak. Jary et al. found conflicting results. In their review in two studies, a significant adjusted increased risk of acute lower respiratory infection due to household air pollution is present, while in four studies they did not find any significant association (Jary et al. 2016).

In the general population, a recent review (Katoto et al. 2021) reported that both acute and chronic exposure to air pollution can have an influence on COVID-19 epidemiology. For acute exposure, $PM_{2.5}$ then PM_{10}, NO_2 and O_3 are pollutants independently associated with COVID-19 incidence and mortality, while for the chronic exposure, a significant association was demonstrated for $PM_{2.5}$ and NO_2. While this association is true is a scientific challenge. In a systematic review published by Maleki and colleagues, the authors report that the scientific literature did not confirm that atmospheric particulate matter pollution enhances the transmission of SARS-CoV-2. Some studies argue that atmospheric PM can enhance the spread operating as a virus carrier. In this sense, other factors need to be considered in the spread of COVID-19 disease, a part from PM concentration, are size of particles in ambient air, weather conditions, wind speed, relative humidity and temperature (Maleki et al. 2021).

9.3 Allergy and Asthma

Pollutants such as particular matter (PM_{10} and $PM_{2.5}$), ozone (O_3), nitric oxide (NO), nitrogen dioxide (NO_2) and sulphur dioxide (SO_2) that are usually present in the polluted air can act on the mucosa of the airways, which causes breathing irritating and causing it to become inflamed. This can, consequently, increase the likelihood that the actual allergens will penetrate the deeper layers of the mucous membranes of the already weakened nose, throat and bronchi and, finally, trigger respiratory allergies and symptoms. Moreover, fine particles can bind to allergens in the atmosphere and thus can facilitate their entry into the respiratory system, up to the bronchi and lungs.

There is an increasing body of evidence according to which climate change has an impact on respiratory allergy and asthma induced by pollen and mould allergens. The reason why this happens is related to the fact that climate change and the global warming have an impact on the quantity, intensity and frequency of rainfall, but also on the frequency of extreme events such as droughts, floods, heat waves, hurricanes and thunderstorms.

As far as allergic respiratory diseases, it is known that pollen and mould allergens are able to promote the release of pro-inflammatory and immunomodulatory mediators, which in turn can accelerate the development of the IgE-mediated sensitization and allergy.

There is evidence that high atmospheric levels of carbon dioxide (CO_2) are associated with the increase of photosynthesis and reproductive effects, and as a consequence, more pollen is produced. Moreover, on the other hand, floods and rainy storms are associated with an increase of mould proliferation, that is linked to the occurrence of respiratory diseases, such as allergic disease (rhinitis) or asthma (D'Amato et al. 2020).

With climate change, mean annual air temperatures are getting hotter in most parts of the world. Since thermometer-based observations began, the year 2015 and the period 2006–2015 were the warmest year and decade on record, respectively. The global average surface temperature has risen at an average rate of 0.07 °C per decade since 1901. During the same period, extreme weather events, such as heat waves, droughts, floods, cyclones and wildfires, have become more and more common (Levi et al. 2018). One of the most important cause of climate change is the emission of pollutants, which also causes mortality and morbidity in particular for cardiovascular and respiratory diseases.

Asthma is a major cause of disability, health resource utilization and poor quality of life for those who are affected. It is the most common chronic disease among children and young adults, particularly because of its early onset (To et al. 2012). In 2016, it was estimated that the prevalence of asthma was more than 330 million people and that this translated into more than 23 million disability adjusted life years, positioning asthma in the 28th place among the leading causes of burden of disease (Arriagada et al. 2019).

Allergic asthma is one of best described asthma phenotypes of primary studies. Allergic sensitization is a strong risk factor for asthma inception and severity in children and in adults. Current therapies can effectively control symptoms and the ongoing inflammatory process but do not affect the underlying, dysregulated immune response. Thus, they are very limited in controlling the progression of the disease (Dhami et al. 2017).

Indoor and outdoor air pollution has been recognized as an important environmental risk factor associated with asthma (Arriagada et al. 2019). Air pollutants reported to adversely impact on health include particulate matter with aerodynamic diameters less than 2.5 ($PM_{2.5}$) and 10 mm (PM_{10}), carbon monoxide (CO), ozone (O_3), nitrogen dioxide (NO_2) and sulphur dioxide (SO_2). Emission of most pollutants in Europe have decreased substantially over the last 20 years, but continue to cause significant mortality and morbidity. Estimates vary, but the WHO has stated over 4.2 million premature deaths from cardiovascular and respiratory diseases and the Lancet Commission reported 9 million premature deaths due to ambient air pollution globally (Sangkharat et al. 2019).

It is important to define the difference between acute exposure and chronic exposure to pollutants.

Acute exposure could increase because of the temperature rise, and the human activities could make more frequent and intense fires; chronic exposure could increase because of the accumulation of polluting gases in earthly atmosphere especially in urban areas.

For the acute exposure (single day lag analysis), there is a very likely positive association for all ages between $PM_{2.5}$ exposition and asthma-related hospital admissions, emergency department visits, asthma-related physician visits (Arriagada et al. 2019). Also ambulance dispatches demonstrate association between all-respiratory and $PM_{2.5}$ and asthma dispatches and NO_2 exposition (Sangkharat et al. 2019). Results for acute exposures were stronger for studies that involved exercise, suggesting that physical activity may amplify the effect of air pollution on acute lung function (Edginton et al. 2019).

For the chronic exposure, Edginton found that increasing levels of average $PM_{2.5}$ and PM_{10} were associated with a statistically significant negative yearly change in FEV1 (forced expiratory volume in the first second) (Edginton et al. 2019). Decreased lung function from long-term exposure to ambient particulate pollution has a number of potential adverse consequences for healthy adults. The effects of acute exposures to particulate air pollution tend to be smaller compared with the effects of exposure over a longer time period. This may be due to a cumulative effect of consistent exposure to high levels of air pollution (Edginton et al. 2019).

In addition, there are significant modification by season in pollutant morbidity. In particular, in warm season (in northern hemisphere from April to October), stronger effect for respiratory diseases are detected for ozone (O_3) and carbon monoxide (CO) while in cold season (from November to March), pooled effect seems to be more pronounced for sulphur dioxide (SO_2), nitrogen dioxide (NO_2) and $PM_{2.5}$ (Bergmann et al. 2020). The effect of morbidity modification by season was found similar in males and females, and there is a slightly higher effect in children and elders. The overall effects of air pollutants on morbidity are being more pronounced in the warm season (Bergmann et al. 2020).

Besides the seasonal effects, the estimate of health burden attributable to climate change is important to include since climate change will lead to altered seasons and changes in temperature; the temperature rise will lead to intensifications of air pollution in several regions (Bergmann et al. 2020).

9.3.1 Chronic Obstructive Pulmonary Diseases (COPD)

In recent years, there has been considerable interest in the health effects of exposure to both short-term fluctuations and long-term air pollution levels, particularly common environmental pollutants including particulate matter (PM), ozone (O_3), carbon monoxide (CO), nitrogen dioxide (NO_2) and sulphur dioxide (SO_2).

Chronic obstructive pulmonary diseases (COPD), such as bronchial asthma, chronic bronchitis and pulmonary emphysema components, are one of the main causes of morbidity and mortality (Schikowski et al. 2010; Schikowski et al. 2014a; Schikowski et al. 2014b).

In the 1960s, an association was established between urban air pollution, symptoms of bronchitis and reduced lung function. The analysis of the Global Burden of Diseases Study suggested that environmental air pollution is the second most common cause of death and disability due to COPD. Few studies have addressed the COPD hypothesis in adults. The causal role of air pollution in the development of COPD is biologically plausible. Oxidative stress and inflammation have been described as consequences of exposure to various air pollutants. Both pulmonary and systemic effects have been observed, and these pathways are likely contributors to COPD-related respiratory diseases.

The causal role of air pollution in the development of COPD is biologically likely. Oxidative stress and inflammation have been described as consequences of exposure to various air pollutants. Both pulmonary and systemic effects have been observed, and these pathways are likely contributors to COPD-related respiratory disease.

European multi-centre study on air pollution and COPD (ESCAPE) estimated that long-term residential exposure to NO_2, PM_{10} and traffic intensity on the nearest major road was positively but not statistically significantly associated with higher COPD prevalence in four European adult cohort studies. The prevalence of COPD was not associated with $PM_{2.5}$. The positive association between traffic intensity on the nearest main road and COPD-defined GOLD reached statistical significance only in women (prevalence and incidence) and non-smokers (incidence).

With the exception of PM, all associations between exposure to air pollutants and the prevalence and incidence of COPD were positive but not statistically significant. SALIA is the only previously published study on air pollution, proximity to traffic and the prevalence of COPD. Results published from the SALIA baseline about 20 years ago demonstrate that the 5-year mean of PM_{10} showed significant associations not only with forced vital capacity and forced expiratory volume in 1 s but also with the probability of having COPD-termed GOLD (stage 1–4): OR 1.68, 95% CI: 1.01–2.78, for 10 μg m—3 PM_{10}. Kan et al. reported that lung function was inversely related to trafficking exposure in women. However, it is unclear whether females are more susceptible to the effects of air pollution than males. An effect of air pollution on lung growth in childhood has been reported, but the link between impaired lung development and COPD in future life has not been established. Similarly, if repeated exacerbations of COPD are considered a cause of disease progression, indirect evidence of a causal role of air pollution on COPD can be affirmed, given the ability of air pollution to trigger flare-ups. Arbex and coauthors examined the influence of the concentration of the main air pollutants on the COPD emergency department visits to the city of Sao Paulo, Brazil. Their discoveries revealed that exposure to PM_{10} and SO_2 showed acute effects regarding visits to the emergency room.

In conclusion, air pollution increases the risk of respiratory mortality, but evidence of impact on lung function and chronic obstructive pulmonary disease (COPD) is less well established.

9.4 Cardiovascular Disease

9.4.1 Myocardial Infarction

In recent years, the scientific evidence in the literature dealing with the effects of climate change on human health is growing. More than any others, cardiovascular diseases are at the forefront of those directly related to increased climate variability in many regions in response to climate change, due to global warming, pollution and so on (De Blois et al. 2015). Here, the consequences that trigger the myocardial infarction or that aggravate it will be described.

Myocardial infarction has become a hot research topic, with many epidemiological studies showing that it is affected by changes in ambient temperature (Andersson et al. 2018; Lee et al. 2014). The most authoritative studies have been undertaken covering various areas of the entire globe, to have a final summary of the phenomenon that includes different types of terrestrial climate. Myocardial infarction (MI or heart attack) is one of the leading causes of cardiovascular mortality. It is due to an atherosclerotic plaque rupture, which leads to partial or complete thrombotic vessel occlusion.

9.4.1.1 The Effect of Temperature

Numerous studies confirm the association between the increase in hospital admissions for myocardial infarction and ambient temperature (Sun et al. 2018; Bhaskaran et al. 2010). When factors like acute stress and anxiety are added to these, the risk increases exponentially (Loughnan et al. 2010). There is a known relationship between temperature and cardiovascular mortality, this relationship is supported by the evidence that links extreme temperatures with changes in blood pressure, blood viscosity and heart rate (Bai et al. 2018). When we talk about temperatures, we refer to both high and low ones. The estimates for the relationship between temperature and the relative risk of MI hospitalization are statistically significant both for a 1 °C increase and for a 1 °C decrease (Versaci et al. 2019). Also, statistical relevance was found in many studies that estimated MI mortality related to a heat wave. It has been noted that the increase in temperature leads to a greater speed of the bloodstream and therefore an increase in blood pressure (Dang et al. 2019). As a result, the decreased supply of oxygen to the tissues could cause an ischaemia which could culminate in a heart attack.

Contrary to what happens with heat, in which the outcome was immediate, with cold it is delayed (Vasconcelos et al. 2013). As a result, exposure to cold and cold spells as well as to heat and heat waves has been associated with an increased risk of myocardial infarction, and this can be explained by some mechanisms (Mohammadi et al. 2018). As for low temperatures, an increase in blood pressure and blood viscosity causes an increase in oxygen demand

and greater cardiac workload (Chen et al. 2019). If these phenomena occur in a heart that is already vulnerable because of multiple pathologies (diabetes mellitus, etc.) or aged people, the risk rises sharply, and the outcome may be inevitable.

9.4.1.2 Atmospheric Pressure and Latitude

As mentioned earlier, exposure to warm and cold too plays an important role in cardiovascular disease: an increase in latitude has been associated with a reduced risk of hospitalization for myocardial infarction (Lam et al. 2018). It has been shown that countries with higher latitude have shown less and weaker effects in relation to exposure to both heat and cold on the risk of hospitalization for myocardial infarction (Madrigano et al. 2013). An explanation of this phenomenon has been found in the adaptive capacity of populations habitually living in a cold climate. An international study in China showed that the heat effects of the northern areas at higher latitude were low.

According to the seasons, atmospheric pressure is one of those factors that strongly affects the possibility of an MI (Furukawa 2019). It has been shown that low pressure or important variations of this and lower winter rainfall greatly increase the risk (Wanitschek et al. 2013).

9.4.1.3 Air Pollution

Poor air quality is recognized as a factor that increases the synergistic effect between temperature and cardiovascular diseases; high pollutant levels cause changes in HR and repolarization parameters that may be precursors of cardiac problems (Chen et al. 2018; Hampel et al. 2010).

Ozone and PM_{10} (particulate matter) are two of the most problematic pollutants for the atmospheric environment, along with nitrogen dioxide (NO_2), sulphur dioxide and carbon monoxide. Ozone is a secondary pollutant that is formed as a result of a series of chemical reactions involving nitrogen oxides (NO_X) and volatile non-methane organic compounds (NMVOC), in the presence of solar radiation and high temperatures. Once ozone has formed, it can remain in the air for several days. Secondary PM_{10}, with aerodynamic diameter below 10 and 2.5 μm, is also produced in the atmosphere by chemical reactions involving various precursor gases, coming from combustion in large industrial plants (SO_2), from the transport sector (NMV and NO_X), from the use of solvents (NMVOC) and from agriculture (NH_3).

9.4.1.4 Conclusion

Eventually, it can be said that the potential impact of climate on health is related to the vulnerability of a specific population and to its ability to adapt and change in response to environmental variations. Furthermore, some diseases are more sensitive to climate stress than others. For an optimal prevention, it is necessary to sensitize world institutions to a more conscientious use of environmental resources, to reduce climate change. Educational programs should be undertaken for the population at risk.

9.5 Heart Failure

Heart failure (HF) is defined as "the inability of the heart to keep up with the demands on it and, specifically, failure of the heart to pump blood with normal efficiency". When this occurs, the heart is unable to provide adequate blood flow to other organs, making sure that secondary insufficiencies are triggered in noble organs such as kidneys, brain and liver (Filippatos and Zannad 2007).

It is known how cardiovascular diseases are affected by an environment subjected to increasingly rapid and extreme climatic events. Among these, heart failure stands out, affecting more and more exposed people who have not had time to adapt. This would mean a peak in the seasonal increase in cases of acute exacerbations and sudden cardiac death related to exposure to cold, even if global temperatures are increasing (De Blois et al. 2015; Vieira et al. 2016).

9.5.1 The Mitochondrial Hypothesis

At the basis of these claims, there is a belief that mitochondria are one of the main targets of environmental toxic substances that can damage mitochondrial morphology, its function and DNA, as this organelle is the primary source for the free radical release upon toxicant exposure (Ekström et al. 2016). By examining the literature on air pollutants that negatively affect mitochondrial morphological and functional activities and how mitochondrial DNA (mtDNA) copies the numerical variation, which reflects the cellular damage induced by the oxidant in the air, we can relate these events with heart failure. These studies conclude that environmental health assessment should focus on the status of cell/circulatory mitochondrial functional copy numbers, which can predict the outcome of cardiovascular diseases. Furthermore, there are alterations of the mitochondrial membrane's potential and changes in the ultrastructure of the cristae. Therefore, the evaluation of functional and morphological mitochondrial changes can be a benchmark to evaluate the impact of environmental pollutants on human health (Boovarahan and Kurian 2018).

9.5.2 The Effect of Temperatures

Heart failure was observed in correlation with low temperature that has prompted the concern on the influence of the seasonality effect. Admissions were observed to peak in autumn and winter (Shiue et al. 2016). The hypothesized mechanism was that an acute reversal in environmental temperature, being too cold or too hot, tends to increase myocardial oxygen consumption and may induce cardiac arrhythmias or angina. Lagged exposure, rather than the exposure at the same day when the cold weather occurred, showed greater effects on HF mortality (Stewart et al. 2019).

As for the heat, the physiological response in patients with HF is abnormal, with a significantly lower flow of blood in the skin and lower cutaneous vascular conductance responses, even if the maximal vasodilatation induced by sodium nitroprusside is not reduced (Münzel et al. 2018a). With more intense heat waves of longer duration, excesses in the mortality and hospitalization of HF patients are to be expected (Cheng et al. 2019; Ponjoan et al. 2017).

9.5.3 Air Pollutants Associations

There are numerous epidemiological and clinical evidences that affirm how $PM_{2.5}$ and PM_{10} are associated with an increase in hospitalizations and mortality especially in people with congestive heart failure and other heart problems (Argacha et al. 2018; Meo and Suraya 2015). Many experimental studies, in support of these statements, demonstrated that air pollution promotes a systemic vascular oxidative stress reaction (Bourdrel et al. 2017). Radical oxygen species induce endothelial dysfunction, monocyte activation and some proatherogenic changes in lipoproteins, which initiate plaque formation (Fiordelisi et al. 2017). Even short-term exposure alone has experienced temporary effects; this is because the increase in systemic blood pressure and vasoconstriction due to short-term exposure to PM could lead to an increase in cardiac postload and the risk of acute decompensated heart failure (Webb et al. 2014).

Acute exposure to diesel exhaust is also followed by an increase in pulmonary vascular resistance and a decrease in pulmonary vessel distensibility at high cardiac output, which may participate in the influence of air pollution on an acute heart failure episode (Tian et al. 2019). Unfiltered dilute diesel engine exhaust (DE) adversely affects cardiac workload in patients with HF (Shah et al. 2013).

9.5.4 Conclusion

Given the worldwide prevalence of exposure to traffic-related air pollution and temperature alterations, these findings are relevant for public health especially in this highly susceptible population.

9.6 Hypertension

9.6.1 Introduction

High blood pressure (BP) now accounts for nearly half of all cardiovascular events and is the leading risk factor for morbidity and mortality worldwide. Nowadays, hypertension most often shows up both for an underlying genetic predisposition and because of unfavourable environmental factors. Colder climates and other seasonal temperature changes, high altitudes, loud noises and air pollutants are each capable of elevating BP values (Giorgini et al. 2017). The most solid evidence is for air pollution: exposure to atmospheric pollutant contributes to increased disability-adjusted life years (DALYs—i.e., the sum of years of potential life lost due to premature mortality and the years of productive life lost due to disability). According to the "Environmental Burden of Disease in European Countries" project report, particulate matter material, along with noise pollution, contributes to more than 75% of the burden of disease attributable to environmental factors (Maria Bruno et al. 2017).

9.6.2 Outdoor Temperatures and Seasonality

The association between seasonal fluctuations and BP was found to be very strong: the most recent systematic review and meta-analysis showed that mean systolic BP was 9 mmHg higher in winter than in summer, and, above 5 °C in outdoor temperature, systolic BP was 6.2 mmHg higher for each 10 °C decrease in temperature. Temperature and seasonality independently affect BP, confirming that outdoor temperature is not a synonym of seasonality (Modesti et al. 2018). Daytime systolic BP resulted negatively affected by temperature, as expected, while seasonality, expressed as the number of daylight hours, mainly affected night-time systolic BP, with more daylight hours correlated with a higher night-time BP, probably due to reduction in sleep quality (Park et al. 2020). However, there are additional complex interrelationships involving several exposure parameters (time of day, durations, indoor vs outdoor temperatures) that determine the full nature of the physiologic responses. The underlying mechanism connecting exposure to cold weather and higher values of BP might be an increase in cold-induced sympathetic nervous system activity, which also leads to increased activity of the renin–angiotensin–aldosterone system, both inducing a BP increase (Brook 2017). Indeed, cold-induced acute arteriolar vasoconstriction is associated with a sympathetically mediated rise in blood pressure, mainly diastolic, and in heart rate and cardiac load. Moreover, cold temperature facilitates platelet aggregation and increases blood viscosity and levels of red blood cell count, plasma cholesterol and plasma fibrinogen (Ikäheimo 2018). However, outdoor temperature is not the only environmental factor accounting for the seasonality of CV diseases: sunlight exposure, perception of cold, changes in dietary and exercise habits, air pollution, circadian rhythm and sleep alterations and conceivably several unknown factors might also play a role in predisposing to CV events during the cold months according to their oscillations (Liu et al. 2015).

9.6.3 Noise Pollution

The WHO estimates that in western European countries, annually 61,000 DALYs are lost due to noise-induced cardiovascular diseases (van Kempen et al. 2018). A large compilation of loud urban and occupational noises has been implicated in increasing BP (e.g., traffic, airplanes, railways, machinery). Consistent evidence further supports that residing in locations chronically affected by loud noise (e.g., nearby road traffic and airports) increases the risk for hypertension in the long term (Yang et al. 2018a; Fu et al. 2017). Whether specific sources are more or less harmful remains to be clarified; however, some studies suggest that night-time exposures have a particularly deleterious impact on increasing BP: 7% more per 10 dB increase in estimated 16-h average traffic noise exposure (Hahad et al. 2019; Münzel et al. 2018c). Important mechanisms seem to be at the base of this association: sympathetic nervous system activity, an increase in circulating stress hormones and activation of the hypothalamic pituitary adrenal axis might lead to endothelial dysfunction, arterial hypertension, development of early atherosclerosis and then to adverse cardiovascular events (Wang et al. 2018; Münzel et al. 2017). Night-time exposure seems capable of disrupting sleep quality and vascular endothelial function, even when an individual is not awoken by the noise (Münzel et al. 2018b). Eventually, the maximal noise intensity threshold deemed safe with regard to BP, and cardiovascular risk remains to be clarified (Münzel et al 2016).

9.6.4 Higher Altitudes

Ascent to higher altitudes increases BP (Bilo et al. 2019). The BP-raising effects can occur over a few days and are independent of the often-concomitant exposure to colder temperatures. The altitude required to pose a threat also remains to be fully clarified. The overall evidence supports different altitudes to present some risks: from 1200 m to 2500 m. The most import biological pathway responsible is likely hypoxia-induced (because of the lower partial pressure of oxygen at higher altitude) activation of the chemoreflex at the level of the carotid body and an ensuing augmentation of sympathetic outflow (Nieuwenhuijsen 2018). Other possible physiologic pathways reported in some studies include increased arterial stiffness, endothelin release and heightened overall blood viscosity.

9.6.5 Air Pollutants

Fine particulate matter less than 2.5 μm air pollution ($PM_{2.5}$), most commonly derived from fossil fuel combustion during modern-day activities (e.g., traffic, power generation, industry), is a leading risk factor for cardiovascular morbidity and mortality (Thompson 2018). Numerous studies across the globe and a recent meta-analysis found that acute exposures are capable of increasing BP (typically by 1–2 μm Hg per 10 mg/m^3) over a period of a few hours to days (Yang et al. 2018b). Numerous biological pathways were found in experimental studies to explain the mechanisms involved in air pollution–mediated BP elevations. The inhalation of $PM_{2.5}$ interacts with a host of pulmonary receptors (transient receptor potential channels) and nerve endings (c-fibres) to initiate nervous system reflex arcs (Salameh et al. 2018). This interaction results in an imbalance in systemic autonomic outflow favouring sympathetic tone. At the same time, the deposition of fine particles in lung airways creates a nidus of pulmonary inflammation. Local mediators including cytokines, active cells and oxidized biological molecules (e.g., phospholipids) spill over into the systemic circulation and thereafter adversely affect the function of the cardiovascular system. Finally, some pro-oxidative constituents of inhaled particles (e.g., nanoparticles, metals, organic compounds) may even be capable of translocating directly into the systemic circulation and mediate direct adverse actions. Together, these responses likely elicit the arterial vasoconstriction and endothelial dysfunction, which are ultimately responsible for the increase in BP (Brook and Kousha 2015). Several personal-level interventions may also reduce the harmful actions of $PM_{2.5}$ exposures include wearing facemasks while outdoors or closing outside windows while in extremely polluted cities and using in-home and automobile cabin high-efficiency particulate arrestance (HEPA) filtration systems (Li et al. 2020).

9.6.6 Other Environmental Factors

A large variety of other environmental factors are also associated with higher BP levels. These factors include persistent organic pollutants, water salinity, strong odours (e.g., nearby farms), metals (i.e., lead, cadmium, mercury, arsenic), chemicals used in plastics (e.g., bisphenol A) and food wraps (e.g., phthalates), health inequalities (low- and middle-income countries) (Rancière et al. 2015; Sarki et al. 2015; da Cunha Martins et al. 2018). Exposure to these elements is imputable of producing several acute and chronic illnesses including hypertension. Nevertheless, pollutants from food: beta-hexachlorocyclohexane, a lipophilic by-product of the production of the insecticide called lindane, has been associated with higher systolic blood pressure among people living close to an industrial area (Argacha et al. 2019).

9.6.7 Conclusion

These considerations remind us that climatic factors can be considered as new potential risk factors of cardiovascular events, and there is an urgent need for large-scale, prospective, community-based and international studies to deeply explore the risk factors to schedule preventive strategies.

References

Ab Manan N, Aizuddin AN, Hod R (2018) Effect of air pollution and hospital admission: a systematic review. Ann Glob Health 84(4):670

Andersson HB, Seth M, Sly J, Bates E, Gurm H (2018) Daily temperature fluctuations and myocardial infarction: implications of global warming on cardiac health. J Am Coll Cardiol 71(11 Supplement):A1152

Argacha JF, Bourdrel T, Van De Borne P (2018) Ecology of the cardiovascular system: a focus on air-related environmental factors. Trends Cardiovasc Med 28(2):112–126

Argacha JF, Mizukami T, Bourdrel T, Bind M-A (2019) Ecology of the cardiovascular system: part II – a focus on non-air related pollutants. Trends Cardiovasc Med 29(5):274–282

Arriagada NB, Joshua AH, Palmera AJ et al (2019) Association between fire smoke fine particulate matter and asthma-related outcomes: systematic review and meta-analysis. Environ Res 179:108777

Bai L, Li Q, Wang J, Lavigne E, Gasparrini A, Copes R et al (2018) Increased coronary heart disease and stroke hospitalisations from ambient temperatures in Ontario. Heart 104(8):673–679

Bergmann S, Li B, Pilot E et al (2020) Effect modification of the short-term effects of air pollution on morbidity by season: a systematic review and meta-analysis. Sci Total Environ 716

Bhaskaran K, Hajat S, Haines A, Herrett E, Wilkinson P, Smeeth L (2010) Short term effects of temperature on risk of myocardial infarction in England and Wales: time series regression analysis of the myocardial Ischaemia National Audit Project (MINAP) registry. BMJ 341:c3823

Bilo G, Caravita S, Torlasco C, Parati G (2019) Blood pressure at high altitude: physiology and clinical implications. Kardiol Pol 77(6):596–603

Boovarahan SR, Kurian GA (2018) Mitochondrial dysfunction: a key player in the pathogenesis of cardiovascular diseases linked to air pollution. Rev Environ Health 33(2):111–122

Bourdrel T, Bind MA, Béjot Y, Morel O, Argacha JF (2017) Cardiovascular effects of air pollution. Arch Cardiovasc Dis 110(11):634–642

Brook RD (2017) The environment and blood pressure. Cardiol Clin 35(2):213–221

Brook RD, Kousha T (2015) Air pollution and emergency department visits for hypertension in Edmonton and Calgary, Canada: a case-crossover study. Am J Hypertens 28(9):1121–1126

Chen H, Li Q, Kaufman JS, Wang J, Copes R, Su Y, Benmarhnia T (2018) Effect of air quality alerts on human health: a regression discontinuity analysis in Toronto, Canada. Lancet Planetary Health 2(1):e19–e26

Chen K, Breitner S, Wolf K, Rai M, Meisinger C, Heier M, KORA Study Group (2019) Projection of temperature-related myocardial infarction in Augsburg, Germany: moving on from the Paris agreement on climate change. Dtsch Arztebl Int 116(31–32):521

Cheng J, Xu Z, Bambrick H, Prescott V, Wang N, Zhang Y et al (2019) Cardiorespiratory effects of heatwaves: a systematic review and meta-analysis of global epidemiological evidence. Environ Res 108610

da Cunha Martins A, Carneiro MFH, Grotto D, Adeyemi JA, Barbosa F (2018) Arsenic, cadmium, and mercury-induced hypertension: mechanisms and epidemiological findings. J Toxicol Environ Health Part B 21(2):61–82

D'Amato G, Chong-Neto HJ, Monge Ortega OP, Vitale C, Ansotegui I, Rosario N, Haahtela T, Galan C, Pawankar R, Murrieta-Aguttes M, Cecchi L, Bergmann C, Ridolo E, Ramon G, Gonzalez Diaz S, D'Amato M, Annesi-Maesano I (2020) The effects of climate change on respiratory allergy and asthma induced by pollen and mold allergens. Allergy 75(9):2219–2228

Dang TAT, Wraith D, Bambrick H, Dung N, Truc TT, Tong S, Dunne MP (2019) Short-term effects of temperature on hospital admissions for acute myocardial infarction: a comparison between two neighbouring climate zones in Vietnam. Environ Res 175:167–177

De Blois J, Kjellstrom T, Agewall S, Ezekowitz JA, Armstrong PW, Atar D (2015) The effects of climate change on cardiac health. Cardiology 131(4):209–217

Dhami S, Kakourou A, Asamoah F et al (2017) Allergen immunotherapy for allergic asthma: a systematic review and meta-analysis. Allergy 72:1825–1848

Doiron D, de Hoogh K, Probst-Hensch N, Fortier I, Cai Y, De Matteis S, Hansell AL (2019) Air pollution, lung function and COPD: results from the population-based UK Biobank study. Eur Respir J 54(1):1802140

Edginton S, O'Sullivan DE, King W et al (2019) Effect of outdoor particulate air pollution on FEV1 in healthy adults: a systematic review and meta-analysis. Occup Environ Med 76:583–591

Ekström A, Brijs J, Clark TD, Gräns A, Jutfelt F, Sandblom E (2016) Cardiac oxygen limitation during an acute thermal challenge in the European perch: effects of chronic environmental warming and experimental hyperoxia. Am J Physiol Regul Integr Comp Physiol 311(2):R440–R449

Enyew HD, Mereta ST, Hailu AB (2021) Biomass fuel use and acute respiratory infection among children younger than 5 years in Ethiopia: a systematic review and meta-analysis. Public Health 10(193):29–40

Filippatos G, Zannad F (2007) An introduction to acute heart failure syndromes: definition and classification. Heart Fail Rev 12(2):87–90

Fiordelisi A, Piscitelli P, Trimarco B, Coscioni E, Iaccarino G, Sorriento D (2017) The mechanisms of air pollution and particulate matter in cardiovascular diseases. Heart Fail Rev 22(3):337–347

Fu W, Wang C, Zou L, Liu Q, Gan Y, Yan S, Song F, Wang Z, Lu Z, Cao S (2017) Association between exposure to noise and risk of hypertension: a meta-analysis of observational epidemiological studies. J Hypertens 35(12):2358–2366

Furukawa Y (2019) Meteorological factors and seasonal variations in the risk of acute myocardial infarction. Int J Cardiol 294:13–14

GBD 2015 Collaborators for Chronic Respiratory Disease (2017) Global, regional and national deaths, prevalence, life suitable for years and years lived with disability for chronic obstructive pulmonary disease and asthma, 1990–2015: a systematic analysis for the 2015 Global Burden of Disease Study. Lancet Respir Med 5:691–670

Giorgini P, Di Giosia P, Petrarca M, Lattanzio F, Stamerra CA, Ferri C (2017) Climate changes and human health: a review of the effect of environmental stressors on cardiovascular diseases across epidemiology and biological mechanisms. Curr Pharm Des 23(22)

Hahad O, Prochaska JH, Daiber A, Münzel T (2019) Environmental noise-induced effects on stress hormones, oxidative stress, and vascular dysfunction: key factors in the relationship between Cerebrocardiovascular and psychological disorders. Oxidative Med Cell Longev 2019:1–13

Hampel R, Schneider A, Brüske I, Zareba W, Cyrys J, Rückerl R, Peters A (2010) Altered cardiac repolarization in association with air pollution and air temperature among myocardial infarction survivors. Environ Health Perspect 118(12):1755–1761

Ikäheimo TM (2018) Cardiovascular diseases, cold exposure and exercise. Temperature 5(2):123–146

Jafta N, Jeena PM, Barregard L, Naidoo RN (2015 May) Childhood tuberculosis and exposure to indoor air pollution: a systematic review and meta-analysis. Int J Tuberc Lung Dis 19(5):596–602

Jary H, Simpson H, Havens D, Manda G, Pope D, Bruce N, Mortimer K (2016) Household air pollution and acute lower respiratory infections in adults: a systematic review. PLoS One 11(12):e0167656

Katoto PDMC, Brand AS, Bakan B, Obadia PM, Kuhangana C, Kayembe-Kitenge T, Kitenge JP, Nkulu CBL, Vanoirbeek J, Nawrot TS, Hoet P, Nemery B (2021) Acute and chronic exposure to air pollution in relation with incidence, prevalence, severity and mortality of COVID-19: a rapid systematic review. Environ Health 20(1):41

La Torre G, Colamesta V, Saulle R, Iarocci G, De Vito C, Mannocci A (2018) Decreasing trends in air pollution in the metropolitan area of Rome: an analysis of 15 years (1999-2013). Ig Sanita Pubbl 74(4):329–336

Lam HCY, Chan JCN, Luk AOY, Chan EYY, Goggins WB (2018) Short-term association between ambient temperature and acute myocardial infarction hospitalizations for diabetes mellitus patients: a time series study. PLoS Med 15(7)

Lee S, Lee E, Park MS, Kwon BY, Kim H, Jung DH et al (2014) Short-term effect of temperature on daily emergency visits for acute myocardial infarction with threshold temperatures. PLoS One 9(4)

Levi M, Kjellstrom T, Baldasseroni A (2018) Impact of climate change on occupational health and productivity: a systematic literature review focusing on workplace heat. Med Lav 109(3):163–179

Li L, Yang A, He X, Liu J, Ma Y, Niu J, Luo B (2020) Indoor air pollution from solid fuels and hypertension: a systematic review and meta-analysis. Environ Pollut 259:113914

Liu C, Yavar Z, Sun Q (2015) Cardiovascular response to thermoregulatory challenges. Am J Phys Heart Circ Phys 309(11):H1793–H1812

Logan WP (1953) Mortality in the London fog incident, 1952. Lancet 1:336–338

Loughnan ME, Nicholls N, Tapper NJ (2010) The effects of summer temperature, age and socioeconomic circumstance on acute myocardial infarction admissions in Melbourne, Australia. Int J Health Geogr 9(1):41

MacIntyre EA, Gehring U, Mölter A, Fuertes E, Klümper C, Krämer U, Quass U, Hoffmann B, Gascon M, Brunekreef B, Koppelman GH, Beelen R, Hoek G, Birk M, de Jongste JC, Smit HA, Cyrys J, Gruzieva O, Korek M, Bergström A, Agius RM, de Vocht F, Simpson A, Porta D, Forastiere F, Badaloni C, Cesaroni G, Esplugues A, Fernández-Somoano A, Lerxundi A, Sunyer J, Cirach M, Nieuwenhuijsen MJ, Pershagen G, Heinrich J (2014) Air pollution and respiratory infections during early childhood: an analysis of 10 European birth cohorts within the ESCAPE Project. Environ Health Perspect 122(1):107–113

Madrigano J, Mittleman MA, Baccarelli A, Goldberg R, Melly S, Von Klot S, Schwartz J (2013) Temperature, myocardial infarction, and mortality: effect modification by individual and area-level characteristics. Epidemiology (Cambridge, Mass.) 24(3):439

Maleki M, Anvari E, Hopke PK, Noorimotlagh Z, Mirzaee SA (2021) An updated systematic review on the association between atmospheric particulate matter pollution and prevalence of SARS-CoV-2. Environ Res 195:110898

Manes M, Salvatori E, La Torre G, Villari P, Vitale M, Biscontini M, Incerti G (2008) Urban green and its relation with air pollution: ecological studies in the metropolitan area of Rome. Ital J Public Health 5(4):278–283

Maria Bruno R, Di Pilla M, Ancona C, Sørensen M, Gesi M, Taddei S, Munzel T, Virdis A (2017) Environmental factors and hypertension. Curr Pharm Des 23(22):3239–3246

Meng Y, Lu Y, Xiang H, Liu S (2021) Short-term effects of ambient air pollution on the incidence of influenza in Wuhan, China: a time-series analysis. Environ Res 192:110327

Meo SA, Suraya F (2015) Effect of environmental air pollution on cardiovascular diseases. Eur Rev Med Pharmacol Sci 19(24):4890–4897

Modesti PA, Rapi S, Rogolino A, Tosi B, Galanti G (2018) Seasonal blood pressure variation: implications for cardiovascular risk stratification. Hypertens Res 41(7):475–482

Mohammadi R, Soori H, Alipour A, Bitaraf E, Khodakarim S (2018) The impact of ambient temperature on acute myocardial infarction admissions in Tehran, Iran. J Therm Biol 73:24–31

Münzel T, Center for Cardiology, University Medical Center Mainz, Mainz, Germany, Sørensen M, and Danish Cancer Society Research Center, Copenhagen, Denmark (2017) Noise pollution and arterial hypertension. Eur Cardiol Rev 12(1):26

Münzel T, Sørensen M, Gori T, Schmidt FP, Rao X, Brook J, Chen LC, Brook RD, Rajagopalan S (2016) Environmental stressors and cardio-metabolic disease: part I–epidemiologic evidence supporting a role for noise and air pollution and effects of mitigation strategies. Eur Heart J:ehw269

Münzel T, Gori T, Al-Kindi S, Deanfield J, Lelieveld J, Daiber A, Rajagopalan S (2018a) Effects of gaseous and solid constituents of air pollution on endothelial function. Eur Heart J 39(38):3543–3550

Münzel T, Schmidt FP, Steven S, Herzog J, Daiber A, Sørensen M (2018b) Environmental noise and the cardiovascular system. J Am Coll Cardiol 71(6):688–697

Münzel T, Sørensen M, Schmidt F, Schmidt E, Steven S, Kröller-Schön S, Daiber A (2018c) The adverse effects of environmental noise exposure on oxidative stress and cardiovascular risk. Antioxid Redox Signal 28(9):873–908

Nieuwenhuijsen MJ (2018) Influence of urban and transport planning and the city environment on cardiovascular disease. Nat Rev Cardiol 15(7):432–438

Nuvolone D, Balzi D, Pepe P, Chini M, Scala D, Giovannini F, Cipriani F, Barchielli A (2013) Ozone short-term exposure and acute coronary events: a multicities study in Tuscany (Italy). Environ Res 126:17–23

Park S, Kario K, Chia Y, Turana Y, Chen C, Buranakitjaroen P, Nailes J, Hoshide S, Siddique S, Sison J, Soenarta AA, Sogunuru GP, Tay JC, Teo BW, Zhang Y, Shin J, Minh H, Tomitani N, Kabutoya T et al (2020) The influence of the ambient temperature on blood pressure and how it will affect the epidemiology of hypertension in Asia. J Clin Hypertens 22(3):438–444

Peel JL, Tolbert PE, Klein M, Metzger KB, Flanders WD, Todd K, Mulholland JA, Ryan PB, Frumkin H (2005) Ambient air pollution and respiratory emergency department visits. Epidemiology 16(2):164–174

Ponjoan A, Blanch J, Alves-Cabratosa L, Martí-Lluch R, Comas-Cufí M, Parramon D et al (2017) Effects of extreme temperatures on cardiovascular emergency hospitalizations in a Mediterranean region: a self-controlled case series study. Environ Health 16(1):32

Rancière F, Lyons JG, Loh VHY, Botton J, Galloway T, Wang T, Shaw JE, Magliano DJ (2015) Bisphenol a and the risk of cardiometabolic disorders: a systematic review with meta-analysis of the epidemiological evidence. Environ Health 14(1):46

Salameh P, Chahine M, Hallit S, Farah R, Zeidan RK, Asmar R, Hosseiny H (2018) Hypertension prevalence and living conditions related to air pollution: results of a national epidemiological study in Lebanon. Environ Sci Pollut Res 25(12):11716–11728

Sangkharat K, Fisher P, Neil Thomas G et al (2019) The impact of air pollutants on ambulance dispatches: a systematic review and meta-analysis of acute effects. Environ Pollut 254:112769

Sarki AM, Nduka CU, Stranges S, Kandala N-B, Uthman OA (2015) Prevalence of hypertension in low- and middle-income countries: a systematic review and meta-analysis. Medicine 94(50):e1959

Schikowski T, Ranft U, Sugiri D et al (2010) Decline in air pollution and change in prevalence in respiratory symptoms and chronic obstructive pulmonary disease in elderly women. Respir Res 11:113

Schikowski T, Adam M, Marcon A, Cai Y, Vierkötter A, Carsin AE, Jacquemin B, Al Kanani Z, Beelen R, Birk M, Bridevaux PO, Brunekeef B, Burney P, Cirach M, Cyrys J, de Hoogh K, de Marco R, de Nazelle A, Declercq C, Forsberg B, Hardy R, Heinrich J, Hoek G, Jarvis D, Keidel D, Kuh D, Kuhlbusch T, Migliore E, Mosler G, Nieuwenhuijsen MJ, Phuleria H, Rochat T, Schindler C, Villani S, Tsai MY, Zemp E, Hansell A, Kauffmann F, Sunyer J, Probst-Hensch N, Krämer U, Künzli N (2014a) Association of ambient air pollution with the prevalence and incidence of COPD. Eur Respir J 44(3):614–626

Schikowski T, Mills IC, Anderson HR, Cohen A, Hansell A, Kauffmann F, Krämer U, Marcon A, Perez L, Sunyer J, Probst-Hensch N, Künzli N (2014b) Ambient air pollution: a cause of COPD? Eur Respir J 43(1):250–263

Schwartz J (2001) Is there harvesting in the association of airborne particles with daily deaths and hospital admissions? Epidemiology 12(1):55–61. https://doi.org/10.1097/00001648-200101000-00010.

Shah AS, Langrish JP, Nair H, McAllister DA, Hunter AL, Donaldson K, Newby DE, Mills NL (2013) Global association of air pollution and heart failure: a systematic review and meta-analysis. Lancet 382(9897):1039–1048

Shiue I, Perkins DR, Bearman N (2016) Relationships of physiologically equivalent temperature and hospital admissions due to I30–I51 other forms of heart disease in Germany in 2009–2011. Environ Sci Pollut Res 23(7):6343–6352

Stewart S, Moholdt TT, Burrell LM, Sliwa K, Mocumbi AO, McMurray JJ et al (2019) Winter peaks in heart failure: an inevitable or preventable consequence of seasonal vulnerability? Card Fail Rev 5(2):83

Sun Z, Chen C, Xu D, Li T (2018) Effects of ambient temperature on myocardial infarction: a systematic review and meta-analysis. Environ Pollut 241:1106–1114

Thompson JE (2018) Airborne particulate matter: human exposure and health effects. J Occup Environ Med 60(5):392–423

Tian Y, Liu H, Wu Y, Si Y, Song J, Cao Y et al (2019) Association between ambient fine particulate pollution and hospital admissions for cause specific cardiovascular disease: time series study in 184 major Chinese cities. Bmj 367:l6572

To T, Stanojevic S, Moores G, Gershon AS, Bateman ED, Cruz AA, Boulet LP (2012) Global asthma prevalence in adults: findings from the cross-sectional world health survey. BMC Public Health 12:204

Tsai SS, Chang CC, Yang CY (2013) Fine particulate air pollution and hospital admissions for chronic obstructive pulmonary disease: a case-crossover study in Taipei. Int J Environ Res Public Health 10(11):6015–6026

van Kempen E, Casas M, Pershagen G, Foraster M (2018) WHO environmental noise guidelines for the European region: a systematic review on environmental noise and cardiovascular and metabolic effects: a summary. Int J Environ Res Public Health 15(2):379

Vanker A, Gie RP, Zar HJ (2017) The association between environmental tobacco smoke exposure and childhood respiratory disease: a review. Expert Rev Respir Med 11(8):661–673

Vasconcelos J, Freire E, Almendra R, Silva GL, Santana P (2013) The impact of winter cold weather on acute myocardial infarctions in Portugal. Environ Pollut 183:14–18

Versaci F, Biondi-Zoccai G, Dei Giudici A, Mariano E, Trivisonno A, Sciarretta S, Scappaticci M (2019) Climate changes and ST-elevation myocardial infarction treated with primary percutaneous coronary angioplasty. Int J Cardiol 294:1–5

Vieira JL, Guimaraes GV, de Andre PA, Saldiva PHN, Bocchi EA (2016) Effects of reducing exposure to air pollution on submaximal cardiopulmonary test in patients with heart failure: analysis of the randomized, double-blind and controlled FILTER-HF trial. Int J Cardiol 215:92–97

Wang D, Zhou M, Li W, Kong W, Wang Z, Guo Y, Zhang X, He M, Guo H, Chen W (2018) Occupational noise exposure and hypertension: the Dongfeng-Tongji cohort study. J Am Soc Hypertens 12(2):71–79.e5

Wanitschek M, Ulmer H, Süssenbacher A, Dörler J, Pachinger O, Alber HF (2013) Warm winter is associated with low incidence of ST elevation myocardial infarctions and less frequent acute coronary angiographies in an alpine country. Herz 38(2):163–170

Webb L, Bambrick H, Tait P, Green D, Alexander L (2014) Effect of ambient temperature on Australian northern territory public hospital admissions for cardiovascular disease among indigenous and non-indigenous populations. Int J Environ Res Public Health 11(2):1942–1959

Yang B-Y, Qian Z, Howard SW, Vaughn MG, Fan S-J, Liu K-K, Dong G-H (2018a) Global association between ambient air pollution and blood pressure: a systematic review and meta-analysis. Environ Pollut 235:576–588

Yang Y, Zhang E, Zhang J, Chen S, Yu G, Liu X, Peng C, Lavin MF, Du Z, Shao H (2018b) Relationship between occupational noise exposure and the risk factors of cardiovascular disease in China: a meta-analysis. Medicine 97(30):e11720

Zanobetti A, Schwartz J, Dockery DW (2000) Airborne particles are a risk factor for hospital admissions for heart and lung disease. Environ Health Perspect 108(11):1071–1077

The Impact of Environmental Alterations on Human Microbiota and Infectious Diseases

10

Barbato Domenico, De Paula Baer Alice, Lia Lorenza, Giada La Torre, Rosario A. Cocchiara, Cristina Sestili, Angela Del Cimmuto, and Giuseppe La Torre

Abstract

Climate change affects the health of human beings through several mechanisms. Changes in the average surface temperature on earth, heat waves, adverse weather events and other environmental alterations can increase mortality and morbidity by increasing the risk of infections carried by food and water. Water-related diseases could occur due to both the deficiency of water and its poor quality. They are generally classified into four groups: waterborne, water-washed, water-based and water-related vector-borne diseases. Food-borne diseases can be achieved either by ingesting the pathogenic microorganism or through the toxins produced by them contained in the food. Furthermore, ecological and environmental changes directly affect the spread of vector-borne diseases that nowadays continue representing some of the major microbial causes of morbidity and mortality in the world.

Environmental alterations associated with climate change have also been proved to influence microbial communities within the human body. This diverse community that colonize us has demonstrated importance on the function of the immune system, food digestion, development of chronic diseases and modulation of brain functioning. It is being affected by elements such as antibiotic use, heavy metals, micro-plastics, organic pollutants, pesticides and food additives. Through all these mechanisms, among yet so many others, climate change is affecting human health.

Keywords

Climate change · Environment · Human microbiota · Waterborne diseases · Vector-borne diseases · Food-borne diseases Pollution

When we think about the microbes that affect the human body, what normally first comes to mind are the pathological microorganisms that colonize humans, causing disease. In the past several decades, research has also focused on them, but more recently, a new and growing interest has emerged towards the human microbiota. The human microbiome is a diverse community of bacteria, archaea, fungi, protozoa and viruses that live on and within human beings. They are responsible for around 1 kg of human weight, outnumber by at least a factor of 10 of our number of cells and colonize virtually all body parts that are exposed to the environment: the genitourinary

B. Domenico (✉) · D. P. B. Alice · L. Lorenza
G. La Torre · R. A. Cocchiara · C. Sestili
A. Del Cimmuto · G. La Torre
Department of Infectious Diseases and Public Health, Sapienza University of Rome, Rome, Italy
e-mail: domenico.barbato@uniroma1.it

tract, eyes, skin and, mostly, the gastrointestinal tract (Grice & Segre 2011).

Many factors influence microbiota composition. In early life, the colonization is thought to begin even before birth, as recent studies have demonstrated the presence of microorganisms in the placenta, amniotic fluid and umbilical cord. Then comes the birthing process, and a series of following factors related to environmental exposures in life, including infant feeding method, diet, use of antibiotics and other medications, geographical location and ethnicity. Throughout life, the microbiota is also influenced by stress and lifecycle stages.

But why is the microbiota important? Gut microbial communities are determinant in the induction, development and performance of the immune system, in addition to its role in food digestion and nutrient synthesis. Modifications in this community composition and richness beyond the natural variation can cause functional dysbiosis, increasing susceptibility to pathogenic infections and the development of chronic diseases. Changes in the microbiome are associated with diseases such as inflammatory bowel disease, obesity, seborrheic dermatitis, metabolic syndrome, asthma, acne, cardiovascular disease, immune-mediated conditions and many neuropsychiatric disorders such as autism, anxiety, schizophrenia, Parkinson's disease and Alzheimer's disease (Bibbò et al. 2016).

The microbiota–gut–brain axis has gained significant attention in the last years (Cresci & Bawden 2015; Cryan et al. 2019). The gut microbiota and the brain communicate and have an impact on each other through the immune system, tryptophan metabolism, the vagus nerve and the enteric nervous system. The gut microbiota responds to stress and stress mediators, and a dysbiotic gut microbiome can induce anxiety and depression. Some microbial metabolites that are produced in the gastrointestinal tract are known to modulate brain functioning, and through the regulation of the vagus nerve, these microorganisms can alter behaviour. New studies are focusing on better understanding these mechanisms as to plan microbial-based intervention and therapeutic strategies for neuropsychiatric disorders (Claus et al. 2016).

As said, human microbiota is influenced by a number of environmental factors. Therefore, it is logical to conclude that climate change, by changing the world we live in, will affect our microbiota and, through this mechanism, among yet so many others, will have an impact on human health. In this section, we will focus on some specific environmental alterations associated with climate change that have been proved to influence microbial communities within the human body. This is a recent field of research, and for sure, many more associations are yet to be fully understood (De Giusti et al. 2019).

10.1 Antibiotics

The overuse of antibiotics and antibiotics resistance are two important concerns of modern days. In addition to pharmaceutical administration of antibiotics, humans can passively ingest those medications through multiple other routes, as high concentrations of various antibiotics are found in the natural environment, including rivers and agricultural soil. However, even with antibiotic consumption being on a rise, human medical application is not where these drugs are most used: two-thirds of the overall antibiotic usage is destined for animal production (de Gunzburg et al. 2018; World Health Organization 2019b).

Modifications of natural ecosystems, due to human activities, can affect the use of antibiotics and the spread of antibiotics resistance (Hernando-Amado et al. 2019). Climate change increases the global biological space in which microorganisms, humans and animals interact, and weather patterns such as El Niño can modify oceanic currents and therefore the world's distribution of pathogens, which may include antibiotic-resistant bacteria (ARBs). Expanding urbanization process results in the destruction of natural habitats, causing loss of animal and plant genetic diversity, a process that is worsen by the anthropogenic selection of a limited range of varieties of economic interest. This implies the homogenization of hosts, which favours the dissemination of antibiotic resistant genes. Climate change is also responsible, through a number of reasons, for increasing migration, that can also lead to the propagation of ARBs. Natural disasters, rising sea levels, drought/floods, scarcity of food and water, are all phenomena that cause not only people, but also animals, to move. Migrating animals and people can carry ARBs, and dissemination might be favoured by poor sanitary conditions and poor access to health services (Wisner & Adams 2002).

Most antibiotics have broad-spectrum activity so they can be used to treat many diseases. However, therapies not only target pathogenic microorganisms but also affect the host microbial communities in the gut, making these medications potentially harmful agents. An increasing number of evidence supports the correlation between antibiotics overuse and the development of many disorders associated with the alteration of gut microbiota.

Previous studies found that antibiotic treatment did not reduce the total number of gut microbiota but altered the proportion of certain species in humans and animals. In addition, antibiotic exposure typically altered the diversity of the microbiome, by either increasing or decreasing diversity. In the West, antibiotic use has been consistently shown to be an environmental risk factor for inflammatory bowel disease. More importantly, the effects of antibiotics on human gut microbiota can persist for several years. Host–microbial interactions are very specific, and antibiotic therapy causes alterations or loss of highly co-evolved processes. Antibiotic-induced changes that are important to microbial regulation of host immunity include loss of bacterial ligands that are recognized by the host, loss of specific bacterial signals and alterations in the metabolites produced by the microbiota such as short-chain fatty acids (Ianiro et al. 2016).

However, it is important to mention that antibiotic treatment not always displays adverse effects on the GM of experimental animals and that, after cessation of treatment, much of microbiota recovers. Nevertheless, it can require several months to fully overcome the damage and some bacterial groups eliminated by treatment do not appear again even several years after discontinuation of treatment.

The effects of antibiotic-induced dysbiosis seem to last longer if they occur early in life, especially in childhood. This happens not only through direct administration of treatment to children but also because antibiotics given to a mother during pregnancy or at birth might have effects on the baby through vertical transmission of the microbiota. Disruption of the immune system development and maturation by altering gut microorganisms in early life might result in autoimmune diseases in later life, such as necrotizing enterocolitis and atopic diseases.

10.2 Heavy Metals

The gastrointestinal microbiota can metabolize a variety of environmental chemicals, directly upon ingestion or after their conjugation by the liver. Reciprocally, environmental chemicals can interfere with the composition and/or metabolic activity of the GI microbiota, with potentially deleterious consequences for the host. As an abundant pollutant in both industrialized and developing countries and increasingly pervasive in terrestrial and aquatic environments, heavy metals are associated with a wide range of toxic effects, including carcinogenesis, oxidative stress, DNA damage and effects on the immune system (Rosenfeld 2017). Heavy metals should also be considered as an important factor in the selection of antibiotic resistant microorganism, as they can coselect for AMR70, stimulate horizontal gene transfer (HGT)71 and modify the dynamics of antibiotics in natural ecosystems.

Studies on animals supplied with contaminated drinking water showed that heavy metals can affect GM, mostly by alteration of the metabolic activity of the gut microbiome as demonstrated by the variation of a number of microbial co-metabolites, such as indole derivatives, observed in urine or faeces the exposed animal. Heavy metal-consuming animals had smaller numbers of *Lachnospiraceae* and larger numbers of *Lactobacillaceae* and *Erysipelotrichaceacae* than control animals.

Arsenic (As) significantly perturbs the microbiome composition and metabolomic profiles in mice. High concentrations of cadmium (Cd) have been observed in aquatic systems, sediment and soil in some countries, especially in developing countries such as China. Cd toxicity is associated with carcinogenesis, hepatotoxicity, oxidative stress and immunotoxicity. There is evidence that cadmium caused a decrease in the populations of all microbial species. Subchronic exposure of mice to a low dose of Cd decreased the relative abundance of *Firmicutes* and *g-Proteobacteria* and increased the relative abundance of Bacteroidetes in the cecum and faeces. These alterations in the composition of the gut microbiome are associated with higher levels of lipopolysaccharide (LPS) in the serum, hepatic inflammation and even energy metabolism dysregulation. Lead (Pb), persistent in air, soil, water, old paint and food, can affect humans through ingestion, inhalation and dermal absorption. Pb poisoning has also been implicated in the development of obesity through its interruption of energy production and other metabolic processes.

Similar to the results obtained with heavy metals, there is discordance across studies as to which gut microbiota are altered by environmental nanoparticle exposure in various animal models and systems that include mice, rats, fruit flies, zebrafish, tilapia, redworms and gut bacteria from a human donor. *Firmicutes* and *Lactobacillus* spp. are the only ones affected in two or more of the studies listed above that represent various taxa (rats, mice, fruit flies, zebrafish and redworms).

10.3 Microplastics

The production of modern plastics has been thriving, with 348 million tons of plastics having been produced in 2017. Plastics are non-biodegradable, so most of the plastics produced in the past still persist in the environment today, and the plastics we manufacture daily will continue to exist for hundreds of years to come. Plastics can fragment into smaller pieces and are referred to as microplastics when particles are smaller than 5 mm. These non-biodegradable particles have been shown to accumulate in marine environments, freshwater habitats and terrestrial soils. The application of wastewater sludge and organic fertilizers to agricultural fields accounts for direct microplastic pollution of soil and potential, indirect pollution of aquatic systems due to runoff. Although freshwater systems are also impacted by runoff, historically, research has been greatly centred around microplastic pollution in marine habitats, with less focus attributed to freshwater and terrestrial ecosystems. However, as publications unveiled microplastic pollution to be pervasive in freshwater environments as well, research focus has shifted to include microplastic pollution in non-marine environment. Rowing evidence suggests microfibers are in the air around us, in our offices and outside in the street. Microplastics have been found in all corners and crevices of the earth, from polar ice to the deep sea and in human lungs and faeces.

A wide spectrum of animals from the wild has been found to ingest microplastics which may be transferred tropically. Although microplastics are most likely to accumulate in the gut, translocation of microplastics to other tissues such as the liver and kidneys, the circulatory system and the gills and accumulation there in have also been observed.

The effects of microplastic ingestion are manifold, and evidence suggests that these include—but are not limited to—the following: alteration of feeding activity, reduction of food assimilation efficiency, stunted growth, negative impacts on reproduction, altered gene expression, oxidative stress and neurotoxicity.

When we focus on its impact on human microbiota, research connecting these two is in its absolute infancy, as only very little scientific work has investigated this issue. Studies have shown that exposure to microplastic particles induced gut dysbiosis, altered bacterial diversity and was even accompanied by further negative health effects.

Microplastics can cause mechanical disruption of the GIT, leading to malnourished individuals, inflammation of the gastrointestinal tract and the breakdown of microplastics to form nanoplastics able to cross biological barriers; moreover, they act as a vector for potential pathogens and foreign, non-core bacteria, leading to competition for limited resources with resident bacteria; harbour chemicals known to disrupt the endocrine system and environmental chemicals such as persistent organic pollutants (POPs), polychlorinated biphenyls (PCBs), polycyclic aromatic hydrocarbons (PAHs) and dichlorodiphenyltrichloroethane (DDT) (Fackelmann & Sommer 2019). These factors may either directly or indirectly induce gut dysbiosis—the most common alterations observed being perturbations of alpha and beta diversity, an increase in potential pathogenic bacteria, a reduction in bacteria characteristic of a healthy gut microbial community, and the rise of negative interactions between essential gut bacteria, which constitute the core microbiome, and potential pathogens. This dysbiosis may involve the immune system, activated by inflammatory processes or pathogens or host hormone signalling, engaged by endocrine disruptors. The disruption of the symbiosis between host and gut microbiome may trigger the onset of (chronic) diseases, deteriorate host health, promote pathogenic infections and alter the gene capacity and expression of gut microbiota, leading to various consequences not yet detailed.

10.4 Organic Pollutants

Persistent organic pollutants (POPs) including organochlorine pesticides, polychlorinated biphenyls (PCBs), polybrominated diphenyl ethers and polycyclic aromatic hydrocarbons (PAHs) are synthetic compounds that have become of concern due to their resistance to degradation, their ability to be transported over long distances by air and water, their potential to accumulate in biological tissues and their potential to cause toxic effects such as immunotoxicity, neurotoxicity, carcinogenicity, mutagenicity and endocrine disruption (Jin et al. 2017; Nadal et al. 2015; Zhang et al. 2015).

Climate change has the potential to alter the distribution and biological effects of POPs, through alterations in environmental variables such as temperature, windspeed, precipitation and solar radiation. Temperature is one of the key parameters able to impact severely the global distribution of POPs—the degradation rates of POPs in the environment can increase up to three times for every 10 °C increment. Cold temperatures can induce their deposition and accumulation in Arctic environmental media. An increase of rainfall can raise POP deposition onto soil, and more frequent storm surges may cause the mobilization of chemicals stored in the soil compartment.

Therefore, global warming is probably influencing the environmental behaviour of POPs and ultimately affecting human exposure (Patz et al. 2000; Reiter 1998; Reiter 2008).

POPs do not usually cause respiratory health effects. In contrast, ingestion and bioaccumulation are routes of concern. Because most exposure to POPs occurs through the diet, the host gastrointestinal tract and gut microbiota are likely to be exposed to and modulated by POPs. In fact, although few studies have been done in the subject, some POPs, mainly 2,3,7,8-tetrachlorodibenzofuran (TCDF), were found to increase gut inflammation, modulated the gut microbiota population, shifting the microbial community from *Firmicutes* towards *Bacteroidetes*, increased bacterial fermentation and had a profound impact on host metabolism in an AHR-dependent manner. These changes in microbiota composition were associated with alterations in bile acid metabolism, and also triggering significant inflammation and metabolic disorders in the host as a result of bacteria.

10.5 Pesticides

Climate change on a global scale will influence local agriculture and increase pest populations, including weeds, insects and insect-borne diseases, therefore likely leading to large increases in pesticide use. Besides, pesticide efficacy is strongly associated with environmental conditions such as rainfall pattern, temperature and CO_2 concentration, another factor that points towards an increase in the use of agrochemicals. In developing countries, which are expected to suffer most from climate change, easily available, biodegradable, low cost and low risk pesticides are needed for low-income peasant farmers and organic farmers, and some countries might even re-introduce or increase the use of banned or restricted pesticides. Such practice will result in a higher risk of elevated exposures of humans to pesticides via residues in their food (Yuan et al. 2019).

Some pesticides, such as organophosphates and carbamates, affect the nervous system. Others may irritate the skin or the eyes, be carcinogens or affect the endocrine system in the body. Pesticides are for the most part persistent, bio-accumulative (become concentrated as they travel through the food chain), affect the physical and chemical properties of soils and toxic, not only to their target organisms, but also for the entire ecosystem and for any living organism. Because of the antimicrobial activity of some pesticides, they have the potential to change the gut microbiome and induce other symptoms in animals.

Pesticides could alter gut microbiota composition and their metabolites, including TMA, bile acids and SCFAs or other metabolites. The changes in the gut microbiota and metabolites could further cause adverse effects on hosts by some known and unknown signalling pathways.

10.5.1 Insecticides

An example of insecticide that can be toxic for non-target organisms is the organophosphate pesticide (OPP), which was found to alter gut microbial populations, composition and energy metabolism in male mice; OPPs were associated with a decrease in the number of beneficial bacteria (*Bifidobacterium* and *Lactobacillus*), accompanied by an increase in the numbers of *Enterococcus* and *Bacteroides*. Other changes in gut microbiota composition were associated with diabetic and obese phenotypes and induced glucose intolerance. Organochlorine pesticides (OCPs) are another kind of insecticide, and a link among serum OCP concentrations, methanobacteriales in the gut, and obesity in the general population has been found. Permethrin (PEM), a pesticide commonly used for residential pest control, was found to reduce the abundance of *Bacteroides*, *Prevotella* and *Porphyromonas*, while increase *Enterobacteriaceae* and *Lactobacillus*. Chlorpyrifos (CPF) is an organophosphate insecticide that also induces higher levels of *Bacteroides*, along with *Enterococcus*, and *Clostridium*, but lower levels of *Lactobacillus* spp. and *Bifidobacterium* spp. in rat intestines. Dysbiosis induced by CPF impairs the mucosal barrier, increases bacterial translocation and stimulates the innate immune system.

10.5.2 Fungicides

The second type of pesticide is the fungicides, biocidal chemical compounds used to kill parasitic fungi or their spores. Examples are carbendazim, imazalil, propamocarb and epoxiconazole. Oral administration of CBZ to mice increased hepatic lipid accumulation and body weight and additionally increased the levels of proinflammatory cytokines, changes that were related to CBZ-induced gut microbiota dysbiosis. In summary, studies have found that fungicides have bad effects on the gut microbiota of mammals or zebrafish and give rise to disorders of the host.

10.5.3 Herbicides

Herbicides, chemical substances commonly known as weed killers, are yet another class of pesticides used to control unwanted plants. As they are commonly and widely used, some residues have been observed in human food, mostly from the residues of herbicides entering water, air and soil. Examples of substances are glyphosate and pentachlorophenol (PCP). As the previous ones, these pesticides were found to have effects on gut microbes in animals and further influence the host.

10.6 Food Additives

Food additives are widely present in food in industrialized regions and can alter the host–microbiota relationship, resulting in an unfavourable gut microbiome shift, intestinal disturbance and inflammation and potentially accounting for the development of diseases such as inflammatory bowel disease, colitis and metabolic syndrome (Defois et al. 2018; Roca-Saavedra et al. 2017).

Non-nutritive sweeteners (NNSs), synthetic compounds such as saccharin, sucralose and aspartame, can provide sweet taste to foods without the high energy content of caloric sugars and are present in diet sodas, cereals and sugar-free desserts. They are normally recommended for people planning to lose weight or individuals suffering from glucose intolerance or type 2 diabetes mellitus. Ironically, chronically feeding NNSs has been related to higher glucose intolerance mediated by alterations in the GM and was associated with increased abundance of bacteria belonged to the genus *Bacteroides* and order *Clostridiales* in the gut. Not only that, but the decrease in beneficial bacteria of GM caused by NNSs was also associated with weight gain in rats.

Dietary emulsifiers (DEs) are present in most processed foods in order to achieve specific textures, prolong shelf-life and freshness. As a factor resulting from industrialization, DEs have been associated with the reduction of GM diversity, interactions and, consequently, increased incidence of metabolic syndrome and other inflammatory diseases in industrialized societies.

Carboxymethylcellulose and polysorbate, two types of DEs, promote bacterial overgrowth and facilitate translocation of bacteria across gut epithelia. Additionally, they reduced the mucus layer thickness and were involved in the onset of intestinal inflammation, obesity and diabetes.

10.7 The Impact of Environmental Alterations on Infectious Diseases

10.7.1 Water-Related Diseases

Water-related diseases are defined as "any significant adverse effects on human health, such as death, disability, illness or disorders, caused directly or indirectly by the condition, or changes in the quantity or quality, of any waters".

Water can cause risks to human health, due to both its deficiency and its poor quality. The limitation or lack of water is particularly a widespread condition mainly due to global climate change; the countries of sub-Saharan Africa are the most affected, but also the countries with a temperate climate, although to a lesser extent (Ahmed et al. 2018). These phenomena determine a lowering of the level of domestic hygiene and, consequently, a possible increase in the risk of the spread of infectious diseases, especially for children, resulting in serious health consequences that are difficult to quantify (Nichols et al. 2018).

The problem of access, distribution and use of safe and drinking water is widely and dramatically present in many developing countries, especially for infectious risks. About 2.2 billion people worldwide do not have access to safe drinking water services, 4.2 billion people do not have safely managed sanitation services. Furthermore, three billion people cannot even wash their hands with soap and water at home, lacking basic handwashing facilities. It is estimated that 1 in 10 people (785 million) do not yet have access to basic services, including the 144 million who drink untreated surface water.

Causes of water-related diseases include human actions (e.g. incorrect disposal of sewage wastes) or weather events. Precipitation pattern and extreme rainfall events affect disease transmission, while water temperature influences the growth and survival of microorganisms. Moreover, drought causes a high concentration of pathogens in the effluents, which not even the water treatment plants are able to contain with consequent contamination of the surface waters.

10.7.1.1 Classification

The pathogens responsible for water-related diseases can be traditional pathogens, new pathogens, emerging pathogens, re-emerging and opportunistic pathogens: bacteria, viruses, protozoa and parasites. According to the type of contact and transmission, these diseases are generally classified into four groups: waterborne, water-washed, water-based and water-related vector-borne diseases (Centers for Disease Control and Prevention 2010; Centers for Disease Control and Prevention 2011; Centers for Disease Control and Prevention 2015; UNICEF 2019; World Health Organization 2019a).

Waterborne Diseases

People can be exposed to water-borne infections through faecal-oral route by ingestion of contaminated drinking water, recreational water or food that has been prepared or washed using water contaminated with pathogens. Flies are also important transmitters of contamination from faeces, in fact it would be good practice to control flies by using nets with insecticides, sprays and using traps or insecticidal baits (David et al. 2013).

It is widely known that the risk most frequently associated with the use of contaminated drinking water is more correlated to the contamination by enteric pathogens that can reach the aquifers and contaminate the surface waters that are subjected to a series of treatments to allow their human consumption. This risk is still very high in developing countries, while in industrialized countries, there has been a decline in diffusion thanks to water treatment and disinfection processes (Kistemann et al. 2012).

Enteric viruses include enterovirus, norovirus, adenovirus, hepatitis E and hepatitis A viruses, rotavirus. Enteric viruses are responsible for a wide range of pathologies, among which the best known are gastroenteritis (diarrhoeal diseases), myocarditis and pericarditis, permanent or transient flaccid paralysis, aseptic meningitis, encephalitis. There is a serious underestimation of their spread in humans and in the environment as they are characterized by mainly an asymptomatic circulation.

Among the viral agents, the hepatitis A virus, spread worldwide both in sporadic and in epidemic forms, is the best known. Purification and chlorination of public water drastically reduce the presence of this virus. Handwashing and the correct disposal of human faeces are the most effective prevention systems, in addition to boiling water, when staying in areas without aqueducts and vaccination.

Regarding bacteria, *Vibrio cholerae* and nontyphoidal *Salmonella* certainly represent the agents of diarrheal diseases that have returned to having a great health impact in recent years. Cholera is a widespread disease in many developing countries, where poor hygiene conditions lead to a high risk of infection. Pandemics of the disease can cyclically spread, affecting especially at-risk countries. Seven pandemics have occurred in the past 200 years, the latest of which caused by the *Vibrio cholerae* serogroup O1 El Tor biotype which still persists in developing countries (Cvjetanovic & Barua 1972). Salmonellosis is the most frequent cause of gastroenteritis, accounting for 50% of gastrointestinal infections.

The most important protozoan infections are amoebic dysentery, cryptosporidiosis and giardiasis. Amebiasis is a ubiquitous disease caused by the parasite *Entamoeba histolytica*. Most infections are asymptomatic, although in some cases they can become clinically important. The intestinal form can assume the characteristics of acute dysentery, with fever and bloody mucoid diarrhoea. The extraintestinal form can involve any organ but mainly affects the liver causing liver abscess.

Cryptosporidium parvum and *Giardia lamblia* produce infectious forms (oocysts and cysts) resistant to the treatments that are performed on water before being distributed in the water supply networks. Giardiasis and cryptosporidiosis are faecal-oral diseases which can pass asymptomatically or cause self-resolving gastroenteritis in immunocompetent subjects. On the other hand, in immunosuppressed people, particularly in AIDS patients, they can become chronic and lethal, especially *Cryptosporidium* infection.

10.7.1.2 Water-Washed Diseases

Water-washed diseases are infections caused by insufficient availability of water with consequent poor personal and household hygiene. The presence of adequate quantities of good quality water represents a good measure of prevention, as well as adequate education of the people about appropriate use of water and sanitation, including basic messages such as correct handwashing with soap (Singh et al. 2011).

This category includes gastrointestinal infections, ophthalmic diseases (trachoma and conjunctivitis), fungal skin diseases such as ringworm, and infections caused or transported by parasites such as lice, mites, fleas and ticks.

Scabies is a highly contagious skin infection caused by *Sarcoptes scabiei*, a skin-burrowing mite, which occurs mainly in vulnerable groups such as young children and the elderly in resource-poor communities. The characteristic symptoms of scabies are an extremely itchy rash and track-like burrows in the skint hat appear on the hands, elbows, knees, breasts, shoulders or elsewhere.

The microorganism responsible for trachoma is *Chlamydia trachomatis* and is the world's leading cause of preventable blindness. Chronic chlamydial conjunctivitis is characterized by exacerbation and improvement phases which is common among children aged 3–6 years in some areas affected by poverty worldwide. Trachoma usually affects both eyes and is very contagious in its early stages and is transmitted by eye-to-eye, hand-to-eye contact, by flies that land on the eyes or by sharing contaminated items (e.g. towels, handkerchiefs, eye make-up) (World Health Organization 2021b).

Water-Based Diseases

Water-based diseases are caused by parasites that spend part of their life cycle within intermediate aquatic host. Human infection can occur through contact or ingestion of infested water. Dracunculiasis and schistosomiasis belong to this group.

Dracunculus medinensis or "guinea-worm" causes dracunculiasis. *D. medinensis* larvae are emitted by the female who comes out of the human skin in contact with the water, they fall into the stagnant water and are ingested by small crustaceans (cyclops). Human infestation occurs by drinking the water from wells and lakes by ingesting the larvae which are subsequently released into the stomach, migrate through the intestine and become adults (World Health Organization 2017). After mating, the females migrate to the subcutaneous tissues, where they grow to become sometimes even a metre long (World Health Organization 2000; World Health Organization 2021a).

Schistosomiasis is caused by several species of flat worms, which can penetrate healthy human skin in contact with water and migrate through the body inside the blood vessels where they lay their eggs.

Symptoms include itchy rashes, fever, aching, diarrhoea, chills, coughs and pains. More serious effects include damage to organs (bladder, liver and kidneys), impairment of the nervous system and, in children, stunted growth and cognitive development.

Infections due to *S. haematobium* and *S. mansoni* are prevalent in sub-Saharan Africa, while those due to *S. japonicum* and *S. mekongi* in South East Asia. The spread of schistosomiasis, both the urinary and intestinal and hepatic forms, is strongly linked to the bad disposal of excrements secondary to the absence of water sources. In order to prevent, it is important to educate the population in endemic areas about how they are transmitted and protected. Therefore, correct disposal of faeces and urine is essential so that the eggs cannot reach the waterways populated by the molluscs (Esrey et al.1991).

Water-Related Vector-Borne Diseases

Water-related vector-borne diseases are infections carried by insects having aquatic immature stages that act as vectors of the disease (Rossati et al. 2016). These insects live near stagnant waters and host the pathogenic microorganisms that are transmitted in humans at the time of the sting: this is what happens for malaria, dengue or African trypanosmosis (sleeping sickness).

In 2018 around 405,000 people died of malaria, 67% of whom were children under the age of 5 years. Around 300 million new cases occur each year, most of them in sub-Saharan Africa. Increasingly intense irrigation projects have increased the spread of this disease and have allowed the proliferation of *Anopheles*, the nocturnal mosquito carriers of the disease.

The species of *Plasmodium* responsible for human malaria are transmitted naturally from man to man by the bite of infected mosquitoes that breed in fresh or brackish waters. A female

Anopheles becomes infected with a blood meal carried out on a subject carrying the sexual forms of the parasite (gametocytes).

Symptoms appear 7, 15 or more days after the infected mosquito bite. They usually consist of fever, often very high, headache, vomiting, diarrhoea, sweating and shaking chills. The pathogenicity of *Plasmodium* is linked to their ability to invade and destroy red blood cells followed by the main symptomatology of the disease, represented by recurrent febrile access and anaemia.

Dengue fever is a viral infection transmitted to humans by mosquito bites that have, in turn, bitten an infected person. The main vector is the mosquito *Aedes aegypti*, although there have been cases reported by *Aedes albopictus*. It is particularly widespread during and after the rainy season in the tropical and subtropical areas of Africa, Southeast Asia, China, India, the Middle East, Latin and Central America, Australia and various areas of the Pacific (World Health Organization 2009). Dengue is an acute, febrile disease with an abrupt onset with fever, intense headache, arthromyalgia, retroorbital pain and skin rash.

African trypanosomiasis is an infection caused by protozoa of the genus *Trypanosoma brucei*, transmitted by the puncture of a tse-tse fly. Where the flies of the *Glossina* group are the main vectors, such as in West and Central Africa, the infection occurs mainly along the waterways in the forests bordering the rivers and ponds.

African trypanosomiasis has three phases: cutaneous, haemolymphatic, central nervous system. Central nervous system involvement causes persistent headache, inability to concentrate, progressive fatigue and apathy, daytime sleepiness, tremor, ataxia and terminal coma (Büscher et al. 2017).

The prevention of all these diseases related to aquatic vectors is based on the health education of the population about the measures for the elimination of the habitats of the larvae and the means of personal protection against day and night habits of mosquitoes, including the use of screens, mosquito nets, protective and repellent clothes and even destroying the environments in which they live, avoiding the indiscriminate destruction of the vegetation.

10.7.2 Food-Borne Diseases

It is widely known that the emerging health risks resulting from climate change are not limited only to excess mortality or morbidity due to heat waves, but concern multiple aspects such as the increased risk of infectious diseases carried by food, water and insect vectors (Bintsis 2017).

Food-borne diseases can be defined as the illness due to the ingestion of spoiled or poisonous food, contaminated by microorganisms or toxicants, which may occur at any stage during food processing from production to consumption.

The growth in the average surface temperature of the earth, the frequency and duration of heat waves, the contamination of internal and coastal waters resulting from the most frequent adverse weather events are all phenomena that influence the prevalence and incidence of infectious diseases transmitted by insect vectors, contaminated water and food with different mechanisms.

The risk factors that contribute to the increase in the incidence of food-related diseases, such as contamination of food and water following flood events, exposure of food to higher temperatures throughout the conservation, distribution and storage chain, social behaviour (public and collective catering especially in the months with warmer temperatures), require targeted monitoring, surveillance and information interventions (Levy et al. 2018).

The warmer average temperatures cause a growth of pathogenic replication, and the prolongation of the hot seasons can increase the possibility of contamination through the handling and use of food. About 32% of food-borne outbreaks in Europe are caused by factors associated with incorrect use of temperatures in food handling.

Currently, more than 250 food-borne diseases are classified, caused by various pathogens: mainly bacteria, but also viruses and parasites. New ones have been identified over the years, some of which are also spreading due to the increase in trade and travel, the use of collective catering and large intensive farms (World Health Organization 2015).

Weather-climatic variables (water and atmosphere temperature, rainfall and floods) can affect the pathogenicity of the six microorganisms recognized as the most frequent pathogens: *Campylobacter*, *Cryptosporidium*, *Listeria*, norovirus, *Salmonella* and *non-choleric Vibrio*.

Generally, the main sources of contaminated food are eggs and egg products, mixed meals, seafood and fish products (European Food Safety Authority (EFSA) et al. 2019). Campylobacteriosis, the most frequently reported zoonotic food infection in humans, shows a clear seasonal pattern. Colonization by *Campylobacter* increases rapidly with rising temperatures. The risk of campylobacteriosis has been positively correlated with average weekly temperatures. Otherwise, the decreasing trend in cases of salmonellosis in humans is probably mainly due to the effective *Salmonella* control programs (highly sensitive to the increase in temperatures) implemented to reduce the prevalence of bacteria in poultry. Cases of salmonella infections increase by 5–10% as each degree increases in weekly temperatures, for temperatures above about 5 °C (Liu et al. 2018).

The effect of temperature is more evident in the week preceding the disease, highlighting the importance of inadequate food preparation and storage at the time of consumption. In fact, poor hygiene, together with poor temperature control during the production, processing, transport, preparation and storage of food, can interact, allowing the multiplication of pathogens. The association between warmer temperatures and diseases highlights that the rates of diseases carried by water and food increase with rising temperatures.

Microscopic filamentous fungi can develop on a wide variety of plants and can lead to the production of highly toxic chemicals, commonly referred to as mycotoxins. The most widespread and studied mycotoxins are the metabolites of some types of mould such as *Aspergillus*, *Penicillium* and *Fusarium*. Contamination due to fungi can occur in almost all stages of the food chain (harvest, storage and transportation). Their colonization and diffusion are favoured by environmental conditions and nutritional components, as well as by other factors such as attacks by insects or weeds. The biosynthesis of mycotoxins is influenced by somewhat peculiar conditions such as the climate and the geographical location of the cultivated plants, the cultivation practices, the deposit and the type of substrate (Assunção et al. 2018).

Harmful algal blooms (HABs), with the consequent production of toxins, can also cause diseases in humans, mainly through the consumption of contaminated molluscs or crustaceans. The warming of sea water can therefore contribute to increase the cases of contaminated molluscs or reef fish and lead to an expansion of the distribution of pathologies such as ciguatera, food poisoning due to ciguatoxin (toxin produced in particular by microalga *Gambierdiscus toxicus*).

Epidemiological data relating to food-borne diseases remain scarce, particularly in developing countries, and even the most evident epidemics are often not recognized or not reported or not properly investigated. The categories of individuals most sensitive to food-borne pathogens are those who have an immature or weakened immune system such as infants and children, pregnant women, the elderly or immunocompromised individuals. In this regard, malnourished infants and children are particularly exposed to food risks and more at risk for the development of serious forms of disease.

In industrialized countries, more than 30% of the population is subject to food-borne infection every year. It is much more complex to estimate the incidence of these infections in developing countries where the high incidence of diarrheal phenomena suggests the presence of a serious problem in food safety (Miraglia et al. 2009).

According to CDC (Centers for Disease Control and Prevention), in the United States each year 48 million people gets sick, 128,000 are hospitalized and 3000 die of food-borne diseases.

In 2018, 5146 food-borne outbreaks occurred in Europe, involving 48,365 people. The main causative agent of epidemic outbreaks was *Salmonella*, responsible for one out of three outbreaks causing the greatest number of cases of illness and hospitalization.

Although *Salmonella* was the most frequently identified causative agent in outbreaks in many countries, in some states epidemic episodes were mainly associated with infection by norovirus, *Campylobacter*, *Trichinella*, bacterial enterotoxins and *Shiga-toxin-producing Escherichia coli* (STEC). Also in 2018, the majority of epidemic outbreaks occurred in the domestic environment while the context that determined the greatest number of epidemic cases was at restaurants, pubs, bars, hotels and other forms of catering.

Symptoms related to food-borne illnesses can be mild and self-limiting, such as nausea, vomiting and diarrhoea, or reach more serious conditions that endanger life-threatening individuals, such as liver failure, brain and neurological disorders, and paralysis.

10.7.2.1 Classification

Contaminated food can have several roles in causing diseases. It can be the indispensable element in the transmission chain: some microorganisms cause diseases only if they reach a microbial load such as to overcome human defences; this load is reached when the microorganisms find in the food the ideal conditions for multiplying (temperature, humidity, acidity, sugar and salt content, presence of oxygen). The harmful action of microorganisms can be achieved either by ingesting the pathogenic microorganism and its toxins with food (food-borne infections and toxicoinfections), or through the only toxins produced by them contained in the food (food-borne intoxication) (De Giusti et al. 2007).

Otherwise, food can be an occasional element in the transmission chain. Contaminated foods can act as simple "vehicles" of the pathogens, but they are not indispensable for the transmission of infections that can also spread in other ways. In these types of infections, microorganisms are highly pathogenic, so a modest microbial load is sufficient, and their multiplication in the food is not necessary before ingestion.

Generally, they are due to contamination of food and water with faecal material from patients or carriers.

These diseases can be caused by bacteria (e.g. *Shigella*), viruses (hepatitis A virus, norovirus), and parasites (Amoeba and Giardia).

Food-Borne Infections

Food-borne infections occur through the ingestion of food contaminated with a high load of live pathogens that are able to reach the host's intestine and adhere to epithelial cells. Microorganisms can cause localized inflammation with diarrheal syndromes, or generalized systemic infections, characterized by diffusion by blood, fever and antibody formation (Cumming et al. 2019).

Examples of food infections are diseases caused by the ingestion of food contaminated by: *Listeria monocytogenes, Shiga-toxin-producing Escherichia coli (STEC), Salmonella spp., Campylobacter* spp.

Salmonella is one of the most commonly spread bacteria that can give rise to a food infection. The main sources of contamination are raw products of animal origin (meat, poultry, eggs, milk products).

The symptoms of salmonellosis are diarrhoea, vomiting and abdominal cramps, but in immunosuppressed individuals, it can also cause very serious conditions.

Shiga toxin-producing Escherichia coli (STEC) strains are enteric pathogens that produce a powerful toxin responsible for serious morbid forms in humans. The best known and most widespread serogroup is O157. The clinical manifestation associated with STEC infection varies from watery diarrhoea, haemorrhagic colitis and uremic haemolytic syndrome.

Human infection is transmitted through the ingestion of contaminated food or water or by direct contact with animals. Among the most at-risk contaminated foods are raw or undercooked meat, unpasteurized milk, cheeses and other unpasteurized milk-based products. Vegetables can also carry the infection through contamination with zootechnical waste. Another route of transmission of STEC infections is oro-faecal route from person to person.

Campylobacter is the most common cause of diarrhoea in the world. It generates fever and abdominal cramps and is found above all in risky raw foods such as poultry and milk.

Listeria monocytogenes is a ubiquitous bacterium, widespread in the environment and is commonly found in the soil, water, vegetation and faeces of numerous animal species. The foods mainly associated with listeriosis infection include fish, raw meat and vegetables, unpasteurized milk and dairy products, processed and prepared foods (ready to eat) including hot dogs, prepacked salads, sandwiches, smoked fish.

Listeriosis can have different clinical forms, from acute febrile gastroenteritis typical of food poisoning, which occurs within a few hours from ingestion and self-limiting in healthy subjects, to invasive or systemic form. Infections contracted during pregnancy can have serious consequences on the foetus (foetal death, abortion, premature birth or congenital listeriosis).

Food-Borne Toxicoinfections

Food-borne toxicoinfections are caused by consumption of foods containing both toxins and bacteria. In this case, the toxicity is given by both the preformed toxins and those produced by live pathogens ingested with the food inside the host. Examples of food poisoning are the syndromes produced by *Bacillus cereus* and *Clostridium perfringens*.

C. perfringens is anaerobic and sporogenic bacterium. It is present in the soil and intestinal tract of humans and animals. Spores are heat resistant and can proliferate in preheated foods, ready to be eaten.

The main causes of *C. perfringens* infection are inadequate hot storage or inadequate cooling of cooked food, especially meat. Symptoms appear 6–24 h after ingesting contaminated food and include abdominal pain and diarrhoea.

Bacillus cereus is a bacterium widespread in the environment, especially in soil and dust. It produces two types of toxins: one, stable to heat, which causes vomiting, another, labile to heat, responsible for a diarrheal form. Transmission occurs through contaminated food that is kept at room temperature for a long time after cooking or has not been quickly and efficiently cooled. This is because *B. cereus* is able to survive in a spore state during the heat treatment of the food and then change into the vegetative form (capable of producing toxins) when conditions become again favourable for its survival.

Food-Borne Intoxications

Food-borne intoxications are caused by the ingestion of a food containing a preformed toxin, produced by bacteria that multiplied in the food prior to consumption. *Clostridium botulinum* and *Staphylococcus aureus* intoxications belong to this group.

Clostridium botulinum lives in the soil, in the absence of oxygen, and produces spores that can resist the external environment even for a long period until they meet conditions suitable for growth. It can produce several toxins, usually designated with letters, from A to F. Toxins A, B, E and F are those responsible for botulism that affects humans.

All stored and processed foods uncooked and with a low degree of acidity (pH above 4.6) can constitute an environment suitable for the growth of botulinum.

The period of onset of symptoms is between 12 and 36 h, and the symptoms include clouded vision or doubled vision, dry mouth, difficulty swallowing, paralysis of the respiratory muscles.

Staphylococcus aureus is a bacterium capable of producing toxins and intoxication occurred by ingesting food contaminated by toxins produced by the bacterium. Generally contaminated foods include cream, cream pastries, milk, preserved meats and fish.

Symptoms usually arise suddenly with severe nausea and vomiting that begin approximately 2–8 h after eating contaminated food.

10.7.2.2 Food Safety Recommendations

WHO developed the Five Keys to Safer Food program to promote safe food handling behaviours and aimed at both food handlers and consumers (World Health Organization 2006). The Five Keys to Safer Food explain the basic principles that every individual should know all over the world to prevent food-borne diseases through easy to adopt and adapt tools.

1. **Keep clean**: wash your hands before touching the food and wash them often while cooking and after going to the toilet; wash and disinfect all work surfaces and materials that come into contact with food.
2. **Separate raw and cooked**: do not use utensils and equipment that come into contact with raw food for other foods, keep the food in closed containers to avoid any contact between raw food and prepared food.
3. **Cook thoroughly**: especially meat, poultry, eggs and seafood; bring the soups and stews to a boil at 70 ° C, check that the meat and poultry are not pink and reheat the cooked and cooled foods.
4. **Keep food at safe temperatures**: do not leave cooked food for more than 2 h at room temperature; refrigerate cooked food and easily perishable foodstuffs at temperatures preferably below 5 °C; keep cooked food warm (above 65 °C) before serving; do not store food too long even in the refrigerator and do not thaw frozen at room temperature.
5. **Use safe water and raw materials**: use only safe or treated water to exclude any contamination; wash fruits and vegetables (especially if eaten raw); do not consume foods beyond their expiration date and prefer foods processed for safety (such as milk pasteurized).

10.7.3 Vector-Borne Diseases

Vector-borne diseases represent a not completely solved public health problem and denote those diseases that are spread by insect vectors. The insects constitute a necessary stage in the transmission of the infection from one person to another or from animal hosts to humans. A large number of viral, bacterial and parasitic pathogens can be transmitted by insect vectors (Cook 1996) (Table 10.1). Moreover, several vector-borne diseases continue representing some of the major microbial causes of morbidity and mortality in the world nowadays. In this regard, suffice it to recall that malaria alone causes million deaths and 273 million cases worldwide each year (World Health Organization 2021c).

The impact of the environmental changes on the spread of vector-borne diseases has been addressed for many years, and several literature reviews have specifically focused on this topic. The study of ecology and the epidemiology of all vector-borne diseases should involve three important aspects:

1. Hosts: human/animal/vector
2. Pathogens: virus, bacteria, parasites
3. Environment

The spread of a vector-borne disease could be influenced by several factors, in particular, the host's ecology and the behaviour; the vector's ecology and the behaviour; the level of susceptibility of the population. The presence of a vector in a specific environmental context does not only depend on climate conditions (temperature, humidity) but also depend on the existence of a suitable habitat (e.g. a lot of vectors cannot survive inside an urban context, their life could be compromised by pollution, pesticides, land use) and on the density of human and animal population with which the vector comes into contact (Paaijmans 2010; Bartlow et al. 2019; Baylis 2017; Gatto et al. 2016).

In the past, most of the cases of recrudescence of a lot of infectious diseases can be attributed to local ecological and environmental changes, which determined an increase in the vector spread or in the likelihood of contact between vector and specific host. There is a great number of drivers of environmental alterations that could have effects on the spread of many vector-borne diseases. Some of these drives are widespread all over the world, having their origin in determinants that can be traced everywhere, such as the temperature increase or the growth of the levels of CO_2 in our atmosphere, while other drivers have local causes but could affect the status of many vector-borne diseases worldwide, for example deforestations, urbanization, chemical pollution, intensive agriculture, trade and travel, water storage and irriga-

Table 10.1 Some examples of vector-borne diseases of clinical importance

Disease	Pathogen	Vector		Reservoir
Dengue	Flavivirus	*Mosquitos*	Aedes albopictus; Aedes aegypti	Human, monkey
Japanese encephalitis	Flavivirus		Culex spp.	Human, pig
West Nile	Flavivirus		Culex spp.	Wild rodents, birds, horse
Yellow fever	Flavivirus		Aedes spp.	Human, monkey
Murray River encephalitis	Flavivirus		Culex annulirostris	Birds
St Louis encephalitis	Flavivirus		Culex spp.	Peridomestic birds
Eastern equine encephalitis	Alphavirus		Mosquito: Aedes, Coquillettidia, Culex spp.	Birds, horses
Rift valley fever	Phlebovirus		Mosquito: Culex spp., Aedes spp.	Domestic ruminants
Chikungunya virus	Alphavirus		Mosquito: Aedes aegypti; Aedes albopictus	Human
Malaria	Plasmodium		Mosquito: Anopheles spp.	Human
Tick-borne encephalitis	Flavivirus	*Ticks*	Tick: Ixodes spp.	Small mammals, rodents
Crimean-Congo haemorrhagic Fever	Nairovirus		Tick: Ixodes spp.	Ticks
Spotted fever	Rickettsia		Tick: Rhipicephalus sanguineus, Dermacentor marginatus	Rodents, dogs, ticks
Lyme disease	*Borrelia burgdorferi*		Tick: Ixodes ricinus, Ixodes persulcatus	Small mammals, birds, reptiles
Anaplasmosis	Anaplasma phagocytophilum		Tick: Ixodes ricinus	Rodents, dogs, ticks
Leishmaniasis	Leishmania	*Phlebotomus*	Phlebotomus	Dogs, foxes, rodents

tion, rains and floods (Sutherst 2004). Since some environmental alterations are universal but have different on a local point of view, it is necessary to address the assessment of the risk of transmission on vector-borne diseases with a multilevel approach, focusing either on the type of infection or on the kind of environmental modifications that could modify geographic distribution of vectors and relative pathogens/diseases.

The *increase of CO_2 concentration* in the atmosphere has a potential effect on the epidemiology of vector-borne diseases. An indirect effect is that plants could improve their water use efficiency, and this can result in larger plants that provide more humid habitat for insect vectors and pathogens. Moreover, higher CO_2 concentrations lead to an increase of ambient temperature and plant biomass and growth of woody vegetation, with an increased longevity of vectors.

Climate change is one of the best-known factors able to influence the geographical distribution of all the conditions that are adequate for vectors and pathogens (European Food Safety Authority (EFSA) et al. 2018; European Food Safety Authority (EFSA) et al. 2020). Pathogens transmitted by vectors and causing vector-borne diseases are usually extremely sensitive to climatic variations because they spend most of their life cycle inside invertebrate hosts, whose temperature is similar to environmental one. Furthermore, climate influences the life cycle of vectors and can alter the reproduction rate of parasites and virus inside vectors and human hosts (Semenza and Menne 2009; Semenza & Suk 2018), in particular temperature increases can reduce the incubation period of these pathogens and the life cycle of vectors, thus enhancing transmission risk of vector-borne disease. Long-term changes in the average temperatures of seasons can also have impact on vector and host animals and could influence human activities and land use, with consequential effects on spatial-temporal distribution of vector-borne diseases and their prevalence (Lindgren et al. 2012). Climate change will be able to influence either vectors or pathogens. The environmental temperature is directly proportional to the level of development of pathogens, but inversely proportional to the survival of some vectors. With the expansion of warm climatic areas, the growth seasons will become longer, causing a quicker development of vectors and pathogens, with an increase in the number of generations per year in cooler areas. But the highest temperature could affect the life spans of the vectors, and it could explain the range expansion of warm climate vectors and pathogens towards suitable temperature areas (Hunter 2003). The evidence suggests that future climate change, if not mitigated, will very likely impact the length of the transmission season and the geographical range of a significant proportion of infectious diseases (Baer De Paula et al. 2019; Caminade et al. 2019; La Torre et al. 2020).

As far as *urbanization* is concerned, there are three factors that can have an impact on the spread of vector-borne diseases: the increased density of people living in the same area; the lack of suitable and healthy infrastructures; poor water supply and bad management of wastewater. Many vectors of different vector-borne diseases (yellow fever, dengue) are able to take advantage from the presence of artificial sources of wastewater in order to thrive and reproduce (Zuo et al. 2018).

In the last decades, *land use* has gone through a period that showed an intensification of agriculture and its spread into new areas where forests existed before, with the consequent need for deforestation and for the distribution of water for irrigation and human consumption. Destruction of habitats, loss of biodiversity, species loss and alteration of existing vector-host–pathogen relationships have been the main consequences of this excessive exploitation of lands. Several types of land use that have an impact on the incidence of vector-borne diseases have been identified, including deforestation, land cultivation, and various water storage, distribution, and irrigation structures and practices. Deforestation and consequent industrial development, residential expansion and agricultural improvement have had well-known effects on the species of some vectors, for example *Anopheles* mosquitoes, and on the rates of transmission of the vector-borne disease. Deforestation, irrigation and water storage are responsible for the development of new more vector-breeding sites and for an increased risk of interaction between human and vectors. The intensification of agriculture determines an increased alteration of the land and local vegetation, increasing the availability of potential vector breeding sites.

Fertilizers, pesticides, industrial toxins and other substances are the main causes of *chemical pollution* that has potential effects on vectors, pathogens and hosts. Industrial and agricultural chemical industries produce thousands of toxic substances each year. These chemical substances have effects on human and animal hormones (Matzrafi 2018). The list of synthetic pyrethroids is of particular concern because of their nature as active endocrine-disrupting chemicals, widely used in agriculture and in impregnated bed nets to prevent malaria: observations of injury of human immune system, in particular in developing countries (where the problem is less kept in attention), could explain the indirect effect on the risk of spread of vector-borne diseases.

References

Ahmed SA, Guerrero Flórez M, Karanis P (2018) The impact of water crises and climate changes on the transmission of protozoan parasites in Africa. Pathog Glob Health 112(6):281–293

Assunção R, Martins C, Viegas S, Viegas C, Jakobsen LS, Pires S, Alvito P (2018) Climate change and the health impact of aflatoxins exposure in Portugal – an overview. Food Addit Contam Part A 35(8):1610–1621. https://doi.org/10.1080/19440049.2018.1447691

Baer De Paula A, Sestili C, Cocchiara RA, Barbato D, Del Cimmuto A, La Torre G (2019) Perception of climate change: validation of a questionnaire in Italy. Clin Ter 170(3):e184–e191

Bartlow AW, Manore C, Xu C, Kaufeld KA, Del Valle S, Ziemann A et al (2019) Forecasting zoonotic infectious disease response to climate change: mosquito vectors and a changing environment. Vet Sci 6(2):40

Baylis M (2017) Potential impact of climate change on emerging vector-borne and other infections in the UK. Environ Health 16(1):45–51

Bibbò S, Ianiro G, Giorgio V, Scaldaferri F, Masucci L, Gasbarrini A, Cammarota G (2016) The role of diet on gut microbiota composition. Eur Rev Med Pharmacol Sci 20(22):4742–4749

Bintsis T (2017) Foodborne pathogens. AIMS Microbiol 3(3):529–563

Büscher P, Cecchi G, Jamonneau V, Priotto G (2017) Human African trypanosomiasis. Lancet 390(10110):2397–2409

Caminade C, McIntyre KM, Jones AE (2019) Impact of recent and future climate change on vector-borne diseases. Ann N Y Acad Sci 1436(1):157

Centers for Disease Control and Prevention (2011a) Estimates of Foodborne Illness in the United States. https://www.cdc.gov/foodborneburden/2011-foodborne-estimates.html

Centers for Disease Control and Prevention (2010b) Hygiene-related diseases "Scabies". https://www.cdc.gov/healthywater/hygiene/disease/scabies.html

Centers for Disease Control and Prevention (2015c). Parasites - amebiasis - entamoeba histolytica Infection. "General Information". https://www.cdc.gov/parasites/amebiasis/general-info.html

Claus SP, Guillou H, Ellero-Simatos S (2016) The gut microbiota: a major player in the toxicity of environmental pollutants? NPJ Biofilms Microbiomes 2:16003. https://doi.org/10.1038/npjbiofilms.2016.3

Cook J, Reich MR. Case 4-1996: paralysis due to schistosomiasis. N Engl J Med. 1996 Jun 6;334(23):1548

Cresci GA, Bawden E (2015) Gut microbiome: what we do and don't know. Nutr Clin Pract 30(6):734–746. https://doi.org/10.1177/0884533615609899

Cryan JF, O'riordan KJ, Cowan CSM, Sandhu KV, Bastiaanssen TFS, Boehme M, Codagnone MG, Cussotto S, Fulling C, Golubeva AV (2019) The microbiota-gut-brain axis. Physiol Rev 99(4):1877–2013. https://doi.org/10.1152/physrev.00018.2018

Cumming O, Arnold BF, Ban R et al (2019) The implications of three major new trials for the effect of water, sanitation and hygiene on childhood diarrhea and stunting: a consensus statement. BMC Med 17(1):173

Cvjetanovic B, Barua D (1972) The seventh pandemic of cholera. Nature 239(5368):137–138

David J-P, Ismail HM, Chandor-Proust A, Paine MJI (2013) Role of cytochrome P450s in insecticide resistance: impact on the control of mosquito-borne diseases and use of insecticides on earth. Phil Trans R Soc B 368:20120429. https://doi.org/10.1098/rstb.2012.0429

De Giusti M, De Medici D, Tufi D, Carolina M, Boccia A (2007) Epidemiology of emerging foodborne pathogens. Ital J Public Health 4(1)

De Giusti M, Barbato D, Lia L, Colamesta V, Lombardi AM, Cacchio D, Villari P, La Torre G. (2019) Collaboration between human and veterinary medicine as a tool to solve public health problems. Lancet Planet Health 3(2):e64–e65

de Gunzburg J, Ghozlane A, Ducher A, Le Chatelier E, Duval X, Ruppé E, Armand-Lefevre L, Sablier-Gallis F, Burdet C, Alavoine L, Chachaty E, Augustin V, Varastet M, Levenez F, Kennedy S, Pons N, Mentré F, Andremont A (2018) Protection of the human gut microbiome from antibiotics. J Infect Dis 217(4):628–636. https://doi.org/10.1093/infdis/jix604

Defois C, Ratel J, Garrait G et al (2018) Food chemicals disrupt human gut microbiota activity and impact intestinal homeostasis as revealed by in vitro systems. Sci Rep 8:11006. https://doi.org/10.1038/s41598-018-29376-9

Esrey SA, Potash JB, Roberts L, Shiff C (1991) Effects of improved water supply and sanitation on ascariasis, diarrhoea, dracunculiasis, hookworm infection, schistosomiasis, and trachoma. Bull World Health Organ 69(5):609–621

European Food Safety Authority (EFSA), Maggiore A, Afonso A, Barrucci F, De Sanctis G (2020) Climate change as a driver of emerging risks for food and feed safety, plant, animal health and nutritional quality. EFSA Support Public 17(6):1881E

European Food Safety Authority and European Centre for Disease Prevention and Control (EFSA and ECDC) (2018) The European Union summary report on trends and sources of zoonoses, zoonotic agents and food-borne outbreaks in 2017. EFSA J 16(12):e05500

European Food Safety Authority and European Centre for Disease Prevention and Control (EFSA and ECDC) (2019) The European Union one health 2018 zoonoses report. EFSA J 17(12):e05926

Fackelmann G, Sommer S (2019) Microplastics and the gut microbiome: How chronically exposed species may suffer from gut dysbiosis. Mar Poll Bull 143:193–203. https://doi.org/10.1016/j.marpolbul.2019.04.030

Gatto MP, Cabella R, Gherardi M (2016) Climate change: the potential impact on occupational exposure to pesticides. Ann Ist Super Sanità 52(3):374–385. https://doi.org/10.4415/ANN_16_03_09

Grice EA, Segre JA (2011) The skin microbiome. Nature reviews. Microbiology 9(4):244–253. https://doi.org/10.1038/nrmicro2537

Hernando-Amado S, Coque TM, Baquero F, Martínez JL (2019) Defining and combating antibiotic resistance from one health and global health perspectives. Nat Microbiol 4(9):1432–1442. https://doi.org/10.1038/s41564-019-0503-9

https://apps.who.int/iris/bitstream/handle/10665/43546/9789241594639_eng.pdf?sequence=1

Hunter PR (2003) Climate change and waterborne and vector-borne disease. J Appl Microbiol 94:37–46

Ianiro G, Tilg H, Gasbarrini A (2016) Antibiotics as deep modulators of gut microbiota: between good and evil. Gut 65(11):1906–1915. https://doi.org/10.1136/gutjnl-2016-312297

Jin Y, Wu S, Zeng Z, Fu Z (2017) Effects of environmental pollutants on gut microbiota. Environ Pol 222:1–9. https://doi.org/10.1016/j.envpol.2016.11.045

Kistemann T, Rechenburg A, Höser C, Schreiber C, Frechen T, Herbst S (2012) Assessing the potential impacts of climate change on food-and waterborne diseases in Europe. Stockholm: ECDC.

La Torre G, De Paula BA, Sestili C, Cocchiara RA, Barbato D, Mannocci A et al (2020) Knowledge and perception about climate change among healthcare professionals and students: a cross-sectional study (Original research). SEEJPH. https://doi.org/10.4119/seejph-3347

Levy K, Smith SM, Carlton EJ (2018) Climate change impacts on waterborne diseases: moving toward designing interventions. Curr Environ Health Rep 5(2):272–282

Lindgren E, Andersson Y, Suk JE, Sudre B, Semenza JC. Public health. Monitoring EU emerging infectious disease risk due to climate change. Science. 2012 Apr 27;336(6080):418–9

Liu H, Whitehouse CA, Li B (2018) Presence and persistence of salmonella in water: the impact on microbial quality of water and food safety. Front Public Health 6:159

Matzrafi M (2018) Climate change exacerbates Pest damage through reduced pesticide efficacy. Pest Manag Sci 75(1):9–13. https://doi.org/10.1002/ps.5121

Miraglia M, Marvin HJP, Kleter GA, Battilani P, Brera C, Coni E et al (2009) Climate change and food safety: an emerging issue with special focus on Europe. Food Chem Toxicol 47(5):1009–1021. https://doi.org/10.1016/j.fct.2009.02.005

Nadal M, Marquès M, Mari M, Domingo JL (2015) Climate change and environmental concentrations of POPs: a review. Environ Res 143:177–185. https://doi.org/10.1016/j.envres.2015.10.012

Nichols G, Lake I, Heaviside C (2018) Climate change and water-related infectious diseases. Atmos 9:385

Paaijmans KP, Imbahale SS, Thomas MB, Takken W. Relevant microclimate for determining the development rate of malaria mosquitoes and possible implications of climate change. Malar J. 2010 Jul 9;9:196

Patz JA, McGeehin MA, Bernard SM, Ebi KL, Epstein PR, Grambsch A et al (2000) The potential health impacts of climate variability and change for the United States: executive summary of the report of the health sector of the US National Assessment. Environ Health Perspect 108(4):367–376

Reiter P (1998) Global-warming and vector-borne disease in temperate regions and at high altitude. Lancet 351(9105):839–840

Reiter P (2008) Climate change and mosquito-borne disease: knowing the horse before hitching the cart. Rev Sci Tech 27(2):383–398

Roca-Saavedra P, Mendez-Vilabrille V, Miranda JM, Nebot C, Cardelle-Cobas A, Franco CM, Cepeda A (2017) Food additives, contaminants and other minor components: effects on human gut microbiota—a review. J Physiol Biochem 74(1):69–83. https://doi.org/10.1007/s13105-017-0564-2

Rosenfeld CS (2017) Gut dysbiosis in animals due to environmental chemical exposures. Front Cell Infect Microbiol 7:396. https://doi.org/10.3389/fcimb.2017.00396

Rossati A, Bargiacchi O, Kroumova V, Zaramella M, Caputo A, Garavelli PL (2016) Climate, environment and transmission of malaria. Infez Med 24(2): 93–104

Semenza JC, Menne B. Climate change and infectious diseases in Europe. Lancet Infect Dis. 2009 Jun;9(6):365–75

Semenza JC, Suk JE (2018) Vector-borne diseases and climate change: a European perspective. FEMS Microbiol Lett 365(2):fnx244

Singh BB, Sharma R, Gill JPS, Aulakh RS, Banga HS (2011) Climate change, zoonoses and India. Rev Sci Techn OIE 30(3):779

Sutherst RW (2004) Global change and human vulnerability to vector-borne diseases. Clin Microbiol Rev 17(1):136–173

UNICEF, W (2019) Progress on household drinking water, sanitation and hygiene 2000–2017. UNICEF Report

Wisner B, Adams J (2002) Environmental health in emergencies and disasters: a practical guide. World Health Organization, Geneva

World Health Organization (2000) Protocol on water and health to the 1992 convention on the protection and use of transboundary watercourses and international lakes. London, 17 June 1999

World Health Organization (2006) Five keys to safer food manual. Switzerland, Geneva

World Health Organization (2009) Dengue guidelines for diagnosis, treatment, prevention and control: new edition. Geneva, World Health Organization

World Health Organization (2015) WHO estimates of the global burden of foodborne diseases: foodborne disease burden epidemiology reference group 2007–2015. World Health Organization, Geneva

World Health Organization. (2017). Guidelines for drinking-water quality, 4th edition

World Health Organization (2019a) Global water, sanitation and hygiene annual report 2018. https://apps.who.int/iris/bitstream/handle/10665/327118/WHO-CED-PHE-WSH-19.147-eng.pdf?ua=1

World Health Organization (2019b) Joint FAO/WHO expert meeting in collaboration with OIE on foodborne antimicrobial resistance: role of the environment, crops and biocides: meeting report

World Health Organization (2021a) Dracunculiasis (guinea-worm disease). https://www.who.int/news-room/fact-sheets/detail/dracunculiasis-(guinea-worm-disease)

World Health Organization (2021b) Trachoma Factsheets. https://www.who.int/news-room/fact-sheets/detail/trachoma

World Health Organization (n.d.-c) World malaria report 2019. https://www.who.int/publications/i/item/world-malaria-report-2019

Yuan X, Pan Z, Jin C, Ni Y, Fu Z, Jin Y (2019) Gut microbiota: an underestimated and unintended recipient for pesticide-induced toxicity. Chemosphere 227:425–434. https://doi.org/10.1016/j.chemosphere.2019.04.088

Zhang L, Nichols RG, Correll J, Murray IA, Tanaka N, Smith PB, Hubbard TD, Sebastian A, Albert I, Hatzakis E, Gonzalez FJ, Perdew GH, Patterson AD (2015) Persistent organic pollutants modify gut microbiota-host metabolic homeostasis in mice through aryl hydrocarbon receptor activation. Environ Health Perspect 123(7):679–688. https://doi.org/10.1289/ehp.1409055

Zuo T, Kamm MA, Colombel J et al (2018) Urbanization and the gut microbiota in health and inflammatory bowel disease. Nat Rev Gastroenterol Hepatol 15:440–452. https://doi.org/10.1038/s41575-018-0003-z

The Relationship Between Environment and Mental Health

Rosario A. Cocchiara, Alice Mannocci, Insa Backhaus, Domitilla Di Thiene, Cristina Sestili, Domenico Barbato, and Giuseppe La Torre

Abstract

Mental health is defined by the World Health Organization as a state of well-being in which the person realizes his/her own abilities, can cope with the stressing situations of life, reaches efficiency in work and is able to contribute to the community he/she belongs to.

In the past years, a deeper understanding about the causes for mental disorders was gained, but in general, they demonstrated to be related to complex interactions between biological and environmental factors. In fact, there is growing evidence that the quality of the environment that surrounds us, both natural and anthropological, impacts on physical and mental health.

The relationship between environmental conditions and mental well-being has long been acknowledged and has recently garnered additional attention in the face of climate change. In this chapter, we will present the main potentially associations between the mental illnesses and heavy metals, the climatic factors and indoor environment, and will give an overview on new psychological effect of ecological crises, such as eco-anxiety, ecological grief and solastalgia.

Finally, this chapter will give an overview on the methods to better face with the prevention and treatment of mental health related to environmental issues.

Keywords

Climate change · Cognitive behavioural therapy · Environment · Eco-anxiety Ecological grief · Mental health · Solastalgia

11.1 Introduction

Mental health is defined by the World Health Organization (WHO 2004) as a state of well-being in which the person realizes his/her own abilities, can cope with the stressing situations of life, reaches efficiency in work and is able to contribute to the community he/she belongs to. Another definition from Keyes, in 2014, includes in mental health well-being, emotional, social and psychological aspects, positive attitudes towards responsibilities and positive social functioning (Keyes 2014).

However, people may experience negative emotions and feelings, such as anger or sadness, but this does not mean that they are not in good mental health. In this regard, an interesting study from Galderisi et al. proposes a new definition of

R. A. Cocchiara · I. Backhaus · D. Di Thiene
C. Sestili · D. Barbato · G. La Torre
Department of Public Health and Infectious Diseases, Sapienza University of Rome, Rome, Italy

A. Mannocci (✉)
Universitas Mercatorum, Rome, Italy
e-mail: alice.mannocci@uniroma1.it

mental health described as "a dynamic state of internal equilibrium which enables individuals to use their abilities in harmony with universal values of society" (Galderisi et al. 2017). The ability to recognize, express and modulate emotions, empathize with others, coping with occurring life events and a positive social functioning does not save the individuals from adverse situations, but furnishes them the right tools to overcome them. Moreover, the condition of equilibrium does not keep itself static, but varies according to life moments or happenings, and challenges the integrity of the individual in implementing changes for a new balance.

In the past years, a deeper understanding about the causes for mental disorders was gained, but in general, they demonstrated to be related to complex interactions between biological and environmental factors. In fact, there is growing evidence that the quality of the environment that surrounds us, both natural and anthropological, impacts on physical and mental health (Di Nardo et al. 2010). From a systematic review published in 2017, evidences about the interaction between urbanicity, socio-ecological environment and mental health were retrieved (Gruebner et al. 2017). Studies indicate that people living within natural environments and green spaces are less likely to fall ill due to depression and that natural green areas help recovery from chronic stress. Furthermore, better concentration is observed in individuals living in natural green environments. On the other side, the urbanization of living areas affects the individual's health through increased levels of pollution and accidents (Correia et al. 2013). Similarly, this process reduces green areas that can contribute in regulating urban microclimate, in moderating extreme temperatures, purifying and filtering air from dust and pollutants, and reducing noise. Studies on anxiety disorder, distress and anger highlight higher rates in urban areas than in rural ones (Sharifi et al. 2015; Lee and Maheswaran 2011).

Besides the rapid climate change our planet is facing, which is having an unpredictable impact on mental health, and represents one of the most urgent public health threats of our times (Gifford & Gifford 2016).

Let us move on now to a deeper analysis about how the environment can affect the mental health of people.

11.2 How Does the Environment Influence Human Health

Humans interact with the environment continuously. These interactions affect quality of life, years of healthy life lived and health disparities. The World Health Organization (WHO) defines environment, as it relates to health, as "all the physical, chemical, and biological factors external to a person and all the related behaviors". The relationship between environmental conditions and mental well-being has long been acknowledged and has recently garnered additional attention in the face of climate change (WHO 2006). This paragraph presents the main potential associations between the mental illnesses and heavy metals, the climatic factors and indoor environment. Close examination of these causes has identified a complex network of factors associated not only with genetic, biological and psychosocial characteristics of the individual but also with environmental determinants present in the area of residence and workplace. One way for the reduction of the mental diseases could be approached interventions on environmental modifiable determinants.

11.2.1 Heavy Metals

The scientific evidences have well established the strong association between the mental diseases and acute intoxications by heavy metals. Among adults, exposure to lead (Pb) has been associated with depression, anxiety, panic disorder, reduced cognitive and response capacity, schizophrenia, Parkinson's disease and, although not as consistently, with Alzheimer's disease. Further, it has been associated with depression and phobic anxiety in women. Exposure to arsenic (As) has been associated with cognitive impairment, depression, anxiety, adjustment problems and Alzheimer's. Whereas cadmium (Cd) has been associated with depression and schizophrenia, manganese (Mn) has been linked to problems in the motor system, memory and poor cognitive performance, hyperactivity and attention deficit disorder. The long-term impact of low-dose exposure is under studied, but the researchers have argued that continuous exposure to heavy metals could give rise to a "silent pandemic" in modern society, one responsible for a subclinical and permanent decrease in the IQ, an increase in school failure, a reduction in productivity and an increased risk of antisocial and criminal behaviour. Recent publications have shown that living in areas with a higher concentration of heavy metals and metalloids in soil is associated with an increased probability of having a mental disorder (Ayuso-Álvareza et al. 2019; Patel et al. 2016; Vigo et al. 2016). These relationships were strengthened in individuals who reported consuming vegetables more than once a day.

11.2.2 Indoor Environment

Maintaining adequate indoor temperature and humidity is necessary to support health and improve quality of life. The relationship between local ambient temperatures and a wide range of mental health outcomes was well investigated, and a quasi-linear relationship between temperatures and mental health was found (Mullins and White 2019a, b). Mental health appears to deteriorate with increased temperatures across the range of temperatures considered. It is in contrast with the physical health measures that respond negatively to both extreme cold and heat. The relationship between mental health and temperatures is closely mirrored by temperature and emotional well-being, violent crime, aggressive behaviours and interpersonal conflict. Moreover, there are secondary causal associations between variation in temperatures and mental health that should be taken into consideration:

- Temperatures impact physical health of self and/or others, which in turn affects mental health.
- Temperatures impact time allocation, which in turn impacts mental health.
- Temperatures impact cognitive function, which in turn impacts mental health.
- Temperatures impact emotional state or emotional regulation, which in turn impacts mental health.
- Temperatures disturb sleep, which in turn affects mental health.

Perceived humidity was also significantly associated to the secondary associations to mental health, especially concerning on sleep quality, although the relationship was less strong than for temperature. In the summer season, measures of absolute humidity affected temperature perception, and measures of temperature affected the perception of humidity (Quinn and Shaman 2017).

11.2.3 Climatic Factors

Different aspects of climate change may affect mental health through direct and indirect pathways, leading to serious mental health problems (Clayton et al. 2001). Different types of extreme weather events appear to relate to high stress levels and consequently to mental health impacts, particularly at onset. The link between extreme anxiety reactions (such as post-traumatic stress disorder or PTSD) and acute weather disasters, such as floods (the most common disasters at global level), heat waves and cyclones, is well established, as are the emergency and other response procedures that are deployed when they occur. The long-term effects of extreme weather events are under studied. However, long-term anxiety and depression, as well as PTSD, increased aggression (in children) and perhaps even suicide have been found to be associated with floods. The scientists are declared that climate change will have significant mental health implications noting the psychological distress and anxiety about the future that may result from acknowledging climate change as a global environmental threat (Watts et al. 2018). The main mental health impacts of climate change will be

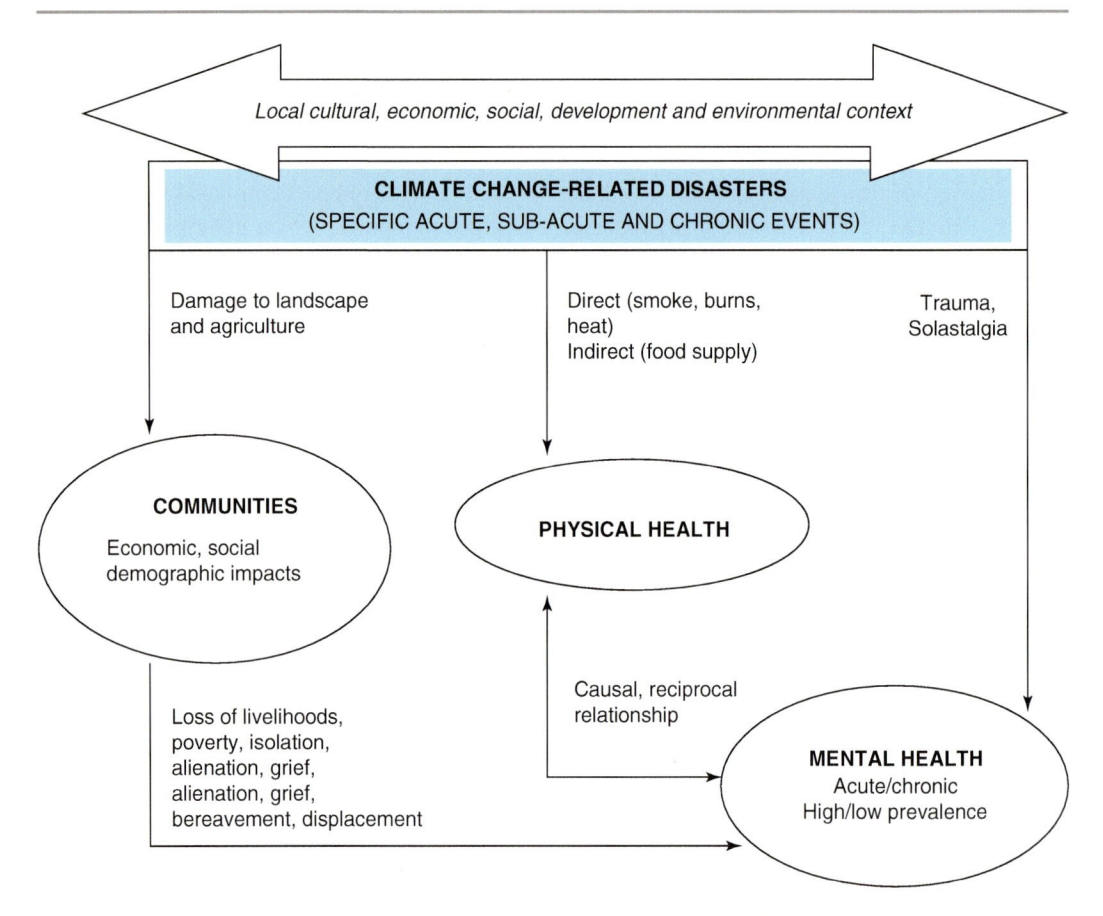

Fig. 11.1 Framework showing putative causal pathways linking climate change and mental health. [Source: Berry, H.L., Bowen, K. & Kjellstrom, T. Climate change and mental health: a causal pathways framework. Int J Public Health 55, 123–132 (2010)]

due to the disruptions that vulnerable communities, in particular, face with regard to the social, economic and environmental determinants of mental health and the future distress and anxiety that climate change may create on an individual level (Fritze et al. 2008). Figure 11.1 shows the framework published by Berry et al. (2010) that explain that climate change will have both direct and indirect effects on mental health. Climate change may affect mental health:

- Directly by exposing people to the psychological trauma based on frequency, severity and duration of adverse weather events/climate-related disasters, climate change including extreme heat exposure, and by destroying landscapes, which diminishes the sense of belonging and solace that people derive from their connectedness to the land.
- Indirectly by two further pathways:
- Physical health: for example, through increased heat stress, injury, disease and disruption to food supply.
- Community well-being: for example, through damage to the economic and, consequently, the social fabric of communities.

11.3 Vulnerability Factors

Climate change is arguably one of the greatest public health threats of our time. A growing body of literature on climate change and mental health suggests that extreme weather events can trigger post-traumatic stress disorder (PTSD), major depressive disorder, anxiety, depression, substance abuse and suicidal ideation (Hayes and Poland 2018).

The extent to which extreme events affect vulnerability is determined by magnitude, duration, timing as well as the sequence of events. Recovery from extreme events can take decades, and some communities may not recover from an event before the next one occurs (Ebi and Bowen 2016). The following chapter provides information on the frequency and intensity of climatic events, their consequences on access to essential infrastructure and sociodemographic differences/social inequalities.

11.3.1 Frequency and Intensity of the Climatic Events

Category 5 Hurricanes such as Hurricane Katrina, Irma and Maria, which have left parts of the Caribbean destructed and hundreds of thousands of people displaced and homeless, are no longer rare (United Nations Office for the Coordination of Humanitarian Affairs 2017). One of the most visible consequences of the climate change is an increase in the intensity and frequency of extreme weather events (IPCC 2014; Seneviratne et al. 2012). It has widely been accepted that global warming and climate change come along with an increase in both the intensity and frequency of extreme weather events (IPCC 2014; Morganstein and Ursano 2020). The Intergovernmental Panel on Climate Change (IPCC) forecasts that risks associated with extreme events will continue to increase as the global mean temperature rises (IPCC 2014). These extreme whether events can include an increase in the frequency and duration of heatwaves, heavy rainfalls and storms (IPCC 2014). Seneviratne et al. (2012) project that droughts will intensify in some seasons in areas such as southern Europe and the Mediterranean region, central Europe, central North America, Central America and Mexico, northeast Brazil, and southern Africa (Seneviratne et al. 2012). Europe, for example, was hit by a heatwave in July 2019, setting all-time high temperature records in Belgium, Germany, Luxembourg, the Netherlands, and the United Kingdom.

11.3.2 Access to Infrastructures and Resources

Extreme weather events can disrupt the serviceability of essential infrastructure such as roads, water treatment, electricity supply grids and essential public health infrastructures (Bell et al. 2018; Deshmukh et al. 2011). A loss of electricity linked to a heatwave, in the United States in 2003, for example, caused hospital emergency generators to shut down, food contamination from loss of refrigeration and increased incidence of total mortality (Beatty et al. 2006). Moreover, natural disasters may also result in infrastructure destruction, increasing the exposure risk to chemical, biological, radiological or nuclear (CBRN) materials. This was the case, for instance, in Japan following the 2011 Tōhoku earthquake and tsunami, which damaged reactors of the Fukushima nuclear power plant and exposed the community to nuclear material (Kumagai and Tanigawa 2018; Morganstein and Ursano 2020). Apart from that, natural disasters can also disturb access to mental health service facilities. The Gulf Coast region of the United States experienced a widespread loss of mental healthcare facilities, treatments and personnel, after Hurricane Katrina (Wang et al. 2007; DeSalvo et al. 2007).

11.3.3 Demographic and Socioeconomic Factors: Are Social Inequalities an Issue?

Although climate change affects everyone, it disproportionately affects the health of vulnerable groups such as children, people of lower socioeconomic status, chronically ill and mobility impaired people as well as pregnant and postpartum women (Cianconi et al. 2020). According to Morganstein and Ursano (2020), increased vulnerability to psychological consequences, following ecological disasters, depends on a variety of factors, including pre-event characteristics (e.g., socioeconomic status), event impact (e.g., event severity) and recovery variables (e.g., social support) (Table 11.1).

Table 11.1 Factors increasing vulnerability

Pre-even characteristics	Event impact	Recovery variables
Socioeconomic status	Duration	Relocation
Age	Severity	Job loss
Gender	Physical injury	Social support and social capital
Social support	Home loss and displacement	Financial stress

Source: Adapted from Morganstein and Ursano (2020)

A lower socioeconomic status, in particular, is often associated with worse health outcomes following disasters (Brown et al. 2013; Morganstein and Ursano 2020). People with fewer financial resources often reside in locations that are less resistant against the effects of ecological disasters, live in much older buildings without proper insulation and often cannot afford to pay for air conditioning and/or heating (Morganstein and Ursano 2020; Sánchez-Guevara Sánchez et al. 2020). A study conducted among residents in Madrid found that almost 23% of households are at risk of energy poverty, meaning that they lack the ability to keep homes warm during the winter and cool during summer (Sánchez-Guevara Sánchez et al. 2020).

Great inequalities also exist in terms of mental health outcomes (Matthews et al. 2019; Tracy et al. 2011). Hurricane Katrina exemplified the increased vulnerability to mental health problems among those of a lower socioeconomic status and minorities (Matthews et al. 2019; Rhodes et al. 2010). Researchers found that while the vast devastation of Hurricane Katrina lead to an overall increase of mental illness among people who survived Katrina, the event was particularly stressful to African American and low-income residents (Rhodes et al. 2010). Rhodes et al. (2010) found that the prevalence of probable serious mental illness doubled among low-income parents exposed to Hurricane Katrina. Mental health outcome may greatly depend on available pre-disaster resources (Sasaki et al. 2019). During Hurricane Katrina, most high- and medium-income families were evacuated and secured places to stay in hotels, or with family and friends in other cities in advance of the storm (Rhodes et al. 2010).

Research also suggests that women are particularly vulnerable to the effects of climate change and extreme weather and that overall these events kill more women than men (World Health Organization 2014). Studies from Europe, for example, have shown that women are at greater risk of dying in heatwaves (Kovats and Hajat 2008).

11.4 Sudden Phenomena with Direct Effects on Human Well-Being

All the anxiety-related diseases (e.g., stress, anxiety, depression, panic attacks, post-traumatic stress disorder (PTSD), sleep disorder) can be related to natural disasters. Catastrophic natural events such as hurricanes, tsunamis, volcanic eruptions earthquakes, floods, heat waves have an immediate impact on human lives and often result not only in the destruction of the physical, biological and social environment of the affected people, but in a longer-term impact on their mental health (Makwana 2019).

Mental health issues caused by disasters, especially in low-income countries are often a neglected area. Along with the social and economic losses, the individuals and communities experience a mental instability which might precipitate PTSD, anxiety and depression in the population. Generally, the disasters are measured by the cost of social and economic damage, but there is no comparison to the emotional sufferings a person undergoes post-disaster (Hackbarth et al. 2012).

Furthermore, it is important to highlight how psychological effects of the disaster have more impact on vulnerable part of the population, especially children, women and dependent elderly population.

Time, individual resilience and post-intervention techniques can help to recover the majority of affected people. However, effective interventions should be given pre-, peri- and post-disaster period to improve the adverse mental health effects of the disaster. Furthermore, great importance has the cultural context of the community and the values of the society to help them to cope with future disasters.

A disaster-induced displacement of entire populations is often associated with an increased risk of mental health and physical pathologies. A study carried out by Jang et al. (2021) was focused on population displaced by natural disasters in Southeast Asia. The researchers had a double approach, using surveillance data by the Emergency Events Database and performing a systematic review of previous studies concerning physical or mental health outcomes. In the period 2004–2017, they found almost 700 disasters, among which mainly earthquakes, floods, storms. All the studies included in the systematic review demonstrated significantly worse mental health outcomes and poor physical health among displaced population compared with non-displaced population.

It is well known that factors of displacement that can have the highest impact are geographic distance from the pre-disaster community, type of post-disaster housing, number of moves post-disaster, and time spent in temporary housing (Hori and Schafer 2010).

It is interesting to underline that unplanned residential displacement due to natural disasters is associated with both infectious diseases, including malaria, measles, diarrhoea, and chronic diseases, such as diabetes, which contributes to a high mortality rate among the displaced population. As far as infectious diseases are concerned, the reason for the increase is connected to the lack of immunity towards new infectious agents or vectors that can be present in the new environment where population are displaced, but also poor water sanitation and overcrowding. As far as mental health is concerned, in case of displacement, an increase of the risk of mental health disorders occurs, including anxiety, depression, perceived stress and post-traumatic stress disorder.

A particular case occurs when in the same period several disasters happen simultaneously. This is the case of the 2011 Fukushima event, in which earthquake, tsunami and nuclear disasters were in rapid succession. On 11 March 2011, an earthquake of very high magnitude (9.0) hit the north-eastern region of Japan. After few minutes, the earthquake was followed by tsunami waves that reached a height of 10 m. Due to this disaster, more than 18,000 people were either killed or missing. Moreover, the populations surrounding the Tokyo Electric Power Company Fukushima Daiichi Nuclear Power Plant (NPP), that is located in the coastal part of Fukushima prefecture, were evacuated mandatorily due to the subsequent disaster due to plant explosions, reactor meltdowns and radioactive material release. Shigemura et al. (2021) carried out a systematic review of the studies on the psychological consequences of the 2011 Fukushima nuclear disaster. They found a large variability of nonspecific psychological distress, depressive symptoms, post-traumatic stress symptoms and anxiety symptoms, that occurred in 8–65% of the population.

The mental health consequences are strictly linked to displacement (i.e., considering that more than 40,000 were displaced), to changing work or school, as well as concerns about the safety of the outdoor environment, food and water.

11.5 Progressive Phenomena with Indirect Effects on Human Well-Being

11.5.1 Solastalgia

This neologism refers to the combination of Latin words solacium (comfort) and pain (algia).

The philosopher Glenn Albrecht coined the term in 2005 to indicate a distress caused by environmental change (Albrecht 2005). In his words is "the homesickness you have when you are still at home," but the change of your environment is disturbing you. Global climate change is the main cause, but other situations such as natural disasters, e.g., volcanic eruptions, or massive destruction of the environment like the destructive coal mining techniques can generate solastalgia. In 2015, the medical journal *The Lancet* included solastalgia as a contributing concept to the impact of Climate Change on Human Health and Well-Being.

A recent scoping review (Galway et al. 2019) stated that solastalgia will be increasingly applied, developed and measured in the future, as one of the key concept in understanding the links between ecosystem health and human health. Further research on this concept will permit to study the cumulative impacts of climatic and environmental change on mental, emotional and social health.

A study published by Eisenman et al. (2015) was focused on the Ecosystems and Vulnerable Populations Perspective on Solastalgia and Psychological Distress After a Wildfire in Arizona, USA, occurred in 2011 (Fig. 11.2). They found that higher solastalgia score was associated with clinically significant psychological distress. Moreover, the distress was also associated with an adverse financial impact of the fire. A low impact of the psychological distress was associated to socio-economic factors, such as an annual household income over $80,000.00 and a higher family functioning score. Dramatic transformation of a landscape by an extreme environmental event, in this case a wildfire, can reduce its value as a source of solace.

11.5.2 Eco-Anxiety

Eco-anxiety was defined by Castelloe as a specific form of anxiety that is related to stress or distress caused by environmental changes, as well as by our knowledge of these changes. However, the diagnosis of eco-anxiety is not specific. In fact, this condition can be presented by insomnia, obsessive thinking, panic attacks, more or less associated to appetite changes strictly related to environmental concerns (Castelloe 2018).

The American Psychological Association defines eco-anxiety a chronic fear of environmental doom. The stress related to the climate changes can affect the mental health of individuals as stated by Glenn Albrecht and others, which coined also the term. Qualitative research provides evidence that some people are deeply affected by feelings of loss, helplessness and frustration due to their inability to feel like they are making difference in stopping climate change. The National Wildlife Federation has estimated that 200 million Americans will eventually experience emotional distress as a result of the effects of climate change (National Wildlife Federation 2011).

Fig. 11.2 The effect of the wall fire on the environment after 10 years

Gislason et al. (2021) in a recent work studied the Interplay between Social and Ecological Determinants of Mental Health for Children and Youth in the Climate Crisis. They carried out a rapid review that demonstrates that climate crisis is associated with increasing levels of mental health distress in children and youth. The main symptoms and signs are represented by feelings of sadness, guilt, changes in sleep and appetite, difficulty concentrating, solastalgia and disconnection from land. The perception of climate change of young people is mainly determined by their social locations and many are dealing with feelings of immense worry and eco-anxiety. According to Gislason et al., the mental health impacts of climate change on children/youth are strictly linked to Social Determinants of Health (SDoH). However, this impact needs to be seen in relation to the Ecological Determinants of Health (EDoH).

11.5.3 Ecological Grief

The ecological grief is defined by Cunsolo and Ellis (2018) as the pain, sadness, grief and suffering that people undergo in experiencing (direct or indirect) the loss of a natural landscape, species or reference environment.

The ecological grief is a pain that situates people within the environments and ecosystems where they live and that are profoundly part of them, which have a deep impact on physical, mental and emotional health, which determine their well-being and which define their identity. This means that in the context of mitigation or adaptation to the risks and problems of climate change, worry, fear and grief must be considered as emotions to be recognized and accepted, as "normal" experiences to be named and narrated, as the first stages of an adaptive process that can allow people to face future challenges, as feeling and experiencing loss means recognizing that we are part of a community.

According to Chalupka et al. (2020), the gradual nature of environmental degradation is able to cause chronic community, familial and individual stress from displacement as well as loss of access to resources, connection to land and place. Evidence suggests that young people are experiencing ecological grief that can lead some to feel angry, frustrated and helpless. Finally we need to recognize also that indirect impacts of climate change events also include climate change exacerbating socioeconomic inequalities and broader social determinants of health during and in the aftermath of disaster events.

11.6 How to Prevent and Treat Mental Health

Between 2009 and 2012, mobilization of people with combined background of psychotherapy and interest in climate and ecological change, converged in creating the Climate Psychology Alliance (CPA). This organization works in the USA: it developed a scientific approach to Climate Psychology which aims to recognize effective strategies for preventing negative effects of environment changes on mental well-being.

The CPA developed a dual approach: firstly, it increases consciousness about the responsibility each individual has within the society, educating to correct habits that could be useful in reducing the impact of our living in the changing of the environment; secondly, it offers support and care towards those individuals that result already in an impairment of their mental health, linked to environment issues. The presence of this team of experts guarantees caring paths and solutions for mental well-being related to climate change (Climate Psychology Alliance 2021).

Conversely, within the European framework, the EMEN project was developed: under the aegis of the European Union, this project answers to the need of a joint interaction between scientific community and policy makers, to implement evidences about mental health, and paying a specific attention to the strengthening of resilience towards climate change issue (EMEN 2020).

In case of emergencies, the capacity to face with people's physical and emotional health is crucial to minimize their impact: disasters driven by natural causes (droughts, earthquakes, floods, hurricanes, tornadoes, volcanic eruptions, etc.); technological disasters like fires, toxic leaks and explosions are all situations in which the capacity of resilience is important to respond to a disaster. Winders et al. (2021) performed a systematic review of the evidence on the effectiveness of interventions provided to first responders to prevent and/or treat the mental health effects in case of emergencies. Among the papers assessed in the review, 72% indicates a positive impact in treating psychiatric symptoms. These interventions (*Critical Incident Stress Debriefing*, meditation, multiple stressor debriefing) were judged cathartic, generally well received, and effective in reducing symptoms of anxiety, hyperarousal and depression post-disaster.

One of the main treatments used in crisis situations is the cognitive behavioural therapy (CBT). This type of treatment, according to Dattilio and Freeman (2004), aims at performing a complete assessment of the person, challenging the person's dysfunctional beliefs, creating options in a cooperative way and, finally, establishing hope. In this case, it is fundamental to understand what are the patient's strengths for coping with the crisis and what is the positive potential of a crisis. These authors proposed a five-stage protocol with the aim of:

1. Developing the therapeutic relationship and establishment of rapport;
2. Assessing the severity of the situation at the beginning of the rapport;
3. Supporting the person to assess and activate his/her strengths and resources;
4. Working in a cooperative way between patient and therapist to develop a plan of positive action;
5. Testing novel behaviours and thoughts.

Interestingly, the CBT can be used in crisis interventions in online psychological counselling. Da Silva et al. (2015) found that in context such as disasters, risk/prevention of suicide and trauma, the most chosen approach and the most efficacious is the CBT, all over the word from Europe (the Netherlands) to Oceania (Australia).

References

Albrecht GA (2005) Solastalgia: a new concept in human health and identity. PAN (Philosophy, Activism, Nature) 3:41–55

Ayuso-Álvareza A, Simón L, Nuñezc O et al (2019) Association between heavy metals and metalloids in topsoil and mental health in the adult population of Spain. Environ Res 179:108784

Beatty ME, Phelps S, Rohner MC, Weisfuse MI (2006) Blackout of 2003: Public health effects and emergency response. Public Health Rep (Washington, D.C.: 1974) 121(1):36–44

Bell JE, Brown CL, Conlon K, Herring S, Kunkel KE, Lawrimore J, Luber G, Schreck C, Smith A, Uejio C (2018) Changes in extreme events and the potential impacts on human health. J Air Waste Manage Assoc 68(4):265–287

Berry HL, Bowen K, Kjellstrom T (2010) Climate change and mental health: a causal pathways framework. Int J Public Health 55(2):123–132

Brown RC, Trapp SK, Berenz EC, Bigdeli TB, Acierno R, Tran TL, Trung LT, Tam NT, Tuan T, Buoi LT, Ha TT, Thach TD, Amstadter AB (2013) Pre-typhoon socio-economic status factors predict post-typhoon psychiatric symptoms in a Vietnamese sample. Soc Psychiatry Psychiatr Epidemiol 48(11):1721–1727

Castelloe M (2018) Coming to terms with ecoanxiety; growing awareness of climate change. Psychol Today. https://www.psychologytoday.com/au/blog/the-me-in-we/201801/coming-terms-ecoanxiety

Chalupka S, Anderko L, Pennea E (2020) Climate change, climate justice, and children's mental health: a generation at risk? Environ Justice 13:10–14

Cianconi P, Betrò S, Janiri L (2020) The impact of climate change on mental health: a systematic descriptive review. Front Psych 11

Clayton S, Manning C, Krygsman K, Speiser M (2001) Mental Health and Our Changing Climate: Impacts, Implications, and Guidance. American Psychological Association and Eco America, Technical Report

Climate Psychology Alliance (2021) Handbook of climate psychology. https://www.climatepsychologyalliance.org/

Correia AW, Peters JL, Levy JI, Melly S, Dominici F (2013) Residential exposure to aircraft noise and hospital admissions for cardiovascular diseases: multi-airport retrospective study. BMJ 347:f5561–f5561

Cunsolo A, Ellis NR (2018) Ecological grief as a mental health response to climate change-related loss. Nat Clim Chang 8(4):275–281

da Silva JA, Siegmund G, Bredemeier J (2015) Crisis interventions in online psychological counseling. Trends Psychiatry Psychother 37(4):171–182

Dattilio MF, Freeman A (2004) Estratégias cognitivo-comportamentais de intervenções em crise, 2nd edn. Artmed, Porto Alegre

DeSalvo KB, Hyre AD, Ompad DC et al (2007) Symptoms of posttraumatic stress disorder in a New

Orleans workforce following Hurricane Katrina. J Urban Health 84(2):142–152

Deshmukh A, Ho Oh E, Hastak M (2011) Impact of flood damaged critical infrastructure on communities and industries. Built Environ Project Asset Manag 1(2):156–175

Di Nardo F, Saulle R, La Torre G (2010) Green areas and health outcomes: a systematic review of the scientific literature. Ital J Public Health 7(4):402–413

Ebi KL, Bowen K (2016) Extreme events as sources of health vulnerability: drought as an example. Weather Climate Extremes 11:95–102

Eisenman D, McCaffrey S, Donatello I, Marshal G (2015) An ecosystems and vulnerable populations perspective on Solastalgia and psychological distress after a wildfire. EcoHealth 12(4):602–610

European Migrant Entrepreneurship Network (EMEN) (2020) A project to build strong support for Europe's migrant entrepreneur. http://emen-project.eu/

Fritze JG, Blashki GA, Burke S, Wiseman J (2008) Hope, despair and transformation: climate change and the promotion of mental health and wellbeing. Int J Ment Health Syst 2:13

Galderisi S, Heinz A, Kastrup M, Beezhold J, Sartorius N (2017) A proposed new definition of mental health. Psychiatr Pol 51(3):407–411. https://doi.org/10.12740/PP/74145

Galway LP, Beery T, Jones-Casey K, Tasala K (2019) Mapping the solastalgia literature: a scoping review study. Int J Environ Res Public Health 16(15):2662

Gifford E, Gifford R (2016) The largely unacknowledged impact of climate change on mental health. Bull At Sci 72:292–297

Gislason MK, Kennedy AM, Witham SM (2021) The interplay between social and ecological determinants of mental health for children and youth in the climate crisis. Int J Environ Res Public Health 18(9):4573

Gruebner O, Rapp MA, Adli M, Kluge U, Galea S, Heinz A (2017) Cities and mental health. Deutsch Arztebl Int 114(8):121–127

Hackbarth M, Pavkov T, Wetchler J, Flannery MJ (2012) Natural disasters: an assessment of family resiliency following hurricane Katrina. Marital Fam Ther 38(2):340–351

Hayes K, Poland B (2018) Addressing mental health in a changing climate: incorporating mental health indicators into climate change and health vulnerability and adaptation assessments. Int J Environ Res Public Health 15(9):1806

Hori M, Schafer MJ (2010) Social costs of displacement in Louisiana after hurricanes Katrina and Rita. Popul Environ 31(1–3):64–86

IPCC (2014) Climate Change 2014: Synthesis Report. Contribution of Working Groups I, II and III to the Fifth Assessment Report of the Intergovernmental Panel on Climate Change. Core Writing Team, Pachauri RK, Meyer LA (eds.). p. 151. https://www.ipcc.ch/report/ar5/syr/

Jang S, Ekyalongo Y, Kim H (2021 Feb) Systematic review of displacement and health impact from natural disasters in Southeast Asia. Disaster Med Public Health Prep 15(1):105–114

Keyes CLM (2014) Mental health as a complete state: how the salutogenic perspective completes the picture. In: Bauer GF, Hämmig O (eds) Bridging occupational, organizational and public health. Springer, Dordrecht, pp 179–192

Kovats RS, Hajat S (2008) Heat stress and public health: a critical review. Annu Rev Public Health 29(1):41–55

Kumagai A, Tanigawa K (2018) Current status of the Fukushima health Managment survey. Radiat Prot Dosim 182(1):31–39

Lee AC, Maheswaran R (2011) The health benefits of urban green spaces: a review of the evidence. J Public Health (Oxf) 33(2):212–222

Makwana N (2019) Disaster and its impact on mental health: a narrative review. J Family Med Prim Care 8(10):3090–3095

Matthews V, Longman J, Berry HL, Passey M, Bennett-Levy J, Morgan GG, Pit S, Rolfe M, Bailie RS (2019) Differential mental health impact six months after extensive river flooding in rural Australia: a cross-sectional analysis through an equity lens. Front Public Health 7

Morganstein JC, Ursano RJ (2020) Ecological disasters and mental health: causes, consequences, and interventions. Front Psych 11

Mullins JT, White C (2019a) Temperature and mental health: evidence from the spectrum of mental health outcomes. Discussion Paper Series IZA DP No. 12603.

Mullins JT, White C (2019b) Temperature and mental health: Evidence from the spectrum of mental health outcomes. J Health Econ 68:102240

National Wildlife Federation (2011) The Psychological Effects of Global Warming on the United States: And Why the U.S. Mental Health Care System is Not Adequately Prepared. National Forum and Research Report, February 2012

Patel V, Chisholm D, Parikh R et al (2016) Addressing the burden of mental, neurological, and substance use disorders: key messages from disease control priorities. Lancet 387:1672–1685

Quinn A, Shaman J (2017 Jul) Health symptoms in relation to temperature, humidity, and self-reported perceptions of climate in New York City residential environments. Int J Biometeorol 61(7):1209–1220

Rhodes J, Chan C, Paxson C, Rouse CE, Waters M, Fussell E (2010) The impact of hurricane Katrina on the mental and physical health of low-income parents in New Orleans. Am J Orthopsychiatry 80(2):237–247

Sánchez-Guevara Sánchez C, Sanz Fernández A, Núñez Peiró M, Gómez MG (2020) Energy poverty in Madrid: data exploitation at the city and district level. Energy Policy 144:111653

Sasaki Y, Aida J, Tsuji T, Koyama S, Tsuboya T, Saito T, Kondo K, Kawachi I (2019) Pre-disaster social support is protective for onset of post-disaster depression: prospective study from the Great East Japan Earthquake & Tsunami. Sci Rep 9(1):19427

Seneviratne SI, Nicholls N, Easterling D, Goodess CM, Kanae S, Kossin J, Luo Y, Marengo J, McInnes K, Rahimi M, Reichstein M, Sorteberg A, Vera C, Zhang X (2012) Changes in climate extremes and their impacts on the natural physical environment. In: Field CB, Barros V, Stocker TF, Qin D, Dokken DJ, Ebi KL, Mastrandrea MD, Mach KJ, Plattner G-K, Allen SK, Tignor M, Midgley PM (eds) Managing the risks of extreme events and disasters to advance climate change adaptation. A Special Report of Working Groups I and II of the Intergovernmental Panel on Climate Change (IPCC). Cambridge University Press, Cambridge/New York, NY, pp 109–230

Sharifi V, Amin-Esmaeili M, Hajebi A et al (2015) Twelve-month prevalence and correlates of psychiatric disorders in Iran: the Iranian mental health survey 2011. Arch Iran Med 18:76–84

Shigemura J, Terayama T, Kurosawa M, Kobayashi Y, Toda H, Nagamine M, Yoshino A (2021) Mental health consequences for survivors of the 2011 Fukushima nuclear disaster: a systematic review. Part 1: psychological consequences. CNS Spectr 26(1):14–29

Tracy M, Norris FH, Galea S (2011) Differences in the determinants of posttraumatic stress disorder and depression after a mass traumatic event. Depress Anxiety 28(8):666–675

United Nations Office for the Coordination of Humanitarian Affairs (2017). HURRICANE SEASON 2017. OCHA. https://www.unocha.org/hurricane-season-2017

Vigo D, Thornicroft G, Atun R (2016) Estimating the true global burden of mental illness. Lancet Psychiatry 3:171–178

Wang PS, Gruber MJ, Powers RE, Schoenbaum M, Speier AH, Wells KB, Kessler RC (2007) Mental health service use among hurricane Katrina survivors in the eight months after the disaster. Psychiatr Ser (Washington, D.C.) 58(11):1403–1411

Watts N, Amann M, Ayeb-Karlsson S, Belesova K, Bouley T, Boykoff M, Byass P, Cai W, Campbell-Lendrum D, Chambers J et al (2018) The Lancet Countdown on health and climate change: from 25 years of inaction to a global transformation for public health. Lancet 391:581–630

Winders WT, Bustamante ND, Garbern SC, Bills C, Coker A, Trehan I, Osei-Ampofo M, Levine AC, GEMLR (2021) Establishing the effectiveness of interventions provided to first responders to prevent and/or treat mental health effects of response to a disaster: a systematic review. Disaster Med Public Health Prep 15(1):115–126. https://doi.org/10.1017/dmp.2019.140. Epub 2020 Feb 14. PMID: 33870882

World Health Organization (2004) Promoting mental health: concepts, emerging evidence, practice (summary report). World Health Organization, Geneva

World Health Organization (2006) Preventing disease through healthy environments. WHO, Geneva. https://www.healthypeople.gov/2020/topics-objectives/topic/environmental-health#one

World Health Organization (2014) Gender, climate change and health (No. 9789241508186). World Health Organization, Geneva. http://www.who.int/gender-equity-rights/knowledge/gender-climate-change-and-health/en/

Planetary Health and Healthcare Workers

12

Giuseppe La Torre, Barbara Dorelli, Alice De Paula Baer, Domenico Barbato, Lorenza Lia, and Maria De Giusti

Abstract

In this chapter, there is a description of the link between planetary health and healthcare workers. The Planetary Health Alliance (PHA) is a consortium of over 240 universities, non-governmental organizations, research institutes, and government bodies from more than 40 countries in the world, whose aim is to understand global climate and environmental change and the impact on population health.

The Planetary Health for Physicians is an initiative that aims to influence health professionals to help them understand the effects of global environmental change on patients and increase education and information available on this issue.

Moreover, the One Health is presented as a worldwide strategy to increase inter-sectorial collaborations in all aspects of healthcare for humans, animals, and the environment, with the aim to "forge co-equal," all-inclusive collaborations among physicians, veterinarians, nurses, and other scientific health and environmentally related disciplines.

Finally, the chapter ends with a systematic review, conducted to summarize the results of all the studies that focused on the collaborations between human medicine and veterinary medicine in Europe with the aim to protect human–animal–environmental health. Literature searches were performed using PubMed and Scopus databases. Forty-Seven articles containing the word One Health were retrieved. The main domains showing integrated approaches were: (a) zoonosis, (b) vaccines, (c) antibiotic resistance, (d) implementations of professional network. The historical background of One Health was consistent with the joint actions and fair collaboration among human, veterinarian, and environmental institutions. In particular, 21 papers deal with accomplished interventions within the One Health approach; 23 articles relate to potential/desirable future collaboration; 3 articles describe historical aspects of the One Health. Although an increasing attention has been recently paid to One Health program, many European countries show lack of cooperation among different sectors. Creating a specific founding system for One Health initiatives could help to overcome all the barriers with the aim of a multidisciplinary European attitude.

G. La Torre (✉) · B. Dorelli · A. D. P. Baer
D. Barbato · L. Lia · M. De Giusti
Department of Public Health and Infectious Diseases,
Sapienza University of Rome, Rome, Italy
e-mail: giuseppe.latorre@uniroma1.it

Keywords

Great transition · Healthcare workers · One Health · Planetary Health Alliance · Planetary health for clinicians

12.1 Planetary Health Alliance

The Planetary Health Alliance (PHA) is a consortium of over 240 universities, nongovernmental organizations, research institutes, and government bodies from more than 40 countries in the world, whose aim is to understand global climate and environmental change and the impact on population health.

In this context, the concept of Anthropocene, i.e., geological era that starting from the beginning of significant human impact on nature, geology, and ecosystems of the earth, is crucial. The Alliance wants to deal not only with climate change but also with declining biodiversity, shortages of arable land and freshwater, pollution, and changing biogeochemical flows (Planetary Health Alliance 2021), through its engagement in advancing planetary health education, policy, and research.

This consortium, launched in 2016 with the support of the Rockefeller Foundation, is based at Harvard University and involves the Harvard University Center for the Environment and the Harvard T.H. Chan School of Public Health. PHA has a precise vision according to which people protect and regenerate earth's natural systems for the future generations.

Elements of the PHA mission include the promotion, mobilization, and leading of a transdisciplinary field of planetary health and its diverse components, such as science, stories, solutions, and communities to achieve the so-called Great Transition, characterized not only by the absence of poverty, war, and environmental destruction, but also attention to egalitarian social and ecological values, attention to increase inter-human connectivity, improve the quality of life of the populations and the planet's health.

To implement PHA mission, the consortium aims at:

– Encouraging and empowering a planetary health community with the interplays of different disciplines and sectors, as well as generations, worldviews, and geographies;
– Facilitating education and training of current and next-generation planetary health practitioners;
– Implementing the collaboration between the main planetary health stakeholders, such as civil society, private sector, general public, and governments;
– Promoting actionable steps that single individuals and whole society can make to reach progress toward planetary health.

Box 12.1: Values of the Public Health Alliance
(adapted from https://www.planetary-healthalliance.org/mission-vision)

Hope: Every dimension of human activity is rich with solutions that can help bring humanity back into balance with our natural systems.

Urgency: PHA is a community that needs to act in urgency, since there is the feeling that the world has only one generation in order to transform society into balance with earth's natural systems.

Science-driven innovation: The scientific evidence on the strict relationship between human well-being and the state of our natural systems is the crucial element for planetary health innovation. PHA has the aim to support the growth of the evidence, to prepare and synthetize it for those who are interested in planetary health, and to disseminate it to policymakers, as well as other potential stakeholder.

Justice, equity, and compassion: PHA believes that justice, equity, and compassion are the driving elements behind a future of planetary health for all. It is important to recognize that next generations and populations with the fewest resources, that are often least responsible for environmental changes, are affected by the health burden of degraded natural systems and suffer from vast disparities and structural inequities worldwide.

Partnership and inclusivity: PHA promotes transdisciplinary, multicultural, and intergenerational partnerships between different communities in order to share ideas, explore different perspectives, and collaborate with the aim of enabling people to act as change-makers.

Humility: It is important to recognize that human beings do not have all the answers, and to achieving the Great Transition will require the interconnection of knowledge coming from different partners.

12.1.1 Planetary Health for Clinicians

For Earth Day on April 22, 2019, the *Lancet* published an article entitled "A Call for Clinicians to Act on Planetary Health" revealing the unprecedented support of the scientific community for a healthier planet (Veidis et al. 2019).

The initiative, under the lead of the PHA, has already been joined by 29 medical associations and international health organizations, including Sapienza University, and aims to raise awareness on the impact that global environmental change has on human health by mobilizing medical communities.

The call to action aims to influence health professionals to help them understand the effects of global environmental change on patients and increase education and information available on this issue. In Fig. 12.1, there is the illustration of the areas of planetary health behavior change and action to share with the patients.

Who are the potential healthcare workers interested? The answer is all the clinicians that provide healthcare for individuals and communities. Among these healthcare workers, we need to consider physicians, physician assistants, nurses, nurse practitioners, midwives, dietitians, traditional healers, and community health workers.

The approach chosen by the clinicians for planetary health aims at (Clinicians for Planetary Health (C4PH) 2021):

- Developing materials for healthcare professionals that encourage individual- and community-level planetary health action, with an emphasis on the benefits that can be derived for both health and environment;
- Making periodic calls to brainstorm, network, and exchange all the available resources useful to increase commitment to planetary health among the healthcare professionals;
- Sharing resources and stories of success through the newest way of communicating, including newsletters, blog series, and podcast.

In this context, since healthcare professionals are among the people most listened to by citizens, it is important to use evidence-based information to guide knowledge increase and behavior change. And among these information, those pertaining lifestyles are really impacting. According to the (American College of Lifestyle Medicine 2021), the lifestyle medicine (Fig. 12.2) can be considered as an evidence-based approach to preventing, treating, and even reversing diseases through the replacement of unhealthy behaviors with positive ones including:

- Eating healthfully: The main concept is that food is medicine. Following this approach, people need to choose predominantly whole, plant-based foods that are rich in fiber and nutrient dense. The Mediterranean diet is the diet based on the following dietetic pattern (Saulle and La Torre 2010): (a) high intake of vegetables, pulses (beans, lentils, etc.), fruit, and cereals; (b) medium-high intake of fish; (c) low intake of meat and saturated fat; d) high intake of unsaturated fat (particularly olive oil); (e) medium-low intake of dairy pro-

- Topic 1: Reconnecting (with community, with nature, green spaces)
- Topic 2: Transport (active transport, public transport, reduce flying)
- Topic 3: Food (eat plants, local consumption, reduce waste)
- Topic 4: Energy (clean energy, reduce use, redesign systems)
- Topic 5: Consumption (fix what you have, buy less, buy ethical, rec)
- Topic 6: Equality and Justice (racial, gender, social)

Fig. 12.1 Areas of planetary health behavior change and action (adapted by Clinicians Community Call 2020)

duces (mainly yogurt and cheese), (f) a moderate intake of wine.
– Being physically active: Regular and consistent physical activity, to be maintained on a daily basis throughout life, is an essential element of a correct lifestyle. According to WHO (2021), the recommendation for adults of age between 18 and 64 years is to make at least 150–300 min of moderate intensity (i.e., walking, heavy cleaning (washing windows, vacuuming, mopping), or light bicycling) or 75–150 min of vigorous intensity activity (i.e., jogging, carrying heavy loads, bicycling fast or aerobics) weekly along with two or more days weekly of strength training.
– Managing stress: Stress can have a double face, on the one hand leading to improve health and productivity, on the other causing

Fig. 12.2 The inter-connections of the elements of the lifestyle medicine

LIFESTYLE MEDICINE FOCUSES ON 6 AREAS TO IMPROVE HEALTH

- Increase PHYSICAL ACTIVITY
- HEALTHFUL EATING of whole, plant-based food
- Develop strategies to MANAGE STRESS
- Avoid risky SUBSTANCES
- Form & maintain RELATIONSHIPS
- Improve your SLEEP

unhealthy conditions, such as anxiety, depression, obesity, cardiovascular diseases. The object of lifestyle medicine in this context is to use techniques that can be useful to improve well-being, including listening to relaxing music, practicing exercise or dance, mindfulness, and yoga (La Torre et al. 2020a, b; Cocchiara et al. 2019), taking time for fun, creative activities or hobbies, taking care of spiritual needs, having time to laugh, using green spaces to influence both the perceived health and the objective physical conditions in a measurable way (Di Nardo et al. 2010).

– Avoiding risky substance abuse: There is strong evidence of the negative health impact related to the use of any addictive substance, including many cancers and cardiovascular diseases. Healthy behaviors that are related to improved population health include tobacco smoking cessation and lowering alcohol intake.

– Adequate sleep: Poor-quality sleep can have an impact on health. There is evidence that suggests a link between poor sleep quality and overweight or obesity in young subjects (Fatima et al. 2016), while in adults poor sleep is also linked to worse well-being and decreasing in performance, productivity, and safety at work (Garbarino et al. 2019). So, there is a strong need to identify dietary, environmental, and coping behaviors to improve sleep quality. Older adults who listened to music experienced significantly better sleep quality than those who did not listen to music (Chen et al. 2021). Moreover, there is evidence that mind–body interventions can be considered as a treatment option for patients with sleep disturbance (Neuendorf et al. 2015).

– Having a strong support system: Social relationships are essential to emotional resiliency. This is not only important for children (Gartland et al. 2019) but also for adults and older people. There is strong evidence that loneliness and social isolation are strongly associated with high level of mortality (Hodgson et al. 2020).

12.2 Knowledge and Attitudes Toward Climate Changes in Healthcare Personnel

Healthcare workers are not only important for the general population because they act as models, but can deeply contribute in making recommendations and supporting favorable policies as they have the expertise to recognize the health consequences related to climate changes, and they have a strong impact on the public opinion (Xie et al. 2018).

Many studies show that the general population, and in particular the healthcare professionals, are sufficiently aware of climate changes and its effects, and mostly could identify individual practices that could help to mitigate its repercussions. La Torre et al. (2020a, b) found significant differences on the amount of information regarding the consequences of global warming among HCW mainly related to the region of residence and to gender, with females having lower odds of giving the correct answers. Most of the participants had already heard about CC, with the main sources of information being TV and school/university. Similar results from this study show were found by a study conducted in China with health professionals (Wei et al. 2014), in which TV also appeared as the main source of information. The importance of mass media is also highlighted in a survey conducted in Bangladesh (Kabir et al. 2016), while the key role of school as a source of information appears in a study made with Iranian students, in which school was the main source, with 38.5% of answers (Yazdanparast et al. 2013). However, it is also important to underline the role of social media in this field. Lewandowsky et al. (2019) put the attention on the role of Internet blogs that became a very useful tool for discussing scientific issues, and CC is now one of the most chosen in the discussions. These authors believe that the use of blogs, and particularly the comment sections of blogs, can play a very important role in disseminating different positions around this issue.

However, we need to recognize that television coverage of public health issues has problems, such as individual selection of information of viewers, journalists' unfamiliarity with the topics, and spread of misinformation (Gollust et al. 2019). Taking this into account, television should be used carefully, and it should be as well important to enhance the key role that educational institutions play, being a more reliable information disseminator.

A research conducted at Yale University showed that, for information about climate change-related health problems, Americans mostly trust their primary care doctor (Leiserowitz et al. 2014). Another study carried out in the USA found that the public health community has an important perspective about climate change that, if shared, could help the public to better understand these issues. Their findings also suggest that the communication should not be focused on the problem of climate change, but on solutions and co-benefits: a healthier future offers environmental benefits (Maibach et al. 2010).

The potential of health professionals as disseminators of information on global warming, according to the results of the Italian study (La Torre et al. 2020a, b), seems to be underused. Concerning the causes of global warming, more than 50% of participants understood that greenhouse gases were CO_2, N_2O, and CH_4, although a significantly amount choose only CO_2. Still, on the matter of greenhouse gases, most of the respondents (92.5%) were aware of human's responsibility on their emissions and on scientists' agreement on the subject, showing a positive consonance between Italian population's knowledge and scientific consensus.

Moreover, in the USA research carried out by Reynolds et al. (2010), more than one third of participants mentioned mainly anthropogenic causes as "things that could cause global warming," such as cars and industries, and 26% specifically mentioned fossil fuel use. Similar reasons were mentioned by the Chinese participants (Wei et al. 2014). However, in a study conducted among nursing students in the USA, 18% of the responders affirmed that natural causes were also primary drivers of global warming, and also in the study made with nursing students in Arab countries, respondents believed that climate change was due to a balance between nature and human causes (Felicilda-Reynaldo et al. 2018). In the Arab region, most of the respondents among nursing faculty said that all presented health-related effects had already increased due to climate change; similar findings were presented by a study conducted in Montana with nursing students (Streich 2014).

12.3 One Health Approach

The concept "One World, One Health" has recently appeared, indicating that the world is increasingly aware of the need to link animal diseases, public health, and environment.

The need for collaboration of the human, animal, and environmental health sectors is motivated by the increase in the emerging human infectious diseases that recognize a zoonotic origin and the public health problem of the resistance of microorganisms to antimicrobial drugs (Sikkema and Koopmans 2016).

One Health is a worldwide strategy to increase inter-sectorial collaborations in all aspects of healthcare for humans, animals, and the environment, with the aim to "forge co-equal," all-inclusive collaborations among physicians, veterinarians, nurses, and other scientific health and environmentally related disciplines (One Health Initiative 2021).

Many examples show how the health of people is related to the health of animals and the environment: Rabies, Salmonellosis, West Nile Virus Fever, Q Fever. In the late nineteenth century, Rudolf Virchow (1821–1902) coined the term "zoonosis" and stated "Between animal and human medicine there are no dividing lines – nor should there be" (Klauder 1958). Zoonotic diseases are a growing concern: approximately 60% of existing human pathogens and over 75% of those that have appeared during the past two decades can be traced back to animals (Dehove 2010).

Greater progress in prevention and control of infectious diseases requires a more direct effort focusing on the complex interplay of human health, the health of animals, and the environment (CDC 2018a, b). This multidisciplinarity involves integration between microbiology and immunology, genetics and genomics, entomology and ecology, and the social sciences, among other disciplines (Dehove 2010).The goal of One Health is to promote the cooperative efforts of different disciplines—working locally, nationally, and globally—to achieve the best health for people, animals, and our environment. Considering that 6 out of every 10 infectious diseases in humans are spread from animals (CDC 2018a, b) the One Health approach, with the synergism between the human and veterinary medicine is considered a winning strategy in the framework of public health practice (Mantovani 2008).

The aim of this study is to perform a systematic review of published scientific articles dealing with collaboration between human and veterinary medicine in Europe.

12.4 Methods

In order to perform this review, we followed the conceptual framework of the PRISMA statement (Preferred Reporting Items Systematic Reviews and Meta-analyses) to evaluate the studies concerning healthcare interventions, multidisciplinary collaboration for active surveillance against zoonotic diseases, and surveillance foodborne disease (Liberati et al. 2009).

The systematic research was conducted using both medical databases Scopus and PubMed (Medline) using the algorithm: "human AND veterinary medicine AND collaboration."

In addition, relevant studies were searched in the references of identified articles. No restrictions of language or years of publication were applied; the geographical set was limited to the UE. Only published articles are included in this systematic review.

Research was initiated on the 24th of July 2017. Scientific papers were managed using the program JabRef 2.8.1.With the aim of investigating the European model, the focus was on the active surveillance against zoonotic diseases (including food-borne diseases).

12.4.1 Study Selection

Two researchers independently selected the identified studies and evaluated the inclusion criteria in title and abstract. Subsequently, they assessed the full-text against the eligibility criteria. Disagreement was resolved by consensus or by a third researcher.

12.4.2 Eligibility Criteria

Studies that took into account a collaboration between human and veterinary medicine in Europe were considered eligible. All study designs were included.

12.4.3 Data Collection Process

Data extracted from the articles included first author name, author's affiliation (membership organization), publication year, journal name, title, type of outcome (see previous paragraph "Eligibility criteria"), nationality, type of publication (study design, letter, viewpoint, etc.), and main results. The analysis was performed with the researchers working in pairs and confronting their output about each paper. Disagreements about the choice of the papers to be included in the review were solved by discussion or by a third researcher.

12.5 Results

12.5.1 Study Selection

The selection of articles, shown in the flowchart, was performed according to the PRISMA statement (Fig. 12.3). Gray literature was not included in the review (7).

Overall 657 papers were found, 416 articles through PubMed, 241 through Scopus. Successively, 139 duplicates and 368 articles that did not meet the inclusion criteria were excluded. The remaining 150 papers were analyzed, and from these, 103 articles with no pertinent full text were removed. For the analysis, 47 papers were finally considered.

12.5.2 Characteristics of the Studies

During the literature search, all the papers were divided into three big macro-areas:

1. *Accomplished interventions*: all actions jointly implemented by human and veterinary medicine
2. *Future perspectives*: planned or desirable but not yet realized jointly interventions by human and veterinary medicine
3. *Historical background*: historical aspects of the "One Health" project and its origin

The characteristics of the included studies are summarized in Table 12.1.

The first macro-area was further divided into four subcategories of interest, on the basis of the main topic of the papers:

1a. *Zoonoses*: diseases that are transmissible from animals to humans through direct contact or though food, water, and environment;
1b. *Vaccines*: biological preparation that improves immunity to a particular disease in humans and animals;
1c. *Networks*: implementations of network of professionals in the field of health and environment;
1d. *Antimicrobial resistance*: it occurs when microorganisms develop resistance mechanism toward drugs used to treat human or animal infection making them ineffective.

The second macro-area was divided into four subcategories of interest:

2a. *Zoonoses*
2b. *Vaccines*

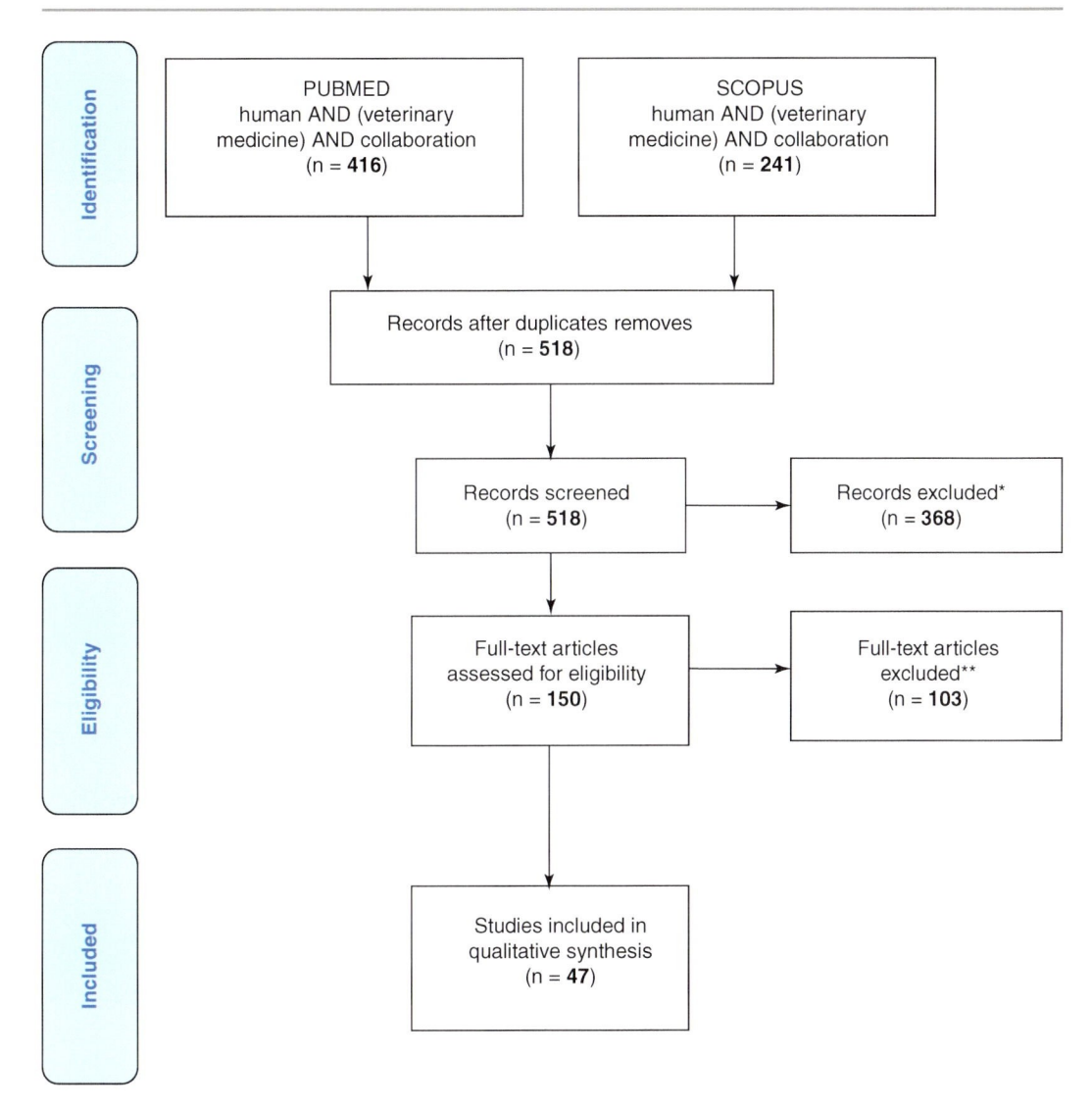

```
Identification:
  PUBMED
  human AND (veterinary
  medicine) AND collaboration
  (n = 416)

  SCOPUS
  human AND (veterinary
  medicine) AND collaboration
  (n = 241)

Screening:
  Records after duplicates removes
  (n = 518)

  Records screened
  (n = 518)            → Records excluded*
                          (n = 368)

Eligibility:
  Full-text articles
  assessed for eligibility
  (n = 150)            → Full-text articles
                          excluded**
                          (n = 103)

Included:
  Studies included in
  qualitative synthesis
  (n = 47)
```

Fig. 12.3 Flow diagram of One Health search strategy. *The papers were removed because they do not consider in the title or the abstract the term "One Health." **The papers were removed because they do not resect the inclusion criteria

2c. *Antimicrobial resistance*

2d. *Generics*: this subcategory includes papers describing hypothetical future collaborations in different areas of intervention.

The first study was published in 1978 and the last one in 2017.

12.5.2.1 Accomplished Interventions

Out of 47 assessed articles, 21 included concrete intervention of collaboration between human and veterinary medicine. Of these 21, 11 articles dealt with zoonosis, 4 vaccines, and 3 networks, while the remaining 3 antimicrobial resistance.

1a. *Zoonoses*

Over the last year, zoonotic pathogens caused (globally?) more than two million deaths and over two billion cases of human illness (CDC, 2018). Focusing on the past decade, more than 40 new zoonotic diseases have emerged. Besides, the ever more rapid

Table 12.1 Summary of the characteristics of the included studies

Author (year)	Realized interventions				Future perspectives				Historical background
	Z	V	N	AMR	Z	V	G	AMR	
Beveridge (1978)							X		
Battelli and Scorziello (1992)	X								
Foley-Nolan et al. (1998)	X								
Schillhorn Van Veen (1998)	X								
Cripps (2000)	X								
Rotivel (2003)	X								
Donaldson and Reynolds (2005)					X				
Michell (2005)							X		
Pawlowski et al. (2005)					X				
Reynolds and Donaldson (2005)		X							
Battelli (2006)	X								
Busani et al. (2006)			X						
Poglayen (2006)					X				
van den Berg et al. (2008)						X			
Friedrich et al. (2008)			X						
Mantovani (2008)									X
Pawlowski (2008)					X				
Gibbs and Anderson (2009)		X							
Kahn et al. (2009)	X								
Hristovski et al. (2010)							X		
Valkanova et al. (2010)	X								
Meisser et al. (2011)							X		
Bartonova (2012)			X						
Bergström et al. (2012)			X						
Paphitou (2013)								X	
Bresalier and Worboys (2014)		X							
Corning (2014)	X								
FVE (2014)							X		
Honey (2014)							X		
Woods and Bresalier (2014)									
Harries (2015)									X
Jarvis (2015)									
Johansen et al. (2015)					X				
McGinn (2015)							X		
Speksnijder (2015)			X						
Stärk et al. (2015)	X								
Wendt et al. (2015)									
Aboul-Enein (2016)		X			X				
Blake (2017)									
Sikkema and Koopmans (2016)					X				X
van de Burgwal et al. (2017)					X				
Wendt et al. (2016)			X						
Cleaveland et al. (2017)					X				
Eussen et al. (2017)									
Gossner et al. (2017)	X								
Mardones et al. (2017)									
Rabinowitz et al. (2017)									

Legend: *Z* zoonosis, *V* vaccines, *N* networks, *AMR* antimicrobial resistance changes at the animal–human–ecosystems interface influence the evolution of pathogens, leading to their possible mutation and recombination, ultimately resulting in increased pathogenicity (Corning 2014).

The oldest study of this specific field was performed by Battelli and Scorziello (1992), who li identified four fields of application for One Health collaboration: (1) applied research: planning and conduct of epidemiological surveys; (2) information and monitoring; (3) planning and management of control measures: experience of collaboration for the brucellosis test was achieved in the Campania region, with joint training courses been delivered and a multidisciplinary group created for the implementation of activities; (4) health education.

Brucellosis was also addressed by Valkanova et al. (2010): these authors studied the spread of brucellosis in Bulgaria from 1950 to 2007. Collaboration of human and veterinary specialists contributed to the fast and proper implementation of countermeasures.

Foley-Nolan et al. (1998) analyzed four cases of Salmonella infection managed by the collaboration between veterinary, public health, medical and environmental health professionals. In the first two cases, communication did not take place early on and disease in animals and humans progressed simultaneously (Foley-Nolan et al. 1998).

In 1998, Schillhorn van Veen (1998) explained the points of convergence between human and veterinary medicine, particularly the pathophysiology (in the 1970s, the research on Feline Leukemia Virus helped to understand the pathophysiology of HIV, due to their similarities), zoonoses (cerebral cysticercosis can be preventable by treating pigs with anthelmintic), zoo-prophylaxis (cowpox and smallpox), and public health (Schillhorn van Veen 1998).

Cripps (2000) faced the problem of veterinary education by redefining the undergraduate course to include: Veterinary Public Health, a veterinary public health master course, interdisciplinary researches, and meetings.

Rotivel (2003) wrote on the eradication of rabies in France in 2001 in terrestrial animals: Rabies control and surveillance are carried out by both human and veterinary authorities. The diagnosis of rabies in animals is carried out in the National Reference Centre for Rabies (NRCR). All the data regarding rabies diagnosis in animals are collected by AFSSA (Agence Française de Sécurité Sanitaire des Aliments). Data concerning human rabies are collected by the NRCR, which depends on the Human Health Authority. They also publish a bulletin annually, which gathers all these data.

In Italy, occupational zoonoses were observed in epidemiological surveys, mainly performed in collaboration with physicians. Health education programs for the control of occupational zoonoses were organized, many of which in cooperation with WHO Collaborating Centers, ISS, Experimental Zoo-prophylactic Institutes, Regional and Local Health Services, and Italian Farmers Union (Battelli et al. 2018).

Other peculiar forms of zoonoses were studied, like the "new" Bovine Amyloidotic Spongiform Encephalopathy (BASE). BASE, localized in the forebrain, has at least the same human health implications as those of classical and more known bovine spongiform encephalopathy (Kahn et al. 2009).

In France, the World Organization for Animal Health (OIE) plays a consistent role in minimizing animal and public health risks attributable to zoonoses. The OIE works in collaboration with the WHO and FAO. Collaboration became apparent in 2003, when a new strain of avian influence (HPAI/H5N1) spread through Asia, Africa, and Europe, and even more in 2009 during the pandemic event implicating a novel H1N1 influenza virus recombinant (Corning 2014).

In Europe, the European Food Safety Authority (EFSA), responsible for animal health and food safety risk assessment, and the European Center for Disease Control (ECDC) are responsible for public health (jointly publish the annual zoonoses report). These agencies collaborate to scientifically support the investigation of translational food-borne and animal diseases outbreaks (e.g., Outbreak of *Schmallenberg* virus) (Stärk et al. 2015).

Gossner et al. (2017) described the human–animal–vector integrated approaches for monitoring and surveillance of West Nile virus (WNV) at national and European levels. A joint approach, involving public health, animal health, and environmental authorities offered the most effective mechanism for tackling WNV transmission. Austria, France, Greece, Italy, and the United Kingdom (UK) implemented different surveillance strategies based on their specific epidemiological situation.

1b. *Vaccines*

Four papers concerned the interventions related to vaccinations in the area of One Health.

Reynolds and Donaldson (2005) described a tangible example of health protection measures for the Avian Influenza in the United Kingdom through the use of regular flu vaccination for poultry workers and other at-risk groups. Examples of more collaborative work on zoonotic diseases were: (1) the development of common standard operating procedures by the veterinary and Public Health Laboratory Test Group for the isolation, identification, and typing of microorganisms and for testing of antibiotic sensitivity; (2) the development of shared databases incorporating phenotypic and molecular information from human and animal sources; (3) surveillance and the training of veterinary and public health staff.

Gibbs and Anderson (2009) showed an Italian experience directed by *Istituto Zooprofilattico Sperimentale delle Venezie*, whose director of the virology department, Ilaria Capua, was directly involved in managing several avian influenza epidemics. In 2000, she developed the DIVA (Differentiating Vaccinated from infected Animals) strategy, based on heterologous vaccination, to fight avian influenza. This strategy obtained the eradication of avian influenza at that time in Italy. Capua deposited her sequences in GenBank and shared it globally.

Bresalier and Worboys (2014) observed the canine distemper vaccine campaign conducted in Britain between 1922 and 1933 and based on collaborations with veterinary professionals, government scientists, the Medical Research Council (MRC), and a commercial pharmaceutical company.

The last article by *Aboul-Enein* et al. (2019) concerned diphtheria. The diphtheria antitoxin (DAT) was developed in the late nineteenth century and played a significant role in the history of Public Health and Vaccinology. Evidence of this comes from diphtheria epidemic in Alaska in 1925, when a coordinated emergency distribution of this life-saving drug by dog-sled kept in worst conditions, led to a profound cultural heritage in the annals of Public Health and Vaccinology (Aboul-Enein et al. 2019).

1c. *Networks*

Three articles on the implementations of network of professionals in the field of One Health were found.

Busani et al. (2006) reviewed the productive cooperation between Public Health and Veterinary institutes in a few countries, such as United Kingdom, Denmark, and Sweden. Inter-professional collaboration in zoonosis research is promoted by the European Union with the Med-Vet-Net network, which comprises 10 countries, with responsibilities for research and provision of advice and consultancy to their respective national governments about animal diseases and welfare, microbiological food safety, and human health risks.

Bartonova (2012) described HENVINET (Health and Environment Network) project, set up to create an interdisciplinary professional network aimed to bridge the communication gap between society and science. It involves 30 organizations from and outside Europe approaching the issue of interdisciplinary collaboration in four ways: (1) the Drivers-Pressures-State-Exposure-Effect-Action framework was used to structure information sharing; (2) interactive online tools were developed to enhance the methods for knowledge evaluation; (3) the measure of scientific agreement was done through quantification methods; (4) a web portal to facilitate collaboration and communication among scientists was developed using open architecture web technology (Bartonova 2012).

Wendt et al. (2015) carried out a literature review and 20 systems were found integrating disparate monitoring and surveillance data from humans and animals. Almost 50% of the systems are composed for surveillance on a word scale (Wendt et al. 2015). This reflects the One Health approach proposed as a global strategy, but most of them operate in North America (n = 6) or Asia (n = 4). Examples of network systems with a global field of interest are: ArboZoonet, Disease Bioportal, GLEWS, Global Food Safety Portal, NBIC, EpiSPIDER. Fifteen systems were set up to survey many diseases (later referred to as broad based) (Conraths et al. 2011).

1d. *Antimicrobial Resistance*

The antimicrobial resistance is one of the most serious problem for human and animal health. Three articles dealt with realized interventions.

Friedrich et al. (2008) describe SeqNet. org, an initiative of 44 laboratories from 25 European countries and one laboratory from Lebanon, founded in 2004. The main objective of the initiative was to establish a European network of excellence for sequence-based genomic typing of microorganisms and to generate unambiguous, easily comparable typing data in electronic form to be used for surveillance of sentinel microorganisms at a national and European level.

Bergström et al. (2012) described the primary epidemic of MRSA infection in horses in Sweden in 2008, and the implemented infection prevention controls, in particular: (1) intensive cleaning and disinfection; (2) isolation of infected horses; (3) collaboration among authorities in animal and human public health; (4) transitory interruption of elective surgery; (5) voluntary MRSA testing of staff; (6) basic hygiene and cleaning policies and staff training. The outbreak has been considered braked, and no new cases occurred for more than a year.

Speksnijder et al. (2015) described the Dutch experience on the management of antibiotic stewardship. In 2011, the Dutch Health Council formulated specific recommendations to prevent the development and spread of antimicrobial resistant bacteria in animal production. The antimicrobial use was reduced by 56% in farm animals in the Netherlands between 2007 and 2012, with the aim to achieve a 70% reduction in 2015.

12.5.2.2 Future Perspectives

Of the 47 articles, 23 included all the articles in which desirable interventions of collaboration between human and veterinary medicine have been proposed for the future, but not been realized yet. Among these, nine articles addressed zoonoses, one vaccines, 11 generic one health topics, and two antimicrobial resistance.

2a. *Zoonoses*

The oldest studies on future perspectives around zoonoses date from 2005. Reynolds and Donaldson (2005) explained that the emergence of new human variant Creutzfeldt–Jakob disease with the potential to link to the epidemic of bovine spongiform encephalopathy in cattle, required collaboration in assessing the potential human health risks. Cooperative relationships were vital to develop: (a) strategic risk assessment and risk management mechanisms at government level; (b) robust surveillance infrastructure; (c) analysis of data on human and animal infections; (d) sharing of these data with all those public health institutions (Donaldson and Reynolds 2005).

Pawlowski et al. (2005) worked on *Tenia solium*, in particular taeniasis and cysticercosis. The authors sustain that the eradication of *T. solium* neurocysticercosis in humans could not be achieved in the early future; however, regional elimination or control of the infection is realistic and urgently needed. The authors observed how the elimination of public health impact of neurocysticercosis in several endemic areas required joint action from medical and veterinary services (Pawlowski et al. 2005). The same author, in 2008, published a study on *Taenia solium* neurocysticercosis and ocular cysticercosis and emphasized the need to control them for medical and economic reasons. In endemic areas, simple operational interventions are proposed, based on oriented chemotherapy (namely niclosamide and praziquantel) using existing healthcare structures and improving collaboration between medical and veterinary services (Pawlowski 2008).

Poglayen (2006) illustrated the role of Veterinary Urban Hygiene (VUH), a cultural revolution in zoonoses approach designed by WHO Expert Consultation (1977). VUH is defined as the activity dealing with health aspects associated with human–animal–environment connections in urban areas (Mantovani 2001). The actual use of animals goes from pet therapy to psychological support in prisons and hospitals, from education in schools to the employment of animals in civil defense. A new domain of public veterinary activity raises the need of strong cooperation among diverse professional categories, such as physicians, biologists, healthcare providers, and others involved in setting animal housing compatible with animal welfare requirement.

Johansen et al. (2015) explained that helminths are mostly diagnosed by methods with low sensitivity and specificity, as well as identification of the eggs in the stool. Standardized test validation, interdisciplinary collaboration, and creation of an international One Health diagnostic platform could significantly contribute to control and elimination of these diseases, sharing best scientific evidences on diagnosis of helminth zoonosis.

Blake and Betson (2017) evidenced the contents of the 2015 Autumn Symposium, focused on One Health. The meeting involved specialists in parasitology and additional complementary experts. Scientists, policy makers, lawyers, and industry representatives were invited to be present at the meeting, promoting and developing One Health understanding with relevance to parasitology (Blake and Betson 2017). Van De Burgwal et al. (2017) explained that the strategies aimed at tackling rabies-specific innovation barriers are important for the development of advances in human rabies treatment and prophylaxis.

To prevent zoonotic diseases, the surveillance plays a major role. Wendt et al. (2016) sustained that in Germany there is no established surveillance system for conducting One Health surveillance. Data about surveillance on zoonoses are collected for the most part separately in different databases for humans or animals: there is a need of integrated surveillance.

Cleaveland et al. (2017) showed that One Health interventions may be more effective and may generate more equitable benefits for human health and livelihoods, in particular in rural areas, than approaches that depend exclusively on treatment of human cases.

2b. *Vaccines*

The review found only one article on the future perspectives of One Health collaboration and vaccines. van den Berg et al. (2008) explained that the final aim of poultry vaccination against H5 and H7 avian influenza subtypes is to stop transmission, in order to achieve efficient control and eradication. The authors sustained the need for a new generation of cost-effective and efficient poultry vaccine that can be applied in the future by mass immunizing methods (spray, drinking water) and induce strong local immunity, in order to control the shedding of the virus.

2c. *Generics*

Eleven articles were found in this subcategory.

The *Beveridge*'s study is the oldest, dated 1978. In this article, he emphasized the need and the importance of the collaboration between the medical and veterinary professions, regarding control of zoonoses; supervising of hygiene of food; detection and prevention of environmental pollution; exchange of research information; supply standardized laboratory animals. According to the author, partnership between these two health professions should be encouraged, for example, by sharing some courses during university education and by joint meetings (Beveridge 1978).

The review by Michell (2005) stated that the creation of the new Medical Research Council Comparative Clinical Science Panel in the United Kingdom will provide the infrastructure and strategic focus to facilitate comparative clinical research and to promote collaboration.

Hristovski et al. (2010) underlined the need of collaboration in One Health concept and how to apply it, for example, by including the concept of One Health in Public Health education and continuous training of health professionals to better cope with the existing and new medical challenges.

The study of Meisser et al. (2011) aimed to investigate the opportunities for implementation of the One Health concept in Switzerland. They had conducted face-to-face interviews with 16 experts in 2010. The following main categories emerged: current cooperation; assessment of the potential of One Health; barriers and bridges; proposals for further action (promotion of interconnections in health and the idea of One Health in the general population and in relevant institutions).

Eussen et al. (2017) focused on the collaboration between human and veterinary healthcare professionals. The administration of 368 questionnaires to health professionals, based on Gaertner and Dovidio's Common Ingroup Identity Model, was conducted to assess the relation between collaboration and common goal. The study's findings indicated that, in order to achieve greater collaboration, it is first and foremost necessary to define a common goal.

Recently, Rabinowitz et al. (2017) affirmed that One Health education efforts in medical schools are in their first stages and are far from veterinary schools, in which One Health represents one of the main parts of their curricula.

There were also reports of meetings, symposium, and conferences. A meeting of European doctors and veterinarians took place in Bruxelles in 2014 and organized jointly by the Federation of Veterinarians of Europe and the Standing Committee of European Doctors focused on 'One Health'. The conclusions emphasized the need for One Health, global strategic approach together with inter-sectorial collaboration and coordination, the establishment of a system for the exchange of information on disease occurrence and early effective surveillance. It is also suggested that political efforts should include the establishment of technical boards at national and international levels and collaboration between medical and veterinary schools (One Health 2014).

A symposium was held in 2014 at the University of Liverpool and brought together researchers, practitioners, policymakers, and business representatives from several universities institutes. The main themes were food safety, natural and social science interactions, antimicrobial resistance, obesity, and emerging of new infectious diseases (Honey 2014).

The Cambridge University One Health Society held a conference in Cambridge in 2015, with the aim of encouraging undergraduates from different backgrounds to engage with the One Health concept (Harries 2015).

Another conference was organized by the Responsible Use of Medicines in Agriculture

Alliance (RUMA) and the Veterinary Medicines Directorate in London in 2015 and focused on the responsible administration of drugs, gathering veterinarians, farmers, producers, doctors, and retailers adopting the One Health approach (Jarvis 2015).

During the International Society of Veterinary Epidemiology and Economics (ISVEE) conference in Mexico in 2015, gaps and needs about One Health were identified. Key messages included: (1) further development of transdisciplinary collaborations; (2) new mechanisms to manage data from different sources; (3) recognition of One Health multi-sectoral priorities; (4) institutionalization of One Health (Mardones et al. 2017).

2d. *Antimicrobial Resistance*

In regard to future perspectives on antimicrobial resistance (AMR), two papers were found. In response to the global Public Health threat posed by AMR, several national and international initiatives were developed in recent years. Paphitou (2013) explained that a holistic view of AMR, as well as intersectorial collaboration between human and veterinary medicine, is required to address the problem. Although the optimally effective and cost-effective strategy to reduce AMR is not known, it should include actions aiming at optimizing antibiotic use, strengthening surveillance and infection control and improving healthcare workers' education. Research efforts to bring new effective antibiotics to patients need to be fostered.

McGinn (2015) stated that the problems such as AMR, which the WHO highlighted as an increasingly serious threat to global public health, require a multidisciplinary approach to resolve them.

12.5.2.3 Historical Background

Of the 50 included articles, only three papers are focused on the historical background of the One Health and its basic concepts. The oldest article was written by Mantovani (2008), who explained that medicine is born as "one," and it could be illustrated like a large tree, whose trunk represents the basic sciences and divides into two branches (human medicine and veterinary medicine) that are connected to each other by a large branch (public health). In the second half of nineteenth century, the WHO set up a program to promote collaboration between the two medicines (Mantovani 2008). Woods and Bresalier (2014) referred to Calvin Schwabe's words, stating that the most significant advances in health had been achieved by figures who had overcome the boundaries between animal and human health, such as Jenner, Virchow, Pasteur, and others, seeking to justify and win support for his One Medicine vision.

Sikkema and Koopmans (2016) declared that the first Western European article reporting the term "One Health" was published in 2008, and since then, the number of publications quickly increased. A lot of new research programs regarding One Health have been funded by EU Research and Innovation Programs since 2007.

By recognizing that human health, animal health, and environmental health are intimately linked, One Health aims at promoting and defending health and well-being of all existing species, by improving cooperation and collaboration between physicians, veterinarians, and all the scientific professionals involved in the health and environmental domains, and by promoting strengths in leadership and management to reach these goals.

From a historic perspective, human and veterinary medicines have been considered a *unicum*, according to the holistic model, dedicated to the health and well-being of humans and animals, and to respect for the environment. Schwabe was the first author who introduced the term "One Medicine" into the scientific literature in 1984 (Schwabe 1984). The concept of One Medicine sets an aim which can be reached by integrating human and veterinary medicine and other branches of science and also by organizing proper training, health education, epidemiological surveillance, public health activities, scientific research, and other activities.

As emerges from the papers that have been reviewed, a One Health approach is strongly required in order to face and control many problems which are emerging or prevailing worldwide. The control of infectious diseases needs a One Health approach, because all the indications show that infectious diseases will continue to have an impact on our health, and the emergence of pathogens will threaten the well-being of people and animals.

Many authors offered several examples of One Health collaborations, which have been achieved through joint efforts regarding: (a) educational initiatives between human medical, veterinary medical schools, and schools of public health and the environment; (b) communications through journals, conferences, allied health networks; (c) assessment, treatment, and prevention of disease transmission and better understanding of cross-species transmission of diseases through comparative medicine and environmental research; (d) cross-species disease surveillance and control efforts in public health; (e) development and evaluation of new diagnostic methods, medicines, and vaccines for the prevention and control of diseases in different species; (f) information and education of political leaders and public sector. Furthermore, numerous organizations have recognized the need for integration of human and animal health, and the health of the environment and have taken steps to develop new programs and form new partnerships to support that integration.

Although in many studies several examples of concrete and pragmatic actions of collaboration between human and veterinary medicine have been presented, the need to make this approach stronger is still a common thought. To reach this purpose, the authors stated that changes of professions related to human and animal health and environment are needed, in both culture and training, as well as a better cooperation between medical and veterinary services. As far as zoonotic diseases are concerned, the authors sustained the need for improving the risk assessment and risk management mechanisms at government level, to strengthen the animal and human health surveillance infrastructure, to analyze data on both human and animal infections, sharing these data with all those institutions which have a role in managing risks and threats to public health. In the field of vaccines, the authors' thought is the need for a new generation of cost-effective and efficient vaccines that can be applied in the future by mass immunizing methods in order to control the shedding of the infectious diseases. At last, the goal to limit the spread of the antimicrobial resistance could be achieved optimizing antibiotic use, strengthening surveillance and infection control, and improving education.

One Health is, in essence, a vision and a plan to guide implementation or to transform the vision into action and, although it represents a challenge, is essential to manage and lead change of this magnitude. Unfortunately, there are significant barriers to the adoption of the One Health concept, perhaps the most important being the need for key leadership (political level or policy makers) to embrace the concept of One Health, to obtain buy-in from medical, veterinarian, industrial, and environmental partners and to make a program of innovation on a global basis. Additional barriers to promote One Health include differences in organization's culture, competing priorities, and absence of resources. Success in embracing the One Health concept will require to break many other barriers, including difficulty in changing ideas and attitudes of healthcare providers from one of "disease care" to one of preventive medicine, increasing specialization in this field, as well as a general insufficiency of awareness and education of physicians, and difficulty getting busy practitioners on board. Only in this way it will be possible to face the serious health threats of all living species and to preserve the same integrity of the ecosystem.

References

Aboul-Enein BH, Puddy WC, Bowser JE. The 1925 Diphtheria Antitoxin Run to Nome - Alaska: A Public Health Illustration of Human-Animal Collaboration. J Med Humanit 2019;40(3):287–96

Aboul-Enein BH, Puddy WC, Bowser JE (2019) The 1925 diphtheria antitoxin run to Nome - Alaska: a public Health illustration of human-animal collaboration. J Med Humanit 40(3):287–296

American College of Lifestyle Medicine (2021) The lifestyle medicine. https://www.lifestylemedicine.org

Battelli G, Scorziello M. Medical-veterinary collaboration in the field of zoonoses. Ann Ig 1992;4(6):395–400.

Bartonova A (2012) How can scientists bring research to use: the HENVINET experience. Environ Health 11(Suppl 1):S2

Battelli G, Scorziello M (1992) Medical-veterinary collaboration in the field of zoonoses. [Article in Italian: Collaborazione medico-veterinaria nel campo delle zoonosi]. Ann Ig 4(6):395–400

Battelli G, Baldelli R, Ghinzelli M, Mantovani A (2018) Occupational zoonoses in animal husbandry and related activities. Ann Ist Super Sanita 42(2):391–396

Bergström K, Nyman G, Widgren S, Johnston C, Grönlund-Andersson U, Ransjö U (2012) Infection prevention and control interventions in the first outbreak of methicillin-resistant *Staphylococcus aureus* infections in an equine hospital in Sweden. Acta Vet Scand 54(1):14

Beveridge WIB (1978) The need for closer collaboration between the medical and veterinary professions. Bull World Health Organ 56(6):849–858

Blake DP, Betson M (2017) One Health: parasites and beyond. Parasitology 144(1):1–6

Bresalier M, Worboys M (2014) 'Saving the lives of our dogs': the development of canine distemper vaccine in interwar Britain. Br J Hist Sci 47(173 Pt 2):305–334

Busani L, Caprioli A, Macrì A, Mantovani A, Scavia G, Seimenis A (2006) Multidisciplinary collaboration in VPH. Ann Ist Super Sanita 42(4):397–496

Centers for Disease Control and Prevention (CDC) (2018a) National Center for Emerging and Zoonotic Infectious Diseases (NCEZID). https://www.cdc.gov/ncezid. Accessed 16 Jan 2021

Centers for Disease Control and Prevention (CDC) (2018b) One Health. https://www.cdc.gov/onehealth/. Accessed 16 Jan 2020

Chen CT, Tung HH, Fang CJ, Wang JL, Ko NY, Chang YJ, Chen YC (2021 Apr) Effect of music therapy on improving sleep quality in older adults: a systematic review and meta-analysis. J Am Geriatr Soc 20

Cleaveland S, Sharp J, Abela-Ridder B et al (2017) One Health contributions towards more effective and equitable approaches to health in low- and middle-income countries. Philos Transact R Soc Lond B Biol Sci 372(1725):20160168

Clinicians Community Call. October 23rd 2020. https://drive.google.com/file/d/1n-NDFAQfHVEiMc3yzgTRB5JETkcSMtRb/view

Clinicians for Planetary Health (C4PH) (2021). https://www.planetaryhealthalliance.org/clinicians-for-planetary-health

Cocchiara RA, Peruzzo M, Mannocci A, Ottolenghi L, Villari P, Polimeni A, Guerra F, La Torre G (2019) The use of yoga to manage stress and burnout in healthcare workers: a systematic review. J Clin Med 8(3):284

Conraths FJ, Schwabenbauer K, Vallat B et al (2011) Animal health in the 21st century-a global challenge. Prev Vet Med 102(2):93–97

Corning S (2014) World organisation for animal Health: strengthening veterinary services for effective one Health collaboration. Rev Sci Tech 33(2):639–650

Cripps P (2000) Veterinary education, zoonoses and public health: a personal perspective. Acta Trop 76(1):77–80

Dehove A (2010) One world, one health. Transbound Emerg Dis 57(1–2):3–6

Di Nardo F, Saulle R, La Torre G (2010) Green areas and health outcomes: a systematic review of the scientific literature. Ital J Public Health 7(4):402–413

Donaldson LJ, Reynolds DJ (2005) Integrated working. Vet Rec 157(22):680–681

Eussen BGM, Schaveling J, Dragt MJ, Blomme RJ (2017) Stimulating collaboration between human and veterinary health care professionals. BMC Vet Res 13(1):174

Fatima Y, Doi SA, Mamun AA (2016) Sleep quality and obesity in young subjects: a meta-analysis. Obes Rev 17(11):1154–1166

Felicilda-Reynaldo RFD, Cruz JP, Alshammari F, Obaid KD, Rady HE, Qtait M et al (2018) Knowledge of and attitudes toward climate change and its effects on health among nursing students: a multi-Arab country study. Nurs Forum 53:179–189

Foley-Nolan C, Buckley J, O'Sullivan E, Cryan B (1998) United front-veterinary and medical collaboration. Ir Med J 91(3):95–96

Friedrich AW, Witte W, De Lencastre H, Hryniewicz W, Scheres J, Westh H (2008) A European laboratory network for sequence-based typing of methicillin-resistant *Staphylococcus aureus* (MRSA) as a communication platform between human and veterinary medicine—an update on SeqNet.org. Euro Surveill 13(19): 18862

Garbarino S, Guglielmi O, Puntoni M, Bragazzi NL, Magnavita N (2019) Sleep quality among police officers: implications and insights from a systematic review and meta-analysis of the literature. Int J Environ Res Public Health 16(5):885

Gartland D, Riggs E, Muyeen S, Giallo R, Afifi TO, MacMillan H, Herrman H, Bulford E, Brown SJ (2019) What factors are associated with resilient outcomes in children exposed to social adversity? A systematic review. BMJ Open 9(4):e024870

Gibbs EP, Anderson TC (2009) 'One World-One Health' and the global challenge of epidemic diseases of viral aetiology. Vet Ital 45(1):35–44

Gollust SE, Fowler EF, Niederdeppe J (2019) Television news coverage of public health issues and implications for public health policy and practice. Annu Rev Public Health 40:167–185

Gossner C, Marrama L, Carson M et al (2017) West Nile virus surveillance in Europe: moving towards an integrated animal-human-vector approach. Euro Surveill 22(18):30526

Harries J (2015) Collaboration on one Health: starting with students. Vet Rec 176(21):538

Hodgson S, Watts I, Fraser S, Roderick P, Dambha-Miller H (2020) Loneliness, social isolation, cardiovascular disease and mortality: a synthesis of the literature and conceptual framework. J R Soc Med 113(5):185–192

Honey L (2014) One Health: time to move on from just talking. Vet Rec 175(4):83

Hristovski M, Cvetkovik A, Cvetkovik I, Dukoska V (2010) Concept of One Health - a new professional imperative. Maced J Med Sci 3(3):229–232

Jarvis S (2015) Working together on the responsible use of medicines. Vet Rec 177(20):511–512

Johansen MV, Lier T, Sithithaworn P (2015) Towards improved diagnosis of neglected zoonotic trematodes using a One Health approach. Acta Trop 141:161–169

Kabir MI, Rahman MB, Smith W, Lusha MA, Azim S, Milton AH (2016) Knowledge and perception about climate change and human health: findings from a baseline survey among vulnerable communities in Bangladesh. BMC Public Health 16:266

Kahn R, Clouser D, Richt J (2009) Emerging infections: a tribute to the one medicine, one health concept. Zoonoses Public Health 56(6–7):407–428

Klauder J (1958) Interrelations of human and veterinary medicine. N Engl J Med 258(4):170–177

La Torre G, De Paula BA, Sestili C, Cocchiara RA, Barbato D, Mannocci A, Del Cimmuto A (2020a) Knowledge and perception about climate change among healthcare professionals and students: a cross-sectional study. S East Eur J Public Health 12:33–47

La Torre G, Raffone A, Peruzzo M, Calabrese L, Cocchiara RA, D'Egidio V, Leggieri PF, Dorelli B, Zaffina S, Mannocci A, Yomin Collaborative Group (2020b) Yoga and mindfulness as a tool for influencing affectivity, anxiety, mental health, and stress among healthcare workers: results of a single-arm clinical trial. J Clin Med 9(4):1037

Leiserowitz A, Maibach E, Roser-Renouf C, Feinberg G, Rosenthal S, Marlon J (2014) Public Perceptions of the Health Consequences of Global Warming 2014. Yale Project on Climate Change Communication, New Haven, CT

Lewandowsky S, Cook J, Fay N, Gignac GE (2019) Science by social media: attitudes towards climate change are mediated by perceived social consensus. Mem Cogn 47:1445–1456

Liberati A, Altman D, Tetzlaff J et al (2009) The PRISMA statement for reporting systematic reviews and meta-analyses of studies that evaluate health care interventions: explanation and elaboration. PLoS Med 6(7):e1000100

Maibach EW, Nisbet M, Baldwin P, Akerlof K, Diao G (2010) Reframing climate change as a public health issue: an exploratory study of public reactions. BMC Public Health 10:299

Mantovani A (2001) Notes on the development of the concept of zoonoses. Hist Med Vet 26(2):41–52

Mantovani A (2008) Human and veterinary medicine: the priority for public health synergies. Vet Ital 44(4):577–582

Mardones F, Hernandez-Jover M, Berezowski J, Lindberg A, Mazet J, Morris R (2017) Veterinary epidemiology: forging a path toward one health. Prev Vet Med 137(Pt B):147–150

McGinn K (2015) Silos in education are the main barrier to collaboration. Vet Rec 176(16):419

Meisser A, Schelling E, Zinsstag J (2011) One health in Switzerland: a visionary concept at a crossroads? Swiss Med Wkly 141:w13201

Michell A (2005) Comparative clinical science: the medicine of the future. Vet J 170(2):153–162

Neuendorf R, Wahbeh H, Chamine I, Yu J, Hutchison K, Oken BS (2015) The effects of mind-body interventions on sleep quality: a systematic review. Evid Based Complement Alternat Med 2015:902708

One Health (2014) collaboration on all levels is key, concludes joint meeting. Vet Rec 174(26):644

One Health Initiative (2021) One World One Medicine One Health. http://onehealthinitiative.com/. Accessed 16 Jan 2020

Paphitou NI (2013) Antimicrobial resistance: action to combat the rising microbial challenges. Int J Antimicrob Agents 42(Suppl):S25–S28

Pawlowski ZS (2008) Control of neurocysticercosis by routine medical and veterinary services. Transact R Soc Trop Med Hyg 102(3):228–232

Pawlowski Z, Allan J, Sarti E (2005) Control of *Taenia solium* taeniasis/cysticercosis: from research towards implementation. Int J Parasitol 35(11):1221–1222

Planetary Health Alliance (2021). https://www.planetaryhealthalliance.org/

Poglayen G (2006) The challenges for surveillance and control of zoonotic diseases in urban areas. Ann Ist Super Sanita 42(4):433–436

Rabinowitz PM, Natterson-Horowitz BJ, Kahn LH, Kock R, Pappaioanou M (2017) Incorporating one health into medical education. BMC Med Educ 17(1):45

Reynolds D, Donaldson L (2005) UK government collaborations to manage threats to animal and human health. BMJ 331(7527):1216–1217

Reynolds TW, Bostrom A, Read D, Morgan MG (2010) Now what do people know about global climate change? Survey studies of educated laypeople. Risk Anal 30:1520–1538

Rotivel Y (2003) Human rabies prophylactics: the French experience. Vaccine 21(7–8):710–715

Saulle R, La Torre G (2010) The Mediterranean diet, recognized by UNESCO as a cultural heritage of humanity. Ital J Public Health 7(4):414–415

Schillhorn van Veen T (1998) One medicine: the dynamic relationship between animal and human medicine in history and at present. Agric Human Values 15(2):115–120

Schwabe C (1984) Veterinary medicine and human health, 3rd edn. Williams and Wilkins, Baltimore, MD

Sikkema R, Koopmans M (2016) One health training and research activities in Western Europe. Infect Ecol Epidemiol 6(1):33703

Speksnijder DC, Mevius DJ, Bruschke CJM, Wagenaar JA (2015) Reduction of veterinary antimicrobial use in the Netherlands. The Dutch success model. Zoonoses Public Health 62(Suppl1):79–87

Stärk K, Arroyo Kuribreña M, Dauphin G et al (2015) One health surveillance – more than a buzz word? Prev Vet Med 120(1):124–130

Streich JL (2014) Nursing faculty's knowledge on health impacts due to climate change. Doctoral thesis. Montana State University-Bozeman, College of Nursing, Bozeman, MT

Valkanova N, Paunov T, Stoyanova K, Romanova H (2010) Problems in anti-epidemic control of brucellosis in Bulgaria. Maced J Med Sci 3(3):268–272

van de Burgwal LHM, Neevel AMG, Pittens CACM, Osterhaus ADME, Rupprecht CE, Claassen E (2017) Barriers to innovation in human rabies prophylaxis and treatment: a causal analysis of insights from key opinion leaders and literature. Zoonoses Public Health 8(64):599–611

van den Berg T, Lambrecht B, Marché S, Steensels M, Van Borm S, Bublot M (2008) Influenza vaccines and vaccination strategies in birds. Comp Immunol Microbiol Infect Dis 31(2–3):121–165

Veidis EM, Myers SS, Almada AA, Golden CD, Clinicians for Planetary Health Working Group (2019) A call for clinicians to act on planetary health. Lancet 393(10185):2021

Wei J, Hansen A, Zhang Y, Li H, Liu Q, Sun Y et al (2014) Perception, attitude and behavior in relation to climate change: a survey among CDC health professionals in Shanxi province, China. Environ Res 134:301–308

Wendt A, Kreienbrock L, Campe A (2015) Zoonotic disease surveillance-inventory of systems integrating human and animal disease information. Zoonoses Public Health 62(1):61–74

Wendt A, Kreienbrock L, Campe A (2016) Joint use of disparate data for the surveillance of zoonoses: a feasibility study for a One Health approach in Germany. Zoonoses Public Health 63(7):50314

Woods A, Bresalier M (2014) One health, many histories. Vet Rec 174(26):650–654

World health Organization (2021) Physical activity-key facts. https://www.who.int/news-room/fact-sheets/detail/physical-activity

Xie E, De Barros EF, Abelsohn A, Stein AT, Haines A (2018) Challenges and opportunities in planetary health for primary care providers. Lancet Planet Health 2:e185–e187

Yazdanparast T, Salehpour S, Masjedi MR, Seyedmehdi SM, Boyes E, Stanisstreet M et al (2013) Global warming: knowledge and views of Iranian students. Acta Med Iran 51:178–184

Endocrine Disruptors and Human Reproduction

13

Francesco Pallotti, Donatella Paoli, and Francesco Lombardo

Abstract

Recent reports regarding lowering fertility rates in industrialized countries are a cause of concern. In general, social and economic factors are involved, together with a large number of medical conditions (genetic abnormalities, endocrine disorders, malignancies, etc.), hazardous lifestyles (smoking, alcohol consumption, unhealthy eating), and environmental factors. In recent years, researchers have focused their attention on a class of substances now known as endocrine disruptors (EDs), a heterogeneous group of substances that can interfere with endogenous hormones. ED-induced hormonal imbalances may then cause disturbances of many physiological functions (such as puberty, breast development, etc.) and induce several pathologies like cancer (prostate and breast), endocrine and neuroendocrine system dysfunctions, cardiometabolic disorders (obesity, cardiovascular diseases, etc.), and hindrances of reproductive functions in both sexes. Various aspects of reproductive function have been investigated in relation to possible interference of several EDs (phthalate esters, bisphenol A, polychlorinated biphenyls, organophosphate pesticides, perfluoroalkylated compounds, in particular) on several key functions such as gonadal ontogenesis, gonadal steroidogenesis, gonocyte proliferation, pubertal development, and adult fertility. Several EDs have also been associated with the onset of urogenital abnormalities (cryptorchidism, anogenital distance, hypospadia) and reproductive diseases, such as polycystic ovary syndrome and endometriosis. Most of the harmful effects of EDCs were observed in experimental conditions on animals, often after exposure to doses higher than those expected from accidental environmental contact and human studies, but results are relatively scant or controversial. Overall, current evidence does not fully explain the impact of EDs on human male reproductive health. This lack of consistency in experimental and in vivo evidence will definitely require further investigations but, in the meantime, we must prudently point toward a positive association between EDC exposure and reproductive system damage.

F. Pallotti · D. Paoli · F. Lombardo (✉)
Laboratory of Seminology—Sperm Bank "Loredana Gandini", Department of Experimental Medicine, "Sapienza" University of Rome, Rome, Italy
e-mail: francesco.pallotti@uniroma1.it;
donatella.paoli@uniroma1.it;
francesco.lombardo@uniroma1.it

Keywords

Endocrine disruptors · Human reproduction
Infertility · Phthalate · Bisphenol A · PCB
Organophosphate pesticides
Perfluorochemicals

13.1 Introduction

Lowering fertility rates are a cause of concern in industrialized countries. These are considered a result of a complex intertwining of medical, social, and economic factors (Den Hond et al. 2015). Frequently, socioeconomic factors result in a postponed pregnancy planning, and the increasing parental age is often associated with worse reproductive outcome (du Fossé et al. 2020). On the other hand, a plethora of medical conditions can be associated with a worsened reproductive potential for both the male and female partners, including genetic abnormalities, endocrine disorders, systemic diseases, congenital or acquired urogenital abnormalities, urogenital tract infections, immunological factors, and malignancies. Additionally, harmful lifestyles (cigarette smoking, alcohol consumption, unhealthy eating, obesity, etc.) and environmental factors can adversely impact the couple's fertility potential (Gianfrilli et al. 2019; Amiri and Tehrani 2020; Corona et al. 2020). It is worth stressing that fertility is a condition characterizing a couple, where the reproductive potential and health of both partners are equally relevant. Besides, a couple can be defined as "infertile" after the failure to achieve a clinical pregnancy after at least 12 months of regular unprotected sexual intercourse. This condition is expected to affect around 15% of couples worldwide but, to further worsen the picture, a trend of increasing number of couples affected by infertility referring to fertility centers has been reported (Mann et al. 2020). In parallel, while roughly one-third of cases can be attributable to a specific condition affecting the male partner, accumulating worldwide literature data in the last decades has brought to a lively discussion on the possibility of a decline in semen quality (Levine et al. 2017). While many literature papers have brought often conflicting evidence of this decline, a recent meta-analysis has supported this possibility (Levine et al. 2017) but, also due to methodological limitations as data heterogeneity, it is currently impossible to ascertain causality with a specific genetic, clinical, or environmental factor. Nonetheless, in recent years, researchers have focused their attention on a class of substances now known as endocrine disruptors (ED).

13.2 Endocrine Disruptors

These compounds are a highly heterogeneous group of substances that, as defined by the Endocrine Society, can interfere with endogenous hormones action at any level (from synthesis to receptor binding, signal transduction, and elimination) (Fig. 13.1) (Gore et al. 2015). This wide definition underlines the fact that a precise pathogenic mechanism is still a matter of debate for many EDs, as causality between exposure and pathological effect is extremely difficult to investigate. In fact, effects of endocrine disruption from these chemicals generally become apparent with a long onset and/or after a prolonged and continuous exposure. Exposure can either be at very low doses, far lower than those expected to cause immediate toxicity, or be the result of the simultaneous interaction of a mixture of chemicals (the so-called cocktail effect) (Kortenkamp 2020). Also, EDs may exert deleterious effects not only in exposed individuals but also in their offspring and, in future generations, through induced epigenetic modifications. Furthermore, these EDs (or ED mixtures) interacting with the individual's hormonal balance can exercise different effects (Barouki 2017; Pallotti et al. 2020). As a consequence, the in vivo biological effects of EDs may show relevant inter-individual variability, thus making it difficult to predict the net effect on a specific organism. Human contact with EDs is almost inevitable as we are literally surrounded by potential sources of contamination. EDs have been detected in water and foods (Su et al. 2020; Mantovani 2016). Industrial products and everyday objects (ranging from cosmetics to plastic tools and toys, etc.) may have been produced with these substances that are generally responsible for their "pleasant" physical characteristics. Even domestic and occupational furnishings and tools generally contain these substances, since they are necessary for several characteristics of the product itself (insulation, fireproof or flame-retardant status, etc.). Under certain conditions (heat, wearing

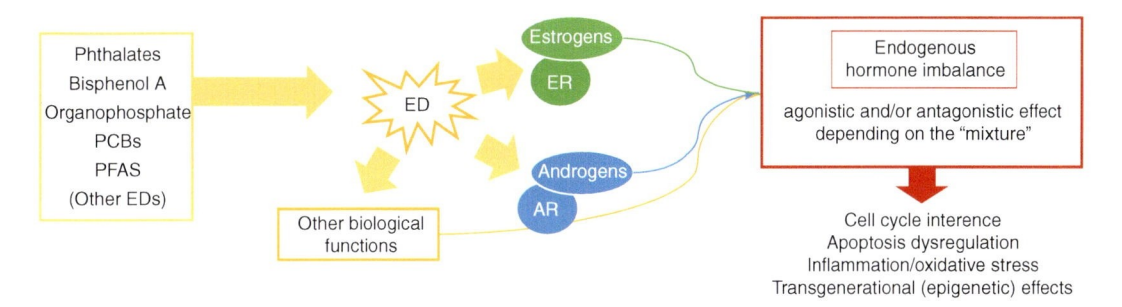

Fig. 13.1 Schematic representation of ED action

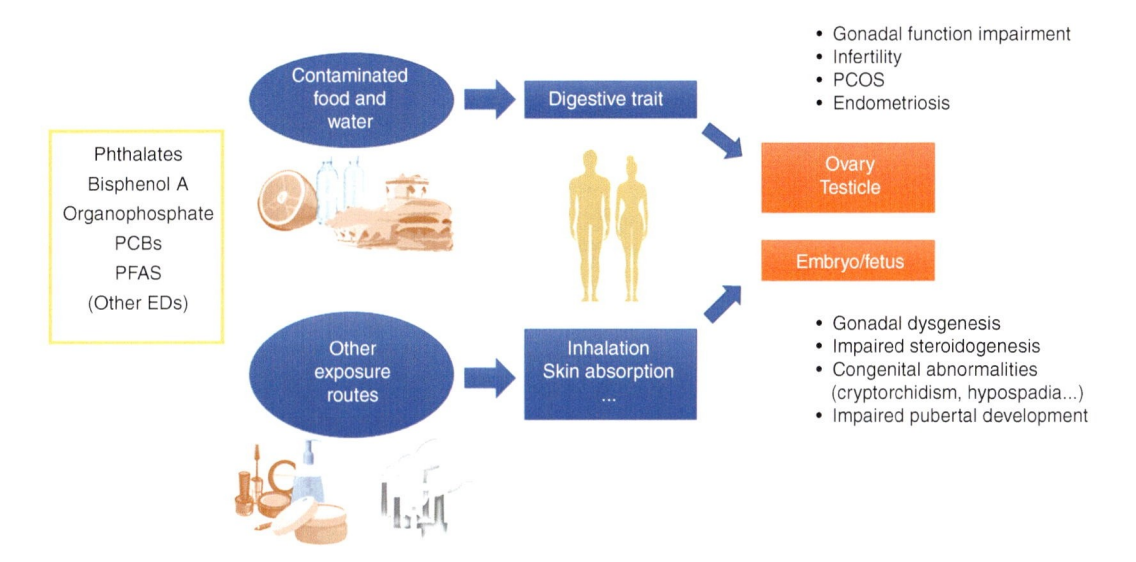

Fig. 13.2 Exposure routes and main reproductive effects of EDs

away, etc.), EDs may leak from the chemical structure of the object and represent a potential source of contact. Therefore, theoretically all routes of exposure are possible, although the most relevant are food and water consumption, inhalation, and direct dermal contact (Fig. 13.2) (Paoli et al. 2020). During fetal and neonatal life, transplacental contamination and breast feeding are also relevant routes of exposure. Once inside the human body, adipose tissue can accumulate EDs, as most of them have some degree of lipophilicity (Cano-Sancho et al. 2020). Lipolysis is then a cause of constant release of ED mixtures from adipocytes, that becomes particularly relevant under certain conditions (such as obesity with dysfunctional adipocytes or weight loss) potentially unleashing deleterious repercussions for endocrine/metabolic diseases or carcinogenesis (Cano-Sancho et al.

2020; Kahn et al. 2020). Another potential effect of EDs may act on an epigenetic level. In general, epigenetic changes are known to modify the level of activation and expression of genes without modifying the genes themselves. This is physiologically achieved through changes in DNA methylation and histone modifications and through the action of small non-coding RNAs. Alterations of these mechanisms have been linked to several pathologies including obesity, CV disease, and cancer. EDs have been linked to potential epigenetic profile shifts and since recent studies reported transmission of these epigenetic shifts from father to child, a transgenerational inheritance of ED modifications is likely (Shi et al. 2020; Pitto et al. 2020; Van Cauwenbergh et al. 2020; Skakkebaek et al. 2016). Consequently, human pre- and postnatal development may be already affected due to prenatal maternal/paternal exposure (Alonso-Magdalena et al. 2016). In the last decades, EDs have been related to disturbances of many physiological functions (such as puberty, breast development, etc.) and to several pathologies like cancer (prostate and breast), endocrine and neuroendocrine system dysfunctions, cardiometabolic disorders (obesity, cardiovascular diseases, etc.), and, of course, hindrances of reproductive functions (Skakkebaek et al. 2016). Human reproduction can be affected by these compounds at several levels as many EDs have estrogenic/antiandrogenic activity, disrupting the activity of the pituitary gonadal axis. Furthermore, there is a growing amount of molecular evidence showing how EDs can involve different pathways and networks after binding or mimicking the actions of either the estrogen receptor (ER) or the androgen receptor (AR), ultimately influencing cell apoptosis, proliferation, differentiation, carcinogenesis, and inflammation (Gore et al. 2015).

13.3 Relevant Classes of EDs in Human Reproduction

Despite exposure to a wide range of EDs has possible impacts, the most studied compounds affecting reproductive functions are polychlorinated biphenyls, organophosphate pesticides, phthalates, bisphenol A, perfluorochemicals.

13.3.1 Polychlorinated Biphenyls

Polychlorinated biphenyls (PCBs) are persistent and lipophilic synthetic compounds constituted by a biphenyl skeleton where hydrogen atoms are replaced by chlorine atoms. In particular, the biphenyl molecule can bind from 2 to 10 chlorine atoms in different positions, resulting in a total of 209 possible congeners (Meeker and Hauser 2010). PCBs are constituents of cutting oils, lubricants, plasticizers, and electrical insulators in transformers and capacitors, but the full range of their potential uses is extremely wide (they could also be used in flame retardants, inks/dyes, sealants/adhesives, and pesticides). Despite their potential, harmful effects on humans from these molecules are well known and led to their ban from developed countries from the 1970s. Unfortunately, due to their extremely long half-lives, PCBs are still distributed worldwide, concentrated, and stored in fatty tissues of animals and humans. In fact, the food chain led to PCB exposure to persist even in countries who discontinued production and banished these products since PCBs present significant levels of bioaccumulation and biomagnification favored by ingestion of contaminated foods (mainly dairy products and fatty fishes and meats). The International Agency for Research on Cancer (IARC) included dioxin-like PCBs in Group 1 of known human carcinogens, based both on strong evidence of the mechanism of carcinogenesis mediated by the aryl hydrocarbon receptor (AhR), which is identical to that of 2,3,7,8-tetrachlorodibenzoparadioxin, and on evidence of carcinogenicity in laboratory animals (Paoli et al. 2015). As for their endocrine disrupting properties, the various congeners of PCBs may show a variety of effects, ranging from estrogenic to anti-estrogenic and/or anti-androgenic effects (Meeker and Hauser 2010).

13.3.2 Organophosphate Pesticides

Organophosphate (OP) compounds are esters of phosphoric acid that know a widespread use as insecticides, herbicides (in agriculture, for example), and flame retardants or plasticizers (in many human activities, including in private households) (Meyer and Bester 2004; Yao et al. 2021). OPs rapidly degrade in the environment, nonetheless more than 40 OP pesticides are classified as moderately or highly hazardous by the US Environmental Protection Agency (EPA) and the WHO Food and Agriculture Organization (Hertz-Picciotto et al. 2018).

Acute toxic activity from exposure to high levels of OPs is exerted toward the nervous system, irreversibly altering the cholinergic nervous transmission (Kaushal et al. 2021).

Conversely, a chronic low-dose exposure to OP pesticides may be associated with a variety of health effects, from neurologic and neurobehavioral hazards to immunotoxicity and carcinogenesis, to endocrine disruption with repercussions on the reproductive axis (Doherty et al. 2019). In particular, alterations of the gonadal hormonal axis (Aguilar-Garduño et al. 2013), testicular damage (Narayana et al. 2006), and impairment of spermatogenesis (Recio-Vega et al. 2008) have been described.

13.3.3 Phthalates

Phthalates are a common name indicating a class of synthetic chemicals, diesters of phthalic acid (1,2-benzenedicarboxylic acid), commonly used in a wide range of industrial, consumer, and personal care products. Several phthalate esters are used in the production of polyvinyl chloride, plastics for food packaging, building materials and medical devices (especially those with a high molecular weight: di-isodecyl phthalate (DIDP), diisononyl phthalate (DiNP), di-2-ethylhexyl phthalate (DEHP), etc.). Other esters, generally low molecular weight phthalates (diethyl phthalate (DEP), dibutyl phthalate (DBP), etc.), can be found within a wide range of personal care products (cosmetics/fragrances, lacquers, and varnishes) and as solvents and plasticizers in cellulose acetate (Giuliani et al. 2020; Nantaba et al. 2021). Phthalates are classified by the European Union's REACH (Registration, Evaluation, Authorization, and Restriction of Chemicals) as dangerous chemicals for reproduction (Ventrice et al. 2013) as they are characterized by a high bioaccumulation deriving from consumption of contaminated food and waters, dermal contact, and through inhalation. After exposure, phthalates are rapidly hydrolyzed into their respective monoesters and conjugated to form the hydrophilic glucuronide conjugate, easily excreted into urine within 24 h, providing a reliable biomarker of exposure (Paoli et al. 2020). Phthalates esters have been found in blood, urine, breast milk, saliva, follicular, and amniotic fluid (Paoli et al. 2020). Phthalates mainly act as anti-androgens, although they may also have weak estrogenic properties (Annamalai and Namasivayam 2015). Many reports have shown that DEHP may cause a variety of male reproductive abnormalities, such as cryptorchidism and hypospadias, partially through its antiandrogen effects (Pallotti et al. 2020).

13.3.4 Bisphenol-A

Bisphenol-A (BPA) is one of the most produced synthetic compounds worldwide as it is involved in the production of polycarbonate plastics and epoxy resins in consumer products from toys to medical tubing and equipment, water pipes, drinking containers, electronic wares, etc. BPA is known to leach from the plastic structure it is held in, and for human exposure, it is especially relevant for BPA leaching from food and beverage containers. Combined with the fact that products containing bisphenol have been present for decades, BPA can be considered nearly ubiquitous, making low-dose human exposure almost inevitable. Exposure to this chemical may occur through oral (food and beverages), respiratory (air dust), and dermal routes (water, cosmetics, etc.) in both humans and animals, and BPA has been detected in many human biological fluids and tissues, including serum, urine, amniotic and follicular fluid (Paoli et al. 2020; Vandenberg et al. 2007). Continuous low-dose exposure to BPA is associated with cardiometabolic diseases, abnormalities of brain development, cancer and hormone dysfunctions, especially of the thyroid and gonadal axes (Pivonello et al. 2020; Abraham and Chakraborty 2020). In particular, endocrine functions disrupted by BPA are secondary to its binding to several receptors, such as estrogen receptor α/β (ER α/β), membrane estrogen receptor (mER), androgen receptor (AR), and thyroid hormone receptor (Siracusa et al. 2018). In addition, BPA may have anti-androgenic and anti-thyroid activity (Kim and Park 2019). Furthermore, as spermatogenesis is under the control of the hypothalamic–pituitary–testicular axis and the thyroid gland, BPA-induced endocrine disruptions are likely to deleteriously affect male semen quality.

13.3.5 Perfluorochemicals

Perfluorochemicals (PFCs) such as perfluorooctanesulfonic acid (PFOS) and perfluorooctanoic acid (PFOA) are synthetic chemicals that have been used in many consumer and industrial wares (protective coatings of carpets and furniture, paper and cloth coatings, polytetrafluoroethylene products, and fire-fighting foam) due to the extreme range of applications of their physical properties (hydrophobicity and lipophobicity, thermal stability). As PFCs are nonbiodegradable and have a long half-life, they are capable of bioaccumulating in the environment and have an extremely long clearance rate in the human body (Foresta et al. 2018). Like previous EDCs, it has been suggested that food, water, and inhalation of airborne particles are the principal human exposure routes (Jian et al. 2017). Despite these compounds can mainly be found bound to albumin in human blood, they are also detectable in seminal fluid, human milk, and tissues (Jian et al. 2017; Cui et al. 2020). PFCs may be associated with a wide range of human diseases, such as thyroid diseases, asthma, behavioral disorders, liver cancer, and immune dysregulation/toxicity, among others (Jian et al. 2017). Also, these compounds may disrupt gonadal hormone axis through both estrogenic- and anti-androgenic-like activities (White et al. 2011; Tian et al. 2019; Di Nisio et al. 2019).

13.4 Impact on Reproduction

13.4.1 Testicular and Ovarian Function

Oogenesis and spermatogenesis are differentiation processes culminating in the selection of an oocyte and the production of motile spermatozoa, respectively. Since both require the complex functional coordination of many endocrine and paracrine factors, alterations arising from the disruption from EDCs have been repeatedly reported (Pallotti et al. 2020; Ge et al. 2019). As discussed before, many EDCs have hormonal activity interfering with both estrogenic and androgenic receptors and altering gonadal functions. BPA and phthalate esters have been widely investigated in relation to ovarian and testicular/spermatogenesis dysfunctions, but relatively robust evidence is limited to animal models, and results can be argued in the sense that experiments often overlook a likely in vivo cocktail effect from different environmental EDCs (Kabir et al. 2015). In vivo analyses of these effects of EDCs on spermatogenesis are difficult to perform, as human EDC-related diseases result of long-term exposure to low concentrations of EDC mixtures and are further complicated by the late onset of the resulting clinical disorders (often many years later or even trans-generational) and a nonlinear dose–response curve of these toxicants (Vandenberg et al. 2012; Welshons et al. 2003). While in vitro evidence focuses on one or a handful of EDCs with a similar activity profile, in everyday life a varied combination of EDCs can act simultaneously and may also involve modulated or antagonistic interactions, making it impossible to predict the net biological effect of the mixture (Lee 2018; Lee et al. 2019). Despite these limitations, we know that dysregulation of androgenic and estrogenic signaling pathways can also arise from an EDC-induced impairment of steroidogenesis, by downregulating the expression of key enzymes such as CYP11A and CYP17A in both testis and ovary, or through the induction of oxidative stress, altering the differentiating spermatozoa DNA compaction and Sertoli cell tight junctions in males (Pallotti et al. 2020; Yeung et al. 2011; Piazza and Urbanetz 2019).

These alterations may potentially hinder both oogenesis and spermatogenesis. Several authors reported relevant histological abnormalities of mouse fetal testis exposed to some phthalate esters (DHP and DCHP) with atrophic and irregular seminiferous tubules with a reduced number of germ cells and several Leydig cells abnormalities (Li et al. 2015; Ahbab and Barlas 2015). These histological abnormalities and even more severe forms of dysgenesis involving Leydig cells differentiation and function (downregulation of enzyme genes related to steroidogenic action and the parallel decrease in INSL3 and CYP11A1) may be due to EDC exposure in critical phases of differentiation (masculinization programming window) (Lara et al. 2017). Male gonocytes might be altered through impairment of the strict interactions with Sertoli cells mediated by EDCs such as phthalate esters, which could also lead to an abnormal cell division and consequent alteration of the seminiferous tubule associated with other peritubular abnormalities (Fisher et al. 2003). In fact, as cellular paracrine signaling might be impaired by the action of EDCs, the fetal Sertoli could fail to control differentiation and function of gonocytes and Leydig cells, resulting in a histologically recognizable dysgenetic gonadal development reflecting (or causing) the EDC-induced steroidogenesis. These alterations will probably lead to impaired adult testicular function, hypospermatogenesis, and infertility, as they have been observed in animal experiments where exposure to BPA corresponded to reduced fetal intratesticular testosterone levels and a reduction in the proliferation and function of adult Leydig cell precursors (which normally only differentiate on puberty), with normal or reduced blood testosterone concentrations and increased LH (Lv et al. 2019).

Animal experiments showed that steroid hormones produced by antral ovarian follicles (testosterone, estrone, estradiol, androstenedione and DHEAS) present a reduction which is dependent from dose and duration of EDCs exposure, corresponding to a reduced follicular growth post-exposure (Wang et al. 2018; Meling et al. 2020) (Mlynarcikova et al. 2014). BPA, for example, has been clearly associated with disruption of steroidogenesis in both theca (inhibition of STAR, CYP450scc, and HSD-3β enzymes) and granulosa cells (inhibition of CYP450) (Bloom et al. 2016; Mahalingam et al. 2017). Evidence for ovarian follicle steroid synthesis in vivo is hard to obtain, but the presence of BPA and phthalate esters in human follicular fluid has been confirmed in women attending an assisted reproduction procedure (Paoli et al. 2020), thus it is likely that enzymes for steroid hormone biosynthesis in human oocytes and granulosa cells are similarly affected as it seems confirmed by scant literature although results still need further validation (Souter et al. 2013; Mok-Lin et al. 2010; Ehrlich et al. 2012).

Also, organochlorine pesticides and PCBs may interfere similarly with hormone synthesis as most animal studies report altered steroidogenesis enzyme expression in ovaries and testes (Rak et al. 2017; Murugesan et al. 2007). Although in vivo studies are extremely rare, these confirm decrease in estradiol in women in association with increasing concentrations of environmental pollutants as dioxins and PCBs. It has to be stressed, however, that as conduction of in vivo studies presents significant difficulties, evidence collected up to date is scant and of low quality.

13.4.2 Gonadal Development

It is clear that, due to the known capability of interfering with hormone synthesis and function, effects of EDCs on embryo gonadal development might be particularly destructive for the human organism. As such, maternal exposure during fetal life appears to be scariest eventuality for the massive risk of developmental alterations EDCs can induce. Since most of these substances present estrogenic/antiandrogenic effects, it is of no surprise that exposure in this phase has been associated with testosterone production hindrance in the male fetus with an increased likelihood of damage to the male reproductive tract development (Conley et al. 2018; Pallotti et al. 2020).

In fact, sex is determined early in fetal life, and reproductive tissue differentiation at this stage is paramount for pubertal development and reproductive life in adulthood. Androgen disruption for the male fetus thus causes disruption of endocrine signaling dedicated to reproductive tissue differentiation and development through the same mechanisms described in the previous paragraph.

In particular, their exposure and the consequent impairment of endocrine signaling lead to a downregulation of the gene expression from the genital tubercle (sonic hedgehog molecules, bone morphogenic proteins, fibroblast growth factors, transforming growth factor β1, and transforming growth factor receptor III) (Zhu et al. 2009). Severity of consequences may vary depending on the unique endocrine balance created between the fetus/mother and the EDCs cocktail, but include malformations (hypospadias, cryptorchidism, testicular/epididymal hypotrophy, deferens and seminal vesicles anomalies, changes to the gubernaculum testis, reduced anogenital distance) impairment of future fertility (altered spermatogenesis, infertility) and an increased risk of testicular cancer during young adulthood (Conley et al. 2018). Congenital urological abnormalities like cryptorchidism might bring severe consequences on adult gonadal function. It is known that testicular descent can roughly be divided in two phases: an early abdominal descent, between gestation weeks 10 and 23, and a final inguinoscrotal descent, from week 28 to birth, mainly ruled by the action of INSL3 and testosterone, respectively. INSL3 acts predominantly in the first phase, while has an adjuvant action of testosterone on the inguinoscrotal phase, when there is regression of the gubernaculum testis that literally "guides" the testicle toward its final position in the scrotum: if there is interference with hormone signaling, abnormalities of descent may occur (Chevalier et al. 2015). There is wide evidence on how phthalate esters and BPA might disrupt INSL3 signaling and, thus, cause cryptorchidism (Wilson et al. 2004; Pathirana et al. 2011; Li et al. 2015; Ahbab and Barlas 2015; Chevalier et al. 2015). Mispositioned testis may suffer, and if it is not treated, spermatogenesis may be disrupted with future infertility and increased incidence of testicular cancer.

Nonetheless, evidence come almost exclusively from animal experiments where pregnant mice are administered with different doses of EDCs (phthalates, bisphenol A, PCBs, etc.) and male offspring show different degrees of urogenital abnormalities (cryptorchidism, hypospadias, reduced anogenital distance) and in certain cases also other non-reproductive organs malformations, especially at higher concentrations (Pallotti et al. 2020). Although we are unlikely to be exposed to these experimental EDC concentrations during everyday life, these experiments pose the theoretical basis for interference of EDCs with human reproduction even during fetal life. It is reasonable to suppose that effects from multiple EDC exposure could co-exist and act in synergy in vivo, manifesting as a continuous spectrum of "exposure disorders." Gonadal dysfunction might be the primary effect suffered from exposure if it happens in the critical stage of embryo development, bringing possible consequences to adult reproductive health.

13.4.3 Pubertal Development

Puberty is a period of complex of endocrine and physical changes, generally about 5–6 years long, which is interposed between late childhood and adulthood and is of critical importance to achieve sexual maturity and development of secondary sex characteristics. Slight differences in puberty timing are present between males and females. In males, puberty occurs later (on average between 9.5 and 13.5 years of age), and the first sign is an increase of testicular volume over the threshold of 4 mL, followed by pubarche within 6 months and penis enlargement usually after 12–18 months. In females, the onset of puberty occurs between 8 and 12 years and is recognizable by the mammary gland development and an increase in height growth speed, while menarche generally follows after a couple of years. All the physical changes occurring during puberty are secondary to an increase of hypothalamic GnRH pulsatility, marking the activation of the hypothalamus–pituitary–gonadal axis, which will then cause an increase of circulating gonadotropins (FSH, LH) and sex steroids (testosterone, estrogens) (Cargnelutti et al. 2020). As it can be expected, endocrine disruption by chemicals may potentially affect both onset and progression in puberty in both sexes. In particular, many papers attempted to investigate the effects of potential interference of EDCs on the onset of puberty in boys and girls, but most evidence comes from observational studies and is unbalanced toward girls' puberty. Several pesticides (PCBs, endosulfan) are markedly and negatively associated with a reduced puberty progression, as measured by Tanner's pubertal stages (Den Hond and Schoeters 2006; Saiyed et al. 2003), suggesting a delay in sexual maturation. A relatively old paper, examining puberty progression in a small number of boys and girls previously exposed to dioxins present in breast milk, detected a delayed breast development in females and a delayed first ejaculation in males (Leijs et al. 2008). Other organochlorine compounds and non-dioxin-like PCBs persisting in the blood of pubertal children are also likely to affect puberty progression but with different effects. A paper confirmed delays in puberty for organochlorine pesticides and dioxin-like compounds; conversely, non-dioxin PCBs tend to anticipate the beginning of puberty (Sergeyev et al. 2017). The final effects on puberty of short-lived EDCs, instead, are more likely to be related to the period of exposure. Prenatal exposure to phthalates and BPA appears to be associated with delayed adrenarche and pubarche onset, while effects on pubertal development of post-natal exposure appear to be milder or absent (Ferguson et al. 2014). Timing of the in utero exposure may also exert different effects, as it has been pointed out that phthalate exposure (DHEP in particular) in the first and second trimesters of pregnancy is associated with increased peripubertal serum estradiol levels, while third trimester exposure seems to affect the onset of pubarche (Watkins et al. 2017).

13.5 Other Reproductive Disease Possibly Linked to EDCs

13.5.1 Polycystic Ovary Syndrome

Polycystic ovary syndrome (PCOS) is currently considered a spectrum of disorders that finds its basis in the various associations of three main conditions: anovulatory cycles, insulin resistance, and hyperandrogenism. This endocrine background is often associated with obesity, dyslipidemia, infertility, and other metabolic disorders. Despite the etiology of this syndrome is not yet fully understood, it is believed that below the different PCOS phenotypes underlie a complex multigenic disorder where epigenetic and environmental influences might have important roles (Escobar-Morreale 2018). Endocrine disrupting chemicals, in association with a genetic susceptibility, might also be involved in PCOS directly and in the associated metabolic diseases indirectly. A precocious exposure to these chemicals, causing abnormal prenatal androgenization, is associated with the development of PCOS. Although epigenetic mechanisms are also thought to be involved, the exact causative mechanisms are still subject of debate (Palioura and Diamanti-Kandarakis 2015; Barrett and Sobolewski 2014). Bisphenol A has been investigated since, in addition to androgenic properties, it also appears to have the capability of altering normal metabolism toward the tendency to weight increase (obesogen) (Barrett and Sobolewski 2014). Most of the evidence once again comes from animal experiments, but it has been observed that female neonatal rats exposed to high concentrations of BPA show development of PCOS-like ovaries in adult life in association with alteration of the pituitary gonadal axis (especially increased testosterone levels) and infertility (Fernández et al. 2009, 2010). In vivo evidence on women is rather scant, but the association between positive correlation between plasma testosterone levels and serum BPA levels in a cohort of PCOS women has been observed, but there was no association with insulin levels (Konieczna et al. 2018). Conversely, another study investigated the role of phthalate esters (DEHP and MEHP) in several adolescents with and without PCOS: while the authors reported no difference in serum phthalates in the two groups, these EDCs were associated significantly with insulin resistance, suggesting the involvement of EDCs in the development of this PCOS feature (Akın et al. 2020). This may underline that EDCs can exert distinct effects on the endocrine balance of the organism and a possible synergistic effect, along with a genetic predisposition, could cause the development of PCOS and its associated disorders.

13.5.2 Endometriosis

Endometriosis is a gynecological condition where endometrial tissue grows outside the uterus and it is a common cause of pelvic pain. Etiology of this disease is still unclear, but ectopic endometrial cells maintain estrogen sensitivity and numerous factors, including EDCs may intervene (M. G. Porpora et al. 2013; Sirohi et al. 2020). PCBs have been described to induce development of endometriosis in female Rhesus monkeys in a dose-dependent manner (Rier et al. 2001). While there is an increased risk of endometriosis in women with higher levels of serum PCBs/p,p'-DDE (Porpora et al. 2009), a causative association still cannot be demonstrated. Similarly, it has been hypothesized that phthalate esters might also be involved in the etiology of endometriosis. While the exact mechanism triggering the development of endometriosis by phthalates is unclear, several experiments showed that phthalate esters might increase viability and proliferation of endometrial cells, suggesting a role in the onset of endometriosis (Kim et al. 2010; Kim et al. 2015). A recent metanalysis of clinical studies found a significant association between the exposure to MEHHP and endometriosis but failed to find associations with other phthalate esters (MEHP, MEP, MBzP, and MEOHP) (Cai et al. 2019). This could be explained by a few confounding factors from the epidemiological study included in the analysis (age, health status, different ethnicity and environment). Nonetheless, even if the evidence is weak, a great attention is needed in monitoring of these substances.

13.6 Conclusions

Current evidence does not fully explain the impact of EDs on human male reproductive health. Most of the harmful effects of EDCs were observed in experimental conditions on animals, often after exposure to doses higher than those expected from accidental environmental contact. Not surprisingly, human study results are relatively scant and controversial. This lack of consistency in experimental and in vivo evidence will definitely require further investigations, but, overall, we must prudently point toward a positive association between EDC exposure and reproductive system damage. Fetal exposure during a precise gonadal developmental window might be particularly destructive and be associated with several degrees of gonadal hormone disruption and the presence of congenital malformations. Xeno-estrogenic effects of several EDCs might impair pubertal development. On the other hand, several in vivo observed effects such as the correlation between EDs and testicular volume, as well as that with testicular cancer, are more uncertain. Although the observed effects may be different among individuals and depending on possible exposure to different cocktails of EDCs, the putative biological link with human reproduction should not be ignored. Further long-term population studies appear necessary in order to identify damaging compounds, clarify sources of exposure, and, possibly, guide in the replacement with harmless substances.

References

Abraham A, Chakraborty P (2020) A review on sources and health impacts of bisphenol A. Rev Environ Health 35(2):201–210

Aguilar-Garduño C, Lacasaña M, Blanco-Muñoz J, Rodríguez-Barranco M, Hernández AF, Bassol S, González-Alzaga B, Cebrián ME (2013) Changes in male hormone profile after occupational organophosphate exposure. A longitudinal study. Toxicology 307(May):55–65

Ahbab MA, Barlas N (2015) Influence of in utero Di-N-hexyl phthalate and dicyclohexyl phthalate on fetal testicular development in rats. Toxicol Lett 233(2):125–137

Akın L, Kendirci M, Narin F, Kurtoğlu S, Hatipoğlu N, Elmalı F (2020) Endocrine disruptors and polycystic ovary syndrome: phthalates. J Clin Res Pediatr Endocrinol 12(4):393–400

Alonso-Magdalena P, Rivera FJ, Guerrero-Bosagna C (2016) Bisphenol-A and metabolic diseases: epigenetic, developmental and transgenerational basis. Environ Epigenetics 2(3):dvw022

Amiri M, Tehrani FR (2020) Potential adverse effects of female and male obesity on fertility: a narrative review. Int J Endocrinol Metab 18(3):e101776

Annamalai J, Namasivayam V (2015) Endocrine disrupting chemicals in the atmosphere: their effects on humans and wildlife. Environ Int 76(March):78–97

Barouki R (2017) Endocrine disruptors: revisiting concepts and dogma in toxicology. C R Biol 340(9-10):410–413

Barrett ES, Sobolewski M (2014) Polycystic ovary syndrome: do endocrine-disrupting chemicals play a role? Semin Reprod Med 32(3):166–176

Bloom MS, Mok-Lin E, Fujimoto VY (2016) Bisphenol A and ovarian steroidogenesis. Fertil Steril 106(4):857–863

Cai W, Yang J, Liu Y, Bi Y, Wang H (2019) Association between phthalate metabolites and risk of endometriosis: a meta-analysis. Int J Environ Res Public Health 16(19):3678. https://doi.org/10.3390/ijerph16193678

Cano-Sancho G, Marchand P, Le Bizec B, Antignac J-P (2020) The challenging use and interpretation of blood biomarkers of exposure related to lipophilic endocrine disrupting chemicals in environmental health studies. Mol Cell Endocrinol 499(January):110606

Cargnelutti, Francesco, Andrea Di Nisio, Francesco Pallotti, Iva Sabovic, Matteo Spaziani, Maria Grazia Tarsitano, Donatella Paoli, and Carlo Foresta. 2020. "Effects of endocrine disruptors on fetal testis development, male puberty, and transition age." Endocrine 72(2):358-374, https://doi.org/10.1007/s12020-020-02436-9.

Van Cauwenbergh O, Di Serafino A, Tytgat J, Soubry A (2020) Transgenerational epigenetic effects from male exposure to endocrine-disrupting compounds: a systematic review on research in mammals. Clin Epigenetics 12(1):65

Chevalier N, Brucker-Davis F, Lahlou N, Coquillard P, Pugeat M, Pacini P, Panaïa-Ferrari P, Wagner-Mahler K, Fénichel P (2015) A negative correlation between insulin-like peptide 3 and bisphenol a in human cord blood suggests an effect of endocrine disruptors on testicular descent during fetal development. Hum Reprod 30(2):447–453

Conley JM, Lambright CS, Evans N, Cardon M, Furr J, Wilson VS, Gray LE Jr (2018) Mixed 'antiandrogenic' chemicals at low individual doses produce reproductive tract malformations in the male rat. Toxicol Sci 164(1):166–178

Corona G, Sansone A, Pallotti F, Ferlin A, Pivonello R, Isidori AM, Maggi M, Jannini EA (2020) People smoke for nicotine, but lose sexual and reproductive health for tar: a narrative review on the effect of cigarette smoking on male sexuality and reproduction. J Endocrinol Investig 43(10):1391–1408

Cui Q, Pan Y, Wang J, Liu H, Yao B, Dai J (2020) Exposure to per- and polyfluoroalkyl substances (PFASs) in serum versus semen and their association with male reproductive hormones. Environ Pollut 266(Pt 2):115330

Den Hond E, Schoeters G (2006) Endocrine disrupters and human puberty. Int J Androl 29(1):264–271; discussion 286–90

Den Hond E, Tournaye H, De Sutter P, Ombelet W, Baeyens W, Covaci A, Cox B, Nawrot TS, Van Larebeke N, D'Hooghe T (2015) Human exposure to endocrine disrupting chemicals and fertility: a case-control study in male subfertility patients. Environ Int 84(November):154–160

Di Nisio A, Sabovic I, Valente U, Tescari S, Rocca MS, Guidolin D, Dall'Acqua S et al (2019) Endocrine disruption of androgenic activity by perfluoroalkyl substances: clinical and experimental evidence. J Clin Endocrinol Metab 104(4):1259–1271

Doherty BT, Hammel SC, Daniels JL, Stapleton HM, Hoffman K (2019) Organophosphate esters: are these flame retardants and plasticizers affecting children's health? Curr Environ Health Rep 6(4):201–213

du Fossé NA, van der Hoorn M-LP, van Lith JMM, le Cessie S, Lashley EELO (2020) Advanced paternal age is associated with an increased risk of spontaneous miscarriage: a systematic review and meta-analysis. Hum Reprod Update 26(5):650–669

Ehrlich S, Williams PL, Missmer SA, Flaws JA, Ye X, Calafat AM, Petrozza JC, Wright D, Hauser R (2012) Urinary bisphenol a concentrations and early reproductive health outcomes among women undergoing IVF. Hum Reprod 27(12):3583–3592

Escobar-Morreale HF (2018) Polycystic ovary syndrome: definition, aetiology, diagnosis and treatment. Nat Rev Endocrinol 14(5):270–284

Ferguson KK, Peterson KE, Lee JM, Mercado-García A, Blank-Goldenberg C, Téllez-Rojo MM, Meeker JD (2014) Prenatal and peripubertal phthalates and bisphenol a in relation to sex hormones and puberty in boys. Reprod Toxicol 47(August):70–76

Fernández M, Bianchi M, Lux-Lantos V, Libertun C (2009) Neonatal exposure to bisphenol a alters reproductive parameters and gonadotropin releasing hormone signaling in female rats. Environ Health Perspect 117(5):757–762

Fernández M, Bourguignon N, Lux-Lantos V, Libertun C (2010) Neonatal exposure to bisphenol a and reproductive and endocrine alterations resembling the polycystic ovarian syndrome in adult rats. Environ Health Perspect 118(9):1217–1222

Fisher JS, Macpherson S, Marchetti N, Sharpe RM (2003) Human 'testicular dysgenesis syndrome': a possible model using in-utero exposure of the rat to dibutyl phthalate. Hum Reprod 18(7):1383–1394

Foresta C, Tescari S, Di Nisio A (2018) Impact of perfluorochemicals on human health and reproduction: a male's perspective. J Endocrinol Investig 41(6):639–645

Ge W, Li L, Dyce PW, De Felici M, Shen W (2019) Establishment and depletion of the ovarian reserve: physiology and impact of environmental chemicals. Cell Mol Life Sci 76(9):1729–1746

Gianfrilli D, Ferlin A, Isidori AM, Garolla A, Maggi M, Pivonello R, Santi D et al (2019) Risk Behaviours and alcohol in adolescence are negatively associated with testicular volume: results from the amico-andrologo survey. Andrology 7(6):769–777

Giuliani A, Zuccarini M, Cichelli A, Khan H, Reale M (2020) Critical review on the presence of phthalates in food and evidence of their biological impact. Int J Environ Res Public Health 17(16):5655. https://doi.org/10.3390/ijerph17165655

Gore AC, Chappell VA, Fenton SE, Flaws JA, Nadal A, Prins GS, Toppari J, Zoeller RT (2015) Executive summary to EDC-2: the endocrine society's second scientific statement on endocrine-disrupting chemicals. Endocr Rev 36(6):593–602

Hertz-Picciotto I, Sass JB, Engel S, Bennett DH, Bradman A, Eskenazi B, Lanphear B, Whyatt R (2018) Organophosphate exposures during pregnancy and child neurodevelopment: recommendations for essential policy reforms. PLoS Med 15(10):e1002671

Jian J-M, Guo Y, Zeng L, Liang-Ying L, Lu X, Wang F, Zeng EY (2017) Global distribution of perfluorochemicals (PFCs) in potential human exposure source- a review. Environ Int 108(November):51–62

Kabir ER, Rahman MS, Rahman I (2015) A review on endocrine disruptors and their possible impacts on human health. Environ Toxicol Pharmacol 40(1):241–258

Kahn LG, Philippat C, Nakayama SF, Slama R, Trasande L (2020) Endocrine-disrupting chemicals: implications for human health. Lancet Diab Endocrinol 8(8):703–718

Kaushal J, Khatri M, Arya SK (2021) A treatise on organophosphate pesticide pollution: current strategies and advancements in their environmental degradation and elimination. Ecotoxicol Environ Saf 207(January):111483

Kim MJ, Park YJ (2019) Bisphenols and thyroid hormone. Endocrinol Metab (Seoul, Korea) 34(4):340–348

Kim Y-H, Kim SH, Lee HW, Chae HD, Kim C-H, Kang BM (2010) Increased viability of endometrial cells by in vitro treatment with Di-(2-ethylhexyl) phthalate. Fertil Steril 94(6):2413–2416

Kim SH, Cho S, Ihm HJ, Young Sang O, Heo S-H, Chun S, Im H, Chae HD, Kim C-H, Kang BM (2015) Possible role of phthalate in the pathogenesis of endometriosis: in vitro, animal, and human data. J Clin Endocrinol Metab 100(12):E1502–E1511

Konieczna A, Rachoń D, Owczarek K, Kubica P, Kowalewska A, Kudłak B, Wasik A, Namieśnik J (2018) Serum bisphenol a concentrations correlate with serum testosterone levels in women with polycystic ovary syndrome. Reprod Toxicol 82(December):32–37

Kortenkamp A (2020) Which chemicals should be grouped together for mixture risk assessments of male reproductive disorders? Mol Cell Endocrinol 499(January):110581

Lara NLM, van den Driesche S, Macpherson S, França LR, Sharpe RM (2017) Dibutyl phthalate induced testicular dysgenesis originates after seminiferous cord formation in rats. Sci Rep 7(1):2521

Lee DH (2018) Evidence of the possible harm of endocrine-disrupting chemicals in humans: ongoing debates and key issues. Endocrinol Metab (Seoul, Korea) 33(1):44–52

Lee D-H, Jacobs DR Jr (2019) New approaches to cope with possible harms of low-dose environmental chemicals. J Epidemiol Community Health 73(3):193–197

Leijs MM, Koppe JG, Olie K, van Aalderen WMC, de Voogt P, Vulsma T, Westra M, ten Tusscher GW (2008) Delayed initiation of breast development in girls with higher prenatal dioxin exposure; a longitudinal cohort study. Chemosphere 73(6):999–1004

Levine H, Jørgensen N, Martino-Andrade A, Mendiola J, Weksler-Derri D, Mindlis I, Pinotti R, Swan SH (2017) Temporal trends in sperm count: a systematic review and meta-regression analysis. Hum Reprod Update 23(6):646–659

Li L, Tiao B, Huina S, Chen Z, Liang Y, Zhang G, Zhu D et al (2015) Inutero exposure to diisononyl phthalate caused testicular dysgenesis of rat fetal testis. Toxicol Lett 232(2):466–474

Lv Y, Li L, Fang Y, Chen P, Wu S, Chen X, Ni C et al (2019) In utero exposure to bisphenol a disrupts fetal testis development in rats. Environ Pollut 246(March):217–224

Mahalingam S, Ther L, Gao L, Wang W, Ziv-Gal A, Flaws JA (2017) The effects of in utero bisphenol a exposure on ovarian follicle numbers and steroidogenesis in the F1 and F2 generations of mice. Reprod Toxicol 74(December):150–157

Mann U, Shiff B, Patel P (2020) Reasons for worldwide decline in male fertility. Curr Opin Urol 30(3):296–301

Mantovani A (2016) Endocrine disrupters and the safety of food chains. Horm Res Paediatr 86(4):279–288

Meeker JD, Hauser R (2010) Exposure to polychlorinated biphenyls (PCBs) and male reproduction. Syst Biol Reprod Med 56(2):122–131

Meling DD, Warner GR, Szumski JR, Gao L, Gonsioroski AV, Rattan S, Flaws JA (2020) The effects of a phthalate metabolite mixture on antral follicle growth and sex steroid synthesis in mice. Toxicol Appl Pharmacol 388(February):114875

Meyer J, Bester K (2004) Organophosphate flame retardants and plasticisers in wastewater treatment plants. J Environ Monit 6(7):599–605

Mlynarcikova A, Fickova M, Scsukova S (2014) Impact of endocrine disruptors on ovarian steroidogenesis. Endocr Regul 48(4):201–224

Mok-Lin E, Ehrlich S, Williams PL, Petrozza J, Wright DL, Calafat AM, Ye X, Hauser R (2010) Urinary bisphenol a concentrations and ovarian response among women undergoing IVF. Int J Androl 33(2):385–393

Murugesan P, Balaganesh M, Balasubramanian K, Arunakaran J (2007) Effects of polychlorinated biphenyl (Aroclor 1254) on steroidogenesis and antioxidant system in cultured adult rat Leydig cells. J Endocrinol 192(2):325–338

Nantaba F, Palm W-U, Wasswa J, Bouwman H, Kylin H, Kümmerer K (2021) Temporal dynamics and ecotoxicological risk assessment of personal care products, phthalate ester plasticizers, and organophosphorus flame retardants in water from Lake Victoria, Uganda. Chemosphere 262(January):127716

Narayana K, Prashanthi N, Nayanatara A, Bairy LK, D'Souza UJA (2006) An organophosphate insecticide methyl parathion (o- O- dimethyl O-4-nitrophenyl phosphorothioate) induces cytotoxic damage and tubular atrophy in the testis despite elevated testosterone level in the rat. J Toxicol Sci 31(3):177–189

Palioura E, Diamanti-Kandarakis E (2015) Polycystic ovary syndrome (PCOS) and endocrine disrupting chemicals (EDCs). Rev Endocr Metab Disord 16(4):365–371

Pallotti Francesco, Marianna Pelloni, Daniele Gianfrilli, Andrea Lenzi, Francesco Lombardo, and Donatella Paoli. 2020. "Mechanisms of testicular disruption from exposure to bisphenol a and phtalates." J Clin Med Res 9 (2):471. https://doi.org/10.3390/jcm9020471.

Paoli D, Giannandrea F, Gallo M, Turci R, Cattaruzza MS, Lombardo F, Lenzi A, Gandini L (2015) Exposure to polychlorinated biphenyls and hexachlorobenzene, semen quality and testicular cancer risk. J Endocrinol Investig 38(7):745–752

Paoli Donatella, Francesco Pallotti, Anna Pia Dima, Elena Albani, Carlo Alviggi, Franco Causio, Carola Conca Dioguardi, et al. 2020. "Phthalates and bisphenol a: presence in blood serum and follicular fluid of Italian women undergoing assisted reproduction techniques." Toxics 8 (4):91. https://doi.org/10.3390/toxics8040091.

Pathirana IN, Kawate N, Tsuji M, Takahashi M, Hatoya S, Inaba T, Tamada H (2011) In vitro effects of estradiol-17β, monobutyl phthalate and mono-(2-ethylhexyl) phthalate on the secretion of testosterone and insulin-like peptide 3 by interstitial cells of scrotal and retained testes in dogs. Theriogenology 76(7):1227–1233

Piazza MJ, Urbanetz AA (2019) Environmental toxins and the impact of other endocrine disrupting chemicals in women's reproductive health. JBRA Assist Reprod 23(2):154–164

Pitto, Letizia, Francesca Gorini, Fabrizio Bianchi, and Elena Guzzolino. 2020. "New insights into mechanisms of endocrine-disrupting chemicals in thyroid diseases: the epigenetic way." Int J Environ Res Public Health 17(21): 7787. https://doi.org/10.3390/ijerph17217787.

Pivonello C, Muscogiuri G, Nardone A, Garifalos F, Provvisiero DP, Verde N, de Angelis C et al (2020) Bisphenol A: an emerging threat to female fertility. Reprod Biol Endocrinol 18(1):22

Porpora MG, Medda E, Abballe A, Bolli S, De Angelis I, di Domenico A, Ferro A et al (2009) Endometriosis and organochlorinated environmental pollutants: a case-control study on Italian women of reproductive age. Environ Health Perspect 117(7):1070–1075

Porpora MG, Resta S, Fuggetta E, Storelli P, Megiorni F, Manganaro L, De Felip E (2013) Role of environmental organochlorinated pollutants in the development of endometriosis. Clin Exp Obstet Gynecol 40(4):565–567

Rak A, Zajda K, Gregoraszczuk EŁ (2017) Endocrine disrupting compounds modulates adiponectin secretion, expression of its receptors and action on steroidogenesis in ovarian follicle. Reprod Toxicol 69(April):204–211

Recio-Vega R, Ocampo-Gómez G, Borja-Aburto VH, Moran-Martínez J, Cebrian-Garcia ME (2008) Organophosphorus pesticide exposure decreases sperm quality: association between sperm parameters and urinary pesticide levels. J Appl Toxicol 28(5):674–680

Rier SE, Turner WE, Martin DC, Morris R, Lucier GW, Clark GC (2001) Serum levels of TCDD and dioxin-like chemicals in rhesus monkeys chronically exposed to dioxin: correlation of increased serum PCB levels with endometriosis. Toxicol Sci 59(1):147–159

Saiyed H, Dewan A, Bhatnagar V, Shenoy U, Shenoy R, Rajmohan H, Patel K et al (2003) Effect of endosulfan on male reproductive development. Environ Health Perspect 111(16):1958–1962

Sergeyev O, Burns JS, Williams PL, Korrick SA, Lee MM, Revich B, Hauser R (2017) The association of peripubertal serum concentrations of organochlorine chemicals and blood Lead with growth and pubertal development in a longitudinal cohort of boys: a review of published results from the Russian Children's study. Rev Environ Health 32(1-2):83–92

Shi, Yanbin, Wen Qi, Qi Xu, Zheng Wang, Xiaolian Cao, Liting Zhou, and Lin Ye. 2020. "The role of epigenetics in the reproductive toxicity of environmental endocrine disruptors." Environ Mol Mutagen, 62(1):78-88. https://doi.org/10.1002/em.22414.

Siracusa JS, Yin L, Measel E, Liang S, Xiaozhong Y (2018) Effects of bisphenol a and its analogs on reproductive health: a mini review. Reprod Toxicol 79(August):96–123

Sirohi D, Al Ramadhani R, Knibbs LD (2020) Environmental exposures to endocrine disrupting chemicals (EDCs) and their role in endometriosis: a systematic literature review. Rev Environ Health 36((1):101–115. https://doi.org/10.1515/reveh-2020-0046

Skakkebaek NE, Meyts ER-D, Buck GM, Louis JT, Andersson A-M, Eisenberg ML, Jensen TK et al (2016) Male reproductive disorders and fertility trends: influences of environment and genetic susceptibility. Physiol Rev 96(1):55–97

Souter I, Smith KW, Dimitriadis I, Ehrlich S, Williams PL, Calafat AM, Hauser R (2013) The association of bisphenol-a urinary concentrations with antral follicle counts and other measures of ovarian reserve in women undergoing infertility treatments. Reprod Toxicol 42(December):224–231

Su C, Cui Y, Liu D, Zhang H, Baninla Y (2020) Endocrine disrupting compounds, pharmaceuticals and personal care products in the aquatic environment of China: which chemicals are the prioritized ones? Sci Total Environ 720(June):137652

Tian M, Huang Q, Wang H, Martin FL, Liu L, Zhang J, Shen H (2019) Biphasic effects of perfluorooctanoic acid on steroidogenesis in mouse Leydig tumour cells. Reprod Toxicol 83(January):54–62

Vandenberg LN, Hauser R, Marcus M, Olea N, Welshons WV (2007) Human exposure to bisphenol a (BPA). Reprod Toxicol 24(2):139–177

Vandenberg, Laura N., Theo Colborn, Tyrone B. Hayes, Jerrold J. Heindel, David R. Jacobs Jr, Duk-Hee Lee, Toshi Shioda, et al. 2012. "Hormones and endocrine-disrupting chemicals: low-dose effects and nonmonotonic dose responses." Endocr Rev 33 (3): 378–455.

Ventrice P, Ventrice D, Russo E, De Sarro G (2013) Phthalates: European regulation, chemistry, pharmacokinetic and related toxicity. Environ Toxicol Pharmacol 36(1):88–96

Wang X, Jiang S-W, Wang L, Sun Y, Xu F, He H, Wang S, Zhang Z, Pan X (2018) Interfering effects of bisphenol a on in vitro growth of preantral follicles and maturation of oocyes. Clin Chim Acta 485(October):119–125

Watkins DJ, Sánchez BN, Téllez-Rojo MM, Lee JM, Mercado-García A, Blank-Goldenberg C, Peterson KE, Meeker JD (2017) Impact of phthalate and BPA exposure during in utero windows of susceptibility on reproductive hormones and sexual maturation in peripubertal males. Environ Health 16(1):69

Welshons WV, Thayer KA, Judy BM, Taylor JA, Curran EM, Frederick S v S (2003) Large effects from small exposures. I. Mechanisms for endocrine-disrupting chemicals with estrogenic activity. Environ Health Perspect 111(8):994–1006

White SS, Fenton SE, Hines EP (2011) Endocrine disrupting properties of perfluorooctanoic acid. J Steroid Biochem Mol Biol 127(1-2):16–26

Wilson VS, Lambright C, Furr J, Ostby J, Wood C, Held G, Earl Gray L Jr (2004) Phthalate ester-induced gubernacular lesions are associated with reduced insl3 gene expression in the fetal rat testis. Toxicol Lett 146(3):207–215

Yao Y, Li M, Pan L, Duan Y, Duan X, Li Y, Sun H (2021) Exposure to organophosphate ester flame retardants and plasticizers during pregnancy: thyroid endocrine disruption and mediation role of oxidative stress. Environ Int 146(January):106215

Yeung BH, Wan HT, Law AY, Wong CK (2011) Endocrine disrupting chemicals: multiple effects on testicular signaling and spermatogenesis. Spermatogenesis 1(3):231–239

Zhu Y-J, Jiang J-T, Ma L, Zhang J, Hong Y, Liao K, Liu Q, Liu G-H (2009) Molecular and toxicologic research in newborn Hypospadiac male rats following in utero exposure to Di-N-butyl phthalate (DBP). Toxicology 260(1-3):120–125

Environmental Factors in the Development of Diabetes Mellitus

14

Caterina Formichi, Andrea Trimarchi, Carla Maccora, Laura Nigi, and Francesco Dotta

Abstract

Diabetes mellitus is a global health issue, with a multifactorial pathogenesis. The increasing prevalence of diabetes requires a careful assessment of risk factors for an adequate preventive strategy. Recently, a great interest has been focused on the environmental factors associated with the onset of the disease, from the best known lifestyle factors, such as diet and physical activity, to environmental pollution and gut microbiome. It is now evident that epigenetic modulation has a pivotal role in determining the phenotype of genetically predisposed individuals in response to environmental factors. Epigenetic changes can occur both in utero and after birth, influencing the onset of chronic diseases, such as diabetes mellitus, in adulthood. Indeed, it has been demonstrated that these new actors in diabetes pathogenesis could represent promising therapeutic targets. Therefore, major efforts are needed to achieve a better understanding of the risk factors influencing the pathogenesis of the disease in order to allow better treatment and prevention strategies.

C. Formichi · A. Trimarchi · C. Maccora · L. Nigi
F. Dotta (✉)
Department of Medicine, Surgery and Neurosciences, University of Siena, Siena, Italy
e-mail: caterina.formichi@unisi.it; francesco.dotta@unisi.it

Keywords

Diabetes mellitus · Environment · Lifestyle Gut microbiome · Epigenetics · Endocrine disruptors · Pollution · Virus

14.1 Environment and Endocrine Diseases

A large body of evidence demonstrates that genetic variability alone cannot explain the highly different risks of developing chronic diseases. Genome-Wide Association Studies (GWAS) revealed a relatively limited causal effect (estimated less than 20%) of genetic susceptibility on phenotypic variance (Guerrero-Bosagna and Skinner 2012).

As a matter of fact, environmental exposure and its interaction with genetic components play an important role in disease development, especially in non-communicable diseases (NCDs), such as cancer, asthma, allergy, cardiovascular and endocrine disease including obesity and diabetes mellitus. However, our current knowledge of environmental factors possibly contributing to the development of NCDs is still limited. It is very likely that the main environmental risk factors are still, at least in part, unknown and their effects on human health are probably underestimated. In addition, the occurrence of a synergic effect of environmental exposure with biologic and/or behavioral aspects regulating the impact on human health should be taken into account (Guerrero-Bosagna and Skinner 2012; Renz et al. 2017).

In 2005, epidemiologist Christopher Wild first proposed the concept of *exposome*, to address global exposure assessment throughout the life course. The exposome is not merely composed by pollutants, but it consists of every exposure an individual experiences, from the origin of life—at the moment of conception—to death (Renz et al. 2017; Wild 2012). Therefore, both the nature of exposures and their temporal variation must be taken into account (Wild 2012). Briefly, three broad categories of non-genetic exposures may be distinguished, according to Wild: (a) internal processes (e.g. metabolism, hormonal balance, morphotype, gut microbiota, aging, etc.), (b) specific external factors (e.g. infections, pollutants, smoking, drugs, etc.), and (c) general external factors (e.g. socio-economic status, urban versus rural environment, climate, etc.) (Wild 2012). Though fascinating, the characterization of exposome represents a real challenge, requiring expensive technologies and a methodological approach taking into account the association of several exposures bearing an impact on human health, and the dynamic nature of the exposome—which implies that its innumerable components must be considered in relation to their changes at any given time point. Indeed, the exposure profile of the same individual varies over time (Renz et al. 2017; Wild 2012).

Studies on exposome are ongoing, all over the world, based on the idea that a better understanding of environmental risk factors and their impact on health could lead to more efficient prevention strategies (Wild 2012). An important tool to integrate GWASs and environment-wide association studies of both individuals and communities in their dynamic environment is represented by the application of high throughput technologies (i.e. "-omics techniques") and computational solutions for the interpretation/analysis of these technology-generated data sets (Renz et al. 2017).

14.1.1 Environmental Epigenetics

The crucial role of environment in shaping phenotypes, during developmental and biological processes, has been long recognized and investigated. However, the basic molecular mechanisms of environmentally induced modulation of long-term gene expression were not fully addressed, until epigenetics studies recently allowed to tackle these issues (Guerrero-Bosagna and Skinner 2012). Epigenetics deals with gene expression regulation, producing inheritable phenotypic changes, without altering the DNA coding sequence (Tiffon 2018). The identification of epigenetic processes provides a clue on genome activity regulation by environmental signals, and on the potential environmental interference with developmental processes involved in determining adult phenotype (Guerrero-Bosagna and Skinner 2012). Epigenetics may play a role in disease pathogenesis through a wide range of phenotypic effects; the set of epigenetic changes produced by environmental exposures has been referred to as "environmental epigenetics" (Guerrero-Bosagna and Skinner 2012; Tiffon 2018).

Briefly, epigenetics provides an additional level of control in gene expression through several epigenetics "marks"—a term describing chemical compounds added to DNA or histones—which induce conformational chromatin changes, either promoting or suppressing gene expression (Tiffon 2018). Indeed, the spatial conformation of chromatin is critical in regulating access of transcriptional factors to DNA: less condensed chromatin (*euchromatin*) is accessible for transcriptional machinery and is associated with greater transcriptional activity, while more tightly packed chromatin (*heterochromatin*) is not actively transcribed. The most studied epigenetic marks are DNA methylation—the process by which methyl groups are added to the cytosine residues in CpG islands to form 5-methylcytosines—usually inducing gene silencing, and histone acetylation—the transfer of an acetyl group on histone tails by histone acetyltransferases—that usually loosens chromatin (Tiffon 2018). Other epigenetic mechanisms include: histone post-translational modifications such as methylation, ubiquitination, sumoylation, phosphorylation, biotinylation, and ADP-ribosylation, which can result in either inhibition or up-regulation of transcriptional activity (Tiffon 2018).

Another type of epigenetic mark is represented by non-coding RNAs (ncRNAs), functional RNA molecules transcribed from DNA but not translated into proteins, which have been proved to modulate several cellular processes (Tiffon 2018). Among ncRNAs, microRNAs (miRNAs) recently gained widespread attention as central regulators of various cellular processes, from cell differentiation and proliferation, to apoptosis (Nigi et al. 2018). MiRNAs are small endogenous non-coding RNAs (19–24 nucleotides) that negatively modulate gene expression by interacting with messenger RNA (mRNA) 3′-UnTranslated Region (UTR) of target gene, leading to mRNA degradation and/or translational repression (Nigi et al. 2018). MiRNAs altered expression have been involved in several human diseases, including metabolic and cardiovascular disorders, neoplasms, and neurological diseases (Nigi et al. 2018). Less known non-coding RNA subtypes worth mentioning are long non-coding RNAs (lncRNAs), transcripts longer than 200 nucleotides, functioning as chromatin remodelers, transcriptional and post-transcriptional regulators, which can recruit chromatin-modifying enzymes to control chromatin states, so as to influence gene expression (Tiffon 2018). Epigenetic marks participate in a complex cross-talk with feed-forward and feed-back loops. Indeed, if on the one hand miRNAs have proved to regulate DNA methylation and histone modifications, on the other hand promoter methylation or histone acetylation can in turn modify miRNA expression (Tiffon 2018).

Environmental and lifestyle factors, such as behavior, nutrition or exposure to pollutants, have been related to epigenetic changes; nutrition, in particular, is an essential in influencing the individual's epigenetic burden, from conception to death (Tiffon 2018). Indeed, a number of studies reported the epigenetic effects of diet on phenotype and susceptibility to diseases through-out life, with special attention to the impact of adverse prenatal nutritional conditions—especially in the first month of gestation—and increased risk of disease. For example, the Dutch Famine Birth Cohort suggested how starvation during pregnancy is associated with adverse health outcomes including, but not limited to, a higher risk of cardiovascular disease, metabolic and cognitive disorders, later in life (Roseboom et al. 2006). On the contrary, calorie restriction, without severe nutritional deprivation, has been shown to promote extended lifespan thanks to an anti-inflammatory effect, through the inhibitory effects on specific genes, needed for the inflammatory pathway regulation (Tiffon 2018). Nutrients can reversibly alter gene expression, either by directly inhibiting epigenetic enzymes or by indirectly affecting epigenetic enzymatic reactions, for instance by altering the availability of substrates, modifying, in turn, the expression of critical genes, affecting health and longevity (Tiffon 2018). Of note, the offspring phenotype is influenced not only by single nutrients, but also by specific dietary patterns. It has been widely demonstrated that Western diets, which is rich in saturated fats, red meats, and refined carbohydrates but low in fresh fruits and vegetables, whole grains, fish, and white meats, are associated with several diseases, including metabolic syndrome and increased risk of cancer (Tiffon 2018). Nutritional epigenetics is a quite recent subfield of epigenetics, aimed at studying how nutrients and bioactive food components can reversibly alter DNA methylation status, histone modifications and, ultimately, gene expression, with a consequent impact on global health (Tiffon 2018). A better understanding of such mechanisms will help to exploit the potential beneficial effects of bioactive food, specific nutrients, and dietary patterns to balance the negative impact of certain environmental exposures (Tiffon 2018).

The genome undergoes methylation and demethylation phases, in response to environmental factors at different stages in development, from parental gametogenesis, through fertilization, embryonic and neonatal growth, to aging (Edwards and Myers 2007). It is worth pointing out how some periods of development, namely prenatal and early postnatal period, are particularly susceptible to the detrimental effects of environmental exposures (Guerrero-Bosagna and Skinner 2012; Tiffon 2018). Exposures during this window of vulnerability are more likely to be critical in disease pathogenesis than adult exposures. Increased cellular and tissue differentiation make adults more resistant to epigenetic changes compared to the high sensitivity of a developing organism (Guerrero-Bosagna and Skinner 2012). Environmental exposures can occur both in somatic and germ cells, influencing the establishment and/or maintenance of specific epigenetic patterns (Guerrero-Bosagna and Skinner 2012). An increasing number of studies on animal models have shown how parental diet and other exposures can influence DNA methylation patterns of the embryo—that could be inherited across generations—and permanently affect postnatal health outcomes (Edwards and Myers 2007). Environmentally induced permanent germline epigenome alterations may supposedly promote epigenetic transgenerational inheritance of these phenotypes, resulting in adult onset disease (Guerrero-Bosagna and Skinner 2012). Indeed, although most environmental factors can produce alterations in the somatic cell epigenome, the transgenerational inheritance involves the occurrence of epigenetic programing during germ line differentiation (Guerrero-Bosagna and Skinner 2012). The study of methylation patterns' inheritance will allow to deepen our knowledge of the mechanisms of transmission to offspring of acquired parental methylation changes (Edwards and Myers 2007).

14.1.2 The Gut Microbiota

Among non-genetic determinants of human disease, in recent years a great deal of attention has been focused on microbiota. The microbiota can be defined as a complex population of microorganisms, including bacteria, viruses, protozoa, and fungi, inhabiting different districts of the human body; more than 70% of the microbiota is located in the gastrointestinal tract in a symbiotic relationship with the host (Pascale et al. 2018).

The human gastrointestinal tract represents one of the largest interfaces in the human body between the host and the outside environment, covering approximately 250–400 m^2 (Thursby and Juge 2017).

Human gastrointestinal tract is massively colonized from birth by a huge number of microbes, referred to as *gut microbiota*, with an estimated ratio of human to bacterial cells ranging from 1:1 to 1:10 (Thursby and Juge 2017). Recent advances in genome sequencing technologies and bioinformatics allowed for more than 2000 species to be identified in humans, and then classified into 12 different phyla, most of which belonging to Proteobacteria, Firmicutes, Actinobacteria, and Bacteroidetes (Thursby and Juge 2017; Tang et al. 2017).

Genetic and environmental factors—such as lifestyle, diet, antibiotics use, illnesses and even the mode of delivery—influence the composition of gut microbiota and are responsible for high inter-individual differences in species and subspecies, in particular in the Firmicutes to Bacteroidetes ratio (Tang et al. 2017). There is increasing evidence that the composition of gut microbiota affects host metabolism and that an altered microbial composition, known as dysbiosis, is involved in a number of pathological conditions (Thursby and Juge 2017; Tang et al. 2017).

These commensal microbes are involved in food digestion, mainly through digestion of complex carbohydrate with production of short-chain fatty acid (SCFAs) or protein fermentation, leading to the production of several metabolites (ammonia, various amines, thiols, phenols, and indoles) (Tang et al. 2017). Apart from the role in digestion, gut microbiota is also needed in a number of metabolic and immune-related processes, such as shaping and maintaining the integrity of the intestinal mucosal barrier, regulation of nutrient uptake and metabolism, production of several biologically active metabolites, support in maturation of immunological tissues—both intestinal mucosal and systemic immune system, and prevention of propagation of pathogenic microorganisms (Thursby and Juge 2017; Tang et al. 2017). The gastrointestinal microbiota is also involved in the de novo synthesis of essential vitamins, which the host cannot produce alone (Thursby and Juge 2017). Gut microbial signals are transmitted across the intestinal epithelium to communicate with host organs and tissues. Signaling molecules can be represented by structural components of microorganisms, such as lipopolysaccharide (LPS) and peptidoglycans, that can trigger several downstream signaling processes, by interacting with host receptors, both at the epithelial cell border, as well as within vasculature, particularly if gut barrier function is compromised. The interaction with host mucosal surface cells is often mediated by pattern recognition receptors (PRR), which recognize pathogen-associated molecular patterns (PAMPs) which, in turn, stimulate and shape host immune response (Tang et al. 2017). Gut microbiota can also influence host processes through bioactive metabolites—trimethylamine/trimethylamine N-oxide, SCFAs, primary and secondary bile acid—that act as endocrine signals in distant organs, directly or through interaction with other endocrine hormones, including ghrelin, leptin, glucagon-like peptide 1 (GLP-1), and peptide YY (PYY) (Tang et al. 2017). Additional bioactive microbial signals have been reported to stimulate the parasympathetic nervous system, therefore impacting metabolic processes (e.g. glucose homeostasis) linked to the development of metabolic syndrome (Tang et al. 2017). SCFAs produced by metabolic activity of gut microbiota during digestion have been reported to play an essential role in cellular processes such as gene expression, chemotaxis, differentiation, proliferation, and apoptosis and regulate metabolic pathways such as hepatic lipid and glucose homeostasis. Moreover, they are involved in appetite regulation and energy intake, both through receptor-mediated mechanisms and in an epigenetic fashion, acting as histone deacetylase inhibitors (Thursby and Juge 2017). The physical presence of the commensal microbiota in the gastrointestinal tract also limits exogenous pathogenic strains expansion, for example, by competing for attachment sites or nutrients, and producing antimicrobial peptides. Therefore, the use of antibiotics, altering the gut commensal microbiota, can lead to pathogens colonization and dissemination (Thursby and Juge 2017).

14.1.3 Endocrine Disruptors

More recently, attention has been focused on chemical disruptors and on their role in human health and disease. As far as endocrine system is concerned, since the beginning of 1940s data regarding the role of the so-called *endocrine disruptors* has been generated and published.

The endocrine system consists of interacting tissues, signaling each other and to distant target organs, through hormonal messengers, whose function is to maintain homeostasis, regulate physiological functions and development of an organism—from the prenatal period to adult age, living in an evolving environment. Hormones, chemical messengers secreted by endocrine tissues, are released into the bloodstream, through which they reach target tissues, where they exert their action by interacting with specific receptors and activating complex signaling pathways (Marty et al. 2011; Kabir et al. 2015). The human being produces more than 50 hormones, and hormonally active molecules (i.e. cytokines and neurotransmitters), allowing communication between different organs and tissues, in order to regulate and coordinate various metabolic processes (Marty et al. 2011; Kabir et al. 2015). However, when external elements (i.e. chemicals or pollutants) trigger a perturbation of the homeostasis exceeding the body's compensation and self-regulation capacities, the resulting imbalance can have serious health consequences (Marty et al. 2011).

Since the 1940s, along with the massive use of anabolic steroids in intensive livestock farming to improve the average weight gain, and of contraceptive agents, such as ethinyl estradiol, noretinodrel, and megestrol acetate, alarming reports describing their potential harmful effects began to circulate (Marty et al. 2011). Therefore, scientific community stressed the need for investigation on the potential side effects of these commonly used compounds (Marty et al. 2011).

Diethylstilbestrol (DES), for example, a compound with strong estrogenic activity, was extensively used for the treatment of habitual abortion, premature delivery, and other complications of pregnancy, until 1971 when the Food and Drug Administration (FDA) forewarned physicians not to prescribe DES to pregnant women due to increased incidence of a rare vaginal cancer in the daughters of women taking DES (Marty et al. 2011); subsequent studies also revealed an increased risk for breast cancer in DES-treated women and possible reproductive tract abnormalities in their offspring (Marty et al. 2011; Giusti et al. 1995).

Other examples of endocrine disruption have been reported in Japan and Taiwan, where researchers described low birth weight and neurological development delays in offspring of women exposed to PCBs and PCDFs, from consumption of contaminated rice oil (Marty et al. 2011; Aoki 2001).

Despite the increasing amount of reports suggesting a potential endocrine toxicity of industrial and pharmaceutical agents, no specific action was taken until the famous meeting, held at Wingspread Conference Center in Racine, Wisconsin, in July 1991, organized by Theo Colborn and coworkers to assess the state of the art about this issue. During the meeting, it was stated that many anthropogenic environmental pollutants, defined as *endocrine disruptors*, have been proven to disrupt endocrine system of both animals and humans (Colborn and Clement 1992; Colborn et al. 1996). Since then, the potential role of environmental chemicals on public health—in particular the effects of long-term exposure to low-dose endocrine disruptors—rapidly raised concerns and attracted significant public attention (Marty et al. 2011; Kabir et al. 2015). U.S. Environmental Protection Agency (EPA) defined an endocrine disrupting chemical (EDC) as "an exogenous agent that interferes with the synthesis, secretion, transport, binding, metabolism or elimination of natural hormones in the body that are responsible for the maintenance of homeostasis, reproduction, development and/or behavior" (Kavlock et al. 1996; Diamanti-Kandarakis et al. 2009). Thus, all chemical compounds or mixtures interfering with the physiological endocrine functions, so as to prevent organism-environment interaction, can be considered as endocrine disruptors (Kabir et al. 2015). Given EDCs ubiquitous presence in present life, endocrine disruption is a major concern for public health, especially in developing countries, and it is mandatory to define strategies to avoid widespread environmental contamination and prevent environmentally induced diseases (Kabir et al. 2015). Raising awareness on how exposures to EDCs are relevant to human health among the different government and policymakers should be promoted, in order to achieve immediate action to ban most of the EDCs chemicals, to implement proper monitoring systems of waste material management and to discourage the use of toxic compounds (Kabir et al. 2015).

Epidemiological evidence suggests that EDCs exposure could be related to increased incidence and prevalence of some diseases over the last 50 years—i.e. breast cancer, prostate cancer, testis cancer, diabetes mellitus, obesity, and subfertility. This increase has been attributed, at least in part, to improvement in diagnosis and in health care, while it is still challenging to assess exposure to EDCs and prove a causal association with human diseases, though time trends data are consistent with such association (Kabir et al. 2015). Indeed, vast geographical differences in incidence of some diseases in more industrialized countries are well documented: in the USA, diabetes prevalence from 1980 to 2011 increased by nearly 170% and the prevalence of obesity in children aged 6–11 years almost tripled from 1980 to 2012 and, likewise, the prevalence has quadrupled over the same period in adolescents (Kabir et al. 2015).

14.1.3.1 EDCs Classification

Natural and synthetic chemicals are extremely heterogeneous and ubiquitously present in the environment. Humans and wildlife population are surrounded by a great variety of natural and synthetic chemicals, so that exposure cannot be avoided (Kabir et al. 2015). Different substances, at different concentrations, may be found in different regions, depending on their availability; the use of a given disruptor, for example, for example, while forbidden in some countries might be allowed in some others (Kabir et al. 2015).

Despite their heterogeneity, different criteria for EDCs classification have been proposed (Table 14.1) (Kabir et al. 2015; Diamanti-Kandarakis et al. 2009; Gore et al. 2014).

Contaminated soil and water are the main source of endocrine disruptors. Disruptors in soil and in ground water are absorbed by microorganisms, plants, and algae, which in turn are eaten by animals and, lastly, by humans. Industrial and pharmaceutical processes are one of the main sources of water and soil pollution, as they result into the production of several EDCs, which are released into waste water and subsequently dispersed in the environment. In addition, some EDCs commonly used in the industrial sector (e.g. pesticide, plastic, BPA, and phthalates) can contribute to environmental pollution, through soil runoff or random discharges into water courses. Wastewater treatment systems available to date, such as flocculation, sedimentation, filtration, and chlorination, unfortunately, cannot guarantee full elimination or removal of these substances from water. Indeed, EDCs can exert their toxic effects on the endocrine system even at low doses; thus, low concentrations of chemicals in watercourses are sufficient to produce endocrine disruption (Kabir et al. 2015; Gore et al. 2014).

Apart from industrial pollution of water and soil, EDCs can also be largely found in agricultural farming. Exogenous sexual steroids are widely used to promote growth and development in farm animals, to help weight gain and to promote feed efficiency, causing high concentrations of these chemicals in the environment, further contaminating soil, water, and air in a particular area as well as in nearby areas (Marty et al. 2011; Kabir et al. 2015). EDCs can be found in many daily use products—and not always included in the chemical compound list—such as personal care products, textile materials, children products, electronic devices, and furniture. Environmental release and consequent human exposure to EDCs raise concerns in the scientific community and for governments (Kabir et al. 2015; Gore et al. 2014). Parabens, phthalates, and glycol ethers are commonly found in soaps and skin care products and can enter the bloodstream after skin absorption. Children are at high risk of poisoning, since blood–brain barrier is not fully developed in childhood, therefore EDCs often present in baby products (i.e. lead, phthalates, cadmium) can easily reach the brain (Kabir et al. 2015; Gore et al. 2014). BFRs are present in a number of commonly used products, from electronic devices to furnishing (i.e. pillows, mat-

Table 14.1 EDCs classifications. *PCBs* polychlorinated biphenyls, *PBBs* polybrominated biphenyls, *BPA* bisphenol A, *DDT* dichlorodiphenyltrichloroethane, *DES* diethylstilbestrol, *PAHs* polycyclic aromatic hydrocarbons

EDCs' classification by occurrence (Kabir et al. 2015, Diamanti-Kandarakis et al. 2009)		EDCs' classification by origin (Kabir et al. 2015)	EDCs' classification by area of use (Kabir et al. 2015)
Natural EDCs (e.g. phytoestrogen)		Hormones (natural—e.g. phytoestrogen—or synthetic—e.g. levothyroxine)	Pesticides
Man-made EDCs	Chemicals (industrial solvents or lubricants and their derivatives, e.g. PCBs or PBBs)	Drugs that can alter endocrine system (e.g. clofibrate)	Chemicals
	Plastics (e.g. BPA)	Industrial or domestic chemicals (e.g. phthalates or PCBs)	Food contact materials (e.g. BPA)
	Plasticizers		
	Pesticides (e.g. DDT or chlorpyrifos)		
	Fungicide (e.g. vinclozolin)	Substances derived from industrial and domestic processes (e.g. dioxins or PAHs)	
	Pharmaceutical agents (e.g. DES)		

tresses, chipboard furniture), from synthetic fabrics, to insulating materials and other building materials (Kabir et al. 2015; Gore et al. 2014). The toxicity of brominate flame retardants (BFRs) is well established. BFRs have a long half-life, so they can persist for a long time, after being released into the environment (Kabir et al. 2015; Gore et al. 2014). Endocrine disruptors can also be found in food and beverages containers and coatings used to protect canned food from pathogens. Bisphenol A (BPA), for example, is commonly used in food lining and can be released into food and subsequently ingested by humans (Kabir et al. 2015; Gore et al. 2014). Commonly used pesticides, such as dichlorodiphenyltrichloroethane (DDT) and chlorpyrifos, have proved to affect human physiology, especially reproductive and neural systems. DDT was once used in agricultural sector, in cattle breeding, in household and in public places; despite having been banned, it is still in use in some countries (Kabir et al. 2015; Gore et al. 2014). DDT has long been recognized as an endocrine disruptor, with potential detrimental effects on several endocrine glands, such as thyroid, endocrine pancreas, gonads, renin-angiotensin system, and neuroendocrine system (Kabir et al. 2015; Gore et al. 2014). Neurotoxicity of chlorpyrifos, an organophosphoric pesticide commonly used as an insecticide against various pests, both in household and agriculture sectors, has been confirmed in several studies (Kabir et al. 2015; Gore et al. 2014).

14.1.3.2 EDCs Absorption and Metabolism

Exposure to EDCs comes from different sources, widely present in the environment. Thus, endocrine disruptors can enter the human body in different ways (Fig. 14.1) (Kabir et al. 2015). In addition, distribution of a given EDC in a specific area is not constant or predictable (Kabir et al. 2015). To make matters worse, exposure to low-level environmental EDCs occurs in an environment in which natural EDCs and/or exposures to pharmaceuticals contribute to the background hormonal milieu (Marty et al. 2011).

EDCs exposure in adults occurs mainly through oral consumption of food and/or water contaminated by endocrine disruptors commonly used in industrial processes that can be found in soil or groundwater (e.g. PCBs, dioxins, and perfluorinated compounds), chemical substances used in food or beverage containers and pesticides residues in vegetables or fruit (Kabir et al. 2015; Gore et al. 2014). Pesticides used in agriculture or vector control (e.g. DDT, chlorpyrifos, vinclozolin) or chemicals used in home furniture, such as brominated flame retardants (BFR), have been shown to enter human body through skin contact and/or by inhalation (Kabir et al. 2015; Gore et al. 2014). Also, EDCs in polluted air can enter human body by inhalation. Skin contact is how some EDCs contained in cosmetics and personal care products (Kabir et al. 2015; Gore et al. 2014). EDCs can be found also in sanitary materials; phthalates, for instance, are used in intravenous tubes and can get into human body through intravenous path (Kabir et al. 2015; Gore et al. 2014). Finally, pregnant and lactating women, previously exposed to EDCs, can transfer EDCs to their babies through placenta and breast milk; additionally, several EDCs are contained in baby products, thus skin contact with these products expose infants to endocrine disruptors (Kabir et al. 2015; Gore et al. 2014).

The metabolism of EDCs in humans is not yet fully characterized, although several data - both in vivo and in vitro - have been published on this subject. It is argued that some metabolites may display a more hazardous disruptive profile or greater storage capacities in human tissues, compared to native compounds (Kabir et al. 2015; WHO (World Health Organization)/UNEP (United Nations Environment Programme) 2013). Non-persistent EDCs usually undergo liver metabolism and are excreted in urine and stool; on the contrary, persistent EDCs can accumulate in human tissues, especially in adipose tissue, from which they are slowly released. Actually, it is widely believed that persistent EDCs may be eliminated in breast milk by breastfeeding women, and some data confirm levels of persistent organic pollutants (POPs) in breast milk far above metabolic capacities (Kabir et al. 2015; WHO (World Health Organization)/UNEP (United Nations Environment Programme) 2013).

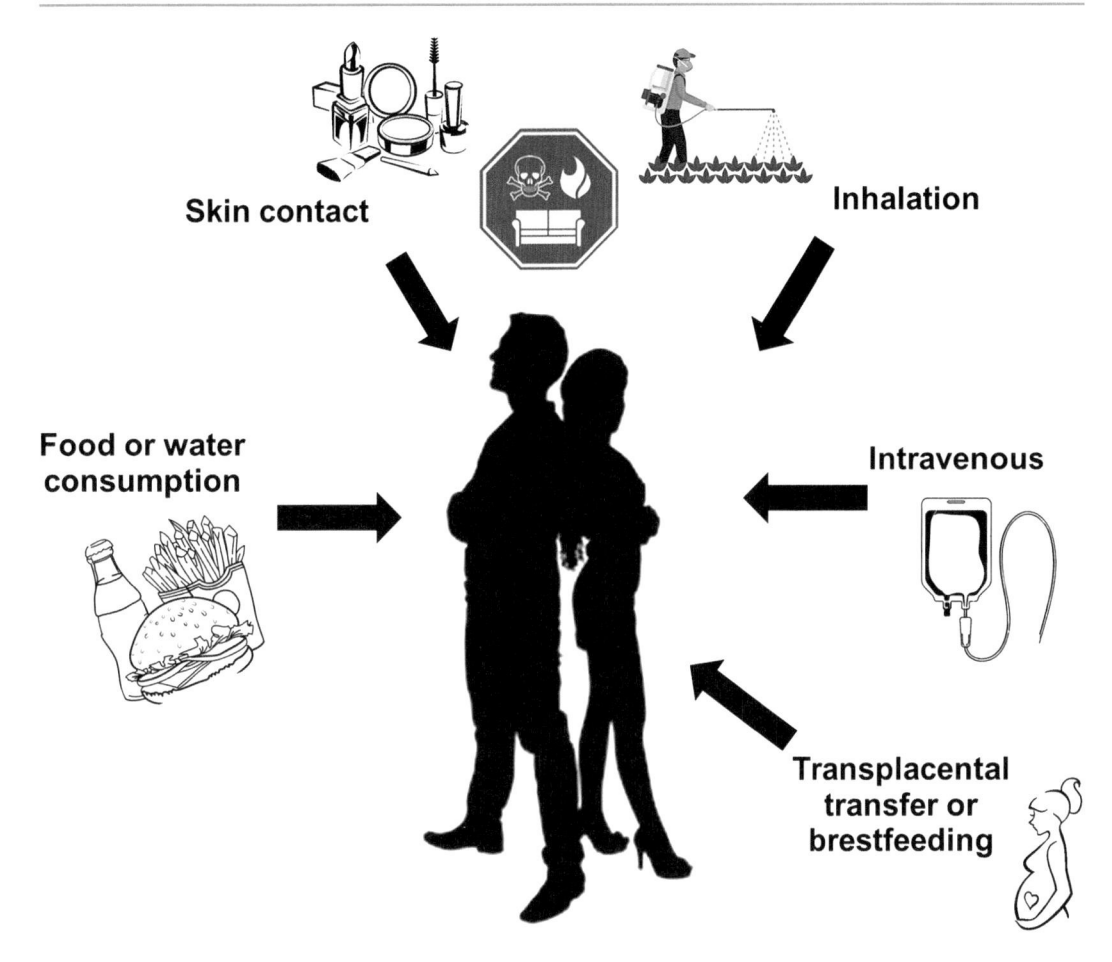

Fig. 14.1 Different ways of human exposure to EDCs

14.1.3.3 Mechanisms of Action and Effects of EDCs

Over the last three decades, many reports on endocrine disruptors, their generation and actions, and how they may lead to biological changes have been published (Kabir et al. 2015; Gore et al. 2014).

As shown in Fig. 14.2, EDCs differ greatly in terms of chemical structure, therefore it is not possible to predict whether a compound possesses endocrine disrupting activity merely on the basis of its structure nor what its mechanism of action is. However, some characteristics, such as the presence of a halogen-free phenolic ring or sulfonic functional groups or the replacement of chlorine or bromine next to OH group of phenolic ring, seem to be required for endocrine disruptive activity (Kabir et al. 2015; Diamanti-Kandarakis et al. 2009; WHO (World Health Organization)/UNEP (United Nations Environment Programme) 2013). Thanks to the rising knowledge about the mechanisms of action of endocrine disruptors, scientists progressed from the initial hypothesis of a direct interaction with nuclear hormone receptors—including estrogen receptors (ERs), androgen receptors (ARs), progesterone receptors, thyroid receptors (TRs), and retinoid receptors—to a much more detailed comprehension of the mechanisms through which EDCs yield their effects (Kabir et al. 2015; Diamanti-Kandarakis et al. 2009).

Endocrine disruptors exhibit the same characteristics as hormones, interfering with their function, subsequently altering endocrine function,

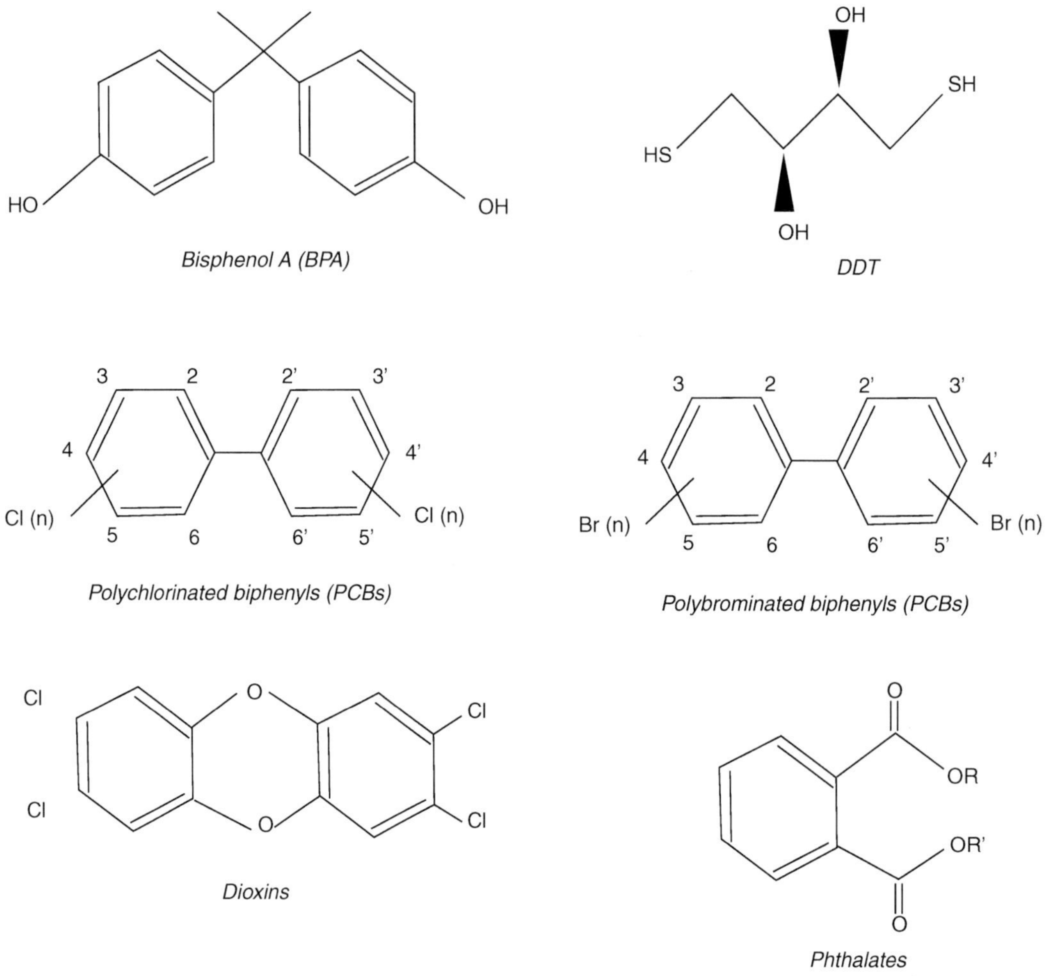

Fig. 14.2 Chemical structure of some common EDCs

such that it leads to adverse effects on human health (Kabir et al. 2015; WHO (World Health Organization)/UNEP (United Nations Environment Programme) 2013). Like hormones, EDCs interact with hormonal receptors, resulting in tissue-specific and age-specific effects (Kabir et al. 2015). As for endogenous hormones, the circulating levels do not always reflect the activity of the chemical substances, which is instead determined by the balance between the free circulating fraction and the protein-bound fraction. Both hormones and EDCs can cause permanent developmental effects. Indeed, sensitivity to EDCs action is much higher during developmental period, requiring lower doses to obtain effects compared to adulthood or aging (Kabir et al. 2015). However, there are some differences between EDCs and hormones: first of all, EDCs can interact with multiple receptors, simultaneously; EDCs can act on membrane receptors as well as nuclear receptors; EDCs can be active at different doses—some are active at low doses, others don't, with different effects at low doses versus high doses; bioaccumulation of EDC has been reported; combination effects of multiple EDCs have been reported (Kabir et al. 2015; WHO (World Health Organization)/UNEP (United Nations Environment Programme) 2013).

Hormonal disruption can arise from different mechanisms of action: direct interaction of EDCs with hormonal receptors, fully or partially imitating endogenous hormones (i.e. sexual hormones or thyroid hormones), and potentially producing overstimulation; receptor antagonism, which prevents the binding of the native hormone and interrupts the hormonal signaling; alteration of liver metabolism of hormones or of hormonal receptors, causing changes in their function; interference with binding proteins and transporters, potentially modifying hormone delivery to its target(s) (Kabir et al. 2015; WHO (World Health Organization)/UNEP (United Nations Environment Programme) 2013).

A number of studies, both in animals and humans, indicate the potential role of EDCs in affecting several endocrine glands, from reproductive systems to thyroid, and metabolic pathways (Kabir et al. 2015). Increasing understanding of EDCs actions and effects has led to a better definition of EDCs-associated endocrine dysfunctions.

The most characterized target for EDCs is thyroid gland, which has been long identified as a preferential target for disruptors. Indeed, several studies found that many EDCs are able to directly disrupt the physiological thyroid function in many phases of thyroid regulation—i.e. iodine uptake, thyroid hormone production, interconversion of thyroid hormones, cellular uptake, cell receptor activation and hormone degradation and elimination (Kabir et al. 2015; Diamanti-Kandarakis et al. 2009). Chemicals have been proven to interfere with the sodium/iodide symporter (NIS), reducing iodide uptake and thyroid hormone synthesis (i.e. perchlorate) and thyroperoxidase (TPO), inhibiting iodide oxidation and transfer to thyroglobulin (i.e. 6-propyl-2-thiouracil (PTU), isoflavones) (Kabir et al. 2015; Diamanti-Kandarakis et al. 2009). Another well-characterized target of endocrine disruptors is the reproductive system.

Of note, some EDCs can produce different dysfunctions in males or in females, acting as estrogens and/or antagonists of androgens. This phenomenon may explain the sexual dimorphism of some human disorders (Kabir et al. 2015; Diamanti-Kandarakis et al. 2009). Female reproductive system disorders potentially associated with EDCs include precocious puberty, polycystic ovary syndrome (PCOS), both further related to metabolic syndrome, menstrual irregularities and subfertility, premature ovarian failure (POF) as well as higher risk of breast and reproductive system cancers (Kabir et al. 2015; Diamanti-Kandarakis et al. 2009). Existing data also suggest that male sexual disorders, such as sperm alterations, hypospadias, and ectopic testis, might be caused by exposure to EDCs, especially during sexual differentiation of the fetus (Kabir et al. 2015; Diamanti-Kandarakis et al. 2009). Some authors reported a gradual decrease of male fertility and impaired sperm quality in several countries, though opinions in this field are conflicting. Nevertheless, variations in sperm quality have been reported, both within and between different countries, and endocrine-active chemicals' impact on sperm quality cannot be ruled out (Kabir et al. 2015; Diamanti-Kandarakis et al. 2009). Hormonal disruption can occur not only as a consequence of the direct effect of EDCs on endocrine glands, but also from disruption of neuroendocrine system, causing interference with hormonal feedback regulation by hypothalamic and pituitary deregulation (Kabir et al., 2015). Neuroendocrine disruption can result in alteration of metabolic rate and sexual differentiation, indirect effect on behavior, cognitive and sensory function disorders, derangement of neurological development as well as in some neuroteratogenic effects (Kabir et al. 2015; Gore 2010). Examples of neuroendocrine disruptors which may affect directly or indirectly endocrine functions include some PCBs, dioxins, pesticides and their related metabolites, heavy metals, synthetic steroids, tamoxifen, and phytoestrogens (Kabir et al. 2015; Diamanti-Kandarakis et al. 2009; Gore 2010). Moreover, exposure to hormonally active compounds may result into disrupted hypothalamic programing, thus negatively affecting individual reproductive success as an adult (Gore 2010).

14.1.3.4 Additional Considerations on EDCs

Although several endocrine disruptive mechanisms have been identified, the exact impact of EDCs on human health and their involvement in human disease has yet to be fully estimated. Indeed, EDCs may be beneficial or detrimental, and can exert different tissue- and age-specific effects (Marty et al. 2011; Kabir et al. 2015). In order to determine their involvement in a given disorder, one should ascertain the rate of exposure to a variety of known and unknown EDCs, personal variations in metabolism and body composition, affecting the half-life, persistence and degradation of EDC in fluids and tissues, as well as genetic background, without underestimating the latency between exposure and disease occurrence and the likelihood of chronic exposure to low amounts of multiple EDCs (Marty et al. 2011; Kabir et al. 2015; Diamanti-Kandarakis et al. 2009).

A comprehensive understanding of EDCs impact on human health must not leave out a meticulous analysis of some controversial aspects pertaining to EDCs exposure. First of all, the age of exposure might determine different effects of EDCs. Indeed, it is well known that developmental period (i.e. pre- and perinatal period) is a phase of great vulnerability, in which the subject is highly sensitive to the harmful effects of EDCs exposure, compared to adulthood; thus, the same EDC might have different effects on newborns and adults. For each individual, the set of diseases developed in adult life is determined by the interaction between genome and environment, including in utero and early postnatal EDCs exposures (Kabir et al. 2015; Diamanti-Kandarakis et al. 2009). Moreover, some EDCs are persistent in the environment, due to their long half-lives; therefore, posing a health risk for future generations—most likely through epigenetic mechanisms— although without actually being directly exposed (Kabir et al. 2015; Diamanti-Kandarakis et al. 2009; WHO (World Health Organization)/UNEP (United Nations Environment Programme) 2013; Preston et al. 2018). Another important determinant of EDCs effects is the latency period between the exposure and the occurrence of a disease. Indeed, it is hard to define the causal relationship between a given EDC and a specific medical condition, because exposure might take place long before the development of a disease (Kabir et al. 2015; Diamanti-Kandarakis et al. 2009). It should also be taken into account that an individual is often exposed to several pollutants, that could exert additive and/or synergistic effects (Kabir et al. 2015; Diamanti-Kandarakis et al. 2009). Additionally, even more concerning is that chemical disruptors may prompt a biological response at very low doses, more so if exposed during the developmental period (Marty et al. 2011; Kabir et al. 2015; Diamanti-Kandarakis et al. 2009). It is also worth noting that endocrine disruptors elicit non-monotonic dose-responses, meaning that biological effects may decrease at increasing doses of the compounds, and chemical potency of an EDC is not directly related to its affinity for a receptor but can be determined by receptor abundance (Kabir et al. 2015; Diamanti-Kandarakis et al. 2009).

14.2 Environmental Factors and Type 1 Diabetes Mellitus

Type 1 diabetes mellitus (T1DM) is a chronic immune-mediated disease characterized by a selective loss and dysfunction of pancreatic insulin-producing beta cells in genetically susceptible individuals (American Diabetes Association 2020a). The clinical onset of disease is preceded by a pre-clinical, prodromal phase ranging from few weeks to many years (Atkinson et al. 2014), during which a progressive decline of functional beta-cell mass (secondary to the immune-mediated inflammatory destructive process) develops, alongside with the appearance, in various titers and combinations, of disease-specific (but non-pathogenic) autoantibodies against beta-cell autoantigens. T1DM can occur at any age, however, there are two peaks in newly diagnosed cases between 5 and 7 years of age and at or near puberty. The incidence of T1DM shows geographical variability and the actual amount is difficult to assert, as T1DM is probably under-diagnosed (Atkinson et al. 2014). However, of note, the incidence has been increasing for the past several decades, with the number of new cases rising by approximately 3% per year in children and adolescents worldwide (Skyler et al.

2017). T1DM is a multifactorial disease in which both genetic predisposition and environmental factors (maybe acting at different time points) promote the triggering of the autoimmune response against pancreatic beta cells (Atkinson et al. 2014). The greatest genetic contribution is conferred by the HLA-DQ locus on the short arm of chromosome 6 (Pociot et al. 2010; Pociot and Lernmark 2016), specifically the DRB1*03-DQB1*0201 (DR3) and DRB1*04-DQB1*0302 (DR4) haplotypes (Mehers and Gillespie 2008). However, only a small proportion (<10%) of subjects with HLA-conferred diabetes susceptibility develops clinical disease (Virtanen and Knip 2003; Knip 2003; Knip and Simell 2012) and, in recent years, the number of children new-diagnosed for T1D with the highest risk genotype has decreased in many populations (Pociot and Lernmark 2016; Knip and Simell 2012; Todd 2010; Gillespie et al. 2004; Hermann et al. 2003). At the same time, the proportion of T1DM patients with low-risk or even protective HLA genotypes has increased (Pociot and Lernmark 2016; Knip and Simell 2012) and T1DM prevalence appears to have increased in populations whose genetic susceptibility for this disease, in previous generations, would have been considered "low" (Atkinson and Chervonsky 2012). Therefore, while genetic susceptibility is believed to be a prerequisite for the development of T1DM, a large body of evidence based on in vitro and animal models, cohort and epidemiological studies and in vivo analysis by now, supports a key role of exogenous triggers in T1DM development (Rewers and Ludvigsson 2016). Supporting the contribution of environment and/or lifestyle on the risk of developing T1DM, studies on twins demonstrated a pairwise concordance of T1DM of <40% among monozygotic twins (Steck and Rewers 2011). Moreover, there is a more than ten-fold difference in the disease incidence among European Caucasians, supporting the concept of high heterogeneity in the geographical distribution of T1DM, mentioned above, and migration studies indicate that the incidence of the disease has increased in population groups who have moved from a low incidence region to another with high incidence (Rewers and Ludvigsson 2016; Knip et al. 2005; Christoffersson et al. 2016). Interestingly, a rising incidence of T1DM in individuals over 50 years of age has been observed and resulted primarily not associated with an increased genetic risk (Gillespie et al. 2004; Hermann et al. 2003; Fourlanos et al. 2008). Finally, a seasonal pattern for both the month of birth as well as the month of T1DM diagnosis has been reported, and in particular, an increase of new-onset T1DM cases during late autumn, winter, and early spring has repeatedly been confirmed in youth (Maahs et al. 2010). All these observations emphasize the influence of environmental factors in T1DM pathogenesis, now believed to have a greater influence on the risk of islet autoimmunity and T1DM than in the past (Maahs et al. 2010; Craig et al. 2019). Over the years, epidemiological studies have shown a significant correlation between the environment and T1DM and a wide range of environmental triggers have been suggested to contribute to the pathogenesis of disease in susceptible individuals. However, in contrast to the relatively rapid advancement in identifying T1DM genes, study and confirmation of environmental factors is challenging for many reasons (Knip et al. 2005). First of all, large numbers of environmental determinants (from different categories) could contribute to the induction of T1DM or, on the contrary, to protection (Rønningen et al. 2015) and, although many environmental factors have been proposed, hardly none of them has been definitely proven as effectively responsible. Moreover, exposures may occur any time before the clinical onset of T1DM disease, from fetal to adult age (Rønningen et al. 2015), and environmental triggers may differ in distinct populations, partly depending on the "genetic background." Finally, the individual risk of developing T1DM in the general population is not very high and pretty variable. Among the environmental factors proposed as potential triggers of T1DM, the following have been thoroughly investigated: viral infections, gut microbiome, dietary factors, toxins, and stress. In this regard, because of the autoimmune process frequently starts in early life, prospective studies on children followed from birth (Rewers and

Ludvigsson 2016) have contributed to identify potential responsible factors of islet autoimmunity (operating in utero, perinatally, or during early childhood) and progression to T1DM. There is a quite number of such studies, so far, prospectively evaluating young children at increased T1DM risk (based on their HLA-DR-DQ genotype or their family history) from birth, such as the Australian Baby Diab Study, the German Baby Diab Study, the Diabetes Autoimmunity Study of the Young (DAISY), the Diabetes Prediction and Prevention Project (DIPP), and the Prospective Assessment in Newborns for Diabetes Autoimmunity (PANDA) (Couper 2001).

14.2.1 Viral Infections and T1DM

Infectious agents are among the most extensively studied potential environmental triggers of T1DM and the hypothesis that viruses could be implicated in the pathogenesis of T1DM was proposed a long time ago. In particular enteroviruses, especially coxsackievirus B-group (CVB), seem to be involved in the initiation and/or acceleration of pancreatic islet autoimmunity, through several mechanisms, not mutually exclusive, including direct destruction of pancreatic beta-cells, bystander activation of autoreactive T-cells, molecular mimicry or viral persistence (Varela-Calvino and Peakman 2003; Bergamin and Dib 2015) (Fig. 14.3).

Enteroviruses are single-strand RNA viruses of the picornaviridae family, which are transmitted by fecal-oral and, less commonly, by respiratory routes. Clinical manifestations are usually mild, making the serologic diagnosis in the acute phase very difficult (Bergamin and Dib 2015). However, after replicating in gastrointestinal or in respiratory mucosa, viruses can spread through the lymphatic system into the circulation (after a brief viremic phase) at secondary replication sites (Roivainen 2006), among which we can hypothesize the pancreas (Bergamin and Dib 2015). Enteroviral infections have been suspected to be involved in the pathogenesis of T1DM since the late 1960s, when (Gamble and Taylor 1969) a seasonal variation in the incidence of T1DM after enterovirus infection was observed (Craig et al. 2013) and a frequency increase of neutralizing antibodies against the CVB4 serotype in new-onset T1DM patients was described (Bergamin and Dib 2015). A viral causality is also supported by the observation of a T1DM spatial and/or temporal clustering (Sheehan et al. 2020), besides a seasonal variation in disease clinical onset (Sheehan et al. 2020), and by various social and demographic factors such as urban/rural status, population density, and socio-economic status (Sheehan et al. 2020). Increasing evidence is leading to demonstrate a causal association between enteroviral infections and T1DM, however, establish viruses as the inducers (or one of the inducers) of T1DM is challenging. The link between infections and autoimmunity

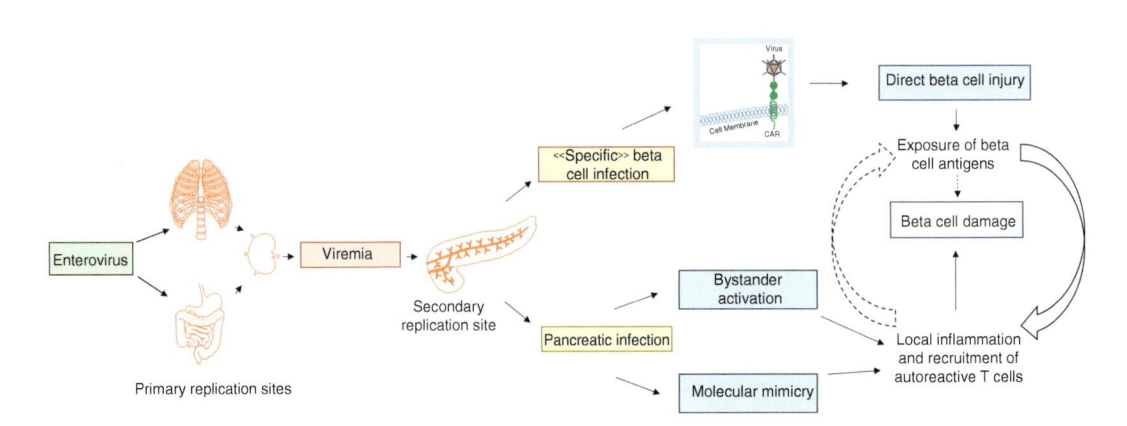

Fig. 14.3 Possible injury mechanisms induced by enterovirus infection in T1DM development

could be multifactorial (Filippi and von Herrath 2005) and several infections may act together or in a temporal sequence to trigger autoimmune process; moreover, the length of time-period between the possible triggering effect(s) and the disease clinical onset (which could also be long/very long) makes difficult to establish a direct relationship. Indeed, what we observe at clinical presentation of T1DM is the "final stage" of an autoimmune/destructive process that may has been going on for a long period of time, making hard to identify viral signatures in serum and pathogens in situ, as viruses which could play a role in the early stages of the process (Bergamin and Dib 2015; Hyöty 2016). Alternatively, viruses could trigger T1DM as a hit-and-run event, leading themselves not always identifiable. Furthermore, the so-called fertile field hypothesis suggests that viral infections make tissues a "fertile ground" for autoreactive lymphocytes to invade and expand, thus favoring the development T1DM, too (In't 2011; Fujinami et al. 2006). Finally, T1DM patients as well as healthy individuals undergo multiple viral infections during their life, and several of these viruses may even protect people from autoimmune diseases (Bergamin and Dib 2015; Rook 2012). Discrepancies in literature about the role of enteroviruses in human T1DM are present and may be due to a variety of effects of different viral strains as well as may be the result of the concomitance of other predisposing environmental factors needed for disease development (Petzold et al. 2015). However, there is an amount of data in favor of the association between enteroviral infections and T1DM. Epidemiological studies identified an increased incidence of T1DM following enterovirus epidemics (Op de Beeck and Eizirik 2016); furthermore, enteroviral RNA has been detected in peripheral blood of patients with new-onset T1DM and serological analysis, particularly for the coxsackievirus-B1 serotype (CVB1) contributed to strengthen a link between enteroviral infections and T1DM (Op de Beeck and Eizirik 2016). Of particular interest, enteroviruses were detected in pancreatic islets of patients with newly diagnosed with T1DM (Krogvold et al. 2015) and CVB4 was isolated from pancreatic tissue of recent-onset T1DM organ donors (Bergamin and Dib 2015; Dotta et al. 2007). In addition, the expression of the viral capsid protein VP1 was detected by immunohistochemistry in islets of T1DM organ donors, specifically within insulin-containing islets, and in islets of pancreatic autopsy specimens from patients with T1DM (but not in exocrine tissue) (Ylipaasto et al. 2004; Richardson et al. 2009), as well as in beta-cells (but not in glucagon secreting alpha-cells) of islets from pancreatic biopsies obtained from recent-onset T1DM patients (Op de Beeck and Eizirik 2016; Krogvold et al. 2015). More in detail, several hypotheses have been proposed to explain how enteroviruses affect T1DM. These viruses seem to have a strong pancreatic tropism (Bergamin and Dib 2015); human islets express the Coxsackie- and adenovirus receptor (Ifie et al. 2018) and beta cells are susceptible to enteroviral infections in vitro (Bergamin and Dib 2015). The specific tropism of CVBs for beta cells is supported by both in vitro as well as in vivo studies showing the ability of these enteroviruses to infect human pancreatic insulin-secreting cells (in particular, CVB4) (Ylipaasto et al. 2004; Frisk and Diderholm 2000), with consequent effects, ranging from functional damage to cell death (Grieco et al. 2012). Enteroviruses have been shown to play cytolytic effects on beta cells (which can expose previously hidden self-components), with several serotypes replicating without apparently destroying cells and to impair beta cell function (Roivainen et al. 2002). Alternatively, beta cell damage may result from a virus-induced inflammatory reaction involving pancreatic tissue. Viral infections can lead to the production of pro-inflammatory cytokines and the activation of (endogenous) antigen presenting cells (APCs), in addition to/rather than direct tissue damage. The inflammation caused by virus infection may be followed by the generation of autoreactive T cells by "bystander activation" or "molecular mimicry" or both (Fujinami et al. 2006). Based on the so-called bystander activation, viral infections may result in the impaired activation of pre-primed autoreactive T cells, which can then initiate the autoimmune process leading to T1DM in genetically predisposed individuals (Grieco et al. 2012), as studied in mice infected with CVB4 and developing clinical diabetes (Bergamin and Dib 2015; Horwitz et al. 1998). According to "molecular mimicry" (Fujinami et al. 2006), instead, the development of an autoimmune disease, including T1DM, can be the result of the activation of a T-cell population against an environmental antigen if the epitope recognized shows sequence or structural similarity with self-protein. However, molecular mimicry is extremely common in nature, suggesting that, rather, an impaired immune response may be crucial for beta-cell damage (Bergamin and Dib 2015). It can be hypothesized also a scenario in which enteroviruses (in particular CVBs) can first spread to the pancreas initiating pancreatic islet infection after internalization through specific receptors [such as human Coxsackie- and adenovirus receptor, hCAR (Ifie et al. 2018)] on beta-cells. Subsequently, viruses may be recognized by innate immune pattern recognition receptors (PRRs), initiating signaling cascades and leading to the production of pro-inflammatory chemokines and cytokines like IFNs-I, with consequent expression of interferon-stimulated genes (ISGs), whose products, directing antiviral and immunoregulatory actions, might limit infection. Even if local inflammation should contribute to eradicate the viral infection, in some genetically susceptible individuals these attempts to eliminate the virus could fail, (maybe for an exaggerated inflammatory response and/or a defective activation of intracellular anti-inflammatory or anti-apoptotic responses) leading to a progression of the inflammatory process and to beta-cell loss (Eizirik et al. 2009) by promoting IFN-I induced activation of autoreactive T-cells. In individuals unable to eliminate the virus, it could remain in beta cells in a slowly replicating form, continuously producing viral RNA and proteins stimulating the innate immune system and perpetuating inflammation and autoimmunity. A chronic/latent infection may be in agreement with the observation that chronic infections are often characteristics of autoimmune diseases. Enteroviruses could establish a chronic/persistent pancreatic infection and acquire genomic deletions, reducing their replication to very low levels (Krogvold et al. 2015; Chapman and Kim 2008; Tracy et al. 2015). A possible mechanism explaining persistence of enteroviruses within pancreatic islets has been suggested by studies of enteroviral myocarditis (Rewers and Ludvigsson 2016): a spontaneous nucleotides deletion in the 5′ non-translated region of the virome led to a low grade persistence of the defective virus, unable to induce cytopathic damage, but able to slowly replicate. In support of this, coxsackieviruses have been shown to persist in the pancreas due to N-terminal deletions in their genome, consistent with a role in maintaining beta cell autoimmunity (Tracy et al. 2015; Yeung et al. 2011).

On the contrary, some authors believe that infections may protect from T1DM autoimmunity. The age of the first exposure to enteroviruses may be critical in determining interaction between virus and the host immune system and possible development of T1DM; in support of this, enterovirus infections during the first year of life have been correlated with protection from the onset of T1DM (Juhela et al. 2000). The supporters of the so-called hygiene hypothesis suggest that children experiencing more infection in childhood are more protected; however, there has been no consensus among researchers (Tracy et al. 2010; Cooke 2009; Ludvigsson et al. 2013). More in detail, the hygiene hypothesis proposes that early exposure to infectious agents in early childhood is necessary for the maturation of the immune response. In the absence of an exposure, there could be a failure of early immune system regulation, leading to the development of autoimmunity in genetically susceptible individuals. Moreover, in support of the rise in the rates of T1DM in highly developed countries, it has also been proposed that this increasing incidence could be due to the loss of environmental protective factors. It has been hypothesized, indeed, that in countries with the highest incidence of T1DM, significant changes in lifestyle and the improvement of sanitary conditions have led to a less exposure to infection, favoring an altered immune response to environmental triggers (Bergamin and Dib 2015) among pregnant women, exposing fetuses and newborns to enteroviral infections (Rewers and Ludvigsson 2016). However, although fascinating, evidence about the hygiene hypothesis in human T1DM is minimal yet (Atkinson and Chervonsky 2012) and not supported by prospective studies.

14.2.2 The Gut Microbiome

It is well known that the microbiome interacts closely with the host immune system, suggesting its strong contribution for providing a mechanistic link between multiple environmental triggers and islet autoimmunity (Craig et al. 2019; Quercia et al. 2014; Hooper et al. 2012). The microbial colonization of the gastrointestinal tract during the first few years of life seems to be fundamental for the development of functional host immune regulation and alterations in either the composition of the microbiota or the host response can result in chronic inflammation (Sommer and Bäckhed 2013). Overgrowth of some microorganisms and reduction/loss of others results in the imbalance of the gut microbial ecosystem, defined as intestinal dysbiosis. Moreover, the composition of the gut microbiome is strongly affected by maternal environment, diet (particularly breastfeeding in early life) use of antibiotics/probiotics (Craig et al. 2019) and, importantly, the pancreas and the gut belong to the same intestinal immune system (Vaarala 2012). Assuming that, in many cases, the autoimmune process leading to T1DM starts during the first few years of life, when the intestinal microbiota develops, we can hypothesize an important implication of the microbiota in T1DM pathogenesis. The role of the gut microbiota in T1DM etiology has been investigated in recent years to clarify its role in disease development and evaluate the possibility of preventive approaches, such as dietary modifications and administration of probiotics (Atkinson and Chervonsky 2012). Dysbiosis of the gut microbiota has been suggested to be associated with islet autoimmunity and T1DM (Zheng et al. 2018).

Findings obtained show that intestinal microbiome of T1DM patients and of subjects at-risk to develop the disease differs from healthy individuals. Individuals with "pre-clinical" T1DM have a gut microbiome characterized by bacteroidetes domination at the phylum level, a scarcity of butyrate-producing bacteria, a reduction of microbial community stability, and of bacterial and functional diversity. All these changes seem to arise after the appearance of autoantibodies that precede the clinical onset of T1DM, suggesting an involvement of the gut microbiota in the progression from islet autoimmunity to clinical disease rather than at the beginning of the autoimmune process (Knip and Siljander 2016). The mechanisms promoting the progression from seroconversion to clinical disease need to be identified (Knip and Siljander 2016); however, important observations are that intestinal butyrate is considered the main energy source for epithelial cells in the colon (Hague et al. 1996) and that it is responsible for the regulation of tight junctions assembly and of gut permeability (Peng et al. 2009). Furthermore, bacteroidetes have been associated with gastrointestinal inflammation and increased intestinal permeability. In T1DM patients the amounts of bacteroidetes have been found significantly higher also, with a firmicutes to bacteroidetes ratio notably decreased, in comparison with healthy subjects (Knip and Siljander 2016). Gut microbiome may contribute by affecting intestinal permeability, molecular mimicry and by modulating both innate as well as adaptive immune system. When the intestinal permeability is increased, intestinal immune system is greater exposed to antigens such as dietary antigens (e.g. proteins and peptides), causing altered immune activation and intestinal inflammation (Vaarala 2008). Moreover, food antigens, intestinal toxins, and infectious factors may translocate from gastrointestinal lumen to intestinal mucosal components, and finally to the pancreatic lymph nodes, trigging and/or promote the progression of the autoimmune process (Bibbò et al. 2017). About molecular mimicry, a number of bacterial proteins have been showed to share similar molecular structure with pancreatic autoantigens and in NOD mice diabetogenic microbes have been observed in the gut, able to induce or speed up the onset of T1DM (Zheng et al. 2018). As for the innate immune system, gut microbiome possesses multiple pathogen-associated molecular patterns, such as LPS, lipoproteins, peptidoglycan, and their nucleic acids. Gut microbes can trigger different toll-like receptors (patter-recognition receptors recognizing pathogen-associated molecular patterns) to induce both pro-diabetogenic as well as anti-diabetogenic signals (Kieser and Kagan 2017). The interaction between gut microbes and adaptive immune system could be important as well for the pathogenesis of T1DM; in regard to this, some particular gut bacteria seem to have the capacity to regulate T-cell subsets and function [e.g. Clostridia seem to have the capability to induce regulatory T cells (Zheng et al. 2018)]. The exact mechanism for bacteria contribution in T1DM is largely unknown, so far, but short-chain fatty acids (SCFAs), which are secreted by gut microbes (resulting reduced in T1D), have been proposed to carry out an important protecting role against T1DM (Mariño et al. 2017). SCFAs may support epithelial cell integrity (as described above for butyrate) as well as the function of adaptive immune cells, e.g. promoting peripheral regulatory T-cell generation; in this way, changes in gut microbiota composition could suppress the function of regulatory T cells, providing a "fertile ground" for dysregulated immune responses and breakdown of immunological tolerance (Zheng et al. 2018; Knip and Siljander 2016). In animal models of autoimmune diabetes an increased intestinal permeability has been described as a feature preceding the clinical onset of T1DM and environmental factors able to modulate permeability seem to be associated with disease development (Vaarala et al. 2008). Both enhanced gut permeability and intestinal inflammation have been linked to the development of T1DM in humans (Vaarala 2012). However, although gut inflammation is considered a precondition for autoimmunity progression and an altered gut barrier function in early life may be causative for an impaired response to antigens, it is unknown whether aberrant responses to gut lumen antigens can induce intestinal inflammation and enhance gut permeability or if the opposite is true. Probably, potential triggers factors are individually insufficient to promote T1DM development in genetically at-risk individuals but a combination of "permissive" gut factors may be necessary (Chia et al. 2017).

14.2.3 Nutritional Factors

Nutritional factors have long been suggested to be involved in the development of islet autoimmunity and T1DM. They could act as antigenic triggers of autoimmunity or as co-factors in the scenario of a gut infection and/or inflammation (Craig et al. 2019). Most investigated dietary factors most investigated include breastfeeding, cow's milk, wheat gluten, and vitamin D (Virtanen and Knip 2003). Despite data suggest possible associations, however, there is a little clear evidence of nutritional factors involvement in the etiology of T1DM. Changes in modern food processing and storage techniques may also contribute (Elliott 2006), as suggested in mouse models in which heat treatment of foods in the presence of sugars (lactose, glucose, and fructose) or ascorbic acid (able to produce glycated products) have showed diabetogenic effects (Peppa et al. 2003). Furthermore, inhibition of advanced glycation product receptors has been shown to inhibit autoimmune diabetes in mice (Chen et al. 2004). Indeed, evidence about protective effects of breastfeeding or early introduction of cow's milk in infants against the development of islet autoimmunity and progression to T1DM in children with beta cell autoantibodies is controversial (Maahs et al. 2010). It may depend, especially for cow milk, on the interplay between a genetic susceptibility with an aberrant mucosal immunity to dietary and other proteins (Craig et al. 2019), along with increased intestinal permeability and exposure to intestinal microbiota (Harrison and Honeyman 1999). In particular, meta-analysis of retrospective studies suggests a small protective effect of breastfeeding and small detrimental effect of the early introduction of cow's milk in infants (Couper 2001). However, it is no possible to draw a conclusion because studies showed discordant results even taking into account some bias, such as lack of distinction between exclusive or partial breastfeeding. Furthermore, the duration of exclusive breastfeeding together with the age of introduction of cows' milk proteins may have influenced the outcomes. Additional confusing factors may be attributable to differences in weaning practices (e.g. weaning to hydrolyzed versus intact protein infant formula) or introduction of cereal food as first infant food rather than infant formula (Chia et al. 2017). The fact that the highest incidence of T1DM worldwide occurs in northern Europe, led to the hypothesis that that low serum concentrations of vitamin D may be a cause for T1D development. Moreover, the seasonality of birth in children with T1DM and the seasonal pattern at the onset of disease could be explained by seasonal variation in vitamin D production from exposure to the sun. Importantly, vitamin D has been examined as a potentially protective factor because it exerts anti-inflammatory effects and it has an active role in the immune system regulation, as well as in metabolic pathways relevant to diabetes. Moreover, polymorphisms in key genes for the metabolism of vitamin D have been described as modulators of T1DM risk (Infante et al. 2019). At this time, existing data suggest that new-onset T1DM patients have lower serum concentrations of this metabolite than healthy controls. However, despite heating interest in vitamin D supplementation to prevent islet autoimmunity and T1DM, there is little supporting evidence from prospective birth cohort studies and trials of vitamin D intake during pregnancy (Atkinson 2012; Dong et al. 2013). Evaluation of the introduction of solid food and cereals in infants as determinants of risk to develop islet autoimmunity as well as clinical T1DM have showed divergent results and differences across studies are in part attributable to country differences in infant nutrition. As an example, the major observational studies, DAISY, BABYDIAB, DIPP, and ABIS mentioned above, have found discordant results about the timing of introduction of any type of cereal (gluten and non-gluten containing) and the risk of islet autoimmunity (Rewers and Ludvigsson 2016; Maahs et al. 2010).

14.2.4 Perinatal and Postnatal Factors

T1DM can be diagnosed at any age, as described above, however, a large body of evidence from longitudinal cohort studies of children at-risk to develop the disease have showed that beta-cell specific autoantibodies can be present from the first year of life and that subjects who develop T1DM at a young age have a more "aggressive" disease. This concept supports the hypothesis that environmental exposures in early life may strongly contribute to T1DM risk, whether related to maternal influences on the fetus during pregnancy, neonatal factors or contributors during infancy and early childhood (Maahs et al. 2010; Craig et al. 2019; Stene and Gal 2013). With particular reference to intrauterine life, non-twin sibling pairs and dizygotic twin pairs both share 50% of their genes, however, individuals with an affected dizygotic twin show a higher risk of T1DM compared to subjects with an affected non-twin sibling, suggesting that the intrauterine environment and/or sharing early life environment influence the risk of developing disease (Stene and Gal 2013). Several perinatal factors have been potentially associated with the development of T1DM in childhood, in particular: maternal-child blood group incompatibility (both AB0 and Rh factor), preeclampsia, maternal enterovirus infections [e.g. rubella, (Stene and Rewers 2012)], neonatal respiratory distress, neonatal infections, gestational age, maternal age at delivery, caesarian section, birth order, and birth weight. It has not been clearly established whether these factors contribute directly to T1DM or how they may be influenced by other currently unknown risk factors (Craig et al. 2019; Atkinson 2012; Stene and Gal 2013). In particular, cesarean section has been suggested to be associated with the risk of progression from islet autoimmunity to clinical T1DM (Bonifacio et al. 2011) and an alteration in the timing and composition of gut microbiota of newborns following caesarean delivery has been hypothesized. Moreover, maternal diet composition during pregnancy has been studied for association with islet autoimmunity and/or clinical T1DM in the offspring in longitudinal birth cohort studies, but no convincing cause–effect relationship has been found (Lamb et al. 2008; Virtanen et al. 2011). An intriguing hypothesis suggests that enteroviral infections (in particular coxsackie viruses) during pregnancy may result in persistent infection with consequent islet autoimmunity in the mother (Rešić Lindehammer et al. 2012) and increased risk for T1DM in the offspring (Rewers and Ludvigsson 2016; Regnell and Lernmark 2017). Environmental factors may interact with each other. Breast milk may protect against enteric infections and enteric infections, in turn, could increase immunity to dietary antigens by increasing intestinal permeability. Another possibility is that alterations of gut mucosal immune function in genetically susceptible individuals may be the basis of the effects of dietary and/or viral proteins on the islet autoimmunity development in early life (Couper 2001). In conclusion, despite some suggestive associations, there is no clear evidence that non-genetic prenatal factors can influence directly T1D development (Stene and Gal 2013). In early infancy, factors contributing to the increase of body's requirement for insulin and subsequent high insulin secretion, such as high birthweight, rapid postnatal growth, overweight, puberty, low physical activity, trauma, infections, and glucose overload might play an important role in development of T1DM, in particular may accelerate the progression to clinical onset of disease in people with beta-cell autoimmunity (Rewers and Ludvigsson 2016; Stene and Gal 2013; Regnell and Lernmark 2017), the so-called beta-cell stress hypothesis. Accordingly, intake of sugar and high glycemic index foods has been proposed to be associated with progression to clinical T1DM in children with islet autoimmunity, but not with islet autoimmunity development per se (Lamb et al. 2015). Rapid longitudinal growth, puberty, low physical activity, and overweight have also been proposed to increase beta-cell stress and promote progression to clinical onset of diabetes in children with islet autoimmunity (Brown and Rother 2008). In particular, excess weight gain may lead to insulin resistance and hyperglycemia in little children, accelerating beta-cell apoptosis directly or by inducing beta-cell specific neo-autoantigens in genetically predisposed individuals. Moreover, rapid growth may increase insulin request promoting beta-cell stress and increasing autoantigens presentation (Wilkin 2001; Rewers 2012). The generation of neo-autoantigens in physiological states related to over secretion of insulin may be induced through post-translational modification of islet proteins (e.g. proinsulin, chromogranin A, islet amyloid polypeptide -IAPP-). Post-translational modifications represent intriguing players, as suggested in other autoimmune diseases such as coeliac disease, multiple sclerosis, rheumatoid arthritis ((Doyle et al. 2014), however, they have only recently become object of systematic studies in T1DM (Rewers and Ludvigsson 2016). Finally, psychological stress (e.g. divorce or family bereavements) has been suggested to increase the probability of both autoantibodies appearance and T1DM development, maybe not only through the increase of insulin resistance (leading to increased demand on the beta-cells) but also via increased cortisol concentrations, directly modulating the immune system (Rewers and Ludvigsson 2016). Prolonged endoplasmic reticulum stress may impair insulin synthesis and cause pancreatic beta-cell apoptosis (Cnop et al. 2012); furthermore, may increase abnormal post-translational modification of endogenous beta-cell proteins (Marre et al. 2015).

14.2.5 Chemical Compounds

Human exposure to environmental chemicals is complex and, while some chemicals may have beneficial effects, others may show detrimental ones in genetically predisposed individuals to autoimmunity. Endocrine disrupting chemicals (EDCs) could be considered environmental triggers for T1DM. EDCs may act as single chemical agent or as chemical mixtures, operating in different temporal windows. We can hypothesize that chemicals may have direct toxic effects on beta-cells, alter the development and the function of the immune system, affect the microbiota, and compromise intestinal permeability (Predieri et al. 2020; Bodin et al. 2015). With particular reference to microbiota, EDCs could be differently metabolized by gut bacteria, modifying absorption, metabolism, distribution, and excretion of these chemicals (Arumugam et al. 2011) with potentially different effects. Moreover, EDCs may contribute to increase or reduce the abundance of specific gut microbial populations and potentially rise the risk for T1DM development, as demonstrated in animal models for several substances (e.g. arsenic, mixture of diethyl phthalate, methylparaben) (Predieri et al. 2020). Moreover, chemical-induced epigenetic modifications, contributing to altered gene expression, are probably involved, with particular reference to in utero effects. However, human studies evaluating the potential causative role of exposure to EDCs on the T1DM pathogenesis are few and showing contradictory results (Bodin et al. 2015). In particular, a strong evidence for a single factor as the major trigger for T1DM development is lacking and the doses of chemicals sufficient to impact the risk of T1DM after exposure are unknown. It can be speculated that several factors could have additive or synergic effects, acting via several mechanisms and/or at different stages in the disease development (Rønningen 2015). Toxins in foods or water are also been proposed to be associated with T1DM development, however, observations are too preliminary and discordant (Atkinson 2012). As an example, a connection between T1DM and water containing nitrates, nitrites or nitrosamines has been observed in some studies (Benson et al. 2010), although no or contradictory associations have been described in others (Rewers and Ludvigsson 2016). Ambient air pollution is another environmental exposure suspected to increase the risk of T1DM development. This cause–effect relationship is supported by a growing body of evidence linking air pollutants with autoimmune mediated diseases and by the fact that systemic oxidative stress following exposure to air pollution, has also been observed in children and adolescents with T1DM (Elten et al. 2020). Few epidemiological as well as case-control studies, examining the association between air pollution (in particular nitrogen oxides, ozone) and the development of T1DM in children have been conducted and preliminary evidence suggests a role for both maternal and early life exposures to ambient air pollution (Elten et al. 2020). Finally, it has been hypothesized that vaccines might trigger autoimmunity, but no association has been observed with islet autoimmunity and T1DM (Rewers and Ludvigsson 2016) and as an example a meta-analysis of 23 studies investigating 16 vaccinations concluded that vaccines do not increase the risk of T1DM in children (Morgan et al. 2016).

14.2.6 Post-Transcriptional Modifications

Several studies (some of them carried out in twins) have found a different methylation status of T1DM (and its complications) strongly associated genes (Esposito et al. 2019). As examples, significant methylation differences have been reported in CpG sites proximal to the transcription start site of the insulin gene between T1DM patients and non-diabetic subjects. Methylation differences have been found also in CpG sites located within the proximal promoter of interleukin 2 receptor alpha gene (involved in regulatory T cells) between T1DM e non-diabetic individuals (Belot et al. 2013) and moreover in CpG sites within the major histocompatibility complex genes between individuals with T1DM at-risk haplotypes compared to controls (Kindt et al. 2018).

14.3 Environmental Factors and Type-2 Diabetes Mellitus

The global prevalence of diabetes mellitus has dramatically increased in the last decades, rising from 4.7% in 1980s to 8.5%, in the adult population, and the number of diabetic people is expected to rise to approximately 700 million by 2045, representing one of the leading causes of death worldwide (International Diabetes Federation 2017; World Health Organization 2017).

Type 2 diabetes mellitus (T2DM) is the most common form of diabetes mellitus, accounting for more than 90–95% of all diabetic forms, and it is characterized by a combination of relative—rather than absolute—insulin deficiency and peripheral insulin resistance (American Diabetes Association 2020a). T2DM is a global public health problem, and a serious issue in developing countries with worrying proportion in children and adolescents.

T2DM is a classic example of multifactorial disease, caused by an interaction of a predisposing genotype with environmental risk factors. T2DM displays a major genetic component, as demonstrated by the higher concordance rate found in monozygotic versus dizygotic twins and the higher incidence in first-degree relatives of T2DM patients compared to the general population (Wu et al. 2014). In addition, segregation studies indicate a polygenic nature of T2DM and GWAS studies identified approximately 75 susceptibility loci related to T2DM (Wu et al. 2014).

Despite the number of genomic loci associated with metabolic diseases identified, the overall effect size is limited and cannot be considered the unique cause of an alarming rise in prevalence of obesity and diabetes over the last decades, thus hinting at a relevant role of the environment (Wu et al. 2014; Tamashiro and Moran 2010).

The increase in T2DM prevalence is strongly related to the increasing proportion of overweight and obesity worldwide. Indeed, the role of the so-called obesogenic environment has been long recognized as a major contributor to the development of T2DM (Wu et al. 2014; Tamashiro and Moran 2010).

However, the broad phenotypic heterogeneity of individuals living in the same "obesogenic" environment has led to the assumption that metabolic risk is determined by a complex gene-environment interaction, therefore under favorable lifestyle and environmental conditions, subjects with predisposing genetic or epigenetic background are at increased risk of T2DM (Fig. 14.4) (Kolb and Martin 2017; Bouret et al. 2015).

14.3.1 Lifestyle Factors

Epidemiological studies suggest that several environmental risk factors contribute to disease development, such as the amount and type of food, sedentary time, physical activity, watching TV, noise, fine dust, sleep quality and duration (i.e. sleep deprivation, sleep disturbance caused by night-time exposure to noise or light, longer sleep duration and day-time napping), shift working, passively or actively smoking, emotional stress, and socio-economic status. Of note, only few environmental or lifestyle factors directly impact β-cell function, while almost all of the above-mentioned risk factors seem to promote an inflammatory status and concurrent insulin resistance (Kolb and Martin 2017).

Recent systematic reviews on the prevention of T2DM in high-risk population agreed that lifestyle intervention can prevent or delay the onset of the disease (Uusitupa et al. 2019). Most of the studies, however, investigated a combination of dietary intervention and physical activity, thus it is not possible to discriminate whether the beneficial metabolic effect depends on diet, weight loss or physical activity alone (Kolb and Martin 2017; Uusitupa et al. 2019). The contribution of lifestyle on diabetes risk has been targeted by several randomized trial, such as Diabetes Prevention Program and the Diabetes Prevention Study, where a lifestyle intervention, including dietary measures and increased physical activity, proved to be associated with a decreased incidence of T2DM by more than 50% after 3–4 years from the intervention, compared to control group (Delahanty et al. 2014; Tuomilehto et al. 2001). The beneficial lifestyle changes granted a sustained risk reduction of T2DM over 10 years of follow-up and the greater the adherence to intervention, the better the long-term prevention (Uusitupa et al. 2019).

Among lifestyle factors linked to T2DM development, diet has been largely investigated.

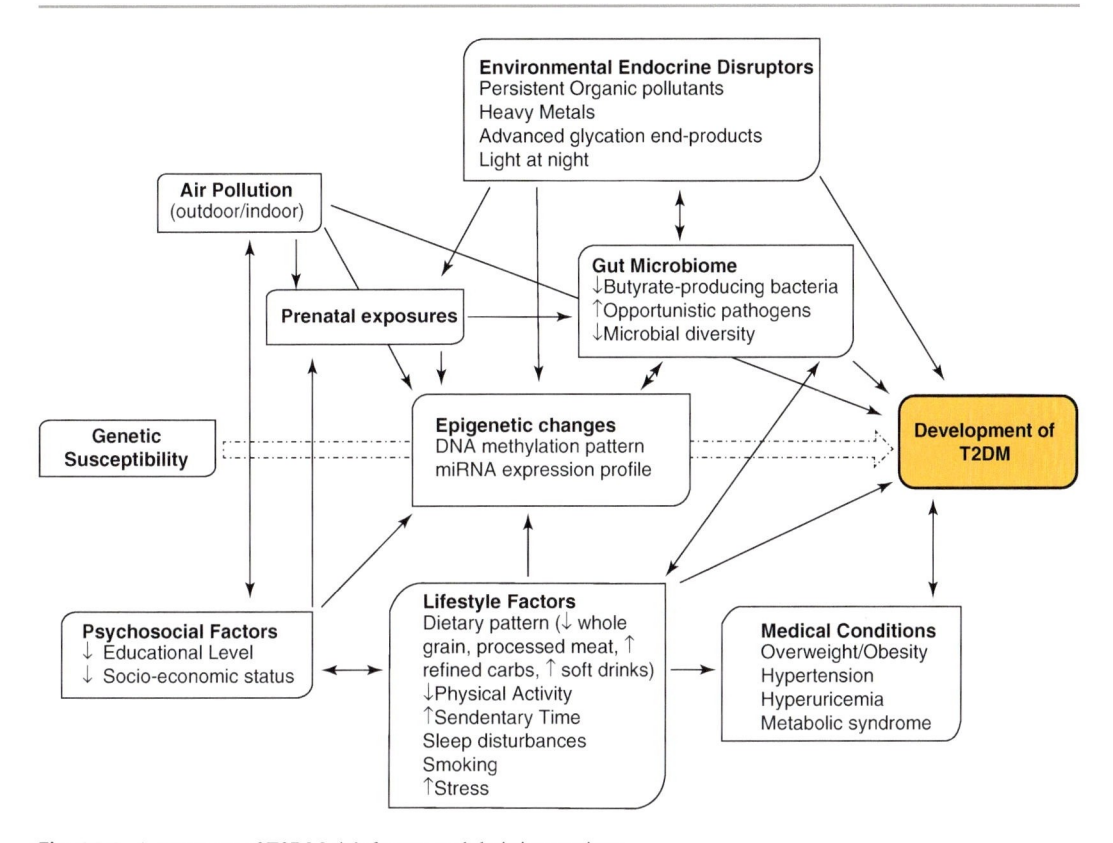

Fig. 14.4 A summary of T2DM risk factors and their interactions

Diet is a modifiable risk factor for T2DM and several observational studies have been conducted to analyze the association between food groups or nutrient consumption and T2DM incidence. In general, studies have shown that a low-fiber diet with a high glycemic index, frequent consumption of processed meat and sugar-sweetened beverages are positively associated with a higher risk of T2DM, and total and saturated fat intake is associated with an increased risk of T2DM independently of BMI, whereas plant food, total dairy products, whole grains, legumes, berries, low energy density food, fish consumption have the opposite effect (Wu et al. 2014; Kolb and Martin 2017; Uusitupa et al. 2019; Bellou et al. 2014). Also, moderate alcohol or coffee consumption have shown a beneficial effect (Kolb and Martin 2017; Bellou et al. 2014). In addition to the impact of individual nutrients and/or foods, several studies have investigated the role of dietary patterns, concluding that Western dietary pattern is associated with increased risk of T2DM, compared to Mediterranean or vegetarian diet (Uusitupa et al. 2019). In a study where no calorie restriction was intended and, on average, less than 1 kg of body weight per person was lost, Mediterranean diet seemed to provide a major reduction of T2DM incidence over 4 years compared to a conventional low fat diet, suggesting the presence of components in the Mediterranean diet that may decrease diabetes risk independently of weight reduction (Salas-Salvadó et al. 2011).

Weight loss obtained by means of a healthy diet combined with regular physical activity is beneficial for the prevention of T2DM. A significant association between anthropometric measures of obesity (e.g. BMI, waist-to-hip ratio, waist circumference) and T2DM has been demonstrated. However, the risk of T2DM onset is not the same for all obese subjects: specifically, increased visceral adiposity has been long recognized as one of the main determinants of insulin resistance, while, on the contrary, peripheral fat accumulation has been linked to a better metabolic profile, as shown by the protective effect of larger hip circumference on T2DM (Bellou et al. 2014).

Based on epidemiological data, all types of recreational or occupational physical activities are inversely related to T2DM risk, with a reduction in relative diabetes risk by approximately 30% and beneficial effect on insulin sensitivity and glycemic control (Kolb and Martin 2017; Bellou et al. 2014). On the contrary, sedentary time and TV watching are strongly associated with obesity and incident diabetes (Kolb and Martin 2017; Bellou et al. 2014).

In addition, available evidence points at an association of lower socio-economic status and educational level with higher risk for T2DM. Lower socio-economic status is associated with higher stress levels, which has been shown to impair neuroendocrine system. Additionally, people with low socio-economic status are more likely to have an unhealthy lifestyle and limited access to healthcare services (Kolb and Martin 2017; Bellou et al. 2014).

Recently, changes in circadian rhythm and sleeping pattern have been linked to a higher risk of developing T2DM. The circadian clock influences several endocrine pathways, and circadian disruption is correlated with obesity, hyperglycemia, hypertension, and elevated triglycerides (Russart and Nelson 2018). The circadian system is highly sensitive to light, therefore aberrant light exposure—especially light at night but also reduced light levels—can disrupt or abolish circadian rhythms (Russart and Nelson 2018). Additionally, studies in subjects working night shifts—who display a higher risk of T2DM compared to people working day shifts—suggest that constant exposure to light at night can alter hormonal signaling, resulting in hypercortisolism, diminished glucose-stimulated insulin secretion and β-cells failure (Russart and Nelson 2018).

14.3.2 The Gut Microbiota

Gut microflora have recently gained considerable attention due to its potential causal role in the development of T2DM and metabolic diseases (Bouret et al. 2015). The observation that germ-free mice, lacking a microbiota, are "protected" from diet-induced obesity and display improved glucose tolerance paved the way to extensive research, focused on the link between gut microbiota and the host metabolism (Sonnenburg and Bäckhed 2016).

Further support to this hypothesis derived from observations that altered microbiota from genetically obese mice or obese people confer an obese phenotype when transferred to lean mice, and insulin-resistant subjects exhibit, in turn, an improvement of insulin sensitivity after receiving fecal microbiota transplant from a lean, insulin-sensitive donor, although the durability of the effect is unknown (Sonnenburg and Bäckhed 2016).

The relationship between gut microbiota and the host is bidirectional and influenced by several factors: on the one hand, the intestinal microbial flora influences processing of nutrients, energy balance, intestinal permeability, metabolic endotoxemia, and inflammation; on the other hand, the profile of the intestinal microbiota changes according to the effect of various factors such as age, food habits, environment, and medications (Upadhyaya and Banerjee 2015). Although affected by genetic and environmental factors, the composition of gut microbiota is rather stable in healthy adult individuals—if no relevant environmental variations arise (Upadhyaya and Banerjee 2015; Brunkwall and Orho-Melande 2017). Nevertheless, an extremely high interindividual variability in gut microbial composition has been recognized (Brunkwall and Orho-Melande 2017).

The composition and function of gut microbiota are largely modulated by diet (Brunkwall and Orho-Melande 2017). Gut microbiota is subjected to rapid and long-term changes in response to changes in diet. Switch between plant- and meat-based diets, for instance, is associated with a change in the composition and function of the microbiota over 1–2 day (Sonnenburg and Bäckhed 2016). However, long-term dietary habits are the prevailing force in shaping the composition of an individual's gut microbiota (Sonnenburg and Bäckhed 2016).

It is worth mentioning that the effect of a particular change in diet is highly variable on different people, because of the individualized composition of gut microbiota (Sonnenburg and Bäckhed 2016). The inverse association between dietary fibers and the incidence of T2DM is largely recognized, and it is widely demonstrated that the consumption of fiber and whole grain increases biodiversity of the human gut microbiota (Brunkwall and Orho-Melande 2017). Moreover, current cumulative evidence suggests a potential disruptive effect of artificial food additives on microbiota, thereby contributing to metabolic impairment (Sonnenburg and Bäckhed 2016). Notably, Western diet is characterized by a low intake of plant-based dietary fibers, which, together with a high amount of nutrients and additives negatively affecting the microbiota, could play a significant role in metabolic diseases (Sonnenburg and Bäckhed 2016). Several studies, investigating the composition of gut microbiota under specific dietary patterns, have shown that *Prevotella* is the predominant species in subjects consuming a large amount of fibers, whereas a decrease in *Bifidobacteria* is associated with high-fat diet, and a predominance of *Bacteroides* is typical of diets rich in animal products (Upadhyaya and Banerjee 2015; Filippo et al. 2010).

The impact of gut microbiota on host metabolism is also dependent on microbial diversity. Several studies suggest that the extent of microbial diversity seems to have a significant impact on metabolic health (Sonnenburg and Bäckhed 2016). Industrialization and the resulting "modern lifestyle" have gone hand in hand with a sharp drop in gut microbial biodiversity (Sonnenburg and Bäckhed 2016). Indeed, the gut microbiota of rural populations showed higher bacterial diversity compared with those of Western populations (Sonnenburg and Bäckhed 2016).

The mechanisms by which alterations in gut microbiota might control an individual predisposition to develop obesity or diabetes have not yet been established. However, it is likely that gut microbial composition regulates nutrient absorption by changing the absorptive gastrointestinal surface, together with inflammatory changes, caused by some dietary pattern (Bouret et al. 2015). Extrinsic factors can induce dysbiosis by altering microbial diversity and function (Upadhyaya and Banerjee 2015). Studies on T2DM patients highlighted a moderate grade of gut dysbiosis in these subjects, characterized by a decrease in butyrate-producing bacteria and an increase in opportunistic pathogens (Tang et al. 2017; Wu et al. 2014). Short-chain fatty acids (SCFAs)—namely butyrate, acetate, and propionate—are end products of microbial fermentation of macronutrients and serve both as energy substrates and signaling molecules (Upadhyaya and Banerjee 2015; Wen and Duffy 2017). SCFAs can activate G-protein-coupled receptors GPR41 and GPR43. Activation of GPR41 induces expression of the anorexigenic enteroendocrine hormone PYY in gut epithelial L-cells, whereas activation of GPR43 in mouse white adipose tissue has proved to suppress insulin-mediated fat accumulation, and to stimulate hepatic and muscular energy expenditure (Tang et al. 2017; Upadhyaya and Banerjee 2015). Activation of GPR41 and GPR43 by SCFAs can also induce GLP-1 secretion, holding metabolic effects in the pancreas, thus contributing to glucose homeostasis regulation, while also having central effects on appetite regulation (Tang et al. 2017; Upadhyaya and Banerjee 2015).

Given the role of gut microbiota on metabolic health, it is not surprising that gastrointestinal microbiota is increasingly regarded as a potential way forward to improve glycemic control in T2DM (Sonnenburg and Bäckhed 2016; Brunkwall and Orho-Melande 2017). Several emerging gut-targeting glucose-lowering treatment strategies (i.e. controlled dietary interventions, use of probiotics) show initial promise, paving the way for microbiota-focused precision nutrition (Sonnenburg and Bäckhed 2016; Brunkwall and Orho-Melande 2017).

14.3.3 Endocrine Disrupting Chemicals (EDCs)

The relationship between EDCs exposure and metabolic impairment implies a pathogenic role of EDCs also in diabetes development (Diamanti-

Kandarakis et al. 2009). Indeed, linked high dioxin levels have been associated to an increased risk of diabetes and impaired glucose metabolism in epidemiological studies (Diamanti-Kandarakis et al. 2009), although most of the research on the effects of EDCs on obesity, diabetes mellitus and diabetes-related metabolic complications are fairly recent (Gore et al. 2015).

Due to the strong relationship between obesity and diabetes, it is difficult to differentiate EDCs involved in the development of obesity from those responsible for the development of diabetes, however, some chemicals, termed "diabetogens," have proved to induce insulin resistance and hyperinsulinemia, by directly targeting insulin- and glucagon-secretory cells, adipocytes, and liver cells (Gore et al. 2015). The so-called diabetogen hypothesis proposes that every EDC that causes insulin resistance, regardless of its obesogenic capacities, may be considered a risk factor for metabolic syndrome and/or T2DM (Alonso-Magdalena et al. 2011).

Diabetogenic action is also associated with altered levels of adiponectin and leptin. Taken together, alterations caused by diabetogenic EDCs—hyperinsulinemia, hyperleptinemia and decreased levels of adiponectin—may impair energy balance and lead to obesity and cardiovascular disease (Gore et al. 2015; Alonso-Magdalena et al. 2011).

In vitro and in vivo studies in animal models suggest harmful effects of EDCs both β- and α-cells in the endocrine pancreas, leading to impairment of glucose and lipid-metabolism, whereas a causal relationship between exposure to chemicals and development of diabetes has not been clearly demonstrated in humans, since epidemiological studies are cross-sectional, and diet represents an important confounding factor (Diamanti-Kandarakis et al. 2009; Gore et al. 2015).

Substantial evidence and prospective studies demonstrated a strong positive association between some persistent organic pollutants (POPs) and T2DM in humans, particularly organochloride pesticides and PCBs (Gore et al. 2014; Magliano et al. 2014). In cellular models, adipocytes seem to represent a reservoir for POPs (La Merrill et al. 2013). POPs accumulation in adipose tissue might represent a protective mechanism to limit their availability and systemic toxicity, however, in the long term, adipose tissue become a source of POPs, which are slowly released into the bloodstream (La Merrill et al. 2013). Lipophilic POPs have proved to induce insulin resistance and decreased expression of insulin-dependent genes involved in lipid homeostasis in differentiated adipocyte cell line (Gore et al. 2015). Furthermore, some POPs decreased glucose uptake by adipocytes and triggered a pro-inflammatory response (Gore et al. 2015; La Merrill et al. 2013). Evidence in adult animals confirms the alteration of plasma insulin levels and glucose tolerance, upon POPs exposure (Gore et al. 2015).

Several cross-sectional studies have linked urinary bisphenol A (BPA) levels with the incidence of T2DM (Shankar and Teppala 2011; Wang et al. 2012). In vitro studies showed that exposure to low dose of BPA decreased insulin sensitivity and glucose utilization and stimulated the release of pro-inflammatory molecules, such as IL-6 and IFN (Valentino et al. 2013), while relevant environmental levels can decrease the release of adiponectin (Hugo et al. 2008). BPA can affect pancreatic β-cell function, both by inducing acute insulin release, through the effect on ATP-sensitive K channels (Soriano et al. 2012) and by increasing glucose-induced insulin biosynthesis (Alonso-Magdalena et al. 2008). Both acute- and long-term effects of BPA have also been demonstrated in mice, with rapid increase in plasma insulin levels and decreased glycaemia after acute injection and hyperinsulinemia, insulin resistance, glucose intolerance, and impaired glucagon secretion upon longer exposures (Diamanti-Kandarakis et al. 2009; Gore et al. 2015).

Among EDCs, arsenic and phthalates that have been associated with T2DM in cross-sectional studies (Gore et al. 2015). Perinatal exposure to phthalates in rats produced important gender- and age-dependent alterations in glucose homeostasis, with worse effects on glucose tolerance in adult females compared with males (Lin et al. 2011). Low doses of arsenic blocked glucose-stimulated insulin secretion in murine islets of Langerhans, without substantial alteration of pancreatic insulin content (Douillet et al. 2013).

It has also been proved that β-cell death and impaired glucose-stimulated insulin secretion from pancreatic β-cells can be induced by other heavy metals, such as inorganic mercury and cadmium, respectively (Chen et al. 2010; Chen et al. 2009).

Finally, exogenous advanced glycation end products (AGEs), contained in tobacco or food cooked at high temperatures and some beverages, cause tissue injury through generation of free radicals, and trigger oxidative stress. Evidences show a correlation between exogenous AGEs exposure and increase in cardio-metabolic risk markers; animal studies suggest that exposure to exogenous AGEs is followed by loss of physiological pancreatic cellular architecture, increased fasting glucose and insulin levels (Diamanti-Kandarakis et al. 2009).

14.3.4 Air Pollution

Increasing evidence highlighted the adverse effects of air pollution on T2DM and a rise in T2DM-related biomarkers, upon increased exposure to air pollutants (Li et al. 2019a). Sources of air pollution are both natural (for instance, volcanic eruptions) and man-made, but the latter is the main source (Kim et al. 2020). We can distinguish "outdoor air pollution," typically produced by industrial processes and combustion of fossil fuels, and "indoor air pollution," produced by domestic appliances (Kim et al. 2020). Air pollution is a complex mixture containing both gases and particles, of which, the best characterized is particulate matter (PM) (Kim et al. 2020). The relationship between air pollution exposure and T2DM development is not linear. Emerging data highlighted sex-, age-, and BMI-specific differences in response to air pollutants, pointing at a higher susceptibility to T2DM under air pollution exposure in elderly, female and obese subjects (Li et al. 2019a). It is also evident that, chemical constituents of air pollution may affect T2DM to varying degrees (Li et al. 2019a). Finally, some authors suggest differences in air pollution composition and effects between countries with different economic levels, although no studies have specifically addressed this issue (Li et al. 2019a; Bowe et al. 2018). However, some data seem to support a higher vulnerability to air pollution in economically disadvantaged populations from underdeveloped and developing countries (Li et al. 2019a; Kim et al. 2020; Bowe et al. 2018).

The mechanisms by which air pollution increases the risk of T2DM are not fully understood. Some studies suggest that the relationship between air pollution and T2DM might be mediated by physical activity, since rates of open-air physical activity is inversely related to increased traffic burden, others attribute the harmful metabolic effect of air pollution to systemic inflammation (Li et al. 2019a).

The inflammatory response to short- and long-term exposure to air pollution has already been reported in in vivo and in vitro studies. After inhalation, PM penetrate deeply into the small airways where they stimulate macrophages and epithelial cells to release pro-inflammatory cytokines such as interleukin (IL)-8, IL-6, monocyte chemotactic protein-1 (MCP-1), macrophage inflammatory protein 2 (MIP2), and tumor necrosis factor-α (TNF-α) (Li et al. 2019a; Kim et al. 2020; Hogg and van Eeden 2009; Pope III et al. 2016). Systemic inflammation is known to have a detrimental impact on glucose metabolism, causing reduced insulin sensitivity, abnormal glucose tolerance, visceral fat inflammation/dysfunction, and increased insulin resistance, which in turn significantly increases the risk of diabetes (Li et al. 2019a; Kim et al. 2020).

Furthermore, the increasing concentration of air pollutants can significantly increase oxidative stress, endoplasmic reticulum stress (ERS), and apoptosis (Li et al. 2019a). Animal studies showed an increased production of reactive oxygen species (ROS) from tissues exposed to air pollutants (Li et al. 2019a). Pancreatic β cells are highly sensitive to ROS because of poor antioxidant capacity and limited availability of antioxidant enzymes (Drews et al. 2010). Thus, oxidative stress results in β-cell dysfunction; indeed, ROS could directly damage islet β-cell and indirectly interfere with the insulin signaling pathway, through NF-κB-mediated inflammatory response (Li et al. 2019a; Drews et al. 2010). Air pollution can also favor ERS, further promoting pancreatic β cells apoptosis, causing impaired insulin secretion and, ultimately, leading to diabetes (Li et al. 2019a).

Air pollution exposure has also been linked to an altered epigenetic state, characterized by increased global methylation (Kim et al. 2020).

14.3.5 Epigenetic Changes in T2DM

As already mentioned, one of the major limit of genetic research on T2DM is the fact that genetic signature explains only 10–15% of the T2DM heritability. Recent technical advances (i.e. the availability of several DNA methylation array) have allowed intensive research on epigenetic modifications involved in age- and lifestyle-related metabolic diseases (Kwak and Park 2016). Several studies have shown altered epigenetic patterns in tissues involved in glucose homeostasis—including pancreas, liver, skeletal muscle, and adipose tissue—from subjects with T2DM, suggesting a pathogenic role of epigenetic in T2DM development (Kwak and Park 2016; Nilsson and Ling 2017). Indeed, available findings suggest tissue-specific epigenetic changes in T2DM, possibly modulated both by environmental factors and genetic factors (Kwak and Park 2016). Epigenetic modifications are highly dynamic and susceptible to environmental factors such as age, obesity, physical activity, and diet (Nilsson and Ling 2017). Diet is actually the main source of methyl groups for methylation reactions (Nilsson and Ling 2017). Reduced circulating levels of folate—important dietary sources of methyl donors—in T2DM patients may contribute to the development of T2DM through reduced DNA methylation of a high number of CpG sites (Nilsson et al. 2015). Also, short-term high-fat diet has been associated to widespread DNA methylation changes partially, reversible after returning to the control diet (Jacobsen et al. 2012). The impact of diet on epigenetic modification is also supported by the association of altered DNA methylation levels and increase body mass index (BMI) (Nilsson and Ling 2017). Even the beneficial effects of physical exercise on metabolism have been linked, at least in part, to epigenetic changes, that have been reported after both acute and long-term physical activity (Nilsson and Ling 2017).

Furthermore, evidence suggests that age affects DNA methylation pattern. Some authors propose that the epigenetic drift seen with aging could explain the increased incidence of chronic metabolic diseases, including T2DM, in the elderly (Nilsson and Ling 2017).

Finally, in the last decade, miRNAs have been addressed as crucial regulators controlling many signaling pathways, including insulin signal transduction (Nigi et al. 2018). Several alterations of miRNAs expression or function, in multiple insulin target organs, have proved relevant for the development of insulin resistance in T2DM, affecting several insulin signaling components, from insulin receptor to intracellular transcription factors (Nigi et al. 2018). Indeed, many studies have shown a direct involvement of miRNAs in β-cell dysfunction and insulin resistance (Nigi et al. 2018). Moreover, growing evidences underlined that miRNAs can be secreted in the bloodstream and exert their regulatory activities also in distant cells from those of origin (Nigi et al. 2018).

14.3.6 The Role of Prenatal Exposures on Disease Development

"Fetal programing" describes the process by which exposure to different environmental factors, during fetal growth, can impact disease susceptibility later in life, through a long lasting impact on metabolic pathways (Nilsson and Ling 2017). Epidemiological studies point at a fundamental influence of intrauterine and early postnatal environments on body weight and energy homeostasis in offspring (Roseboom et al. 2006; Tamashiro and Moran 2010).

Initial evidence of the effect of prenatal exposure in humans were derived from studies of infants who were in utero during periods of famine, providing direct evidence that maternal undernutrition during pregnancy predispose to long-term risk of T2DM in postnatal life (Roseboom et al. 2006; Bouret et al. 2015). Available data seems to indicate that the third trimester is the most vulnerable developmental period in terms of long-term regulation of glucose homeostasis, for individuals exposed during late gestation were most affected (Bouret et al. 2015). Interestingly, the risk of obesity and cardiovascular disease was instead more evident in subjects exposed to undernutrition during early gestation (Bouret et al. 2015). Since modern lifestyle and Western diet have led to a two-fold increase in the incidence of maternal overweight and obesity (to over 45%) over the last two decades, an increasing number of studies focused on the risk of poor metabolic health in the offspring, following parental overnutrition and/or obesity during the periconceptual period (Tamashiro and Moran 2010; Bouret et al. 2015; Fernandez-Twinn et al. 2019). Thus, both fetal under- and overnutrition seem to increase the risk of metabolic syndrome in adulthood (Tamashiro and Moran 2010; Fernandez-Twinn et al. 2019). The relationship between birth weight and adult fat mass has developed into a "U" shaped curve, thus individuals with either extremely low or high birth weight may be more susceptible to increased fat mass and metabolic impairment, in childhood and/or adulthood (Tamashiro and Moran 2010).

Until recently, most studies have focused on potential unfavorable consequences of the maternal nutritional status; however, although limited, emerging data suggest that also paternal nutritional status can have detrimental consequences in the offspring (Bouret et al. 2015; Fernandez-Twinn et al. 2019). Animal models provide evidence that paternal obesity at conception is associated with metabolic dysfunction in the offspring, characterized by obesity, glucose intolerance, and insulin resistance, and the effects of paternal diet-induced obesity on offspring adiposity and insulin sensitivity seems to persist for at least two generations (Bouret et al. 2015).

Also preterm delivery seems to be associated to a higher risk of adult onset T2DM, probably through intrauterine growth restriction. Preterm newborns, indeed, usually display low birth weight and they have higher propensity to develop impaired glucose metabolism in later life, which, in turn, increases the risk of T2DM (Bellou et al. 2014).

Additional prenatal exposures potentially associated with adult onset T2DM include parental smoking, infections, and other environmental pollutants (Fernandez-Twinn et al. 2019). Moreover, prenatal stress—originating from work and social environments—has also been shown to increase susceptibility to obesity through chronic activation of the hypothalamic-pituitary-adrenal (HPA) axis, causing hypercortisolism, which in turn has been associated to the development of features related to the metabolic syndrome (Tamashiro and Moran 2010; Fernandez-Twinn et al. 2019).

Epigenetic changes provide a reasonable mechanism for "metabolic programing" in utero, a mechanism through which prenatal exposures could result in long-term metabolic alterations (Tamashiro and Moran 2010). Indeed, increasing evidence highlights how specific in utero exposures can have an impact on offspring epigenotype by modifying the DNA methylation pattern (Fernandez-Twinn et al. 2019). Studies demonstrating epigenetic changes in T2DM- and obesity-associated candidate genes are limited (Bouret et al. 2015) and additional data are needed to demonstrate a causal role for epigenetic variation in mediating the effects of prenatal exposures on poor offspring metabolic health (Fernandez-Twinn et al. 2019). The phenotypic effects of epigenetic changes during intrauterine life may be evident only in adulthood in response to a predisposing environment after birth, for instance the consumption of energy dense diets (Bouret et al. 2015; Nilsson and Ling 2017).

Pre- and perinatal environment can influence adult health also by affecting gut microbiome. Dysbiosis in newborns of obese women has been implicated childhood inflammatory diseases, non-alcoholic fatty liver disease (NAFLD), and increased obesity risk (Fernandez-Twinn et al. 2019).

Several studies have suggested that the composition of the gut microbial flora is different in newborns delivered by cesarean section from that of infants born by vaginal delivery, the former being more prone to obesity and/or diabetes compared to the latter (Wen and Duffy 2017). Infant feeding is another important determinant of gut microbial composition. Obese mothers seem to produce milk containing different, less-diverse bacteria, than milk from normal-weight mothers (Wen and Duffy 2017).

14.4 Environmental Factors and Gestational Diabetes

According to the recommendations of World Health Organization, hyperglycemia during pregnancies is classified as "diabetes mellitus in pregnancy" and "gestational diabetes mellitus" (GDM) (World Health Organization 2013). The latter is defined as "diabetes diagnosed in the second or third trimester of pregnancy that was not clearly overt diabetes prior to gestation" (American Diabetes Association 2020a). The prevalence in the world is increasing and approximately 2–6% of all pregnancies are complicated by GDM, suggesting lifestyle and environmental exposures may contribute to the development of this condition (Tang et al. 2020). Given the health outcomes for both mother and fetus, earlier diagnosis and treatment are crucial: since the diagnosis is often made at a late stage of pregnancy, it is important to study the pathophysiology of GDM and to find out more specific and precise markers of this disease (Ehrlich et al. 2016). GDM pathophysiology is still not clearly defined: the main and the most plausible mechanism is the inadequate beta-cell reaction to peripheral insulin resistance, as it happens during second and third trimesters of gestation; in fact, during physiological pregnancy, a compensatory increased insulin resistance, associated with hypertrophy and/or hyperplasia of beta-cells and with more insulin secretion, occurs (Sebastiani et al. 2017).

The initial evaluation of a pregnant woman must include two steps: the screening of overt type 1 or type 2 diabetes and that for GDM. In the first case, the same criteria valid outside of pregnancy are used for diagnosis. It is also recommended to screen for GDM on the basis of specific risk factors, including positive first-degree familiarity for diabetes, previous GDM, previous pregnancies fetal macrosomia, overweight/obesity (BMI $\geq$25 kg/m^2), age $\geq$ 35 years, high-risk ethnic groups: these women should perform a 75 g OGTT at week 24th–28th of gestation. In high-risk women (BMI $\geq$30 kg /m^2, previous GDM and fasting glucose 100–125 mg/dl early in pregnancy) early screening is required, with 75 g OGTT at week 16th–18th, to be repeated, if negative, at week 24th–28th (Associazione Medici Diabetologi, Società Italiana di Diabetologia 2018). As a consequence, treatment cannot start before the late third trimester, a period of high vulnerability for fetal morbidity and mortality. Therefore, an early screening performed during the first or second trimester of pregnancy could be of great help to avoid pregnancy adverse outcomes planning out an effective treatment to normalize glucose levels: in fact, there are greater maternal and fetal risks not only related to the degree of hyperglycemia but to chronic complications and comorbidities of diabetes too (American Diabetes Association 2020b; Guarino et al. 2018). In general, adverse outcomes of diabetes in pregnancy include spontaneous abortion, need for cesarean delivery, large for gestational age infant, >90th percentile for neonatal skin-fold thickness, neonatal hypoglycemia, hyperbilirubinemia, preeclampsia, macrosomia, and neonatal respiratory distress syndrome (American Diabetes Association 2020b; Bellavia et al. 2019). Not only overt GDM could be dangerous for the mother and the newborn: also increased blood glucose levels, although below the clinical threshold for GDM diagnosis, may also increase the potential risk of negative consequences such as macrosomia, preeclampsia, and perinatal depression (Bellavia et al. 2019). Furthermore, the clinical and metabolic effects of GDM on mother and offspring do not disappear after delivery: women diagnosed with GDM have an increased risk by sevenfold of subsequent T2DM—specifically, an incidence of 60% at

10 years from diagnosis of GDM, progressively increasing thereafter, without signs of plateau—and related cardiovascular diseases (CVD). Additionally, higher rate of obesity and metabolic syndrome have been demonstrated in women with prior GDM, further increasing the risk of CVD (Guarino et al. 2018).

14.4.1 Environment and Pregnancy

The increasing prevalence of obesity and metabolic syndrome over the last decades might be attributed to an interaction between multiple environmental and genetic factors not only in adults but also during childhood: children of mothers previously affected by GDM have increased likelihood of developing CVD in their life, and it is possible that they will have increased rates of childhood obesity and impaired glucose tolerance during adolescence. In addition to the possible role played by a suboptimal milieu in utero due to maternal diet and lifestyle, increasing attention has been focused on the possible actions of EDCs in the development of metabolic diseases and other endocrine conditions, especially during delicate moments of women's life such as pregnancy and delivery (Filardi et al. 2020; Li et al. 2019b).

Exposure to industrial chemicals and to other environmental toxins can occur in various ways: for example, pregnant women can be exposed through air and water pollution, contamination of personal care products, and cosmetics or foods (Tang et al. 2020; Filardi et al. 2020). A study conducted in the USA reported that almost 100% of pregnant women were found to have detectable levels of several EDCs in the bloodstream and in urine samples (certain polychlorinated biphenyls, organochlorine pesticides, perfluorinated compounds, phenols, polybrominated diphenyl ethers, phthalates, polycyclic aromatic hydrocarbons, and perchlorate) (Woodruff et al. 2011). Given that endocrine disorders such as diabetes mellitus, infertility, and obesity are multifactorial diseases, it is right to consider among risk factors also industrial chemicals and other environmental toxins; in particular, some phthalates/bisphenol A (BPA) could be involved in development of obesity and glucose metabolism disorders (Martínez-Ibarra et al. 2019; Zlatnik 2016). BPA and phthalates have been proved to target different pathophysiological pathways of GDM: they have been linked to weight gain, insulin resistance, and pancreatic beta-cell dysfunction (Filardi et al. 2020). As mentioned above, attention is increasingly being focused on pregnancy: in fact, high levels of phthalates and BPA in urine have been associated to a negative impact on both mothers and children's health (preterm delivery and metabolic disorders) (Martínez-Ibarra et al. 2019). Furthermore, BPA is metabolized and eliminated at a low rate in fetuses and children and for this reason they have significantly high levels of BPA in blood and urine (Filardi et al. 2020).

The timing of life in which exposure to endocrine disruptors occurs is very important, both during embryogenesis and during fetal life, when organ development takes place. Awareness of these susceptibility windows should alert women's health care providers of the importance of safeguarding women of child-bearing age from toxins exposure that could impact their future health or fertility, as well as the physical wellbeing of a fetus or a breastfeeding baby (Zlatnik 2016).

14.4.1.1 The Mother and EDCs: Short- and Long-Term Effects

In a cohort of 378 pregnant women evaluated in Charleston (South Carolina), in over 93% of urine samples collected during the second trimester of gestation there were detectable concentrations of eight phthalate metabolites (Wenzel et al. 2018). In another sample of 1274 pregnant women, BPA and eleven phthalate metabolites were measured in urine samples alongside blood dosage of four metals, including lead, cadmium, mercury, and arsenic, both evaluations made during the first trimester: no association of increased risk of GDM or impaired glucose tolerance (IGT) diagnosis in relation to phthalate or BPA or other metals exposure was found and a statistical evidence of the association between blood arsenic concentration with GDM and GDM/IGT combined, but not IGT alone, was observed. Several hypotheses have been raised on the mechanism linking diabetes to arsenic: impaired insulin-dependent glucose uptake and glucose-stimulated insulin secretion, oxidative stress, up-regulation of inflammatory markers (tumor necrosis factor alpha and IL-6), inhibition of peroxisome proliferator-activated receptor-γ (PPAR γ) and alteration of methylation patterns of diabetes-related genes are just some of the candidates (Shapiro et al. 2015).

An association between phthalate exposure and the risk to develop GDM or IGT during pregnancy has been previously highlighted (Shaffer et al. 2019). In different studies, it is reported that mono-ethyl phthalate (MEP), the prevalent urinary metabolite of phthalate, can reach a median concentration of 30 mcg/L in pregnant women (Filardi et al. 2020).

As mentioned above, some ethnic groups, a higher BMI value and gestational weight gain can be considered as risk factors for GDM. In a sample group of 347 pregnant women (Lifecodes cohort study), a positive association was found between obesity measures and maternal urinary MEP (Bellavia et al. 2017). In a subsequent analysis of the Lifecodes cohort study, the association between first and second trimester BPA urinary levels and glucose levels stratifying by BMI categories was investigated: no clear relationship emerged when considering the entire group of women, but there was a positive association between the concentrations of BPA at both trimesters and increased glucose levels in the overweight/obese women (Bellavia et al. 2018).

Risk factors for GDM include also high-risk ethnic groups (South Asia, the Middle East, the Caribbean) (Associazione Medici Diabetologi, Società Italiana di Diabetologia 2018): a study conducted using race-stratification analysis found a strong race-specific association in Asians between several phthalates and blood glucose. It has been hypothesized that the pathogenic mechanism behind this phenomenon is linked to genetic factors. Indeed, in Asian groups, a low prevalence of PPARγ 2 polymorphism Pro12Ala was detected and this polymorphism is associated with decreased risk of diabetes. Consequently, Asian pregnant women may have greater vulnerability to the effects of exposure to this class of EDC, including diabetes and insulin resistance, since it has been suggested that PPARγ as a link between insulin resistance and phthalates effects (Shaffer et al. 2019).

Data generated in animal models highlighted that pregnant mice treated with a subcutaneous injection of BPA had significantly higher weight gain and greater values of plasma insulin, leptin, triglyceride, and glycerol as recorded 4 months after delivery (Alonso-Magdalena et al. 2010) and also impaired glucose control compared to controls (Alonso-Magdalena et al. 2015). These effects can be explained with the BPA action on insulin signaling pathways in peripheral tissues such as liver and adipose tissue: in a mouse model, BPA inhibits the phosphorylation of Akt, with subsequent alteration of the glucose metabolism (Filardi et al. 2020; Alonso-Magdalena et al. 2010).

Contact with phthalates can also involve other aspects of pregnancy such as blood pressure control: after mono-benzyl phthalate (MBzP) exposure, higher maternal diastolic blood pressure may be encountered, in addition to gestational hypertension, preeclampsia, eclampsia or HELLP syndrome whose three main components are hemolysis, elevated liver enzymes and low platelet count (Zlatnik 2016).

From the data presented so far, it emerges that increasing attention has been placed on the effects of BPA and phthalates on pregnant women. With regard to parabens we still don't have much data available (Bellavia et al. 2019). In a study conducted in the USA, the U.S. general population from the 2005–2006 National Health and Nutrition Examination Survey, many different parabens such as methylparaben and propylparaben were found in 99.1% and 92.7% of the participants, respectively, and butylparaben in 47% (Calafat et al. 2010). These endocrine disruptors have actions similar to BPA on glucose control through estrogen-dependent signaling: its role on glucose metabolism can occur also in pregnant women, increasing insulin-resistant state and even GDM risk. In a cohort of women at high risk of GDM from a fertility clinic, during the first trimester, butylparaben and propylparaben urinary concentrations were, respectively, positively and negatively associated with glucose levels, even after considering the effects of potential confounders (Bellavia et al. 2019).

The following pathogenic mechanisms of the relationship between paraben exposure and blood glucose levels have been hypothesized:

- Parabens can stimulate glucocorticoid receptors activity;
- Parabens can function as a thyroid hormone receptor;
- Parabens act as peroxisome proliferator–activated receptor agonist, resulting in impaired glucose metabolism through alterations of adipocyte production and insulin resistance;
- Butylparaben is able to increase plasma leptin concentration;
- Propylparaben has the ability to affect mitochondrial actions (Bellavia et al. 2019).

In contrast, in another study, no statistically significant association between parabens and GDM in a population of pregnant women was found. Moreover, paraben concentrations were not significantly different across pre-pregnancy BMI, although in the sub-group of overweight/obese women there was a positive association between GDM onset and slightly higher levels of Prp urinary paraben. These data could be explained by the relationship between estrogen hormonal imbalance, obesity, and GDM: indeed, obesity is a risk factor for GDM and some studies reported that non-pregnant women with higher BMI had higher level of bioavailable estrogen; consequently, overweight/obese women may be more sensitive to exogenous estrogen compounds such as parabens (Li et al. 2019b).

However, many mechanisms of the etiopathogenesis remain unclear and further researches are still needed to investigate and understand how parabens can impact glucose metabolism in clinical practice and especially during pregnancy (Bellavia et al. 2019).

14.4.1.2 The Newborn and EDCs: Short- and Long-Term Effects

Available data regarding phthalates levels in newborns provide conflicting information: some evidence report similar concentrations in children and mothers, some others two-or three fold lower concentrations in offspring and some others found higher levels of urinary phthalates in children compared to adults (Filardi et al. 2020).

EDCs can be transferred to child and fetus in different ways, such as through circulation and amniotic fluid or also via lactation and may affect fetal growth and the length of pregnancy (Filardi et al. 2020). In a cohort composed by 476 pregnant women, sex-specific associations between biomarkers of maternal phenol exposure during pregnancy and reduced fetal growth were observed (Ferguson et al. 2018); in another study a positive association was found between phthalate exposure at the third trimester of gestation and the risk of preterm birth (Broe et al. 2019).

In addition to short-term effects, great attention was paid to long-term outcomes during prenatal and early life: for instance, a positive association between higher BPA concentration and accelerated growth in 2–5 years old aged children was observed (Braun et al. 2014). Moreover, pesticides can have a similar effect: exposure during early pregnancy to this kind of EDCs may increase the risk of developing metabolic diseases such as obesity, diabetes, dyslipidemia, and hypertension during childhood (González-Casanova et al. 2020).

Further research is needed to better understand the real impact of EDCs in increasing the risk of GDM and to identify the short- and long-term effects on the offspring in order to assess and to develop more accurate prevention strategies (Filardi et al. 2020).

14.4.2 microRNAs as Prognostic and Diagnostic Markers of EDCs' Effects

In GDM, increasing evidence has been provided in support of the potential role of microRNAs biomarkers to predict and to recommend the most suitable and targeted therapy, to set the best glycemic targets and to better understand new pathogenic mechanisms (Ehrlich et al. 2016; Sebastiani et al. 2017). Moreover, microRNAs have a role as potent regulators of cell survival and proliferation and as mediators of tissue cross-talk.

Eventually, they orchestrate beta-cell compensatory response and beta-cell function in diabetes. Taking into account all these functions, microRNAs could almost be considered as "novel hormones." Interestingly, microRNAs can be detected in many biological fluids including serum or plasma, associated with proteins or with extracellular vesicles (EVs), classified as exosomes or microvesicles, according to their size, cell/tissue of origin, and function (Guarino et al. 2018). Of note, EVs and exosomes play several roles during pregnancy, from regulation of immune responses to maternal metabolic adaptation to gestation and can even participate to the cross-talk between placenta and other organs/tissues. Furthermore, the shift of non-coding microRNAs via exosomes in the maternal circulation can have a regulatory effect on other cells (Ehrlich et al. 2016; Sebastiani et al. 2017; Associazione Medici Diabetologi, Società Italiana di Diabetologia 2018; American Diabetes Association 2020b; Guarino et al. 2018).

Deregulation of microRNA expression has also been associated with GDM (Guarino et al. 2018): for example, plasma miR-330-3p could be helpful in identifying GDM patients with potential worse gestational diabetes outcome, since this microRNA may directly be transferred from plasma to beta-cells thus modulating key target genes involved in proliferation, differentiation, and insulin secretion (Sebastiani et al. 2017). Additional data have been generated in a study conducted in pregnant women with and without GDM, at 16–19 weeks of pregnancy, to assess plasma levels of microRNAs. These two groups had different profile of expression of plasma microRNA and the major targets of these differentially expressed microRNAs were associated with insulin resistance and poor pregnancy outcomes (Zhu et al. 2015).

New evidence suggesting a link between microRNA expression, EDCs and GDM outcome are emerging, although this relationship has only recently been explored: the hypothesis is that EDCs, in particular environmental phenols, may induce exosome signaling from the placenta and may have an effect on GDM development (Ehrlich et al. 2016). In a study conducted in a group of 179 women in the Harvard Epigenetic Birth Cohort and the Predictors of Preeclampsia Study, concentrations of phenols and parabens were found to be significantly associated with differential expression of miR-142-3p, miR15a-5p, and miR-185 in placenta (LaRocca et al. 2016) and in another study an elevated expression of miR-146a was observed in BPA-treated immortalized cytotrophoblast cell lines (Avissar-Whiting et al. 2010). In a study conducted in a cohort of Mexican pregnant women, higher levels of miR-9-5p, miR-29a-3p, and miR-330-3p were detected in the group of patients with GDM with respect to non-diabetic women and phthalates were identified in 97–100% of urine samples, on the other hand BPA in 40%. Of note, these three microRNAs are involved in the regulation of glucose metabolism and homeostasis. In addition, a positive correlation between adjusted urinary mono-benzyl phthalate levels and miR-16-5p expression levels and between adjusted mono (2-ethyl hexyl) phthalate concentrations and miR-29a-3p expression levels was found; a negative correlation between unadjusted and adjusted mono-n-butyl phthalate concentrations and miR-29a-3p expression levels and between unadjusted mono-isobutyl phthalate concentrations and miR-29a-3p expression levels was encountered. From these results, it is possible to imagine how the dosage of circulating microRNAs could be useful and applicable in prevention strategies and in improving the quality of life of pregnant women, limiting their exposure to endocrine disruptors and environmental pollutants (Martínez-Ibarra et al. 2019).

Although additional studies are needed in order to clarify the possible use of microRNA in clinical practice, the measurement of circulating levels of these markers may help in deciphering the mechanisms and pathophysiology of GDM and of its complications, to guide the choice of a target therapy in GDM (Sebastiani et al. 2017).

14.4.3 Air Pollutants and GDM

The link between air pollutants and GDM has not been extensively studied. Prolonged exposure to air pollutants from fetal life to childhood has resulted into increased prevalence of hypertension, obesity, and metabolic disorders and, in addition, into a greater risk of cardiovascular disease in future generations. Moreover, exposure to air pollution often has effects on pregnancy outcomes with an increased risk of stillbirth, low birth weight, preterm birth and raising propensity towards developing future CVD risk in the offspring, such as elevated blood pressure (Kim et al. 2020). Most studies assessed the exposure during first and second trimesters; however, only few studies were performed before the conception, which is a time when exposure to EDCs has effects on offspring (Tang et al. 2020). Air pollution can be defined as a heterogeneous set of solid and gaseous components among which it is possible to include lead, fine particulate matter (PM), carbon monoxide, nitrogen dioxide, sulfur dioxide, and ozone. PMs (particles ≤2.5 μm in diameter [PM2.5]), together with nitrogen dioxide (NO_2) and airborne polycyclic aromatic hydrocarbons, are known to have endocrine disrupting effects and have been linked with obesity and GDM (Kim et al. 2020). About the exposure to residential levels of PM2.5 and NO_2, mothers' contact to NO_2 in the first trimester and to PM2.5 in the second trimester was associated with increased risk of developing GDM (Choe et al. 2019). From a recent review, a positive association between exposure to NO_2, nitrogen oxide (NOx), and sulfur dioxide (SO_2) and the second trimester exposure to PM2.5 with the increased risk of GDM was noted (Tang et al. 2020). Moreover, during the first trimester, PM10 exposure could increase the risk of gestational hypertension, such as SO2 exposure is related with increased risk of GDM and there is also a positive association between PM2.5 and preeclampsia. This evidence could be explained by the more sensitive window in the early pregnancy (Bai et al. 2020). Even the period before pregnancy is sensitive to the effect of environmental pollutants, in fact it has been found that an exposure to PM2.5 within 3 months before conception increases the risk of development of GDM and higher values of fasting glycaemia (Zhang et al. 2020). Furthermore, PM10, SO_2, and O_3 in prepregnancy were positively associated with the risk of GDM (Yao et al. 2020).

The biological mechanisms involved in the correlation between air pollution and GDM are not completely understood: oxidative stress, endothelial dysfunction, dyslipidemia or various inflammation patterns, such as both systemic and peripheral inflammation, as it takes place at the level of the adipose tissue, are some of the proposed pathways regarding the association between air pollution and metabolic diseases (Tang et al. 2020; Yao et al. 2020). Interestingly, PM2.5 exposure during pregnancy can increase insulin resistance (Tang et al. 2020; Zhang et al. 2020), down-regulate the expression of glucose transporter 2 in pancreatic beta-cells and therefore impaired glucose metabolism in GDM animal models. Furthermore, PM2.5 exposure before pregnancy could induce pathological changes on beta-cells, including periductal inflammation. Also SO_2, NO_2, and NOx could cause phenomena such as inflammation, dysfunction, and damage of pancreatic beta-cells (Zhang et al. 2020). In the end, placental translocation of particles following maternal exposure on the fetal side of human placenta can be considered as a direct mechanism of action (Melody et al. 2020).

References

Alonso-Magdalena P, Ropero AB, Carrera MP, Cederroth CR, Baquié M, Gauthier BR et al (2008) Pancreatic insulin content regulation by the estrogen receptor ER alpha. PLoS One 3:e2069. https://doi.org/10.1371/journal.pone.0002069

Alonso-Magdalena P, Vieira E, Soriano S, Menes L, Burks D, Quesada I et al (2010) Bisphenol A exposure during pregnancy disrupts glucose homeostasis in mothers and adult male offspring. Environ Health Perspect 118:1243–1250. https://doi.org/10.1289/ehp.1001993

Alonso-Magdalena P, Quesada I, Nadal A (2011) Endocrine disruptors in the etiology of type 2 diabetes mellitus. Nat Rev Endocrinol 7:346–353. https://doi.org/10.1038/nrendo.2011.56

Alonso-Magdalena P, García-Arévalo M, Quesada I, Nadal A (2015) Bisphenol-A treatment during pregnancy in mice: a new window of susceptibility for the development of diabetes in mothers later in life. Endocrinology 156:1659–1670. https://doi.org/10.1210/en.2014-1952

American Diabetes Association (2020a) Classification and diagnosis of diabetes: standards of medical care in diabetes - 2020. Diabetes Care 43(Suppl 1):S14–S31. https://doi.org/10.2337/dc20-S002

American Diabetes Association (2020b) Management of diabetes in pregnancy: standards of medical care in diabetes. 2020 Diabetes Care 43(Suppl 1):S183–S192. https://doi.org/10.2337/dc20-S014

Aoki Y (2001) Polychlorinated biphenyls, polychlorinated dibenzo-p-dioxins, and polychlorinated dibenzofurans as endocrine disrupters—what we have learned from Yusho disease. Environ Res 86:2–11. https://doi.org/10.1006/enrs.2001.4244

Arumugam M, Raes J, Pelletier E, Le Paslier D, Yamada T, Mende DR et al (2011) Enterotypes of the human gut microbiome. Nature 473:174–180. https://doi.org/10.1038/nature09944

Associazione Medici Diabetologi, Società Italiana di Diabetologia (2018) Standard italiani per la cura del diabete mellito 2018. http://www.siditalia.it/clinica/standard-di-cura-amd-sid. Accessed 27 Apr 2018

Atkinson MA (2012) The pathogenesis and natural history of type 1 diabetes. Cold Spring Harb Perspect Med 2:a007641. https://doi.org/10.1101/cshperspect.a007641

Atkinson MA, Chervonsky A (2012) Does the gut microbiota have a role in type 1 diabetes? Early evidence from humans and animal models of the disease. Diabetologia 55:2868–2877. https://doi.org/10.1007/s00125-012-2672-4

Atkinson MA, Eisenbarth GS, Michels AW (2014) Type 1 diabetes. Lancet 383:69–82. https://doi.org/10.1016/S0140-6736(13)60591-7

Avissar-Whiting M, Veiga KR, Uhl KM, Maccani MA, Gagne LA, Moen EL et al (2010) Bisphenol A exposure leads to specific microRNA alterations in placental cells. Reprod Toxicol 29:401–406. https://doi.org/10.1016/j.reprotox.2010.04.004

Bai W, Li Y, Niu Y, Ding Y, Yu X, Zhu B et al (2020) Association between ambient air pollution and pregnancy complications: A systematic review and meta-analysis of cohort studies. Environ Res 185:109471. https://doi.org/10.1016/j.envres.2020.109471

Bellavia A, Hauser R, Seely EW, Meeker JD, Ferguson KK, McElrath TF et al (2017) Urinary phthalate metabolite concentrations and maternal weight during early pregnancy. Int J Hyg Environ Health 220:1347–1355. https://doi.org/10.1016/j.ijheh.2017.09.005

Bellavia A, Cantonwine DE, Meeker JD, Hauser R, Seely EW, McElrath TF et al (2018) Pregnancy urinary bisphenol-A concentrations and glucose levels across BMI categories. Environ Int 113:35–41. https://doi.org/10.1016/j.envint.2018.01.012

Bellavia A, Chiu YH, Brown FM, Mínguez-Alarcón L, Ford JB, Keller M, ; the EARTH Study Team et al. Urinary concentrations of parabens mixture and pregnancy glucose levels among women from a fertility clinic. Environ Res 2019;168:389–396. https://doi.org/10.1016/j.envres.2018.10.009

Bellou V, Belbasis L, Tzoulaki I, Evangelou E (2014) Risk factors for type 2 diabetes mellitus: an exposure-wide umbrella review of meta-analyses. PLoS One 13:e0194127. https://doi.org/10.1371/journal.pone.0194127

Belot M-P, Fradin D, Mai N, Le Fur S, Zélénika D, Kerr-Conte J et al (2013) CpG methylation changes within the IL2RA promoter in type 1 diabetes of childhood onset. PLoS One 8:e68093. https://doi.org/10.1371/journal.pone.0068093

Benson VS, Vanleeuwen JA, Taylor J, Somers GS, McKinney PA, Van Til L (2010) Type 1 diabetes mellitus and components in drinking water and diet: a population-based, case-control study in Prince Edward Island. Canada J Am Coll Nutr 29:612–624. https://doi.org/10.1080/07315724.2010.10719900

Bergamin CS, Dib SA (2015) Enterovirus and type 1 diabetes: what is the matter? World J Diabetes 6:828–839. https://doi.org/10.4239/wjd.v6.i6.828

Bibbò S, Dore MP, Pes GM, Delitala G, Delitala AP (2017) Is there a role for gut microbiota in type 1 diabetes pathogenesis? Ann Med 49:11–22. https://doi.org/10.1080/07853890.2016.1222449

Bodin J, Stene LC, Nygaard UC (2015) Can exposure to environmental chemicals increase the risk of diabetes type 1 development? Biomed Res Int 2015:208947. https://doi.org/10.1155/2015/208947

Bonifacio E, Warncke K, Winkler C, Wallner M, Ziegler AG (2011) Cesarean section and interferon-induced helicase gene polymorphisms combine to increase childhood type 1 diabetes risk. Diabetes 60:3300–3306. https://doi.org/10.2337/db11-0729

Bouret S, Levin BE, Ozanne SE (2015) Gene-environment interactions controlling energy and glucose homeostasis and the developmental origins of obesity. Physiol Rev 95:47–82. https://doi.org/10.1152/physrev.00007.2014

Bowe B, Xie Y, Li T, Yan Y, Xian H, Al-Aly Z (2018) The 2016 global and national burden of diabetes mellitus attributable to PM2.5 air pollution. Lancet Planet Health 2:e301–e312. https://doi.org/10.1016/S2542-5196(18)30140-2

Braun JM, Lanphear BP, Calafat AM, Deria S, Khoury J, Howe CJ et al (2014) Early-life bisphenol A exposure and child body mass index: a prospective cohort study. Environ Health Perspect 122:1239–1245. https://doi.org/10.1289/ehp.1408258

Broe A, Pottegård A, Hallas J, Ahern TP, Lamont RF, Damkier P (2019) Phthalate exposure from drugs during pregnancy and possible risk of preterm birth and small for gestational age. Eur J Obstet Gynecol Reprod Biol 240:293–299. https://doi.org/10.1016/j.ejogrb.2019.07.023

Brown RJ, Rother KI (2008) Effects of beta-cell rest on beta-cell function: a review of clinical and preclinical data. Pediatr Diabetes 9:14–22. https://doi.org/10.1111/j.1399-5448.2007.00272.x

Brunkwall L, Orho-Melander M (2017) The gut microbiome as a target for prevention and treatment of hyperglycaemia in type 2 diabetes: from current human evidence to future possibilities. Diabetologia 60:943–951. https://doi.org/10.1007/s00125-017-4278-3

Calafat AM, Ye X, Wong LY, Bishop AM, Needham LL (2010) Urinary concentrations of four parabens in the US population: NHANES 2005–2006. Environ Health Perspect 118:679–685. https://doi.org/10.1289/ehp.0901560

Chapman NM, Kim KS (2008) Persistent coxsackievirus infection: enterovirus persistence in chronic myocarditis and dilated cardiomyopathy. Curr Top Microbiol Immunol 323:275–292. https://doi.org/10.1007/978-3-540-75546-3_13

Chen Y, Yan SS, Colgan J, Zhang HP, Luban J, Schmidt AM et al (2004) Blockade of late stages of autoimmune diabetes by inhibition of the receptor for advanced glycation end products. J Immunol 173:1399–1405. https://doi.org/10.4049/jimmunol.173.2.1399

Chen YW, Yang CY, Huang CF, Hung DZ, Leung YM, Liu SH (2009) Heavy metals, islet function and diabetes development. Islets 1:169–176. https://doi.org/10.4161/isl.1.3.9262

Chen YW, Huang CF, Yang CY, Yen CC, Tsai KS, Liu SH (2010) Inorganic mercury causes pancreatic beta-cell death via the oxidative stress-induced apoptotic and necrotic pathways. Toxicol Appl Pharmacol 243:323–331. https://doi.org/10.1016/j.taap.2009.11.024

Chia JSJ, McRae JL, Kukuljan S, Woodford K, Elliott RB, Swinburn B et al (2017) A1 beta-casein milk protein and other environmental pre-disposing factors for type 1 diabetes. Nutr Diabetes 7:e274. https://doi.org/10.1038/nutd.2017.16

Choe SA, Eliot MN, Savitz DA, Wellenius GA (2019) Ambient air pollution during pregnancy and risk of gestational diabetes in New York City. Environ Res 175:414–420. https://doi.org/10.1016/j.envres.2019.04.030

Christoffersson G, Rodriguez-Calvo T, von Herrath M (2016) Recent advances in understanding type 1 diabetes. F1000Res 5:F1000 Faculty Rev-110. https://doi.org/10.12688/f1000research.7356.1

Cnop M, Foufelle F, Velloso LA (2012) Endoplasmic reticulum stress, obesity and diabetes. Trends Mol Med 18:59–68. https://doi.org/10.1016/j.molmed.2011.07.010

Colborn T, Clement C (1992) Chemically-induced alterations in sexual and functional development: the wildlife/human connection. Princeton Scientific Publishing Co., Princeton, NJ

Colborn T, Dumanoski D, Myers JP (1996) Our stolen future: are we threatening our fertility, intelligence, and survival? a scientific detective story. Dutton, New York

Cooke A (2009) Review series on helminths, immune modulation and the hygiene hypothesis: how might infection modulate the onset of type 1 diabetes? Immunology 126:12–17. https://doi.org/10.1111/j.1365-2567.2008.03009.x

Couper JJ (2001) Environmental triggers of type 1 diabetes. J Paediatr Child Health 37:218–220. https://doi.org/10.1046/j.1440-1754.2001.00658.x

Craig ME, Nair S, Stein H, Rawlinson WD (2013) Viruses and type 1 diabetes: a new look at an old story. Pediatr Diabetes 14:149–158. https://doi.org/10.1111/pedi.12033

Craig ME, Kim KW, Isaacs SR, Penno MA, Hamilton-Williams EE, Couper JJ et al (2019) Early-life factors contributing to type 1 diabetes. Diabetologia 62:1823–1834. https://doi.org/10.1007/s00125-019-4942-x

Delahanty LM, Pan Q, Jablonski KA, Aroda VR, Watson KE, Bray GA, Diabetes Prevention Program Research Group et al (2014) Effects of weight loss, weight cycling, and weight loss maintenance on diabetes incidence and change in cardiometabolic traits in the diabetes prevention program. Diabetes Care 37:2738–2745. https://doi.org/10.2337/dc14-0018

Diamanti-Kandarakis E, Bourguignon JP, Giudice LC, Hauser R, Prins GS, Soto AM et al (2009) Endocrine-disrupting chemicals: an endocrine society scientific statement. Endocr Rev 30:293–342. https://doi.org/10.1210/er.2009-0002

Dong JY, Zhang WG, Chen JJ, Zhang ZL, Han SF, Qin LQ (2013) Vitamin D intake and risk of type 1 diabetes: a meta-analysis of observational studies. Nutrients 5:3551–3562. https://doi.org/10.3390/nu5093551

Dotta F, Censini S, van Halteren AGS, Marselli L, Masini M, Dionisi S et al (2007) Coxsackie B4 virus infection of beta cells and natural killer cell insulitis in recent-onset type 1 diabetic patients. Proc Natl Acad Sci U S A 104:5115–5120. https://doi.org/10.1073/pnas.0700442104

Douillet C, Currier J, Saunders J, Bodnar WM, Matoušek T, Stýblo M (2013) Methylated trivalent arsenicals are potent inhibitors of glucose stimulated insulin secretion by murine pancreatic islets. Toxicol Appl Pharmacol 267:11–15. https://doi.org/10.1016/j.taap.2012.12.007

Doyle HA, Yang M-L, Raycroft MT, Gee RJ, Mamula MJ (2014) Autoantigens: novel forms and presentation to the immune system. Autoimmunity 47:220–233. https://doi.org/10.3109/08916934.2013.850495

Drews G, Krippeit-Drews P, Düfer M (2010) Oxidative stress and beta-cell dysfunction. Pflugers Arch 460:703–718. https://doi.org/10.1007/s00424-010-0862-9

Edwards TM, Myers JP (2007) Environmental exposures and gene regulation in disease etiology. Environ Health Perspect 115:1264–1270. https://doi.org/10.1289/ehp.9951

Ehrlich S, Lambers D, Baccarelli A, Khoury J, Macaluso M, Ho SM (2016) Endocrine disruptors: a potential risk factor for gestational diabetes mellitus. Am J Perinatol 33:1313–1318. https://doi.org/10.1055/s-0036-1586500

Eizirik DL, Colli ML, Ortis F (2009) The role of inflammation in insulitis and beta-cell loss in type 1 diabetes. Nat Rev Endocrinol 5:219–226. https://doi.org/10.1038/nrendo.2009.21

Elliott RB (2006) Diabetes—a man made disease. Med Hypotheses 67:388–391. https://doi.org/10.1016/j.mehy.2005.11.047

Elten M, Donelle J, Lima I, Burnett RT, Weichenthal S, David M et al (2020) Ambient air pollution and incidence of early-onset paediatric type 1 diabetes: a retrospective population-based cohort study. Environ Res 184:109291. https://doi.org/10.1016/j.envres.2020.109291

Esposito S, Toni G, Tascini G, Santi E, Berioli MG, Principi N (2019) Environmental factors associated with type 1 diabetes. Front Endocrinol 10:592. https://doi.org/10.3389/fendo.2019.00592

Ferguson KK, Meeker JD, Cantonwine DE, Mukherjee B, Pace GG, Weller D et al (2018) Environmental phenol associations with ultrasound and delivery measures of fetal growth. Environ Int 112:243–250. https://doi.org/10.1016/j.envint.2017.12.011

Fernandez-Twinn DS, Hjort L, Novakovic B, Ozanne SE, Saffery R (2019) Intrauterine programming of obesity and type 2 diabetes. Diabetologia 62:1789–1801. https://doi.org/10.1007/s00125-019-4951-9

Filardi T, Panimolle F, Lenzi A, Morano S (2020) Bisphenol A and phthalates in diet: an emerging link with pregnancy complications. Nutrients 12:525. https://doi.org/10.3390/nu12020525

Filippi C, von Herrath M (2005) How viral infections affect the autoimmune process leading to type 1 diabetes. Cell Immunol 233:125–132. https://doi.org/10.1016/j.cellimm.2005.04.009

Filippo CD, Cavalieri D, Paola MD, Ramazzotti M, Poullet JB, Massart S et al (2010) Impact of diet in shaping gut microbiota revealed by a comparative study in children from Europe and rural Africa. PNAS 107:14691–14696. https://doi.org/10.1073/pnas.1005963107

Fourlanos S, Varney MD, Tait BD, Morahan G, Honeyman MC, Colman PG et al (2008) The rising incidence of type 1 diabetes is accounted for by cases with lower-risk human leukocyte antigen genotypes. Diabetes Care 31:1546–1549. https://doi.org/10.2337/dc08-0239

Frisk G, Diderholm H (2000) Tissue culture of isolated human pancreatic islets infected with different strains of coxsackievirus B4: assessment of virus replication and effects on islet morphology and insulin release. Int J Exp Diabetes Res 1:165–175. https://doi.org/10.1155/edr.2000.165

Fujinami RS, von Herrath MG, Christen U, Whitton JL (2006) Molecular mimicry, bystander activation, or viral persistence: infections and autoimmune disease. Clin Microbiol Rev 19:80–94. https://doi.org/10.1128/CMR.19.1.80-94.2006

Gamble DR, Taylor KW (1969) Seasonal incidence of diabetes mellitus. Br Med J 3:631–633. https://doi.org/10.1136/bmj.3.5671.631

Gillespie KM, Bain SC, Barnett AH, Bingley PJ, Christie MR, Gill GV et al (2004) The rising incidence of childhood type 1 diabetes and reduced contribution of high-risk HLA haplotypes. Lancet 364:1699–1700. https://doi.org/10.1016/S0140-6736(04)17357-1

Giusti RM, Iwamoto K, Hatch EE (1995) Diethylstilbestrol revisited: a review of the long-term health effects. Ann Intern Med 122:778–788. https://doi.org/10.7326/0003-4819-122-10-199505150-00008

González-Casanova JE, Pertuz-Cruz SL, Caicedo-Ortega NH, Rojas-Gomez DM (2020) Adipogenesis regulation and endocrine disruptors: emerging insights in obesity. Biomed Res Int 2020:7453786. https://doi.org/10.1155/2020/7453786

Gore AC (2010) Neuroendocrine targets of endocrine disruptors. Hormones (Athens) 9:16–27. https://doi.org/10.14310/horm.2002.1249

Gore AC, Crews D, Doan LL, La Merrill M, Patisaul H, Zota A (2014) Introduction to endocrine disrupting chemicals (EDCs); Joint endocrine society–IPEN initiative. Endocrine Society, Washington, DC

Gore AC, Chappell VA, Fenton SE, Flaws JA, Nadal A, Prins GS et al (2015) EDC-2: the endocrine society's second scientific statement on endocrine-disrupting chemicals. Endocr Rev 36:E1–E150. https://doi.org/10.1210/er.2015-1010

Grieco FA, Sebastiani G, Spagnuolo I, Patti A, Dotta F (2012) Immunology in the clinic review series; focus on type 1 diabetes and viruses: how viral infections modulate beta cell function. Clin Exp Immunol 168:24–29. https://doi.org/10.1111/j.1365-2249.2011.04556.x

Guarino E, Delli Poggi C, Grieco GE, Cenci V, Ceccarelli E, Crisci I et al (2018) Circulating MicroRNAs as biomarkers of gestational diabetes mellitus: updates and perspectives. Int J Endocrinol 2018:6380463. https://doi.org/10.1155/2018/6380463

Guerrero-Bosagna C, Skinner MK (2012) Environmentally induced epigenetic transgenerational inheritance of phenotype and disease. Mol Cell Endocrinol 354:3–8. https://doi.org/10.1016/j.mce.2011.10.004

Hague A, Butt AJ, Paraskeva C (1996) The role of butyrate in human colonic epithelial cells: an energy source or inducer of differentiation and apoptosis? Proc Nutr Soc 55:937–943. https://doi.org/10.1079/pns19960090

Harrison LC, Honeyman MC (1999) Cow's milk and type 1 diabetes: the real debate is about mucosal immune function. Diabetes 48:1501–1507. https://doi.org/10.2337/diabetes.48.8.1501

Hermann R, Knip M, Veijola R, Simell O, Laine AP, Akerblom HK, ; FinnDiane Study Group et al. Temporal changes in the frequencies of HLA genotypes in patients with type 1 diabetes--indication of an increased environmental pressure? Diabetologia 2003;46:420–425. https://doi.org/10.1007/s00125-003-1045-4

Hogg JC, van Eeden S (2009) Pulmonary and systemic response to atmospheric pollution. Respirology 14:336–346. https://doi.org/10.1111/j.1440-1843.2009.01497.x

Hooper LV, Littman DR, Macpherson AJ (2012) Interactions between the microbiota and the immune system. Science 336:1268–1273. https://doi.org/10.1126/science.1223490

Horwitz MS, Bradley LM, Harbertson J, Krahl T, Lee J, Sarvetnick N (1998) Diabetes induced by Coxsackie virus: initiation by bystander damage and not molecular mimicry. Nat Med 4:781–785. https://doi.org/10.1038/nm0798-781

Hugo ER, Brandebourg TD, Woo JG, Loftus J, Alexander JW, Ben-Jonathan N (2008) Bisphenol A at environmentally relevant doses inhibits adiponectin release from human adipose tissue explants and adipocytes. Environ Health Perspect 116:1642–1647. https://doi.org/10.1289/ehp.11537

Hyöty H (2016) Viruses in type 1 diabetes. Pediatr Diabetes 17:56–64. https://doi.org/10.1111/pedi.12370

Ifie E, Russell MA, Dhayal S, Leete P, Sebastiani G, Nigi L et al (2018) Unexpected subcellular distribution of a specific isoform of the Coxsackie and adenovirus receptor, CAR-SIV, in human pancreatic beta cells. Diabetologia 61:2344–2355. https://doi.org/10.1007/s00125-018-4704-1

In't VP (2011) Insulitis in the human endocrine pancreas: does a viral infection lead to inflammation and beta cell replication? Diabetologia 54:2220–2222. https://doi.org/10.1007/s00125-011-2224-3

Infante M, Ricordi C, Sanchez J, Clare-Salzler MJ, Padilla N, Fuenmayor V et al (2019) Influence of vitamin D on islet autoimmunity and beta-cell function in type 1 diabetes. Nutrients 11:2185. https://doi.org/10.3390/nu11092185

International Diabetes Federation (ed) (2017) IDF Atlas, 8th edn. International Diabetes Federation, Brussels. https://www.idf.org/e-library/epidemiology-research/diabetes-atlas/134-idf-diabetes-atlas-8th-edition.html. Accessed 27 Jun 2020

Jacobsen SC, Brøns C, Bork-Jensen J, Ribel-Madsen R, Yang B, Lara E et al (2012) Effects of short-term high-fat overfeeding on genome-wide DNA methylation in the skeletal muscle of healthy young men. Diabetologia 55:3341–3349. https://doi.org/10.1007/s00125-012-2717-8

Juhela S, Hyöty H, Roivainen M, Härkönen T, Putto-Laurila A, Simell O et al (2000) T-cell responses to enterovirus antigens in children with type 1 diabetes. Diabetes 49:1308–1313. https://doi.org/10.2337/diabetes.49.8.1308

Kabir ER, Rahman MS, Rahman I (2015) A review on endocrine disruptors and their possible impacts on human health. Environ Toxicol Pharmacol 40:241–258. https://doi.org/10.1016/j.etap.2015.06.009

Kavlock RJ, Daston GP, DeRosa C, Fenner-Crisp P, Gray LE, Kaattari S et al (1996) Research needs for the risk assessment of health and environmental effect of endocrine disruptors: a report of the USEPA-sponsored workshop. Environ Health Perspect 104(Suppl 4):715–740. https://doi.org/10.1289/ehp.96104s4715

Kieser KJ, Kagan JC (2017) Multireceptor detection of individual bacterial products by the innate immune system. Nat Rev Immunol 17:376–390. https://doi.org/10.1038/nri.2017.25

Kim JB, Prunicki M, Haddad F, Dant C, Sampath V, Patel R et al (2020) Cumulative lifetime burden of cardiovascular disease from early exposure to air pollution. J Am Heart Assoc 9:e014944. https://doi.org/10.1161/JAHA.119.014944

Kindt ASD, Fuerst RW, Knoop J, Laimighofer M, Telieps T, Hippich M et al (2018) Allele-specific methylation of type 1 diabetes susceptibility genes. J Autoimmun 89:63–74. https://doi.org/10.1016/j.jaut.2017.11.008

Knip M (2003) Environmental triggers and determinants of beta-cell autoimmunity and type 1 diabetes. Rev Endocr Metab Disord 4:213–223. https://doi.org/10.1023/a:1025121510678

Knip M, Siljander H (2016) The role of the intestinal microbiota in type 1 diabetes mellitus. Nat Rev Endocrinol 12:154–167. https://doi.org/10.1038/nrendo.2015.218

Knip M, Simell O (2012) Environmental triggers of type 1 diabetes. Cold Spring Harb Perspect Med 2:a007690. https://doi.org/10.1101/cshperspect.a007690

Knip M, Veijola R, Virtanen SM, Hyöty H, Vaarala O, Akerblom HK (2005) Environmental triggers and determinants of type 1 diabetes. Diabetes 54(Suppl 2):S125–S136. https://doi.org/10.2337/diabetes.54.suppl_2.s125

Kolb H, Martin S (2017) Environmental/lifestyle factors in the pathogenesis and prevention of type 2 diabetes. BMC Med 15:131. https://doi.org/10.1186/s12916-017-0901-x

Krogvold L, Edwin B, Buanes T, Frisk G, Skog O, Anagandula M et al (2015) Detection of a low-grade enteroviral infection in the islets of langerhans of living patients newly diagnosed with type 1 diabetes. Diabetes 64:1682–1687. https://doi.org/10.2337/db14-1370

Kwak SH, Park KS (2016) Recent progress in genetic and epigenetic research on type 2 diabetes. Exp Mol Med 48:e220. https://doi.org/10.1038/emm.2016.7

La Merrill M, Emond C, Kim MJ, Antignac JP, Le Bizec B, Clément K et al (2013) Toxicological function of adipose tissue: focus on persistent organic pollutants. Environ Health Perspect 121:162–169. https://doi.org/10.1289/ehp.1205485

Lamb MM, Myers MA, Barriga K, Zimmet PZ, Rewers M, Norris JM (2008) Maternal diet during pregnancy and islet autoimmunity in offspring. Pediatr Diabetes 9:135–141. https://doi.org/10.1111/j.1399-5448.2007.00311.x

Lamb MM, Frederiksen B, Seifert JA, Kroehl M, Rewers M, Norris JM (2015) Sugar intake is associated with progression from islet autoimmunity to type 1 diabetes: the diabetes autoimmunity study in the young. Diabetologia 58:2027–2034. https://doi.org/10.1007/s00125-015-3657-x

LaRocca J, Binder AM, McElrath TF, Michels KB (2016) First-trimester urine concentrations of phthalate metabolites and phenols and placenta miRNA expression in a cohort of U.S. women. Environ Health Perspect 124:380–387. https://doi.org/10.1289/ehp.1408409

Li Y, Xu L, Shan Z, Teng W, Cheng H (2019a) Association between air pollution and type 2 diabetes: an updated review of the literature. Ther Adv Endocrinol Metab 10:1–15. https://doi.org/10.1177/2042018819897046

Li Y, Xu S, Li Y, Zhang B, Huo W, Zhu Y et al (2019b) Association between urinary parabens and gestational diabetes mellitus across prepregnancy body mass index categories. Environ Res 170:1519. https://doi.org/10.1016/j.envres.2018.12.028

Lin Y, Wei J, Li Y, Chen J, Zhou Z, Song L et al (2011) Developmental exposure to di(2-ethylhexyl) phthalate impairs endocrine pancreas and leads to long-term adverse effects on glucose homeostasis in the rat. Am J Physiol Endocrinol Metab 301:E527–E538. https://doi.org/10.1152/ajpendo.00233.2011

Ludvigsson J, Holmqvist BM, Samuelsson ULF (2013) Does modern high standard life style cause type 1 diabetes in children? Diabetes Metab Res Rev 29:161–165. https://doi.org/10.1002/dmrr.2377

Maahs DM, West NA, Lawrence JM, Mayer-Davis EJ (2010) Epidemiology of type 1 diabetes. Endocrinol Metab Clin N Am 39:481–497. https://doi.org/10.1016/j.ecl.2010.05.011

Magliano DJ, Loh VH, Harding JL, Botton J, Shaw JE (2014) Persistent organic pollutants and diabetes: a review of the epidemiological evidence. Diabetes Metab 40:1–14. https://doi.org/10.1016/j.diabet.2013.09.006

Mariño E, Richards JL, McLeod KH, Stanley D, Yap YA, Knight J et al (2017) Gut microbial metabolites limit the frequency of autoimmune T cells and protect against type 1 diabetes. Nat Immunol 18:552–562. https://doi.org/10.1038/ni.3713

Marre ML, James EA, Piganelli JD (2015) β cell ER stress and the implications for immunogenicity in type 1 diabetes. Front Cell Dev Biol 3:67. https://doi.org/10.3389/fcell.2015.00067

Martínez-Ibarra A, Martínez-Razo LD, Vázquez-Martínez ER, Martínez-Cruz N, Flores-Ramírez R, García-Gómez E et al (2019) Unhealthy levels of phthalates and bisphenol A in mexican pregnant women with gestational diabetes and its association to altered expression of miRNAs involved with metabolic disease. Int J Mol Sci 20:3343. https://doi.org/10.3390/ijms20133343

Marty MS, Carney EW, Rowlands JC (2011) Endocrine disruption: historical perspectives and its impact on the future of toxicology testing. Toxicol Sci 120:S93–S108. https://doi.org/10.1093/toxsci/kfq329

Mehers KL, Gillespie KM (2008) The genetic basis for type 1 diabetes. Br Med Bull 88:115–129. https://doi.org/10.1093/bmb/ldn045

Melody SM, Wills K, Knibbs LD, Ford J, Venn A, Johnston F (2020) Maternal Exposure to Ambient Air Pollution and Pregnancy Complications in Victoria. Aust Int J Environ Res Public Health 17:2572. https://doi.org/10.3390/ijerph17072572

Morgan E, Halliday SR, Campbell GR, Cardwell CR, Patterson CC (2016) Vaccinations and childhood type 1 diabetes mellitus: a meta-analysis of observational studies. Diabetologia 59:237–243. https://doi.org/10.1007/s00125-015-3800-8

Nigi L, Grieco GE, Ventriglia G, Brusco N, Mancarella F, Formichi C et al (2018) MicroRNAs as regulators of insulin signaling: research updates and potential therapeutic perspectives in type 2 diabetes. Int J Mol Sci 19:3705. https://doi.org/10.3390/ijms19123705

Nilsson E, Ling C (2017) DNA methylation links genetics, fetal environment, and an unhealthy lifestyle to the development of type 2 diabetes. Clin Epigenetics 9:105. https://doi.org/10.1186/s13148-017-0399-2

Nilsson E, Matte A, Perfilyev A, de Mello VD, Käkelä P, Pihlajamäki J et al (2015) Epigenetic alterations in human liver from subjects with type 2 diabetes in parallel with reduced folate levels. J Clin Endocrinol Metab 100:E1491–E1501. https://doi.org/10.1210/jc.2015-3204

Op de Beeck A, Eizirik DL (2016) Viral infections in type 1 diabetes mellitus — why the β cells? Nat Rev Endocrinol 12:263–273. https://doi.org/10.1038/nrendo.2016.30

Pascale A, Marchesi N, Marelli C, Coppola A, Luzi L, Govoni S et al (2018) Microbiota and metabolic diseases. Endocrine 61:357–371. https://doi.org/10.1007/s12020-018-1605-5

Peng L, Li ZR, Green RS, Holzman IR, Lin J (2009) Butyrate enhances the intestinal barrier by facilitating tight junction assembly via activation of AMP-activated protein kinase in Caco-2 cell monolayers. J Nutr 139:1619–1625. https://doi.org/10.3945/jn.109.104638

Peppa M, He C, Hattori M, McEvoy R, Zheng F, Vlassara H (2003) Fetal or neonatal low-glycotoxin environment prevents autoimmune diabetes in NOD mice. Diabetes 52:1441–1448. https://doi.org/10.2337/diabetes.52.6.1441

Petzold A, Solimena M, Knoch KP (2015) Mechanisms of beta cell dysfunction associated with viral infection. Curr Diab Rep 15:73. https://doi.org/10.1007/s11892-015-0654-x

Pociot F, Lernmark Å (2016) Genetic risk factors for type 1 diabetes. Lancet 387:2331–2339. https://doi.org/10.1016/S0140-6736(16)30582-7

Pociot F, Akolkar B, Concannon P, Erlich HA, Julier C, Morahan G et al (2010) Genetics of type 1 diabetes: what's next? Diabetes 59:1561–1571. https://doi.org/10.2337/db10-0076

Pope CA III, Bhatnagar A, McCracken JP, Abplanalp W, Conklin DJ, O'Toole T (2016) Exposure to fine particulate air pollution is associated with endothelial injury and systemic inflammation. Circ Res 119:1204–1214. https://doi.org/10.1161/CIRCRESAHA.116.309279

Predieri B, Bruzzi P, Bigi E, Ciancia S, Madeo SF, Lucaccioni L et al (2020) Endocrine disrupting chemicals and type 1 diabetes. Int J Mol Sci 21:2937. https://doi.org/10.3390/ijms21082937

Preston JD, Reynolds LJ, Pearson KJ (2018) Developmental origins of health span and life span: a mini review. Gerontology 64:237–245. https://doi.org/10.1159/000485506

Quercia S, Candela M, Giuliani C, Turroni S, Luiselli D, Rampelli S et al (2014) From lifetime to evolution: timescales of human gut microbiota adaptation. Front Microbiol 5:587. https://doi.org/10.3389/fmicb.2014.00587

Regnell SE, Lernmark Å (2017) Early prediction of autoimmune (type 1) diabetes. Diabetologia 60:1370–1381. https://doi.org/10.1007/s00125-017-4308-1

Renz H, Holt PG, Inouye M, Logan AC, Prescott SL, Sly PD (2017) An exposome perspective: early-life events and immune development in a changing world. J Allergy Clin Immunol 140:24–40. https://doi.org/10.1016/j.jaci.2017.05.015

Rešić Lindehammer S, Honkanen H, Nix WA, Oikarinen M, Lynch KF, Jönsson I et al (2012) Seroconversion to islet autoantibodies after enterovirus infection in early pregnancy. Viral Immunol 25:254–261. https://doi.org/10.1089/vim.2012.0022

Rewers M (2012) The fallacy of reduction. Pediatr Diabetes 13:340–343. https://doi.org/10.1111/j.1399-5448.2011.00832.x

Rewers M, Ludvigsson J (2016) Environmental risk factors for type 1 diabetes. Lancet 387:2340–2348. https://doi.org/10.1016/S0140-6736(16)30507-4

Richardson SJ, Willcox A, Bone AJ, Foulis AK, Morgan NG (2009) The prevalence of enteroviral capsid protein vp1 immunostaining in pancreatic islets in human type 1 diabetes. Diabetologia 52:1143–1151. https://doi.org/10.1007/s00125-009-1276-0

Roivainen M (2006) Enteroviruses: new findings on the role of enteroviruses in type 1 diabetes. Int J Biochem Cell Biol 38:721–725. https://doi.org/10.1016/j.biocel.2005.08.019

Roivainen M, Ylipaasto P, Savolainen C, Galama J, Hovi T, Otonkoski T (2002) Functional impairment and killing of human beta cells by enteroviruses: the capacity is shared by a wide range of serotypes, but the extent is a characteristic of individual virus strains. Diabetologia 45:693–702. https://doi.org/10.1007/s00125-002-0805-x

Rønningen KS (2015) Environmental Trigger(s) of type 1 diabetes: why so difficult to identify? Hindawi Publishing Corporation BioMed Research International, p 321656. https://doi.org/10.1155/2015/321656

Rønningen KS, Norris JM, Knip M (2015) Environmental trigger(s) of type 1 diabetes: why is it so difficult to identify? Biomed Res Int 2015:847906. https://doi.org/10.1155/2015/847906

Rook GA (2012) Hygiene hypothesis and autoimmune diseases. Clin Rev Allergy Immunol 42:5–15. https://doi.org/10.1007/s12016-011-8285-8

Roseboom T, de Rooij S, Painter R (2006) The Dutch famine and its long-term consequences for adult health. Early Hum Dev 82:485–491. https://doi.org/10.1016/j.earlhumdev.2006.07.001

Russart KLG, Nelson RJ (2018) Light at night as an environmental endocrine disruptor. Physiol Behav 190:82–89. https://doi.org/10.1016/j.physbeh.2017.08.029

Salas-Salvadó J, Bulló M, Babio N, Martínez-González MÁ, Ibarrola-Jurado N, Basora J, PREDIMED Study Investigators et al (2011) Reduction in the incidence of type 2 diabetes with the Mediterranean diet: results of the PREDIMED-Reus nutrition intervention randomized trial. Diabetes Care 34:14–19. https://doi.org/10.2337/dc10-1288

Sebastiani G, Guarino E, Grieco GE, Formichi C, Delli Poggi C, Ceccarelli E et al (2017) Circulating microRNA (miRNA) expression profiling in plasma of patients with gestational diabetes mellitus reveals upregulation of miRNA miR-330-3p. Front Endocrinol (Lausanne) 8:345. https://doi.org/10.3389/fendo.2017.00345

Shaffer RM, Ferguson KK, Sheppard L, James-Todd T, Butts S, Chandrasekaran S, the TIDES Study Team et al (2019) Maternal urinary phthalate metabolites in relation to gestational diabetes and glucose intolerance during pregnancy. Environ Int 123:588–596. https://doi.org/10.1016/j.envint.2018.12.021

Shankar A, Teppala S (2011) Relationship between urinary bisphenol A levels and diabetes mellitus. J Clin Endocrinol Metab 96:3822–3826. https://doi.org/10.1210/jc.2011-1682

Shapiro GD, Dodds L, Arbuckle TE, Ashley-Martin J, Fraser W, Fisher M et al (2015) Exposure to phthalates, bisphenol A and metals in pregnancy and the association with impaired glucose tolerance and gestational diabetes mellitus: the MIREC study. Environ Int 83:63–71. https://doi.org/10.1016/j.envint.2015.05.016

Sheehan A, Freni Sterrantino A, Fecht D, Elliott P, Hodgson S (2020) Childhood type 1 diabetes: an environment-wide association study across England. Diabetologia 63:964–976. https://doi.org/10.1007/s00125-020-05087-7

Skyler JS, Bakris GL, Bonifacio E, Darsow T, Eckel RH, Groop L et al (2017) Differentiation of diabetes by pathophysiology, natural history, and prognosis. Diabetes 66:241–255. https://doi.org/10.2337/db16-0806

Sommer F, Bäckhed F (2013) The gut microbiota—masters of host development and physiology. Nat Rev Microbiol 11:227–238. https://doi.org/10.1038/nrmicro2974

Sonnenburg JL, Bäckhed F (2016) Diet–microbiota interactions as moderators of human metabolism. Nature 535:56–64. https://doi.org/10.1038/nature18846

Soriano S, Alonso-Magdalena P, García-Arévalo M, Novials A, Muhammed SJ, Salehi A et al (2012) Rapid insulinotropic action of low doses of bisphenol-A on mouse and human islets of Langerhans: role of estrogen receptor β. PLoS One 7:e31109. https://doi.org/10.1371/journal.pone.0031109

Steck AK, Rewers MJ (2011) Genetics of type 1 diabetes. Clin Chem 57:176–185. https://doi.org/10.1373/clinchem.2010.148221

Stene LC, Gal EAM (2013) The prenatal environment and type 1 diabetes. Diabetologia 56:1888–1897. https://doi.org/10.100//s00125-013-2929-6

Stene LC, Rewers M (2012) Immunology in the clinic review series; focus on type 1 diabetes and viruses: the enterovirus link to type 1 diabetes: critical review of human studies. Clin Exp Immunol 168:12–23. https://doi.org/10.1111/j.1365-2249.2011.04555.x

Tamashiro KL, Moran TH (2010) Perinatal environment and its influences on metabolic programming of offspring. Physiol Behav 100:560–566. https://doi.org/10.1016/j.physbeh.2010.04.008

Tang WH, Kitai T, Hazen SL (2017) Gut Microbiota in Cardiovascular Health and Disease. Circ Res 120:1183–1196. https://doi.org/10.1161/CIRCRESAHA.117.309715

Tang X, Zhou J-B, Luo F, Han Y, Heianza Y, Cardoso M et al (2020) Air pollution and gestational diabetes mellitus: evidence from cohort studies. BMJ Open Diabetes Res Care 8:e000937. https://doi.org/10.1136/bmjdrc-2019-000937

Thursby E, Juge N (2017) Introduction to the human gut microbiota. Biochem J 474:1823–1836. https://doi.org/10.1042/BCJ20160510

Tiffon C (2018) The impact of nutrition and environmental epigenetics on human health and disease. Int J Mol Sci 19:3425. https://doi.org/10.3390/ijms19113425

Todd JA (2010) Etiology of type 1 diabetes. Immunity 32:457–467. https://doi.org/10.1016/j.immuni.2010.04.001

Tracy S, Drescher KM, Jackson JD, Kim K, Kono K (2010) Enteroviruses, type 1 diabetes and hygiene: a complex relationship. Rev Med Virol 20:106–116. https://doi.org/10.1002/rmv.639

Tracy S, Smithee S, Alhazmi A, Chapman N (2015) Coxsackievirus can persist in murine pancreas by deletion of 5′ terminal genomic sequences. J Med Virol 87:240–247. https://doi.org/10.1002/jmv.24039

Tuomilehto J, Lindström J, Eriksson JG, Valle TT, Hämäläinen H, Ilanne-Parikka P, Finnish Diabetes Prevention Study Group et al (2001) Prevention of type 2 diabetes mellitus by changes in lifestyle among subjects with impaired glucose tolerance. N Engl J Med 344:1343–1350. https://doi.org/10.1056/NEJM200105033441801

Upadhyaya S, Banerjee G (2015) Type 2 diabetes and gut microbiome: at the intersection of known and unknown. Gut Microbes 6:85–92. https://doi.org/10.1080/19490976.2015.1024918

Uusitupa M, Khan TA, Viguiliouk E, Kahleova H, Rivellese AA, Hermansen K et al (2019) Prevention of type 2 diabetes by lifestyle changes: a systematic review and meta-analysis. Nutrients 11:2611. https://doi.org/10.3390/nu11112611

Vaarala O (2008) Leaking gut in type 1 diabetes. Curr Opin Gastroenterol 24:701–706. https://doi.org/10.1097/MOG.0b013e32830e6d98

Vaarala O (2012) Is the origin of type 1 diabetes in the gut? Immunol Cell Biol 90:271–276. https://doi.org/10.1038/icb.2011.115

Vaarala O, Atkinson MA, Neu J (2008) The "perfect storm" for type 1 diabetes: the complex interplay between intestinal microbiota, gut permeability, and mucosal immunity. Diabetes 57:2555–2562. https://doi.org/10.2337/db08-0331

Valentino R, D'Esposito V, Passaretti F, Liotti A, Cabaro S, Longo M et al (2013) Bisphenol-A impairs insulin action and up-regulates inflammatory pathways in human subcutaneous adipocytes and 3T3–L1 cells. PLoS One 8:e82099. https://doi.org/10.1371/journal.pone.0082099

Varela-Calvino R, Peakman M (2003) Enteroviruses and type 1 diabetes. Diabetes Metab Res Rev 19:431–441. https://doi.org/10.1002/dmrr.407

Virtanen SM, Knip M (2003) Nutritional risk predictors of beta cell autoimmunity and type 1 diabetes at a young age. Am J Clin Nutr 78:1053–1067. https://doi.org/10.1093/ajcn/78.6.1053

Virtanen SM, Uusitalo L, Kenward MG, Nevalainen J, Uusitalo U, Kronberg-Kippilä C et al (2011) Maternal food consumption during pregnancy and risk of advanced β-cell autoimmunity in the offspring. Pediatr Diabetes 12:95–99. https://doi.org/10.1111/j.1399-5448.2010.00668.x

Wang T, Li M, Chen B, Xu M, Xu Y, Huang Y et al (2012) Urinary bisphenol A (BPA) concentration associates with obesity and insulin resistance. J Clin Endocrinol Metab 97:E223–E227. https://doi.org/10.1210/jc.2011-1989

Wen L, Duffy A (2017) Factors Influencing the gut microbiota, inflammation, and type 2 diabetes. J Nutr 147:1468S–1475S. https://doi.org/10.3945/jn.116.240754

Wenzel AG, Brock JW, Cruze L, Newman RB, Unal ER, Wolf BJ et al (2018) Prevalence and predictors of phthalate exposure in pregnant women in Charleston, SC. Chemosphere 193:394–402. https://doi.org/10.1016/j.chemosphere.2017.11.019

WHO (World Health Organization)/UNEP (United Nations Environment Programme) (2013) The state of the science of endocrine disrupting chemicals - 2012. In: Bergman Å, Heindel JJ, Jobling S, Kidd KA, Zoeller RT (eds) . UNEP/WHO, Geneva. http://www.who.int/ceh/publications/endocrine/en/index.html. Accessed 24 May 2020

Wild CP (2012) The exposome: from concept to utility. Int J Epidemiol 41:24–32. https://doi.org/10.1093/ije/dyr236

Wilkin TJ (2001) The accelerator hypothesis: weight gain as the missing link between type I and type II diabetes. Diabetologia 44:914–922. https://doi.org/10.1007/s001250100548

Woodruff TJ, Zota AR, Schwartz JM (2011) Environmental chemicals in pregnant women in the United States: NHANES 2003-2004. Environ Health Perspect 119:878–885. https://doi.org/10.1289/ehp.1002727

World Health Organization (2013) Diagnostic criteria and classification of hyperglycaemia first detected in pregnancy. WHO guidelines approved by the guidelines review committee. World Health Organization, Geneva. https://www.who.int/diabetes/publications/Hyperglycaemia_In_Pregnancy/en/. Accessed 6 Jun 2020

World Health Organization (2017) Global Report on Diabetes 2016. WHO, Geneva. https://www.who.int/diabetes/global-report/en. Accessed 27 Jun 2020

Wu Y, Ding Y, Tanaka Y, Zhang W (2014) Risk factors contributing to type 2 diabetes and recent advances in the treatment and prevention. Int J Med Sci 11:1185–1200. https://doi.org/10.7150/ijms.10001

Yao M, Liu Y, Jin D, Yin W, Ma S, Tao R et al (2020) Relationship between temporal distribution of air pollution exposure and glucose homeostasis during pregnancy. Environ Res 185:109456. https://doi.org/10.1016/j.envres.2020.109456

Yeung WC, Rawlinson WD, Craig ME (2011) Enterovirus infection and type 1 diabetes mellitus: systematic review and meta-analysis of observational molecular studies. BMJ 342:d35. https://doi.org/10.1136/bmj.d35

Ylipaasto P, Klingel K, Lindberg AM, Otonkoski T, Kandolf R, Hovi T et al (2004) Enterovirus infection in human pancreatic islet cells, islet tropism in vivo and receptor involvement in cultured islet beta cells. Diabetologia 47:225–239. https://doi.org/10.1007/s00125-003-1297-z

Zhang M, Wang X, Yang X, Dong T, Hu W, Guan Q et al (2020) Increased risk of gestational diabetes mellitus in women with higher prepregnancy ambient PM2.5 exposure. Sci Total Environ 730:138982. https://doi.org/10.1016/j.scitotenv.2020.138982

Zheng P, Li Z, Zhou Z (2018) Gut microbiome in type 1 diabetes: A comprehensive review. Diabetes Metab Res Rev 34:e3043. https://doi.org/10.1002/dmrr.3043

Zhu Y, Tian F, Li H, Zhou Y, Lu J, Ge Q (2015) Profiling maternal plasma microRNA expression in early pregnancy to predict gestational diabetes mellitus. Int J Gynaecol Obstet 130:49–53. https://doi.org/10.1016/j.ijgo.2015.01.010

Zlatnik MG (2016) Endocrine-disrupting chemicals and reproductive health. J Midwifery Womens Health 61:442–455. https://doi.org/10.1111/jmwh.12500

Some Landscape and Healthcare Considerations Comparing European Union and Indian Federation

15

Vittorio Ingegnoli and Elena Giglio

Abstract

Background: Springer Nature, including this book in the series "SGD3 Good Health and Well-Being" (Sustainable Development Goals), suggested a further chapter focused on the relationship between health and the high-/low-income and developing countries' gap. Even if this argument is not the main goal of our studies, we propose some considerations on the comparison between European Union and India.

Theory and Method: These two nations present many similarities, e.g., the number of states, the territorial surface, a common ancient origin of their population (Indo-Europeans) but even sharp differences, as in population number, urbanized people, income pro capita, life expectancy, and cancer incidence. As underlined by Gandhi, the peculiar character of India is to be structured in villages, while Europe is structured in cities. To better understand these differences, the environmental conditions must be considered. Applying the principles and methods of Landscape Bionomics to the period 1880–2010, we will see that the main environmental parameters remained constant in Europe, while in India they were changing decidedly.

Findings: The diagnostic evaluation HH/BTC shows the opposite directions of the land transformations in these countries, where the changes in India were 3.6 times wider than in Europe. Given these differences, you can note that the increase of cancer incidence in Europe has been four times stronger than that in India, where its prevalence resulted nearly 1/10 than in Europe. On the contrary, the increase of type 2 diabetes was higher in India, even in low-income people.

Discussion and Conclusion: In conclusion, it is not correct to compare these two nations following economic parameters: it is necessary to consider also ecological and health aims. The life expectancy gap in 1980 was of 20 years, but in 2015, with a healthcare expenditure from 2.0 to 3.6% (+1.6) of GNP, India gained 15.2 years, while to add only 10.0 years, the EU needed an expenditure from 5.5 to 9.5% (+4.0) of GNP.

V. Ingegnoli (✉)
Landscape Bionomics and Planetary Health, Department of Environmental Sciences and Policy, University of Milan, Milan, Italy

Planetary Health Alliance, Harvard University, USA, and SIPNEI (Psycho-Neuro-Endocrine-Immunology Italian Society), Rome, Italy
e-mail: vittorio.ingegnoli@guest.unimi.it

E. Giglio
Human and Environment relationships PhD, Teacher of Geography at High School, Milan, Italy

Planetary Health Alliance (member of), Education, Boston, USA

Keywords

European Union · EU · India · Environmental parameters · Health/environment relationships

15.1 Similarities and Differences Between European Union and Indian Federation

15.1.1 European Union and India: A Concise Frame Comparison

Recently, Springer Nature referees included this book in the series "SGD3 Good Health and Well-Being" (Sustainable Development Goals Books). They suggested a further chapter focused on the relationship between health and the high-/low-income and developing countries' gap. Even if this argument is not the main goal of our studies, we propose some considerations on the comparison between European Union and Indian Federal Republic, focusing on their ecological characters and some health conditions. Data for EU have been collected from the European Environmental Agency (EEA).

These two nations present many similarities, for instance:

(a) A common ancient origin of their population (Indo-Europeans): even if researchers still do not agree on which one was the original Indo-European homeland, new researches (Kristiansen 2020; India State-Level Disease Burden Initiative Diabetes Collaborators 2018) put it in Anatolia about VIII millennia ago. Within the three to five successive millennia, Indo-Europeans expanded in Europe (but only in present Balkans, Italy, Austria, Germany, France) and in Asia (Iran, Indian Region). Previously it was talked about the steppes of Ukraine, Russia, or Kazakhstan (Balter 2015; Roser et al. 2019).
(b) The same language groups (WHO-Europe 2017). The Indo-European languages are a family of related languages that today are largely spoken in the Americas, Europe, and also Western and Southern Asia. Languages such as Spanish, French, Portuguese, and Italian are all descended from Latin; similarly, Indo-European languages are derived from a hypothetical language known as Proto-Indo-European (no longer spoken).
(c) A similar number of states, 27 in European Union and 28 in India Federal Republic.
(d) A comparable extension, with a territorial surface of 4.23 km^2 in EU-27 and 3.3 million in India Federal Republic.
(e) A similar geographical frame, as India represents the most part of the Indian sub-Continent and European Union quite the same for the European Continent.

However, even sharp differences are known, for instance:

(a) The population number, being the Indian about 2.5 times the European.
(b) The urbanized people, being EU 2.3 times more urbanized.
(c) The income per-capita, pair to $2010 in India vs. $43,700 in the EU (2018) (Ministry of Home Affairs, Government of India 2011; Tian et al. 2014).
(d) Life expectancy, 69 years in India vs. 78 years in EU (2015) (Sapna and Prabhanshu 2019)
(e) Cancer incidence, 356/100,000 in EU vs. 35–40/100,000 in India (Steliarova-Foucher et al. 2014).

We can add an interesting observation: studies on mitochondrial DNA, Y-chromosome, and autosomal loci confirmed an influence of west Eurasians (Europeans) in the genetic composition of Indians. However, as underlined by (Smith and Mallat 2019), the Indian population differs in the physiology from western populations; the YY-paradox suggests that Indians have approximately three times the fat percent of a western individual with a similar body mass index (BMI).

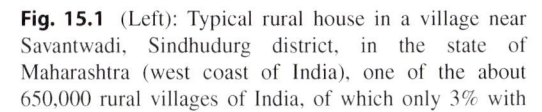

Fig. 15.1 (Left): Typical rural house in a village near Savantwadi, Sindhudurg district, in the state of Maharashtra (west coast of India), one of the about 650,000 rural villages of India, of which only 3% with above 5000 inhabitants. To the right, traditional shops in a district capital village of Rajasthan. Sketches by V. Ingegnoli (1983–1984)

As underlined by Mahatma Gandhi (Gandhi 1962; Gray and Atkinson 2003), the peculiar character of India is to be structured in villages, while Europe is structured in cities (Fig. 15.1).

The Mahatma wrote: "If the village perishes, India will perish too. It will be no more India. Her own mission in the world will get lost." And again: "Small countries like England or Italy may afford to urbanize their systems. A big country like America with a very sparse population, perhaps, cannot do otherwise. But one would think that a big country, with a teeming population with an ancient rural tradition which has hitherto answered its purpose, need not, must not copy the Western model." If, even today, the structure of India is eminently based on rural landscapes, this is due to the presence of about 650,000 rural villages (Violatti 2014): so, in India, the rural population is 66.5% vs. only 33.5% of urban population (2015).

15.1.2 European Union and India: A Few Words on Urbanization

On the other hand, the Indian cities are very crowded, and in 2018 Mumbai city (603.4 km^2) counted 18.2 million inhabitants, whereas Milan city (181.2 km^2, nearly 1/3 of Mumbai) counted only 1.38 million (i.e., 301.6 ab/ha vs. 76.2 ab/ha). Consequently, many peripheral Indian belts are shantytowns. Table 15.1 provides the comparison between the 10 main metropolitan areas in India and Europe. Note that the total population of the first 10 Indian metropolis is 2.7 times the total of EU ones, but they represent only 9.03% of the total Indian population.

The dynamics of urbanization is yet similar both in EU and India, as we see in Fig. 15.2, having increased about 14–15% in the last 60 years. The percent increase is higher in India because the starting level was 1/3 of the European one. As we will see forward ahead, in Tables 15.4 and 15.5, the total urbanized surface in India remains lower than in Europe: in the last 130 years, it passed from 0.7% in 1880 to 2.7% in 2010, while in Europe from 2.3% to 4.7% (within the same period).

Let us remember the interpretation of Gandhi (Gray and Atkinson 2003) comparing cities and villages: "There are two schools of thought current in the world. One wants to divide the world into cities and the other into villages. The village civilization and the city civilization are totally different things. One depends on machinery and industrialization, and the other on handicrafts. We have given preference to the latter. After all, this industrialization and large-scale production are only of comparatively recent growth."

Table 15.1 Comparison between the main metropolitan areas in India and Europe (EU) (2019)

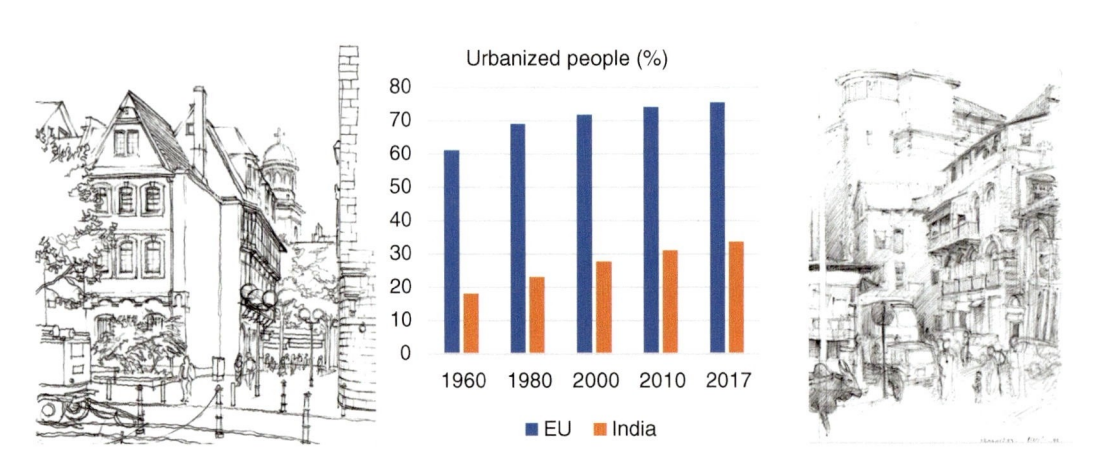

	The largest metropolitan areas in India	Population (million)	The largest metropolitan areas in the European Union	Population (million)	
1	Delhi	29.62			
2	Mumbai	23.36			
3	Kolkata	17.56			
4	Bangalore	13.71			
5	Chennai	11.32			
			Paris	11.02	1
6	Pune	7.76			
7	Ahmadabad	7.41			
			Essen-Dusseldorf	6.13	2
			Madrid	6.03	3
8	Surat	5.81			
			Milan	4.91	4
			Barcelona	4.59	5
9	Lucknow	4.58			
10	Jaipur	4.17			
			Naples	3.57	6
			Athens	3.33	7
			Rome	3.19	8
			Cologne-Bonn	2.09	9
			Hamburg	1.98	10
	Total	**125.29**	**Total**	**46.82**	
	Population India	*1387.20*	*Population Europe*	*427.10*	
	Ratio (%)	***9.03***	***Ratio (%)***	***10.96***	

Fig. 15.2 The increase of urbanization in Europe (Frankfurt) and India (Mumbai, both by V. Ingegnoli, 1984). Note that the increase of India's urbanized population is 187% vs. 123% of the European one

To confirm this assertion, we present Tables 15.2 and 15.3, showing the complex ecological and bionomic parameters which characterize urban/suburban vs. rural/agricultural landscape types (Ingegnoli 2015).

However, we may add another observation comparing the two sketches (Fig. 15.2) of the German city of Frankfurt and the Indian city of Mumbai, in their historical centers: the European city is extremely ordered, while the Indian one is

Table 15.2 Ranges of bionomic *normality* in temperate landscape types in which prevail human activities

Bionomic parameters	URB	SUB.T	SUB.R	AGR	AGR. PRT
RNT, forest area (%)	4.0–8.0	5.0–12.0	8.0–25	10.0–30	20–55
CBSt, Concise Bionomic State of forests	16.5–22.5	18.5–25.5	22.5–30.5	23.5–32.0	27.5–37.5
Allochthonous forest formations (%)	2–5.0	1–4.0	1–2.0	1–2.0	0–1.5
pCA, potential core area (%)	5–10.0	10.0–20	20–35	35–50	50–65
CON, connection ($\alpha + \gamma$)	0.50-0.65	0.55-0.70	0.60-0.75	0.65-0.85	0.70-0.90
Network efficiency	0.6-0.8	0.6-0.8	0.7-0.9	0.8-1.2	1.0-1.8
PRD, agricultural areas (%)	5.0–15	10.0–35	30–55	45–75	35–60
RSD, urbanized areas (%)	25–50	10.0–25	8.0–15	4–9.0	2–6.0
SBS, industrial and transport areas (%)	10.0–20	15–50	7.0–12	4–7.0	2–4.0
HH, human habitat (%)	85–95	80–92	70–86	60–82	42–68
HS/HS* bionomic carrying capacity	0.1-0.5	0.4-1.0	0.6-3.0	1.5-9.0	3.0–12.0
BTC, LU (Mcal/m^2/a)	0.40-0.90	0.60–1.20	0.80–1.60	1.00–2.00	1.70-3.50
CBSt, Concise Bionomic State of the entire territory	4.0–12.0	6.0–16	8.0–20	10.0–24	12.0–30
g-LM, general landscape metastability	3.5-6.0	5.5-8.0	7.5–15	14–24	20–30
LTpE (Landscape Type Evaluation) = 10 (HNd/HU) + BTC$_{A+F}$	0.35–3.0	2.7-6.3	5.7–11.0	10.2–24	22–40
IFF, river functionality	120–180	120–180	180–200	180–200	200–250
Low noise areas (<60 dB) %	40–50	35—45	45–55	50–60	55–70

Landscape types: *URB* urban, *SUB-T* suburban technologic, *SUB-R* suburban rural, *AGR* agricultural, *AGR-PRT* protective agricultural. See Chap. 3 for acronym details

Table 15.3 Ranges of bionomic *normality* in temperate landscape types in which prevail natural processes

Bionomic parameters	Agr-Syl-Tour	AGR.FOR	For-Tour	FOR.SN	FOR.NAT
RNT, forest area (%)	25–65	40–75	45–80	60–85	65–95
CBSt, Concise Bionomic State of forests	28–38	29.5–40.5	30–41	33.5–45.5	>39.5
Allochthonous forest formations (%)	0–1	0–1	<0.5	0	0
pCA, potential core area (%)	45–70	60–75	65–80	75–85	>85
CON, connection ($\alpha + \gamma$)	0.65–0.95	0.75–1.00	0.7–1.0	0.8–1.1	0.8–1.1
Network efficiency	1.1–2.0	1.5–2.0	1.2–2.1	1.6–2.2	>1.8
PRD, agricultural areas (%)	15–40	20–50	10.0–25.0	5.0–20	3.0–6.0
RSD, urbanized areas (%)	2–7.0	1–3.0	0.5–2.0	<1.5	<1
SBS, industrial and transport areas (%)	1.5–3.0	1–2.0	<1	<1	<0.5
HH, human habitat (%)	20–40	22–48	10.0–25	5.0–20	0–12
HS/HS* bionomic carrying capacity	1.2–12.0	3.5–15	1.5–15	4.0–18	>4.5
BTC, UdP (Mcal/m^2/a)	1.8–4.5	3.00–5.00	3.5–6.0	4.50–7.00	6.0–8.5
CBSt, Concise Bionomic State of the entire territory	14–34	16–38	18–42	20–46	>45
g-LM, general landscape metastability	27–37	32–43	40–47	48–52	52–55
LTpE (Landscape Type Evaluation) = 10 (HNd/HU) + BTC$_{A+F}$	34–55	45–88	65–140	100–245	>170
IFF, river functionality	200–270	250–270	250–300	261–300	261–300
Low noise areas (<60 dB) %	70–80	65–75	75–85	80–90	>90

Landscape types: *Agr-Syl-Tour* agro-sylvo-touristic, *AGR-FOR* agro-forest, *For-Tour* forest-touristic, *FOR-SN* forest-seminatural, *FOR-NAT* forest-natural marked by the people's personalization up to the most unthinkable details.

15.2 Theory and Methodology

15.2.1 Landscape Bionomics Principle and Methods: A Short Recall

The most important environmental alterations of the last 125 years can be resumed in forest destruction, and agrarian and urban landscapes increase. This process has many implications with all the other components, as listed by Planetary Health Alliance (PHA), from biodiversity loss to climate change. Moreover, these alterations are related to the failing of "green revolution," which reduced the increase of agrarian areas, but destroyed rural landscapes. To better understand the cited differences between India and EU, the environmental conditions must be more considered.

The new discipline of Bionomics,[1] proposed by Ingegnoli widening Biology through the Theory of Complex Systems (so, enhancing the spatio-temporal-information scale interdependence too), recognizes the hole biological spectrum as constituted of living entities, thus the landscape too. This fact allows to reach a diagnostic evaluation of a territorial unit, putting the basis to geo-health/human-health relation.

The theoretical corpus of this new discipline is quite complex, so we presented here only an extreme synthesis of the most important principles and functions that are needed to understand the applications used in this chapter, in accordance with what already written in Chap. 3. A deepening knowledge of LB will be found in the books of Ingegnoli (Ingegnoli 2011; Ingegnoli 2015; Ingegnoli 2019).

1. The principles needed to find a new ecological discipline, more systemic, able to really upgrade the old conventional concepts, are:
   (a) *Life on earth is organized in a hierarchy of types of hyper-complex systems (living entities)* strictly integrated to each other, which *cannot exist/be without their own environment*.
   (b) Each land-biological system (e.g., Ecotope, Landscape Unit, Eco-Region) is a peculiar biological level, i.e., the *"ecological units" of the territory are living entities* composed by the integration of natural and human systems.
   (c) The ambiguous meaning of the concept of ecosystem must be reconsidered and scaled down.
   (d) Through opportune constraints, the processes of the upper spatio-temporal-information scale explain the significance of the examined level, while the inferior scale processes explain its origin (Allen and Hoekstra 1992): so, as demonstrated by the Principle of Emergent Properties, a top-down criterion of observation and study is needed.
   (e) The environmental balance must follow non-equilibrium thermodynamics and irreversible processes (time arrow).
   (f) *Landscape physiology and pathology* are to be studied through a *qualitative-quantitative clinical diagnostic approach*.
   (g) Systemic functions and indicators, of both LB and medicine, integrated into peculiar systemic diagnoses are needed.
2. The key-state functions are (see also Chap. 3, Sect. 3.3.1):
   (a) **Human habitat (%)**, that is, the surface evaluation (% of landscape unit) of the

---

[1] Note that the name Landscape Bionomics derives by the first book *Landscape Ecology: A Widening Foundation* (Ingegnoli 2011) with the foreword by Richard Forman (Harvard), in which Ingegnoli re-founded this discipline as Biological-Integrated Landscape Ecology, then synthetized in Landscape Bionomics (LB) (Ingegnoli 2015, 2019).

human ability to affect and limit the self-regulation capability of natural systems. Ecologically speaking, the HH cannot be the entire territorial (geographical) surface: it is limited to the human ecotopes and landscape units (LU) (e.g., urban, industrial, and rural areas) and to the semi-human ones (e.g., semi-agricultural, plantations, ponds, managed woods). The NH are the natural ecotopes and landscape units, with dominance of natural components and biological processes, capable of normal self-regulation.

(b) **Biological territorial capacity of vegetation (BTC, Mcal/m²/year),** led by the *physiology of vegetation* to the concept of *latent capacity of homeostasis* of a phytocoenosis (Ingegnoli 2002; 2011; 2015; 2019; Ingegnoli and Giglio 2005; Ingegnoli and Pignatti 2007; Ingegnoli et al. 2017). It can be studied by integrating all the following concepts and parameters: the concept of resistance stability; the type of vegetation community and its metabolic data (biomass, net or gross primary production, respiration, B, NP, GP R); their metabolic relations R/GP (respiration/gross production) and their order relations R/B (respiration/biomass) = dS/S (antithermic maintenance). Evaluations regard the degree of the relative metabolic capacity of principal vegetation communities and the degree of the relative antithermic (i.e., order) maintenance of the same main vegetation communities. Therefore, the BTC function is very important because it is systemic and can evaluate the flux of energy available to maintain the order reached by a complex system.

(c) **Standard habitat (SH, m²/ab),** that is, the state function called *vital space per capita* (Ingegnoli 2002, 2011, 2015, 2019; Ingegnoli and Pignatti 2007), intended as the set of portions of the landscape apparatuses (LA)[2] within the examined landscape units (LU) indispensable for an organism to survive, better known as standard habitat per capita (SH). In the case of human populations, we will have SH_{HH}, that is, an SH referred to the human habitat (HH):

SH_{HH} = (HGL + PRD + RES + SBS + PRT) areas/N° of people [m²/inhabitant]

The connected Minimum Theoretical Standard Habitat per capita (SH*) is the state function estimated in the dependence of the main agrarian crops and the needed Kcal/capita. Finally, the ratio SH/SH*, named bionomic carrying capacity (σ) of a LU, is the state function able to evaluate the self-sufficiency of the human habitat (HH), a basilar question for sustainability and ecological territorial planning.

15.3 Findings: Landscape Transformations and Health

15.3.1 Forest Evaluation and Climatic Differences

Present study was done at a very wide geographic scale (1:12.5 millions), so it acts as a frame.

In EU the total forest cover is 1,820,000 km², which is 41.5% of the total area. Between 1990 and 2015, the area covered by forests and woodlands increased by 90,000 km² (3600 km²/year). The studies on European forests have been sum-

---

[2] A Landscape Apparatus (L-Ap) is a functional system of landscape elements (even not connected), performing a specific physiologic function within the landscape unit (see also Sect. 3.3.1). HGL = hydrogeologic LA; PRD = productive LA; RES = residential LA; SBS = subsidiary LA; PRT = protective LA.

Table 15.4 Main Indian forests BTC estimation

Main Indian types of forest	Land cover × 1000 km²	High BTC	Low BTC	BTC Mean	%	BTC
Evergreen forest						
Tropical evergreen	95.5	12.00	8.00	10	72.51	7.25
Himalayan conifer	13.4	9.00	6.50	7.75	10.18	0.79
Pine forest	22.8	8.00	6.00	7	17.31	1.21
	131.7					**9.25**
Deciduous forest						
Tropical moist	207.6	10.00	8.00	9	45.11	4.06
Tropical dry	217.7	9.00	7.00	8	47.31	3.78
Mountain temperate	34.9	8.00	6.50	7.25	7.58	0.55
	460.2					**8.39**
Other forests	50	8.00	7.00	7.5		**7.50**

marized in Chap. 5, in Tables 5.6 and 5.7, obviously more detailed.

The total forest cover in India is 708,273 km², which is 21.54% of the total area of the country (2018). Between 2015 and 2017, India has added 6778 km² of forest cover and extended 1243 km² of tree cover, despite populations and livestock pressures.

Remembering the surveys in India made by Ingegnoli 30 years ago on the main landscapes, and the data from State Forest Report and Sudhacar et al. (Ingegnoli 2002; The World Bank 2017), the BTC evaluation has been summarized in Table 15.4. Sudhacar's study found 29 land use/land cover classes of natural vegetation including 14 forest types and seven scrub types. Hybrid classification approach has been used for the classification of forest types.

The predominant forest types of India are tropical dry deciduous and tropical moist deciduous. Of the total forest cover, tropical dry deciduous forests occupy an area of 217,713 km² (34.80%) followed by 207,649 km² (33.19%) under tropical moist deciduous forests, 48,295 km² (7.72%) under tropical semi-evergreen forests and 47,192 km² (7.54%) under tropical wet evergreen forests. Tropical evergreen forests and tropical moist forests may reach the higher level of BTC, locally even >12.0 Mcal/m²/year. However, forest cover in India is reducing, as we will see in Table 15.8.

From a macro-climatic point of view, EU and India show a similar latitudinal extension but shifted toward North in Europe (about 34.4°–70.0°N) and South in India (8.4°–37.6°N) (Fig. 15.3). Only the South margins of Spain and Greece and the isles of Sicily and Cyprus present subtropical characters in EU. Only the North margins of Himalaya present cold semi-arid characters in India. Mediterranean climate is present in Punjab. India has also a desertic area in Rajasthan. We cannot forget to remember the crucial presence of monsoons in India.

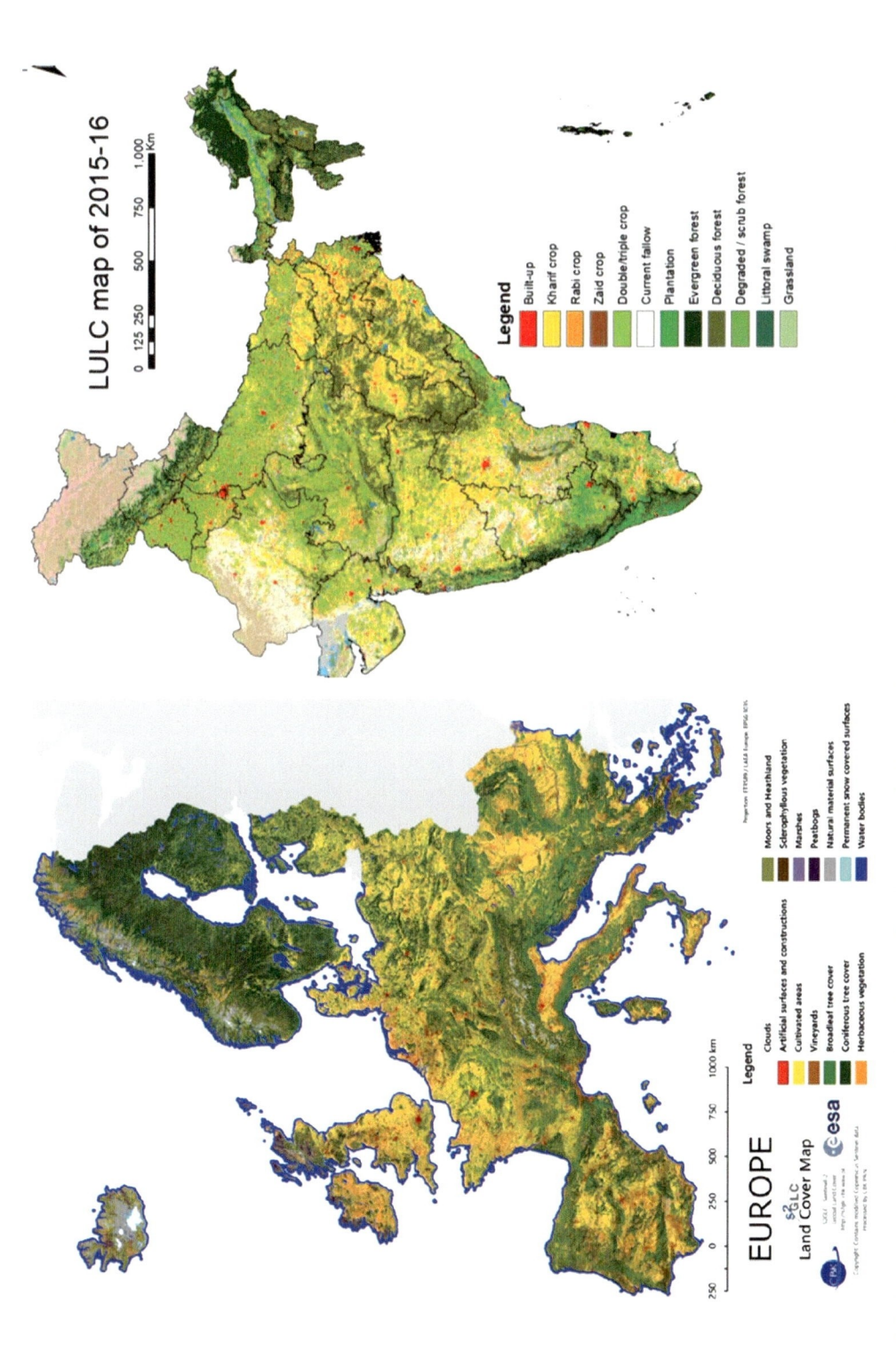

15 Some Landscape and Healthcare Considerations Comparing European Union and Indian Federation

Fig. 15.3 Land cover/land use maps of North-Central-West Europe and India. Latitudinal extension is similar but shifted toward North in Europe (about 34.4°–70.0°N) and South in India (8.4°–37.6°N). Only the South margin of Spain, Sicily, and South of Greece and Cyprus present subtropical characters in EU

15.3.2 Low-Income Countries and Bionomic Principles

In 1985, classifications of countries were related mostly to the values of incomes per capita: so "low-income developing countries" were those with incomes per-capita below $400; "middle-income developing countries" were those with incomes per-capita between $400 and $4000 (data from Encyclopedia Britannica). But economy cannot be the dominant field in the definition of developing/developed countries: also advanced ecology must be recalled, especially if we want to study their public health.

The first relation between human health and the environment can be found in the vital space per capita (SH) and especially the consequent carrying capacity (SH/SH*). This function implies to refer to the concept of population density, and it permits to underline the importance to follow the principles of landscape bionomics, because the geographic density of a population is an ecological nonsense, as it considers also desert and inhospiTable areas. Even Odum (1971) indicated to avoid the use of geographic density. So, we must use the inverse of SH related to the human habitat (HH), evaluating an *ecological density (ab/ha)*.

As expressed in Table 15.5, SH/SH* changes with the climatic belts, consequently tropical countries need a smaller vital space per capita and register a huge amount of population.

In Table 15.6, we show that the differences between geographic vs. ecological density of the

Table 15.5 Evaluation of the theoretical minimum standard habitat per capita (SH*) related to the main types of ecological regions

Ecoregion types (*f*)	Boreal	Temperate	Sub-tropical	Tropical
Kcal/day per capita	3100	2850	2600	2350
Yearly cultures BTC	1.12	1.20	1.27	1.40
SH-PRD, m^2	1228	1050	903	735
SH-PRT BTC	5.6	6.0	6.0	7.4
Deduced BTC (r)	2.35	2.2	2.3	2.35
SH-RSD, m^2	110	105	100	95
SH-SBS, m^2	84	80	76	72
Mean BTC (PRD-RSD-SBS)	1.37	1.46	1.53	1.66
Needed SH-PRT, m^2	429	241	193	112
Resulting SH*, m^2	1851	1476	1272	1014

Table 15.6 Comparison between geographic and ecological density of population in EU-27

Nations (2010)	Europe 27	India 28	France	Tanzania
Surface (km^2)*	4,230,000	3,270,000	543,000	886,000
Population (million inhabitants)[a]	490	1.300	65	47
Human habitat (%)	50.1	53.1	52.3	41.0
Carrying capacity	2.56	1.19	2.1	6.92
Geo. density (ab/km^2)	*116*	*364*	*119*	*54*
Ecol. density (ab/km^2)	*231*	*791*	*229*	*130*
Ecol. density/Geo. density ratio	*2.0*	*2.17*	*1.92*	*2.38*
Max population (million)[b]	*1.325*	***1469***	*179*	*309*

[a]Not considering overseas territories, protectorates, etc. EU-27 in 2010 re-elaborated including Croatia and excluding U.K
[b]The population x carrying capacity (×0.95)

population depend mainly on the levels of human habitat (HH), which generally varies from 25% to 60% of the entire territory. Note that these analyses allow to estimate the maximum amount of population without altering the ecological and bionomic state of the country.

As shown in Chap. 5, the European nation presenting the maximum HH (1.68 above the average) also presents the maximum cancer onset (1.39 above the average). Note that in Table 15.6, the maximum HH = 53.1% pertains to India, a country having HH = 42.7% only 20 years ago (a difference of 24.3%): this increase must be contained and rested!

Thus, we will deepen in the next paragraph the importance of adding the bionomic ones to the geographic parameters (land cover, land use).

15.3.3 Environmental Transformations

Following the BTC evaluations, it was possible to reach a quite correct average value both for EU and India. Applying the principles and methods of landscape bionomics and the main land cover/land use of cropland, grassland, forest, and urbanized to the period 1880–2010, we elaborated Tables 15.7 and 15.8. There we will see that in the period 1880–2010, in Europe, the main environmental parameters remained near constant, while in India they were changing decidedly. Agrarian landscapes mark the main differences (Verma 2017): a small loss of 15% in Europe due to land abandonment and urbanization, but with a compensation in forest re-grown, vs. a strong agrarian increase in India (+77%) due to the pop-

Table 15.7 Geographic and bionomic dynamics in countries pertaining to European Union, 1880–2010

Europe, 27–4.2 million km²	1880	1920	1960	2000	2010
Cropland (%)	33.70	35.82	34.30	32.47	28.50
Grassland (%)	24.80	24.04	25.00	22.74	24.20
Forest L (%)	27.20	26.8	28.00	28.80	32.00
Urbanized L (%)	2.30	2.7	3.90	4.74	4.75
Population (million inhabitant)	*189.3*	*284.8*	*384.6*	*472.7*	*489.6*
Human habitat (%)	52.70	54	54.10	52.50	50.10
BTC (Mcal/m²/year)	2.31	2.28	2.31	2.36	2.45
Carrying capacity SH/SH*	2.51	2.40	2.28	2.50	2.56

Table 15.8 Geographic and bionomic dynamics in India, 1880–2010

India, 28–3.27 million km²	1880	1920	1960	2000	2010
Cropland (%)	27.5	30.6	36.7	40.4	48.8
Grassland (%)	13	13	12	8.3	5.4
Forest L (%)	26.6	25.4	22	20.8	20.1
Urbanized L (%)	0.7	0.8	1.3	2.2	2.7
Population (million inhabitant)	205	240	420	1.020	1.190
Human habitat (%)	31.8	34.65	39.7	42.78	46
BTC (Mcal/m²/year)	2.9	2.85	2.59	2.43	2.36
Carrying capacity SH/SH*	1.9	1.8	1.5	1.10	1.15

Land Transformations in India and Europe (1880-2015)

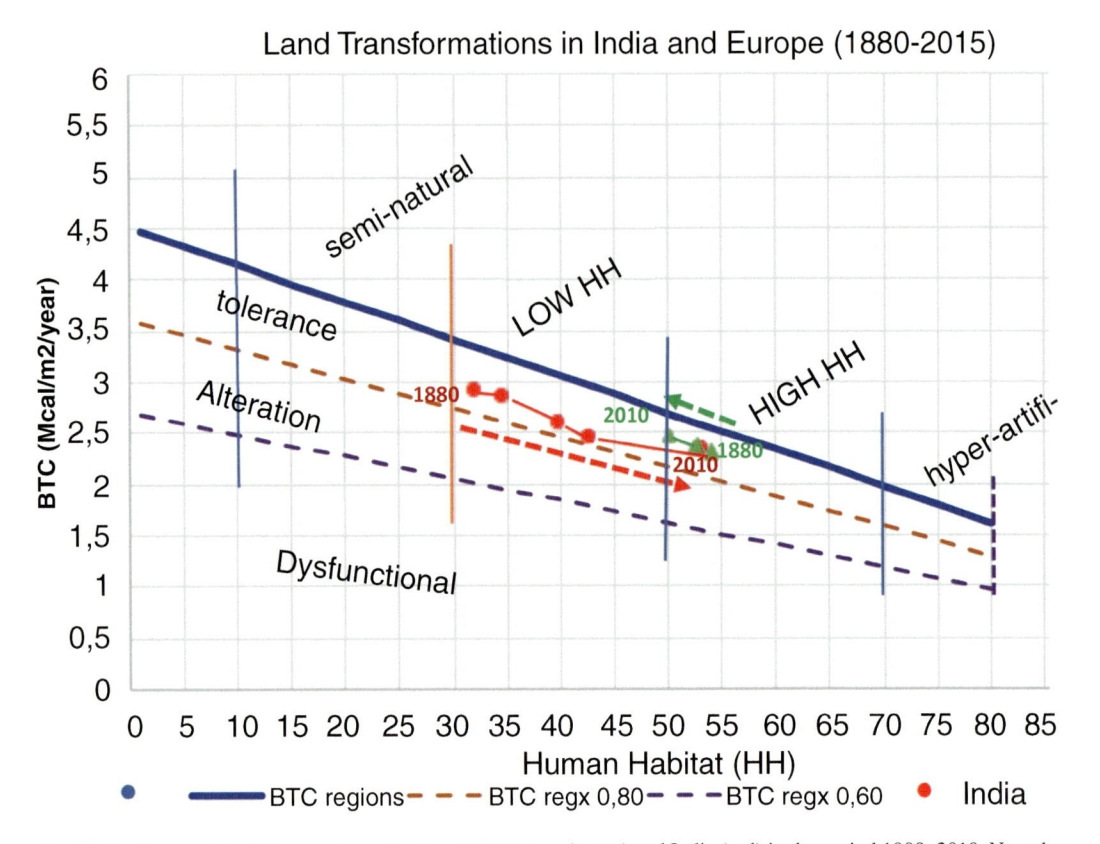

Fig. 15.4 The model of diagnostic evaluation HH/BTC (Chap. 3) at regional scale shows the opposite directions (dashed lines) of the land transformations in Europe (green) and India (red) in the period 1880–2010. Note that the changes in India were 3.6 times wider than in Europe

ulation growth, which decreased also the grassland (−60%) and the forest (−25%).

Consequently, the human habitat (HH) presents a little decrease in Europe from 52.7 to 50.1 (−5%), while in India the increase is strong: HH passing from 31.8 to 46.0 (+45%). However, the most important environmental parameter is the BTC, with a small increase in Europe (+6%) and a strong decrease in India (−19%). The carrying capacity in Europe remained near constant, but in India, SH/SH* = 1.15 (2010) is approaching to the pathologic limit of 0.88–0.92,[3] the country threshold of *heterotrophy*. Remember that India suffers from malnutrition even today (about 15–20% of young people).

The diagnostic evaluation HH/BTC shows (Fig. 15.4) the *opposite* directions of the land transformations in Europe and India. Note that the changes in India were 3.6 times wider than in Europe (Verma 2017); however, both the nations are moving within the tolerance range (80% of the normal ecological state, blue line). While Europe remains in the section of high HH countries, today India has reached the level of Europe: this signifies a similar environmental alteration. We have to underline that these alterations may produce different damages in Europe and India, because Europe is mainly in a temperate belt while India is in a tropical one, so desertification processes may be threatening.

---
[3]Considering 10% of tolerance.

15.4 Some Environment/Cancer Relations

Given these differences and similarities, let us measure some parameters capable to indicate a peculiar aspect of environment/health relation: the cancer incidence growth, which is strongly increasing all over the world and especially in Europe (Steliarova-Foucher et al. 2014; Sudhacar et al. 2015; World Data Atlas 2018). The parallel increase of urbanization, people growth, and aging can be measured (Table 15.6) through ecological density (inhabitants/km^2) and life expectancy (years). From this Table, it is possible to note that while in Europe the strong increase of cancer incidence (712%) is supported by a robust increase of both ecological density (230%) and aging (188%), in India the cancer incidence presents a limited growth (172%) vs. a very strong growth of the other two parameters: ecological density (402%) and life expectancy (291%). Consequently, we need to plot all these data with the bionomics parameters of the previous Tables 15.7 and 15.8, to compare the general environmental states with Table 15.9.

In Fig. 15.5, we can see that the main environmental parameters of the European Union have been near constant since 1880: HH (brown), BTC (green), and carrying capacity (dotted orange). The ecological density of the population (red) and life expectancy (blue) present evident increases. The heavy growth of cancer incidence (7 times) is ineviTable due to high HH, quite low BTC, crowded urbanization, and population aging; it means that, in a quite altered environment (high HH in Fig. 15.4), a process like this is expected.

The same main environmental parameters are plotted also in the analysis of India (Fig. 15.6). Here the dynamics of the analyzed landscape functions present a continuous change. The increase of human habitat (HH) is underlined by the decrease of biological territorial capacity (BTC) of vegetation and the carrying capacity SH/SH*, while the ecological density (red) is 2.8 times the European one. Life expectancy is lower than the European, but with a stronger increment.

The growth of cancer cases in India is limited to 1.7 times and 1/10 of the European increase. The evident proportionality with the increase of human habitat (HH) shows a marked relation with this environmental parameter, which is reinforced by the bionomic alteration of agrarian landscapes. As underlined in the introduction, the rural landscapes are crucial in India, so a process like this can be expected.

However, as noted by Steliarova-Foucher et al. 2014, cancer screening, which is being considered by the Government of India, is known to increase the incidence while reducing mortality. All these factors will lead to further increases in India's future cancer burden.

Table 15.9 Ecological density, life expectancy, and cancer incidence in EU-27 and India in the period 1880–2010

	1880	1920	1960	2000	2010
EU ecological density (ab/km^2)	102	147	198	216	231
EU life expectancy (year)	41	52	67.5	74	77
EU cancer (cases/100,000)	50	80	170	293	356
India, ecological density (ab/km^2)	*197*	*212*	*323*	*729*	*791*
India, life expectancy (year)	*23*	*25*	*41.5*	*62.5*	*67*
India, cancer (cases/100,000)	*19.5*	*22*	*23.2*	*30.7*	*33.5*

Life expectancy and Cancer incidence in Europe

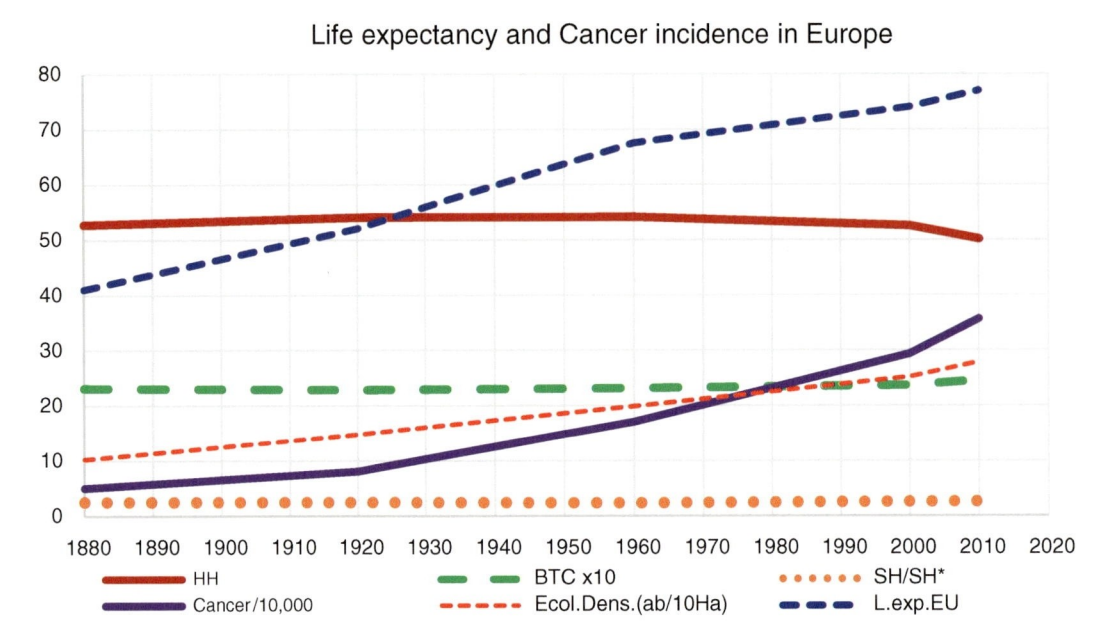

Fig. 15.5 The main environmental parameters of the European Union have been near constant since 1880, as we can see: HH (brown), BTC (green), and carrying capacity (dotted orange). Evident increases present the ecological density of the population (red) and life expectancy (blue). The heavy growth of cancer incidence (7 times) is inevitably due to high HH, quite low BTC, crowded urbanization, and population aging

Human Habitat and Cancer incidence in India

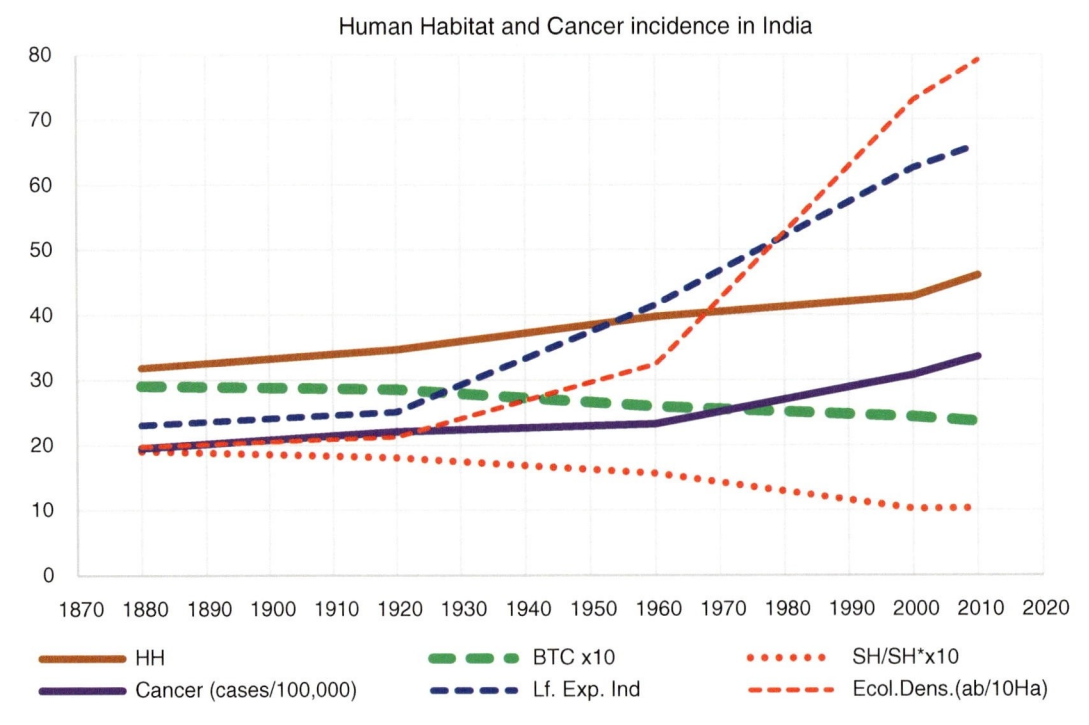

Fig. 15.6 The main environmental parameters of India between 1880 and 2010. The dynamics of the analyzed land functions presents a continuous change. The increase of human habitat (brown) is underlined by the decrease of BTC (dashed green) and carrying capacity SH/SH* (dotted orange), while the ecological density (red) is 2.8 times the European one. Life expectancy is lower than the European, but with a stronger increment. The growth of cancer is limited to 1.7 times and however 1/10 of the European increase

15.5 Some Environment/Diabetes Relations

Today a wide increase in type 2 diabetes is becoming increasingly evident, so that many scientists write of epidemics. In this case, contrary to cancer incidence, the growth of type 2 diabetes incidence is higher in India than in Europe (India State of Forest Report 2015). As plotted in Fig. 15.7, the comparison is immediate (blue bars) in both urban and rural environments. Furthermore, the difference between low- and high-income is not so wide: 6.1% vs. 8.2%.

The problem is that, in India, High-income class (i.e., people with income per-capita > 12,375.00 $) counts 16 million people, so the prevalence of type 2 diabetes may interest 1.3 million (1/1000), while low-income class (<1,000.00 $) counts 684 million people, so the prevalence of diabetes interests 41.7 million people. Note that diabetes is known to be linked with the strong growth of overweight people, consequently, to be a problem of upper classes.

However, it is not the case, because (Fig. 15.8) in the last 10–15 years even the low-income group is arriving to 75–76% of high-income group of overweight persons. As high-income group (1.3%) presents 10% people being overweight, low-income group (56.1%) presents 7–7.5%: so, we will have in India 1.6 million and 47.9 million overweight, respectively, justifying the prevalence of diabetes, but also the difficulties in its healthcare support.

The reasons for the growth of diabetes incidence are manifold and can also influence the low-income people:

1. Increase of caloric food, due to industrialized agriculture
2. Diet modification in modern criteria
3. Increase of urbanization and consequent food artificialization
4. Reduction of physical activity

WHO suggested this etiology, but recalling what was exposed in Chap. 5, we must add at least three reasons:

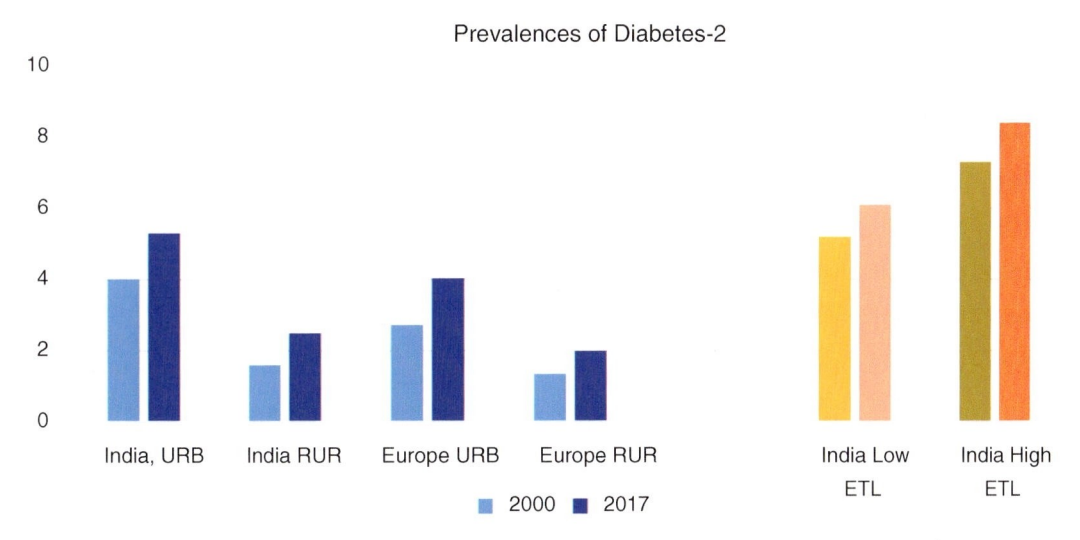

Fig. 15.7 Comparison between the prevalence of diabetes in India and European Union. Note the differences related to urban and rural landscapes, in both India and Europe, and the increase in only 17 years. In the same period, the difference between low and high economic levels (ETL) of population could be meaningful

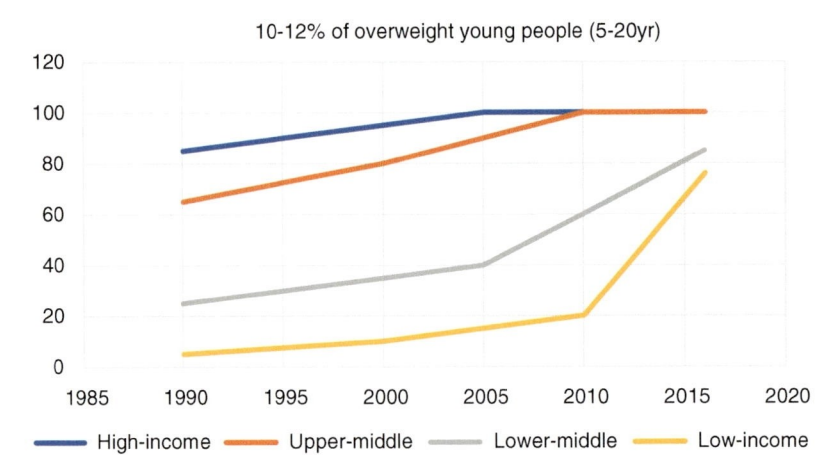

Fig. 15.8 Data from World Data Atlas (2018) show that young people (5–20 years) from high-income nations since 2003 present at least 10–12% of people overweight, followed by high-middle income. But it is impressive that even lower-middle and low-income young people have a strong increase, going to reach the situation of high-income people in few years

(a) Gut microbiota (GM) alterations, due to agrarian landscape banalization (Shetty et al. 2013)
(b) Inter-scale relationship disruptions, due to agrarian landscape industrialization
(c) Environmental toxicants are acting as endocrine-disrupting chemicals (EDCs) (Sargis and Simmons 2019)

The gut *microbiome dysbiosis* may reshape intestinal barrier functions and host metabolic and signaling paths directly or indirectly related to the insulin resistance. Thousands of the metabolites derived from microbes interact with the epithelial, hepatic, and cardiac cell receptors that regulate host physiology, as expressed by Sapna and Prabhanshu (2019). Xenobiotics (e.g., dietary components, antibiotics, and anti-inflammatory nonsteroidal drugs) strongly affect the gut microbial composition and promote dysbiosis. Any change in the gut microbiome can shift the host metabolism toward increased energy harvest during diabetes and obesity.

See Sect 3.4.2 and Fig. 3.14. Bertolaso (2013) remembers what matters the relationships among the components of a system and how such interactions reflect or emerge from the inter-level regulatory processes that are typically the object of inquiry in life sciences.

EDCs implicated in diabetes pathogenesis include various inorganic and organic molecules of natural and synthetic origin, including arsenic, bisphenol A, phthalates, polychlorinated biphenyls, and organochlorine pesticides. As underlined by Sargis and Simmons (2019), evidence implicates EDC exposures across the lifespan in metabolic dysfunction; moreover, specific developmental windows exhibit enhanced sensitivity to EDC-induced metabolic disruption, with potential impacts across generations. Importantly, differential exposures to diabetogenic EDCs also contribute to racial/ethnic and economic disparities. Despite these emerging links, clinical practice guidelines fail to address this underappreciated diabetes risk factor.

As Gandhi underlined clearly, India is a mainly rural nation, so the alteration of agrarian

15 Some Landscape and Healthcare Considerations Comparing European Union and Indian Federation

Life expectancy and Healthcare expend./GDP 1980–2015

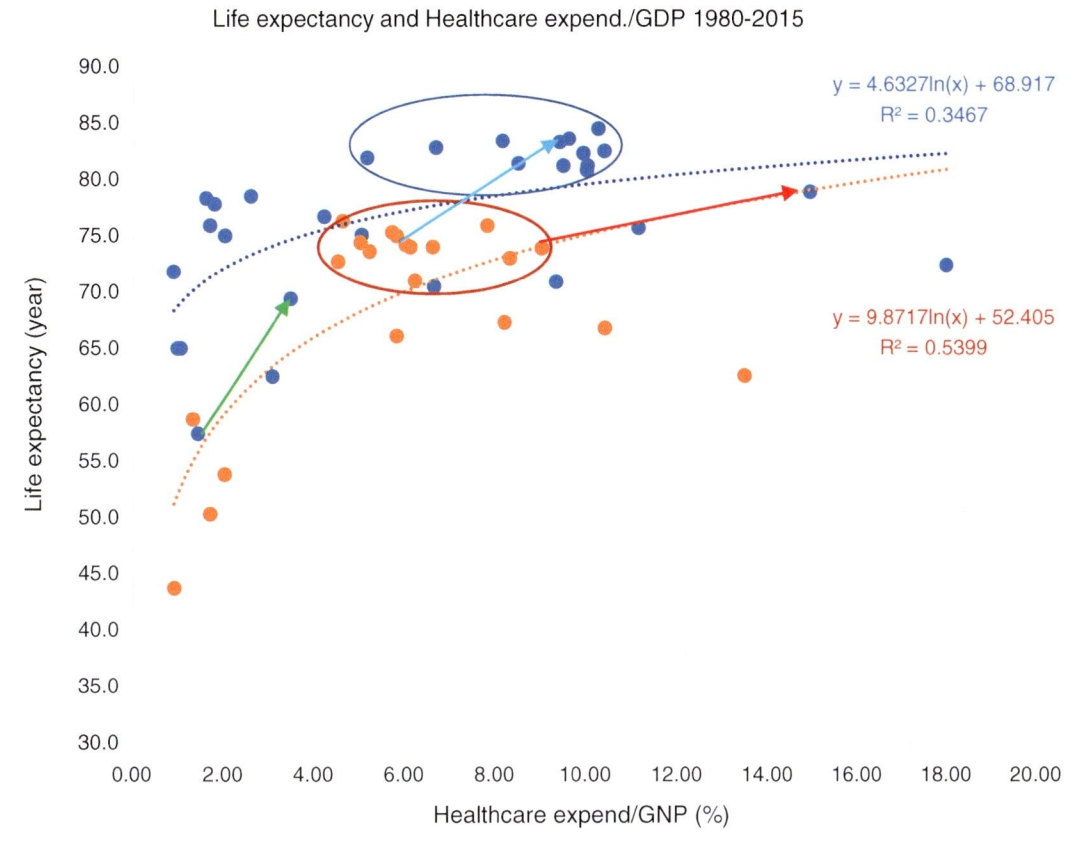

$y = 4.6327\ln(x) + 68.917$
$R^2 = 0.3467$

$y = 9.8717\ln(x) + 52.405$
$R^2 = 0.5399$

Fig. 15.9 Life expectancy and healthcare expenditure HE/GNP per capita in 1980 (red) and 2015 (blue) for the 30 nations of Table 15.10. Even if the trend curves are logarithmic, the nation's reaching the highest life expectancy show an increase proportional to their ratio HE/GNP not exceeding 11%. Note that the growth of HE/GNP from 2.0% to 3.6% led India to add 15.2 years, while to add only 5.0 years the USA growth of HE/GNP passed from 9.0% to 14.9%. Environmental factors (e.g., urbanization and food) must be more considered

landscapes results in real danger even for human health.

15.6 Discussion and Conclusion

Even from these few considerations of the different situations between a high-income group of nations (European Union) and a low-middle one (India), it appears that it is not correct to compare these two types of societies following economic parameters: ecological and health aims must be considered too.

The life expectancy gap in 1980 was of 20 years, but in 2015, with a healthcare expenditure from 2.0 to 3.6% (+1.6) of GNP, India gained 15.2 years (green arrow), while the USA needed an expenditure from 9.0 to 14.4% (+4.4) of GNP to add only 5.0 years (red arrow). This process is plotted in Fig. 15.9, where 30 nations of high-, middle-, and low-income have been studied in the mentioned two times, 1980 and 2015 (Table 15.10) (Ministry of Home Affairs, Government of India 2011; Sapna and Prabhanshu 2019; Tian et al. 2014) [35]. An average condition, between the two extremes of India and the

Table 15.10 Life expectancy and HE/GDP in the examined nations

	HE/GNP, %	LExpt, 2015
Australia	9.40	83.3
Austria	8.50	81.4
Brazil	11.12	75.7
Canada	9.92	82.3
China	4.21	76.7
Cote d'Ivoire	1.42	57.4
Denmark	10.00	80.8
Egypt	0.89	71.8
Ethiopia	1.04	65
France	10.38	82.5
Germany	10.01	81.2
Greece	6.64	70.5
India	3.47	69.4
Italy	9.60	83.6
Japan	10.25	84.5
Mexico	5.03	75.07
Poland	2.58	78.5
Portugal	5.16	81.9
Romania	1.69	75.9
Russia	17.95	72.4
Saudi Arabia	2.02	75
South Africa	3.06	62.5
South Korea	6.68	82.8
Spain	8.15	83.4
Sweden	9.32	70.9
Tanzania	0.97	65
Turkey	1.60	78.3
UK	9.48	81.2
United Arab Emirates	1.79	77.8
USA	14.90	78.9

Note: *HE* healthcare expend, *GNP* gross national product, *LExpt* life expectancy

USA, may be done by Italy (blue arrow), which gained 10 years with expenditure from 6.0 to 9.6 of GNP (+3.6).

The two best conditions in 1980 and 2015 are shown in the ellipses (red > 70 years; blue > 80). The passage between these levels is from about 4.0–9.0% to 4.5–10.5% of HE/GNP. So, to reach the best life expectancy, the curve has to increase, but not over a reasonable limit. The crucial thing is to develop the health organization (Odum 1971) that is compatible with the nations' environmental/social structures (Table 15.11): for their good functioning, this subdivision should be related to the study of the environmental health of the proper spatial scale, that is, punctual, local, territorial (divided in landscape units and landscape) (Table 15.12).

Table 15.11 Articulation of the Rural Indian Organizations in the three main sections of education, health, and social development. Note the presence of Health Missions

Sarva Shiksha Abiyaan (Education)	National Rural Health Mission (Health)	National Rural Livelihood Mission (Rural Development)
District	District	District
Taluka (sub-district)	Health Sub-district	Blocks
Cluster	Community Health Centers	Gram Panchayat
School	Primary Health Centers	Villages
–	Sub-centers	–

Table 15.12 Proper landscape subsystems and their related surface range

Scale	Biotic reference	Living entity	Peculiar systems	Usual surface (km^2)
Regional	Biome	Ecoregion	L. system (LS)	5000–100,000
Territorial	System of community	*Landscape*	*Landscape (L)*	500–10,000
			Complex LU (c-LU)	50–1000
			Landscape unit (LU)	5–100
			Ecotope (ECT)	0.5–10.0
Local	Community	Ecocoenotope	Tessera (TS)	0.005–1.0

References

WHO-Europe (2017) Incidence of cancer per 100,000. In: European health information gateway

Allen TFH, Hoekstra TW (1992) Toward a unified ecology. Columbia University Press, New York

Balter M (2015) Mysterious Indo-European homeland may have been in the steppes of Ukraine and Russia. https://doi.org/10.1126/science.aaa7858

Bertolaso M (2013) How science works: Choosing levels of explanation in biological sciences. Aracne, Roma

Gandhi MK (1962) Village Swaraj. Publ. Jitendra T, Desai, Ahmedabad

Gray RD, Atkinson QD (2003) Language-tree divergence times support the Anatolian theory of Indo-European origin. Nature 426:435–439. https://doi.org/10.1038/nature02029

India State of Forest Report (2015) Important characteristics of different forest type groups

India State-Level Disease Burden Initiative Diabetes Collaborators (2018) The increasing burden of diabetes and variations among the states of India: The Global Burden of Disease Study 1990–2016. Lancet Glob Health 6:e1352–e1362

Ingegnoli V (2002) Landscape ecology: a widening foundation. Springer, Berlin/New York, pp XXIII–357

Ingegnoli V (2011) Bionomia del paesaggio. L'ecologia del paesaggio biologico-integrata per la formazione di un "medico" dei sistemi ecologici. Springer-Verlag, Milano, pp XX–340

Ingegnoli V (2015) Landscape bionomics. Biological-integrated landscape ecology. Springer, Heidelberg/Milan/New York, pp XXIV–431

Ingegnoli V (2019) Infrastrutture ecologiche e Diagnosi dell'ambiente. Bonizzi, Campana, Cordini (Eds), Il Governo dei Parchi, Aracne Ed. Roma, pp. 173–214. ISBN 978-88-255-2726-I. https://doi.org/10.4399/97888255272618.

Ingegnoli V, Giglio E (2005) Ecologia del Paesaggio: manuale per conservare, gestire e pianificare l'ambiente. Sistemi editoriali SE, Napoli, pp 685–XVI

Ingegnoli V, Pignatti S (2007) The impact of the widened Landscape Ecology on Vegetation Science: towards the new paradigm. Springer, Rendiconti Lincei Scienze Fisiche e Naturali, s. IX, vol. XVIII, pp. 89–122

Ingegnoli V, Bocchi S, Giglio E (2017) Landscape bionomics: a systemic approach to understand and govern territorial development. WSEAS Trans Environ Dev 13:189–195

Kristiansen K (2020) The archaeology of Proto-Indo-European and Proto-Anatolian: locating the split. In: Serangeli M, Olander T (eds) Dispersals and diversification: linguistic and archaeological perspectives on the early-stages of Indo-European. BRILL, Leiden-Boston

Ministry of Home Affairs, Government of India (2011) Villages/Towns Directory

Odum EP (1971) Fundamental of ecology. Saunders, Philadelphia, PA

Roser M, Ortiz-Ospina E, Ritchie H (2019) Life expectancy. Our World in Data

Sapna S, Prabhanshu T (2019) Gut microbiome and type 2 diabetes: where we are and where to go? J Natl Biochem 63:101–108

Sargis RM, Simmons RA (2019) Environmental neglect: endocrine disruptors as underappreciated but potentially modifiable diabetes risk factors. Diabetologia 62(10):1811–1822

Shetty SA et al (2013) Opportunities and challenges for gut microbiome studies in the Indian population. Microbiome 1:24. http://www.microbiomejournal.com/content/1/1/24

Smith RD, Mallat MK (2019) History of the growing burden of cancer in India: from antiquity to the 21st century. JCO Global Oncol 5:1–15

Steliarova-Foucher E, O'Callaghan M, Ferlay J, Masuyer E, Rosso S, Forman D, Bray F, Comber H (2014) The European cancer observatory: a new data resource. Eur J Cancer 2015(51):1131–1143

Sudhacar RC, Diwacar PG, Chandra SJ, Vinay KD (2015) Nationwide classification of forest types of India using remote sensing and GIS. Environ Monit Assess 187(12)

The World Bank (2017) Current Health Expenditure per Capita (current US $). License CC-BY4

Tian H, Banger K, Bo T, Dadhwal VK (2014) History of land use in India during 1880–2010: large-scale land transformations reconstructed from satellite data and historical archives. Global Planet Change 121:78–88

Verma R (2017) India unclear how many villages it has, and why that matters. India Spend

Violatti C (2014) Indo-European languages. Ancient history encyclopedia

World Data Atlas (2018) India current expenditure on health per capita

Printed by Printforce, the Netherlands